Lecture Notes in Physics

Edited by J. Ehlers, München, K. Hepp, Zürich, and
H. A. Weidenmüller, Heidelberg
Managing Editor: W. Beiglböck, Heidelberg

31

Transport Phenomena

Sitges International School of
Statistical Mechanics, June 1974
Sitges, Barcelona/Spain
Director: L. Garrido

Edited by G. Kirczenow and J. Marro
University of Oxford, 12 Parks Road
Oxford/England
Belfer School of Sciences
2495 Amsterdam Ave.
New York 10033, USA

Springer-Verlag
Berlin · Heidelberg · New York 1974

ISBN 3-540-06955-0 Springer-Verlag Berlin · Heidelberg · New York
ISBN 0-387-06955-0 Springer-Verlag New York · Heidelberg · Berlin

 Library of Congress Catalog Card Number 74-17900. Printed in Germany.

Offsetprinting and bookbinding: Julius Beltz, Hemsbach/Bergstr.

In memoriam

Prof. L. Rosenfeld

PREFACE

The 1974 Sitges International School of Statistical Mechanics is the third to be held in Spain on this subject. The first took place at Mallorca during the summer 1969 and the second at Sitges in 1972.

The aim of the School is mainly to direct the interest of professors and students of Spanish Universities towards this field of research, besides contributing to the scientific formation of many students from other countries, and facilitating personal contacts for future collaboration between physicists.

I should like to thank all those Institutions who helped to make this School a reality; in particular:

Presidencia de las Cortes Españolas
The Embassy of Japan
Ayuntamiento de Barcelona
Diputación de Barcelona
Rectorado de la Universidad de Barcelona
Ayuntamiento de Sitges
Patronato "Alfonso X el Sabio"

I extend my acknowledgement to:

The Steering Committee;
Dr. L. Navarro for his collaboration on the organization of the School;
Prof. Jones for his great assistance on the edition of the Proceedings;
I also wish to thank my wife for all her continuous help.

L. Garrido
Director del Instituto de Física Teórica
Consejo Superior de Investigaciones Científicas

CONTENTS

SOME SIMPLE REMARKS ON THE BASIS OF TRANSPORT THEORY — SIR R. PEIERLS

I. Introduction 2
II. The most naive transport problem 2
III. Angle dependent scattering; many collision times 8
IV. A still more general transport collision time 10
V. How valid is the Boltzmann equation? 12
VI. The situation is better than it seems 14
VII. There are still reservations 17
VIII. Pauli's remark and off-diagonal elements 18
IX. Summary of limits of validity 20
X. Extension to many-body problem 20
XI. Neglect of correlations 21
XII. Conservation laws 22
XIII. Limit on collision rate 23
XIV. Omission of off-diagonal terms. Rigorous derivations 24
XV. Illustrations. Lattice thermal conductivity 26
XVI. Content of Boltzmann equation 27
XVII. Exponential behaviour 29
XVIII. The use of modern methods 30
XIX. Conclusions 32
References 33

ENTROPY, DYNAMICS AND SCATTERING THEORY — I. PRIGOGINE, F. MAYNÉ

I. Introduction 35
II. The Mac Kean model 36
III. Irreversibility as a symmetry breaking 41
IV. Star unitary transformation 45
V. Construction of the Λ - transformation 49
VI. Potential scattering 57
VII. Concluding remarks 71
References 73

RESPONSE, RELAXATION AND FLUCTUATION — R. KUBO

I.	Introduction	75
II.	Classical Brownian motion and its generalizations	76
III.	Rice's method (harmonic analysis)	79
IV.	Direct integration, path integral representation	80
V.	Stochastic Liouville equation	85
VI.	Retarded friction, fluctuation-dissipation theorems	87
VII.	Force correlations	93
VIII.	Some examples	97
IX.	Some comments	105
X.	Damping-theoretical method	113
XI.	Concluding remarks	119
	Notes and references	122

FLUCTUATING HYDRODYNAMICS AND RENORMALIZATION OF SUSCEPTIBILITIES AND TRANSPORT COEFFICIENTS — P. MAZUR

I.	General introduction	126
II.	On the critical behaviour of the dielectric constant for a non-polar fluid	130
III.	Renormalization of the diffusion coefficient in a fluctuating fluid	144
	References	154

IRREVERSIBILITY OF THE TRANSPORT EQUATIONS — J. BIEL

I.	Introduction	156
II.	General remarks on irreversibility	158
III.	The irreversibility of the Boltzmann transport equation	171
IV.	The irreversibility of other equations	191
	References	201

ERGODIC THEORY AND STATISTICAL MECHANICS — J. L. LEBOWITZ

I.	Introduction	203
II.	Ergodicity and ensemble densities	207
III.	Systems of oscillators and the Kam theorem	213
IV.	Mixing	216
V.	K- and Bernoulli systems	223
VI.	Ergodic properties and spectrum of the induced unitary transformation	230
VII.	Infinite systems	231
	References	234

CORRELATION FUNCTIONS IN HEISENBERG MAGNETS — M. DE LEENER

I. Introduction 238
II. Neutron scattering experiments and spin correlation functions 243
III. Some general properties of the spin correlation functions 248
IV. Low temperature theory 251
V. High temperature theory 261
VI. The critical region 274
References 287

ON THE ENSKOG HARD-SPHERE KINETIC EQUATION AND THE TRANSPORT PHENOMENA OF DENSE SIMPLE GASES — M. G. VELARDE

I. Introduction: The hard-sphere model interaction 289
II. From the Boltzmann approach to the Enskog equation 294
III. Hydrodynamic equations and the (new Enskog) collisional (or potential) transfer 300
IV. Solution of the Enskog equation for practical purposes 305
V. Transport coefficients from the Enskog equation 310
VI. Comparison with experimental data 316
VII. The square-well fluid 327
VIII. Final comments 330
References 336

WHAT CAN ONE LEARN FROM LORENTZ MODELS? — E. H. HAUGE

I. Models 338
II. From kinetic theory to hydrodynamics 340
III. Higher density effects 349
IV. Rigorous results 362
References 366

CONDUCTIVITY IN A MAGNETIC FIELD — R. B. STINCHCOMBE

I. Introduction 369
II. Derivation of the Boltzmann equation in a magnetic field 376
III. Solution of the Boltzmann equation 387
IV. Quantum effects 395
V. Collisions between carriers 401
VI. Collisions with phonons 405
VII. Concluding remarks 411
References 412

TRANSPORT PROPERTIES IN GASES IN THE PRESENCE OF EXTERNAL FIELDS — J. BEENAKKER

I. Introduction 414

II. The non-equilibrium polarizations 426

III. The limitation of the one moment description 441

IV. The effective cross sections and their behaviour 449

V. Field effects in the rarefied gas regime 456

References 466

TRANSPORT PROPERTIES OF DILUTE GASES WITH INTERNAL STRUCTURE — R. F. SNIDER

I. Introduction 470

II. The role of free molecular motion 472

III. Field dependence of the viscosity 479

IV. On the Boltzmann equation for molecules with internal structure 496

V. Collision integrals of the linearized W-S equation 505

References 516

LECTURERS

Prof. J.J.M. BEENAKKER, Leiden, Holland
Dr. J. BIEL, Valencia, Spain
Dr. M. DE LEENER, Brussels, Belgium
Prof. E.H. HAUGE, Trondheim, Norway
Prof. R. KUBO, Tokyo, Japan
Prof. J.L. LEBOWITZ, New York, USA
Dr. F. MAYNE, Brussels, Belgium
Prof. P. MAZUR, Leiden, Holland
Prof. R. PEIERLS, Oxford, Great Britain
Prof. I. PRIGOGINE, Brussels, Belgium
Prof. R.F. SNIDER, British Columbia, Canada
Dr. R.B. STINCHCOMBE, Oxford, Great Britain
Dr. M.G. VELARDE, Madrid, Spain

PARTICIPANTS

Prof. Z. ALEXANDROWICZ, Rehovot, Israel
Mr. J. ALONSO, Valladolid, Spain
Dr. J.R. BARKER, Warwick, Great Britain
Dr. M. BERRONDO, Mexico, Mexico
Dr. D. BEDEAUX, Leiden, Holland
Prof. A. BERNALTE, Bilbao, Spain
Mr. D. BICHSEL, Geneva, Switzerland
Dr. G.K. BIRKNER, Cologne, Germany
Prof. B. BOSCO, Florence, Italy
Prof. L.J. BOYA, Zaragoza, Spain
Dr. E. BRAUN, Mexico, Mexico
Dr. J.J. BREY, Sevilla, Spain
Prof. C. CALVO, Hamilton, Canada
Dr. R. CANAL, Barcelona, Spain
Mr. V. CANIVELL, Barcelona, Spain
Mr. K. CARNEIRO, A.E.C. Risö, Denmark
Mr. C. CASANOVA, Valladolid, Spain
Prof. J. CASAS, Barcelona, Spain

Dr. A. CHAMORRO, Bilbao, Spain
Mr. M. CIEPLAK, Warsaw, Poland
Dr. N. CLAVAGUERA, Barcelona, Spain
Dr. P. CLIPPE, Liège, Belgium
Dr. C. COHEN, Rehovot, Israel
Mr. A. CORDOBA, Sevilla, Spain
Dr. M. CRISAN, Geneva, Switzerland
Prof. R.C. DESAI, Toronto, Canada
Prof. J. DESTRY, Montréal, Canada
Dr. Ing. C. DEUTSCH, Orsay, France
Mr. C. DUFOUR, Mons, Belgium
Mr. S. FALL, Oxford, Great Britain
Prof. B.U. FELDERHOF, London, Great Britain
Prof. J. FERREIRA DA SILVA, Oporto, Portugal
Dr. A. FLORES, Mexico, Mexico
Dr. D. FRENKEL, Amsterdam, Holland
Mr. L.J. GALLEGO, Santiago de Compostela, Spain
Dr. O. GIJZEMAN, Amsterdam, Holland
Mr. A. GIL, Barcelona, Spain
Prof. M. GITTERMAN, Ramat-Gan, Israel
Dr. J.J. GONZALEZ, Trondheim, Norway
Prof. M. GRMELA, Montréal, Canada
Miss F. GUYON, C.N.R.S. France
Prof. R. HAAG, Hamburg, Germany
Dr. F.Y. HANSEN, Lyngby, Denmark
Dr. Ing. P. HOFFMANN, Montreuil, France
Mr. P. v. HOYNINGEN, Zürich, Switzerland
Mr. W. HUYBRECHTS, Antwerp, Belgium
Mr. J.L. IBAÑEZ, Madrid, Spain
Dr. J.J. ICAZA, Bilbao, Spain
Prof. R.B. Jones, London, Great Britain
Dr. R. KAPRAL, Toronto, Canada
Dr. P. KOCEVAR, Graz, Austria
Dr. M. LEAL, Valladolid, Spain
Mr. D. LONGREE, Brussels, Belgium
Dr. Ing. K. LUCAS, Bochum, Germany
Mr. M. MALLOL, Barcelona, Spain
Mr. E. MARSCH, Kiel, Germany
Mr. J.M. MASSAGUER, Barcelona, Spain
Prof. S. MIRACLE-SOLE, Zaragoza, Spain
Dr. M.T. MORA, Barcelona, Spain
Dr. J.A. MADARIAGA, Bilbao, Spain

Miss N. MORELL, Barcelona, Spain
Dr. M. MURMANN, Heidelberg, Germany
Mr. M. NAPIORKOWSKI, Warsaw, Poland
Mr. J. PARRA, Barcelona, Spain
Prof. R. PATHRIA, Waterloo, Canada
Mr. F. PEGORARO, Pisa, Italy
Mrs.A. PELLET, Marseille, France
Prof. V. PEREZ-VILLAR, Santiago de Compostela, Spain
Mrs. E.B. POHLMEYER, Hamburg, Germany
Prof. J. POP-JORDANOV, Belgrade, Yugoslavia
Mr. A. REY, Sevilla, Spain
Prof. F. RICCI, Rome, Italy
Mr. P.A. RIKVOLD, Oslo, Norway
Mr. J.R. RODRIGUEZ, Santiago de Compostela, Spain
Mr. L. RULL, Sevilla, Spain
Dr. M.T. SACCHI, Cagliari, Italy
Prof. J.A. DE SAJA, Valladolid, Spain
Prof. J. SALMON, Paris, France
Mr. M. SAN MIGUEL, Barcelona, Spain
Prof. E. SANTOS-CORCHERO, Valladolid, Spain
Mr. I. DE SCHEPPER, Nijmegen, Holland
Mr. P. SEGLAR, Barcelona, Spain
Dr. S.K. SHARMA, Warwick, Great Britain
Dr. B. SHIZGAL, Vancouver, Canada
Mr. M. SHLESINGER, Rochester, USA
Mr. V. ŠKARKA, Belgrade, Yugoslavia
Prof. J. SOUSA, Oporto, Portugal
Dr. N. SZABO, Geneva, Switzerland
Prof. M.J. TELLO, Granada, Spain
Miss C. TORRENT, Barcelona, Spain
Mr. A. VAZQUEZ, Barcelona, Spain
Miss C. ZARAGOZA, Valencia, Spain

DIRECTOR

Prof. L. GARRIDO, Barcelona, Spain

EDITORS

Mr. G. KIRCZENOW, Oxford, Great Britain
Dr. J. MARRO, New York, USA

SECRETARIES

Mrs. R. CHESTER, Edinburgh, Great Britain
Mrs. S. PASHLEY, Edinburgh, Great Britain

SOME SIMPLE REMARKS ON THE BASIS OF TRANSPORT THEORY

Sir Rudolf Peierls,

University of Oxford
Oxford, England

(Lecture notes taken by V. Canivell, Nuria Morell, J. Parra, M. San-Miguel, Carmen Torrent)

I. INTRODUCTION
II. THE MOST NAIVE TRANSPORT PROBLEM
III. ANGLE DEPENDENT SCATTERING; MANY COLLISION TIMES
IV. A STILL MORE GENERAL TRANSPORT COLLISION TIME
V. HOW VALID IS THE BOLTZMANN EQUATION?
VI. THE SITUATION IS BETTER THAN IT SEEMS
VII. THERE ARE STILL RESERVATIONS
VIII. PAULI'S REMARK AND OFF-DIAGONAL ELEMENTS
IX. SUMMARY OF LIMITS OF VALIDITY
X. EXTENSION TO MANY-BODY PROBLEM
XI. NEGLECT OF CORRELATIONS
XII. CONSERVATION LAWS
XIII. LIMIT ON COLLISION RATE
XIV. OMISSION OF OFF-DIAGONAL TERMS. RIGOROUS DERIVATIONS
XV. ILLUSTRATIONS. LATTICE THERMAL CONDUCTIVITY
XVI. CONTENT OF BOLTZMANN EQUATION
XVII. EXPONENTIAL BEHAVIOUR
XVIII. THE USE OF MODERN METHODS
XIX. CONCLUSIONS
REFERENCES

I. INTRODUCTION

In this set of lectures I shall talk about very simple things, namely the basic principles of transport theory. Much of what I have to say is quite old, and in many of the problems my approach may appear to you old-fashioned. The reason is that such quantities as the electric conductivity of copper and the thermal conductivity of diamond stayed the same for many years, and some theoretical understanding was reached some time ago. Although old, it is not necessarily wrong. Sometimes there is the attitude that just because a method has been used a long time it must be worse than a more elegant one discovered yesterday. I will show you some examples where new techniques give a wrong answer, and the old ones happen to be right. That is my excuse for looking at things in a rather simple and old-fashioned way.

I will start looking at a few typical problems in the most elementary way, just to see the basic physics clearly, and then show how these very simple ideas can be used in cases where this approach is adequate, and point out other problems in which more refinement and deeper insight is necessary.

II. THE MOST NAIVE TRANSPORT PROBLEM

The first transport problems properly handled in physics were quantities like viscosity or thermal conductivity of gases, studied by people like Boltzmann and Maxwell. We could start from these oldest examples but there is a simpler situation still, and that is the model of electrons in metals in which we imagine the electrons scattered by impurities and irregularities in the medium. The difference between this and the gas is that we can imagine the impurities fixed and therefore are dealing with a one-body problem, while in the gas each collision involves two particles, both having statistical properties. The electron model is no different in principle but makes the considerations simpler.

Consider a situation in which an electron, described by its momentum $\underline{p}$, can be scattered by some centre to a new momentum $\underline{p}'$. We assume that there is a differential cross-section $\sigma(\underline{p}, \underline{p}')$ for this process. The scattering probability which will determine the electron behaviour is $w(\underline{p}, \underline{p}') = v \sum_i N_i \sigma_i$, where the sum is over the possible different kinds of scattering centres, each of density N_i ; v is the electron velocity p/m. We assume elastic collisions and therefore the scattering probability will contain a delta function in the energy:

$$w(\underline{p}, \underline{p}') = W(\underline{p}, \underline{p}') \, \delta(E - E') \qquad (1)$$

We can follow Boltzmann in writing the Boltzmann equation which governs the rate of change of the numbers of electrons in a state. Then, if $n(\underline{p}, \underline{r})$ is the density of electrons in momentum and coordinate space, we have:

$$\frac{\partial n(\underline{p},\underline{r})}{\partial t} = - \underline{v}.\frac{\partial n}{\partial \underline{r}} - \underline{F}.\frac{\partial n}{\partial \underline{p}} + \int d^3 \underline{p}' \, w(\underline{p}, \underline{p}') \left[n(\underline{p}') - n(\underline{p}) \right] \qquad (2)$$

where the first term is the effect of the motion; the second the effect of the acceleration due to the force acting on the electron. The integral is the collision term, in which the first part represents what is called the scattering-in from the state $\underline{p}'$ into the state $\underline{p}$, and the other one is the scattering-out from the state $\underline{p}$ into the state $\underline{p}'$. We have used the law of detailed balancing, which is connected with the symmetry of the underlying mechanical process against time reversal, so that

$$w(\underline{p}, \underline{p}') = w(\underline{p}', \underline{p}) \; . \qquad (3)$$

There are problems like conductivity in an external magnetic field, which do not leave this symmetry. This makes no substantial change in the results of the application of statistical mechanics except that one has to use more care in writing down the expressions. For simplicity I shall not treat such problems.

One may wonder how it is that we get here an equation which describes irreversible processes. It is well known that this equation gives sensible answers, for example, for the electric conductivity which is antisymmetric under time reversal, because in the ordinary process of conduction the charge flows from the higher to the lower potential. If the time were reversed it should flow from the lower potential to the higher potential and the conductivity would have a negative sign.

So, in going from the reversible equations of mechanics to this Boltzmann equation, we have already smuggled in the irreversible behaviour in some place. This place is the stosszahlansatz of Boltzmann.

His statement about the number of collisions is basically the following: I consider some target of cross-section σ_{ab} for the process under consideration:

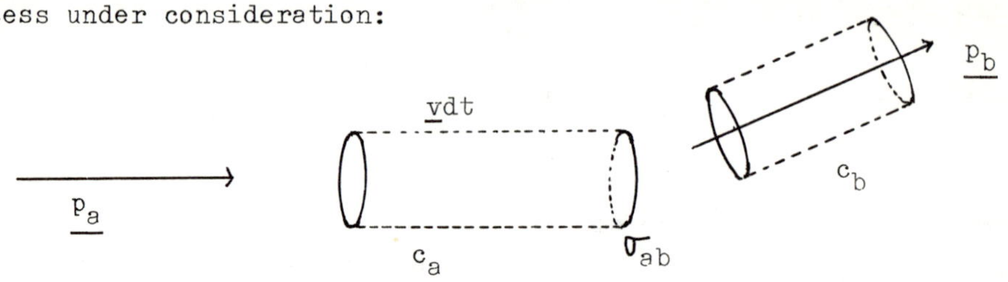

Evidently the number of particles of type a which make the appropriate collision in a time dt is the number of particles contained in a cylinder of cross-section σ_{ab} and length $v_a dt$, where $\underline{v}_a$ is the velocity of the particles. That is $\rho_a \sigma_{ab} v_a dt$. This is exact. The important statement is that the density of particles of type a, ρ_a, in that cylinder, is assumed the same as the density of particles of type a anywhere in the system, because the particles in that cylinder "do not yet know" that they are going to collide, and the only special thing about this cylinder is that it is leading to a collision.

Now by making this statement we introduce a sense of time into the problem. If we consider the inverse reaction, the same cross-section applies to the inverse collision as long as we deal with elastic collisions, by the law of detailed balancing. Therefore to describe the inverse collision we would have to assume that the density of particles with momentum $-p_b$ in the cylinder c_b is the same as elsewhere. Now of course, if we consider collisions in the inverse case this is true, but for the outgoing particles in the direct case this is not true because the cylinder c_b contains the particles that have just made a collision and their number is determined not by what happens in this second cylinder but by the incoming particles that have collided.

Of course in statistical equilibrium this does not matter. But if we are not in equilibrium, if we have for example a current flowing in the direction p_a, then the density of "a" electrons is higher than that of b electrons. After the collision, they are scattered into the direction b, and the density of outgoing b electrons in the cylinder, ρ_b, is therefore greater than elsewhere in the gas.

In accepting this stosszahlansatz we make an assumption which destroys reversibility. We are allowed to do this physically because any experimental situation always arises by bringing the system at some time into some given state, and then allowing it to behave as it likes subsequently. We are concerned with an initial-value problem. But we are incapable of conducting an experiment in which, given a condition at some time, we can observe what the system did in the past. The reason for this is another question, but what we see is that we need a preferred sense in time in practical physical situations.

I mention this because in any theoretical treatment of transport problems, it is important to realize at what point the irreversibility has been incorporated. If it has not been incorporated, the treatment is wrong. A description of the situation which preserves the reversibility in time is bound to give the answer zero or infinity for any conductivity. If we do not see clearly where the irreversibility is introduced, we do not clearly understand what we are doing.

Let us return to our equation and consider the collision term:

$$\int d^3\underline{p}'\, w(\underline{p}, \underline{p}')\, [n(\underline{p}') - n(\underline{p})] \quad .$$

I can do the integration in the second term, and the factor of $n(\underline{p})$, which is of the dimension of an inverse time, can be called the inverse of a "collision time", τ , so the integral becomes:

$$\int d^3\underline{p}'\; w(\underline{p}, \underline{p}')n(\underline{p}') - \frac{1}{\tau}\, n(\underline{p}) \quad . \tag{4}$$

Now, there are situations in which one can simplify the first term also, and I shall start by looking into these situations, although they are quite exceptional in practice.

Let us assume that we are dealing with free electrons so that, for a given energy, W in (1) depends on $\underline{p}'$ only through its direction. Now let us assume that W is independent of the direction of $\underline{p}'$. In that case we are able to integrate also the first part of the collision term, and we are left simply with

$$\frac{1}{\tau}\, (\bar{n} - n) \quad , \tag{5}$$

where $\bar{n}$ is the average of n over directions. This simplest case is applicable to a classical model of hard-sphere scattering, or in quantum mechanics to scattering centres small compared to the wavelength. Then the scattering cross-section is isotropic.

Let us consider as an example two cases:

(a) There is no field and no gradient. Then the Boltzmann equation becomes

$$\frac{\partial n}{\partial t} = \frac{1}{\tau}(\bar{n} - n) \quad . \tag{6}$$

By averaging over all directions, $\frac{\partial \bar{n}}{\partial t} = 0,$ (7)

so that we can write $\frac{\partial}{\partial t}(n - \bar{n}) = -\frac{1}{\tau}(n - \bar{n})$ (8)

which has the solution $n - \bar{n} = F(\Omega)\, e^{-t/\tau}$ (9)

(Ω stands for the direction in space) .

In this trivial model any deviation from equilibrium dies out exponentially and the exponent is always the same for any angular distribution. Of course we are talking here about deviations in the angular distribution. If the energy distribution is wrong then the mechanism we are discussing will not put it right, because we are considering only elastic collisions.

One talks about τ as the "relaxation time". Already in this primitive model we have two relaxation times, one for angular deviations and another for deviations in the energy distribution. In our model the latter is infinite.

(b) As a second case we shall consider an external electric field along the z-axis. The Boltzmann equation now reads:

$$\frac{\partial n}{\partial t} = \frac{1}{\tau}(\bar{n} - n) - e\,\mathcal{E}_z \frac{\partial n}{\partial p_z} \tag{10}$$

In this form it is still a complicated equation to discuss because it is non-linear, in the sense that the deviation from equilibrium will grow with the electric field, and it is here multiplied by the electric field, so that there is a quadratic term. In practice we are nearly always interested in fields weak enough to neglect terms higher than linear in the fields; in other words, we consider the region in which Ohm's law is valid. In this case we can replace $\frac{\partial n}{\partial p_z}$ by $\frac{\partial n_o}{\partial p_z}$ where n_o is the distribution in the absence of the external field, in statistical equilibrium. Then:

$$\frac{\partial n}{\partial t} = \frac{1}{\tau}(\bar{n} - n) - e\,\mathcal{E}_z \frac{\partial n_o}{\partial p_z} \quad . \tag{11}$$

We have in fact to make this approximation if we want to describe a stationary state, because if we keep terms quadratic in $\mathcal{E}$ there will not, in fact, be any stationary states. The reason is that we then include the Joule heat, which will make the electrons hotter and hotter, and this excess of energy cannot in our model be removed, at least by the collisions, so that no steady state is possible. Of course, one can either include in the model some process that can remove this excess energy or take into account that the metal has surfaces that can conduct it away, but this would lead to a complicated discussion.

In our approximation we can have a steady state, so that: $\frac{\partial n}{\partial t} = 0$. n_0 will be a function of the energy only: $n_0 = f(E)$ and therefore

$$\frac{\partial n_0}{\partial p_z} = v_2 \frac{\partial f}{\partial E} \quad . \tag{12}$$

Now our equation is easy to solve and we have:

$$n = f + e\tau \mathcal{E}_z v_z \frac{\partial f}{\partial E} \quad . \tag{13}$$

So we have solved our first transport problem.
We can now immediately compute the current density:

$$j = \int d^3p \; e \, v_z \, n = \rho_e e^2 \tau \langle v_z^2 \rangle \mathcal{E}_z \int \frac{\partial f}{\partial E} E^{\frac{1}{2}} dE \Big/ \int f E^{\frac{1}{2}} dE =$$

$$= \rho_e e^2 \frac{\tau}{m} \mathcal{E}_z \tag{14}$$

where ρ_e is the spatial electron density and the average of v_z^2 is taken over the surface of constant energy. This formula applies both to classical Boltzmann statistics or Fermi statistics. In other words, it would apply both to semiconductors and to metals. In the case of a metal $-\frac{\partial f}{\partial E}$ is nearly a delta function; it has a steep maximum at the Fermi surface, and, therefore the average of v_z^2 is the average over the Fermi surface.

It is not a priori evident whether our derivation is right for a Fermi gas, because we have not taken the Pauli principle into account. The number of collisions from state $\underline{p}'$ to $\underline{p}$ is proportional not only to the number of electrons in the initial state, but also to the number of vacancies in the final state $[1 - n(p)]$ and similarly in the other term: but the quadratic terms cancel out in the elastic case, which we are considering here

$$n(\underline{p}') \left[1 - n(\underline{p})\right] - n(\underline{p}) \left[1 - n(\underline{p}')\right] = n(\underline{p}') - n(\underline{p}) \quad .$$

III. ANGLE DEPENDENT SCATTERING; MANY COLLISION TIMES

Let us see what happens if we take a slightly more realistic case, in which the scattering probability depends on the final direction. We assume we are in an isotropic medium. Then we can say that

$$W(\underline{p}, \underline{p}') = W(\cos \Theta) \tag{15}$$

where Θ is the angle between $\underline{p}$ and $\underline{p}'$. In other words, the collision probability depends only on the relative direction of the initial and final momenta.

Again this will be true in a gas, but it is never exact in a metal because of the crystalline structure. However, the present approximation is somewhat more realistic than the previous model.

The Boltzmann equation now becomes an integral equation. Fortunately we can exploit the isotropy to give us an immediate solution. As we are in an isotropic medium, the energy can depend only on the magnitude of the momentum, so that the energy surface is a sphere, and the collision term in (2) can be written as

$$\text{const.} \int d\Omega' \, W(\cos \Theta) \left[n(\underline{p}') - n(\underline{p})\right] \tag{16}$$

where $d\Omega$ is the element of solid angle, and $\underline{p}'$ differs from $\underline{p}$ only in direction.

We simplify this expression by using the spherical symmetry:

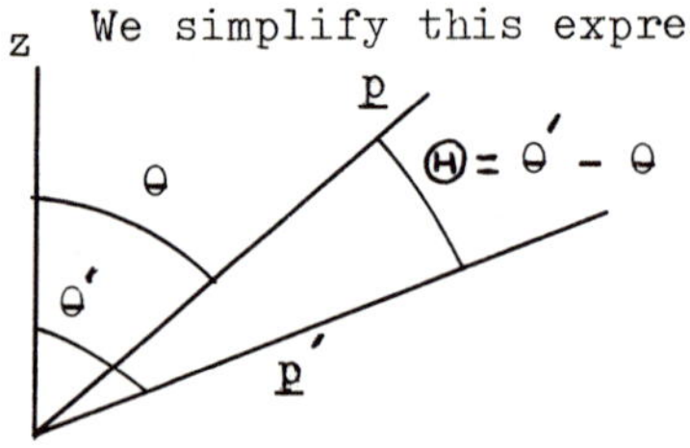

Expand $n(\underline{p})$ in spherical harmonics:

$$n(\underline{p}) = \sum_{\ell m} n_{\ell m} Y_{\ell m}(\theta, \phi) \tag{17}$$

Also, we can expand $W(\cos \Theta)$ in terms of Legendre polynomials

$$W(\cos \Theta) = \sum_{\ell} \frac{w_\ell}{(2\ell+1)} P_\ell (\cos \Theta) \quad . \tag{18}$$

(The factor $(2\ell+1)^{-1}$ is put in for convenience).

Using the composition theorem to express P_ℓ as a sum of products of spherical harmonics in both directions, the integral in (16) is easily done, and the result for the original collision integral is then:

$$\text{const.} \sum_{\ell m} (w_0 - w_\ell) n_{\ell m} Y_{\ell m} \quad . \tag{19}$$

From this it is immediate that, if only collisions are present

$$\frac{dn_{\ell m}}{dt} = -\frac{1}{\tau_\ell} n_{\ell m} \quad \text{where} \quad \frac{1}{\tau_\ell} = (w_0 - w_\ell). \tag{20}$$

In particular, $1/\tau_0 = 0$; there is no change in n_0 because that represents the total number of electrons at the given energy. We are still able to maintain the concept of a collision time, but this now depends on the nature of the deviation from equilibrium which we consider.

However, in the driving term of the Boltzmann equation, we have something proportional to $e\,\mathcal{E}_z v_z$. In an isotropic medium v_z is proportional to $\cos\theta$, a first-order spherical harmonic. The relevant collision time for the disturbance caused by an external field is therefore τ_1. This is sometimes called the transport collision time, and multiplied by the velocity it gives the transport mean free path.

The transport collision rate is given by:

$$\frac{1}{\tau_1} \simeq \int W(\cos\Theta)\,(1 - \cos\Theta)\,d\Omega \quad . \tag{21}$$

This result says that small-angle collisions carry less weight because they are not well able to remove the preferred direction due to the acceleration in the external field. For small angles, the weight is proportional to Θ^2. This is also evident: if we represent the electron momentum by a point on a sphere of constant momentum, small-angle collision will displace the point on the surface in a kind of Brownian motion, so that the displacement is proportional to the square root of the number of collisions. The number of collisions required to get us away from the initial direction is therefore proportional to the inverse square of the angle in each collision.

This fact has been responsible for building up more confidence in the concept of a collision time than was justified. In particular we can apply the same kind of considerations to the thermal conductivity. In this case we have a different driving force, but its directional dependence is also proportional to a velocity component. It is therefore again an $l = 1$ deviation from equilibrium. If we compare the electric and thermal conductivities, we see they depend on the same collision time τ_1.

Therefore all results are the same in that comparison, as if we were dealing with hard sphere collisions and only had one collision time. In particular we get the Wiedemann-Franz law, which says that, at least at high temperatures where we can neglect the quantum nature of the lattice vibrations, the ratio of electric- to thermal conductivity is the same for all metals and varies inversely as the temperature.

IV. A STILL MORE GENERAL TRANSPORT COLLISION TIME

We have not completely justified its empirical success, because metals are really not isotropic. However we can go a little further; when we look for a stationary distribution in an electric field and in a temperature gradient we have:

$$0 = \frac{\partial n}{\partial t} = \int d^3\underline{p}'\, W(\cos\Theta)\,[n(\underline{p}') - n(\underline{p})]\;\delta(E-E') - e\,\underline{\mathcal{E}}.\underline{v}\,\frac{\partial n_0}{\partial E} - \underline{v}.\frac{\partial n}{\partial T}\frac{\partial T}{\partial \underline{r}} \qquad (22)$$

where the first term is the collision term and the last two can be written together as

$$\underline{v}.\underline{F}\,[E(\underline{p})]\;. \qquad (23)$$

As long as we consider elastic collisions, where the equation contains a given energy at a time, we have only to solve this integral equation on the energy surface with the velocity on the right-hand side, which is now a complicated function on the surface.

The solution of this one equation gives us both electric and thermal conductivity, so, it is no wonder that we arrive at the same ratio as when we dealt with hard-sphere collisions. We must not use this agreement as evidence that we are dealing with the case of a clearly defined collision time. This makes our work easy in one sense because we can understand the simple empirical law of Wiedemann and Franz, but it makes it harder in the sense that it is not possible to use this empirical law to draw any conclusions about the collision mechanism.

However, this situation is not general; if we are interested in the Hall effect or in magneto-resistance, we have a magnetic field present as well as an electric field. We are looking for an effect that is proportional to the product of both the electric and magnetic fields. We therefore have to consider a second-order term, which is no longer of the form of a single spherical harmonic. The details of the shape of the energy surface and of the collision mechanisms come in.

Indeed if, in the classical situation which we are now discussing, there was a single collision time, there could be no magnetoresistance. The Hall field which develops makes the equipotential surfaces tilted:

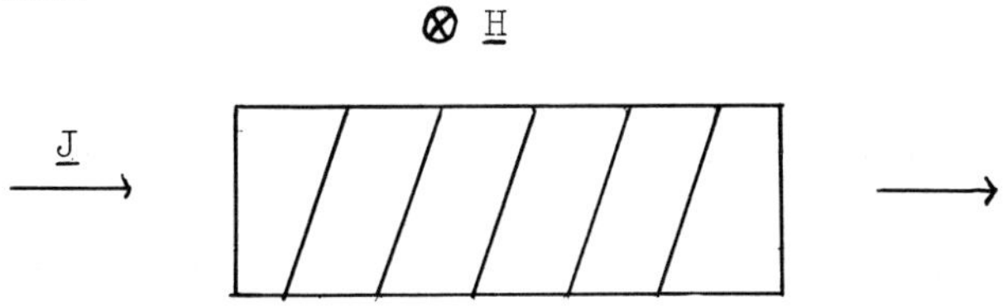

The force required to cancel out the transverse force on the electrons is the same for each electron because the collision time is the same for all. This just cancels the effect of the magnetic force in the longitudinal direction and there is no magnetoresistance.

This was found for the first time by Sommerfeld, to his great surprise. He found a small correction, because all the electrons do not all have exactly the same energy. The Fermi distribution falls off over an energy interval of the order of kT, and within that fall-off range, the electrons that contribute to the conductivity have their energy spread out by an amount kT. If the collision time depends on the energy, this gives a small variation and there will be a small magnetoresistance.since the collision times are no longer equal.

I recall a personal experience. In the early days of the theory of metals I found some results about the variation of magnetoresistance with magnetic field, without determining the magnitude, and I agreed to talk about this at a conference. To my distress I found on the eve of the conference that in my model (which used a single collision time) the magnitude of the effect was zero.

We emphasize the fact that collision time, while intuitively a very useful concept, has to be handled with care. In many situations the assumption of a single collision time can lead us astray.

V. HOW VALID IS THE BOLTZMANN EQUATION?

So far we have looked at these problems in a very simple way. The first thing we should do is to ask in what circumstances this simple picture is adequate, in classical physics.

The electron travels in straight lines between collisions,

the time interval between them being on the average a collision time.

The duration of the collision τ_c classically depends just on the size of the object with which the electron collides. Our momentum distribution is valid between collisions rather than inside the scattering centre. We do not try to describe what is going on inside, and we only consider the asymptotic initial and final states at each collision. There are two reasons why we should assume that $\tau_c \ll \tau$.

The first reason is that if they were comparable, the electron might find itself at the same time under the influence of two scattering centres and our calculations would not be valid.

Secondly, in evaluating the average of some physical quantity we neglect the contribution from inside the scattering centre. While the electron is inside the scattering centre, its properties differ from what they are outside. So if we are dealing with a situation where the fraction of time spent inside the scattering centres is appreciable, then our description is wrong.

In practice τ_c is comparable with τ only in very dirty metals, or in liquids, where the "lattice" is imperfect everywhere. In such cases transport theory becomes very much harder.

Consider now the changes caused by the quantum theory, apart from the Pauli principle, which we have already mentioned. In this new situation we have to describe the scattering by each centre by quantum mechanics. We take the same equations as before, but for the scattering probability we take the solution from the Schrödinger equation

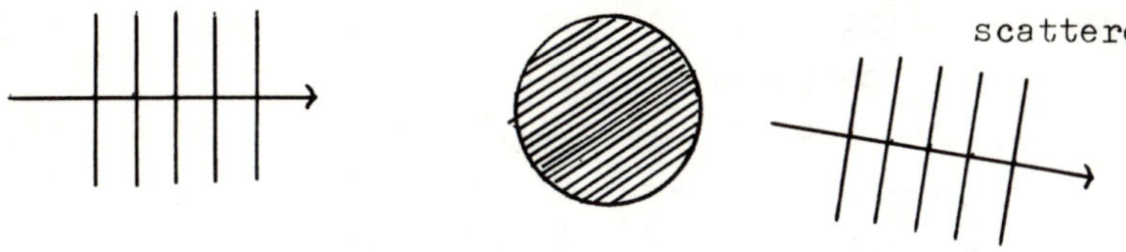

The normal formula for the scattering amplitude refers to the asymptotic form. This is valid at a distance much greater than the wavelength λ, so in using it we assume that the distance between scattering centres, r, is very much larger than λ. If a second scatterer is placed at a distance r comparable to λ then we have new diffraction phenomena, or as one sometimes says "off-energy-shell scattering".

This can be brought out in another way, if we derive the scattering probability by time-dependent perturbation theory. We write:

$$\psi(\underline{r}) = \sum a_k u_k \qquad (24)$$

where the u_k may be plane waves. To the leading order:

$$|a_k|^2 = \sum_{k'} |W_{kk'}|^2 \frac{1-\cos\Delta}{\Delta^2} |a_{k'}|^2 = \sum_{k'} |W_{kk'}|^2 |a_{k'}|^2 D(\Delta) \qquad (25)$$

where $\Delta = \dfrac{E_k - E_{k'}}{\hbar}$

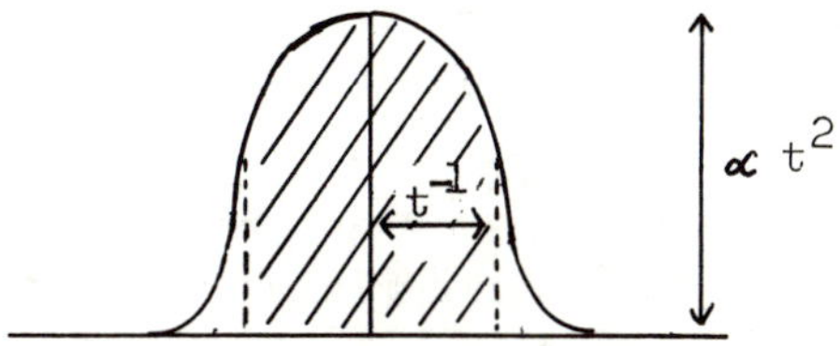

$D(\Delta)$ is a sharp peak, of a height of the order $\hbar^2 t^2$ and a width of the order $\hbar/t$. The area is therefore proportional to t. If the other factors can be treated as constant over an energy range of $\hbar/t$, we obtain an expression of the form

$$\text{const. } t\, |a_{k'}|^2 |W_{kk'}|^2 \delta(E - E')$$

as in Fermi's "golden rule".

This rule is implied in our use of scattering theory, and it is seen to depend on the variation of the factors in (25) within an energy range of $\hbar/t$ being small. But t must be smaller than the collision time, to justify the consideration of a single collision. So $\hbar/\tau$ must be small compared to the relevant energy.

Now in metals the electron distribution function drops off over an energy range of kT, and since this function appears in one of the factors in (25) our derivation seems to require that

$$\hbar/\tau \ll kT \tag{26}$$

In fact this inequality is frequently violated in metals.

VI. THE SITUATION IS BETTER THAN IT SEEMS

Fortunately, while our derivation requires the condition (26), the result is valid subject only to the condition

$$\hbar/\tau \ll E_F \tag{27}$$

where E_F is the Fermi energy. In metals, (27) is much weaker than (26), and is usually satisfied. In semiconductors we cannot improve on (26).

The argument for replacing (26) is quite subtle. We shall use for this purpose the fluctuation-dissipation theorem, which was first applied to conductivity by Kubo (1956, 1958), obtaining an expression for the conductivity in closed form, though involving quantities which are not easy to evaluate.

I shall give a derivation of the Kubo formula for the case of impurity scattering in which we may, as before, consider one electron at a time, and work in terms of single-particle dynamics. I am here following a paper by Greenwood (1958).

In terms of the density matrix ρ,

$$\rho = \rho_0 + \rho_1 \tag{28}$$

where ρ_0 is that of statistical equilibrium, and ρ_1 small of first order in the applied electric field $\underline{\mathcal{E}}$, the current density is

$$j_z = \mathrm{Tr}\,(ev_z\,\rho_1)\,. \tag{29}$$

The equation of motion for ρ is

$$\dot{\rho}_1 = {}^{i}/_{\hbar}\,[H, \rho_1] + {}^{i}/_{\hbar}\,[V, \rho_0] \quad . \tag{30}$$

Here H is the Hamiltonian including the effect of the scattering centres, and V is the potential of the external electric field. I have neglected ρ_1 in the last term since its contribution would be of second order.

To fix the solution of the time-dependent equation (30) we assume that the field has grown gradually since $t = -\infty$ by a factor $e^{\epsilon t}$ with infinitesimal ϵ, and that ρ_1 also was zero at $-\infty$. This is where we violate time-reversal symmetry.

Now ρ_1 also has a time dependence as $e^{\epsilon t}$. (30) can now be written as

$$\epsilon\rho_1 - {}^{i}/_{\hbar}\,[H, \rho_1] = {}^{i}/_{\hbar}\,[V, \rho_0] \quad . \tag{31}$$

We use the representation in which H is diagonal and label its eigenstates by α, β .....

$$\left[\epsilon - {}^{i}/_{\hbar}(E_\alpha - E_\beta)\right] \langle\alpha|\,\rho_1\,|\beta\rangle = e\,\mathcal{E}_z \frac{\langle\alpha|\,v_z|\beta\rangle}{E_\alpha - E_\beta}(f_\alpha - f_\beta) \tag{32}$$

On the right-hand side we have eliminated the operator z, which appears in $V = -e\,\mathcal{E}_z z$, and which does not go well with cyclic boundary conditions, in favour of the velocity by the equation $\dot{z} = v_z$ in the Heisenberg form; f_α is the Fermi function at energy E_α.

The solution of (32) is

$$\langle\alpha|\,\rho_1|\beta\rangle = i\hbar e\,\mathcal{E}_z \frac{\langle\alpha|v_z|\beta\rangle}{E_\alpha - E_\beta + i\hbar\epsilon}\;\frac{f_\alpha - f_\beta}{E_\alpha - E_\beta} \quad . \tag{33}$$

The current density per electron is now, from (29)

$$j_z = i\hbar e^2\,\mathcal{E}_z \sum_{\alpha,\beta} \frac{|\langle\alpha|v_z|\beta\rangle|^2}{E_\alpha - E_\beta + i\hbar\epsilon}\;\frac{f_\alpha - f_\beta}{E_\alpha - E_\beta} \quad . \tag{34}$$

Here every factor is symmetric in α and β, except the first denominator, which would change sign on interchanging α with β except for the small term $i\hbar\epsilon$. We therefore replace the expression by its average, with α , β interchanged, to give

$$\frac{ihe^2}{2}\mathcal{E}_z\sum_{\alpha,\beta}|\langle\alpha|v_z|\beta\rangle|^2 \; \frac{f_\alpha - f_\beta}{E_\alpha - E_\beta} \left\{\frac{1}{E_\alpha - E_\beta + i\hbar\epsilon} - \frac{1}{E_\alpha - E_\beta - i\hbar\epsilon}\right\} \tag{35}$$

We are interested in a macroscopic system in which the energy spectrum is practically continuous. Summing, for example over all states β with energies in a small interval dE_β , and then integrating, we pass above the pole in one term in the last bracket and below the pole of the other, so the difference reduces just to the residue at the pole

$$j = \pi\hbar e^2\,\mathcal{E}_z\sum|\langle\alpha|v_z|\beta\rangle|^2 \left(\frac{\partial f}{\partial E}\right)_{E_\alpha}\delta(E_\alpha - E_\beta)\;. \tag{36}$$

This is essentially the Kubo formula for the case in hand, in the representation used.

Note that (36) makes sense only in the macroscopic limit, with a continuous spectrum. Otherwise, if there are no pairs of states of exactly the same energy, coupled by matrix elements of v , (36) is zero, and if there are, they give terms in $\delta(0)$, which is infinite. For a large but finite system we should not take ϵ in (35) to go to zero, but remain larger than the spacing between levels, so that the δ-function in (36) is replaced by a finite peak covering many levels.

For our present purpose it is important that (36) is of the form

$$j = \mathcal{E}\int dE\; F(E)\left(-\frac{\partial f}{\partial E}\right) \tag{37}$$

where $F(E)$ is some function of the energy and f the Fermi function. $F(E)$ is then a kind of mobility, which will vary slowly with energy. The temperature enters only through $(-\frac{\partial f}{\partial E})$, which is a steep maximum near E_F of unit area, so that (37) is practically

$$\mathcal{E}F(E_F) \tag{38}$$

as long as kT is small compared to the energy range over which t vanishes. If $F(E)$ behaves like a power of E, (38) remains valid up to temperatures of the order of E_F/k. At these temperatures the expansion we have used is valid subject only to (27), and since

(38) no longer contains the temperature, the result is independent of T, and does not require (26).

This argument was given long ago by Landau, but never published by him. His starting-point was that it must be possible to express the conductivity in the form (37), and all we have done is to show explicitly that this is the case.

VII. THERE ARE STILL RESERVATIONS

Note that we rely essentially on the smoothness of F(E). If F(E) contained terms with rapid oscillations, the argument would fail, and the limit of our approximation would go towards (26). In metals there is no reason why there should be such oscillations, but it is dangerous in physics to assume the absence of a phenomenon because we know of no reason to expect it.

As an example of the situation that might arise, we may think of metals in a magnetic field. Here the quantization of the electron orbits results in a discrete spectrum with a spacing of $2\mu B$ (μ = Bohr magneton, B magnetic intensity) and if this is not small compared to kT, one gets oscillations in the susceptibility (de Haas-van Alphen effect). Here the temperature dependence is dominated by a factor of the form

$$e^{-\text{const}\, kT/\mu B} \qquad (39)$$

which vanishes when $\frac{\mu B}{kT} \to 0$ but cannot be expanded in a Taylor series in that quantity. Unfortunately this type of behaviour is not uncommon in statistical mechanics, and therefore looking for corrections involving higher powers of the small quantity under consideration is not always safe.

Many authors, including my student van Wieringen (1954), have looked at the next term in perturbation theory and its effect on the Boltzmann equation, and have found corrections of the relative magnitude $\hbar/\tau E_F$, not $\hbar/\tau kT$. This is comforting, but does not rule out exponential corrections, similar to (39). These are just what one might expect in cases in which a rapid oscillatory behaviour of F(E) invalidates Landau's argument, based on the form (37).

There is another, much more trivial sense in which perturbation theory may be inadequate. Our use of perturbation theory implies the use of first-order Born approximation for the scattering from a single centre. For a strong short-range force, for example a small hard sphere, this is invalid, although the scattering cross-section may be quite small. In that case it is easy to determine the exact scattering cross-section of a single centre from the appropriate Schrödinger equation and use it in Boltzmann's equation. It is on this basis that (27) (or in some cases perhaps (26)) represents the true limit of validity.

VIII. PAULI'S REMARK AND OFF-DIAGONAL ELEMENTS

There is one further aspect of the usual procedure which requires justification. In deriving the collision term of the Boltzmann equation, we ask, at any time, for the rate of change, over a small time interval, of the probability of the electron being in a state p. This is given by $|a_p|^2$, where a_p is an expansion of the electron wave function in terms of eigenfunctions of the Hamiltonian in the perfect lattice.

In deriving Fermi's golden rule, one assumes that the system is initially in a single such eigenstate, but as a result of the interaction there will be a complicated superposition

$$\psi = \sum a_p u_p \tag{40}$$

of the eigenstates u_p.

The probability $|a_p|^2$ at time $t + dt$ will contain not only squares of the form $|a_{p'}|^2$ at time t but also products of the form $a_{p'}^* a_{p''}$. This was first pointed out by Pauli (1928), who suggested averaging over the phases of the eigenstates at each step.

In another language, we can describe the ensemble of electrons in terms of a density matrix $\langle p | \rho | p' \rangle$, of which the diagonal elements give the probabilities or occupation numbers. Pauli's method involves following the time evolution of ρ over a small time interval, assuming it diagonal at t, and omitting the off-diagonal elements that may have developed at $t + dt$.

The merit of Pauli's paper lies in bringing out clearly the need for a step of this kind in the usual procedure. Since then there have been many papers to show how to avoid, or justify, this omission of off-diagonal terms.

For a more rigorous treatment, we might try to start from the Kubo formula, e.g. (36), and exploit the fact that the collision rate τ^{-1} is small. Unfortunately, the direct use of perturbation theory on this expression is useless. The point is that we are dealing with an expression for the conductivity, which, in the absence of collisions, would be infinite. An expansion from infinity as starting-point is not sensible.

I knew of the existence of a formula of the type of (36) many years ago, but regarded it as academic because I could not evaluate it even for the problems which I knew were soluble by the Boltzmann method. This did not do justice to the result, because it can easily be extended to oscillating fields. In that case nothing singular happens in the absence of collisions, and if the frequency ω of the field satisfies the condition

$$\omega\tau \gg 1 \tag{41}$$

a direct application of perturbation theory to the Kubo formula works well.

The static case is, in this sense, the most difficult. One attempt to give a valid derivation is found in the paper by Greenwood (1958) which has already been mentioned.

For this purpose he writes the density matrix in terms of eigenstates of the Hamiltonian H_0 containing only the kinetic energy of the electrons and the potential of the perfect lattice, but not that of the scattering centres.

By perturbation theory one then expresses the rate of change of the diagonal elements of ρ in terms of the off-diagonal ones. In the equation for the rate of change of the off-diagonal elements there occur both diagonal, and other off-diagonal elements. We assume the diagonal ones to be dominant, and can then eliminate the off-diagonal ones, to obtain an equation of motion containing only diagonal elements of ρ. This turns out to be the usual Boltzmann equation. It is now possible to examine approximately the error caused by omitting off-diagonal terms in the second equation, by going one step further, and Greenwood finds that the error is small provided (27) is satisfied.

In this argument, which is suggestive, rather than rigorous, it is important to use the spatial uniformity of the problem. This is connected with the stochastic uniformity of the scattering potential. If we consider the stochastic average of the potential $W(\underline{r})$ of the scattering centres, it is just a constant which can be absorbed in the lattice potential. A quadratic quantity like $\langle W(\underline{r}_1)\, W(\underline{r}_2)\rangle$ taken as a stochastic average, will depend only on $\underline{r}_1 - \underline{r}_2$. This results in a suppression of off-diagonal elements in $\underline{p}$. If the potential were exactly uniform, ρ would be exactly diagonal in $\underline{p}$. With a potential which is stochastically uniform, the diagonal elements dominate.

IX. SUMMARY OF LIMITS OF VALIDITY

So far I have restricted myself to systems in which we were essentially concerned with a one-body statistical problem. One of the conclusions reached was that the simple treatment by the Boltzmann equation was limited to systems with not too great a collision frequency. It is clear that there must be such a limitation, because, if the particles spend an appreciable fraction of their time under the influence of a scattering centre, the correct equilibrium equation of state is given by a Boltzmann factor $\exp\{\beta(T + V)\}$ (or the corresponding Fermi function) where T is the kinetic energy, and V the potential of the scatterers. To describe the latter, we must know the spatial correlations of the particles with the scatterers, for which there is no room in Boltzmann's equation. If we cannot describe statistical equilibrium, we clearly cannot expect to deal satisfactorily with transport problems.

Another way of looking at the same difficulty is that, if $\hbar/\tau > E_F$, the interaction with the irregularities is comparable with the main electron Hamiltonian. In other words, it is no longer possible to divide the Hamiltonian into an "unperturbed" part, which has simple eigenstates and a "perturbation" which is small. The same forces which dominate the dynamics, are responsible for the approach to equilibrium or for the limited response to an external field.

This is typically the situation in a liquid or in an amorphous substance, where it is no longer possible to set up an exact theoretical description, although one can obtain qualitative results in many practical cases.

X. EXTENSION TO MANY-BODY PROBLEM

Let us now remove the limitation to one-body problems, and consider binary collisions. To fix the ideas, consider collisions between free fermions. Let the momenta of the incident particles be $\underline{p}_1$, $\underline{p}_2$ and the final states $\underline{p}_1'$, $\underline{p}_2'$. Then if v_{12} is the relative velocity of approach and, for elastic collisions, also the velocity of separation, and $w(\underline{p}_1, \underline{p}_2; \underline{p}_1', p_2')$ again the differential collision cross-section, the number of collisions specified is

$$v_{12}\, w(\underline{p}_1\underline{p}_2;\, \underline{p}_1'\, \underline{p}_2')\, n(p_1)n(p_2)\, [1 - n(p_1')]\, [1 - n(p')] \qquad (42)$$

where the last factors again account for the number of particles in the initial states, and the number of vacancies in the final states. The difference between the rates for the specified collision and its inverse, which determines the rate of change in the occupation of one of the states in question, is

$$v_{12}\, w(\underline{p}_1\, \underline{p}_2;\, \underline{p}_1'\, \underline{p}_2') \Big\{ n(\underline{p}_1)\, n(\underline{p}_2)\, [1 - n(\underline{p}_1')][1 - n(\underline{p}_2)$$
$$- n(p_1')\, n(p_2')\, [1 - n(\underline{p}_1)]\, [1 - n(p_2)] \Big\} \qquad (43)$$

If we were dealing with bosons, $1 - n$ would in each case be replaced by $1 + n$.

XI. NEGLECT OF CORRELATIONS

(43) should strictly be read as giving the probability of an individual transition, minus that of the inverse transition with the $n(p)$ being 0 or 1. In finding the mean rates of transition, we should average the expression, thus forming an average of terms like

$$\langle n\, (\underline{p}_1)\, n(\underline{p}_2) \rangle \qquad (44)$$

and of higher powers. This involves the two and three-particle distributions, and if we look for an equation for their change with time, we require four and five particle distribution. This would lead us to an infinite hierarchy of linked equations.

Fortunately, this is unnecessary in the dilute system, because we can then assume, with Boltzmann, that the many-body distribution functions factorize, e.g.

$$\langle n(\underline{p}_1)\, n(\underline{p}_2) \rangle \;=\; \langle n(\underline{p}_1) \rangle \langle n(\underline{p}_2) \rangle\,. \qquad (45)$$

We have already discussed the "stosszahlansatz" which implies (45) <u>before</u> any collision. We noted also that <u>after</u> each collision the particles will be correlated. If a given pair of particles which had just collided, were likely to collide again, our assumption (45) would therefore be unjustified. However, in a dilute system the mean free path is much longer than the average distance between the particles. Before the two given particles are likely to collide again, they will have many opportunities of colliding with others, and this will wipe out the correlation. Evidently this would not apply in a dense system, in particular in which a pair of neighbouring molecules may make a large number of collisions with each other interspersed only each time by one collision with another.

We conclude that, for the dilute system, we may treat each of the $n(p)$ in (43) as an average.

XII. CONSERVATION LAWS

Then (43) should vanish in equilibrium. Indeed, the bracket can be expressed, in an obvious shorthand notation, as

$$n_1 n_2 n_1' n_2' \left\{ \frac{1-n_1'}{n_1'} \; \frac{1-n_2'}{n_2'} - \frac{1-n_1}{n_1} \; \frac{1-n_2}{n_2} \right\} \tag{46}$$

For the Fermi distribution

$$\frac{1-n}{n} = e^{(E-E_F)/kT} \tag{47}$$

so that the bracket in (46) is

$$e^{-2E_F/kT} \left\{ e^{(E_1' + E_2')/kT} - e^{(E_1+E_2)/kT} \right\} . \tag{48}$$

But for elastic collisions the cross-section w vanishes unless

$$E_1' + E_2' = E_1 + E_2 . \tag{49}$$

In general, there are other distributions which are not affected by the collisions. In a gas the collisions conserve momentum,

$$\underline{p}_1' + \underline{p}_2' = \underline{p}_1 + \underline{p}_2 \tag{50}$$

and therefore (44) will vanish for a distribution for which there is an exponential function of momentum multiplying the right-hand side of (47). This is a Fermi gas with an arbitrary drift velocity.

The most general distribution which will be stationary in the presence of collisions is therefore of the form

$$n(\underline{p}) = \left\{ e^{(E(\underline{p})+\underline{u}.\underline{p} - E_F)/kT} + 1 \right\} \qquad (51)$$

where T and E_F are arbitrary quantities and $\underline{u}$ an arbitrary vector. These parameters reflect the conservation of energy, particle number, and momentum. In general it is important to know what conservation laws hold, in order to understand the effect of the collisions.

XIII. LIMIT ON COLLISION RATE

The next question is how to derive the collision cross-section w. Here we can be brief, because the considerations are similar to those in the one-body problem. It may be convenient to express the collision cross-section in terms of first-order Born approximation when this is justified, but when it is not, as for strong short-range interactions, one can use the exact solution of the two-body collision problem instead.

Assuming this is done when necessary, the only remaining condition for the validity of the Boltzmann equation is a limit on the collision rate $1/\tau$.

A system will certainly be dense, and a treatment by Boltzmann's equation invalid, if $\hbar/\tau$ exceeds E_F. There may, however, be some difficulty already when $\hbar/\tau$ exceeds kT, since the previous reasoning which allowed us to replace (26) by (27), depended strongly on the absence of energy exchange. (In metals at temperatures above the Debye temperature Θ the energy exchange in electron-phonon collisions is at most $k\Theta$, i.e., smaller than kT. It is likely that such collisions can be regarded as changing the electron energy by a negligible amount, so that the Landau argument remains valid.)

In addition, we have seen that there are several different definitions of collision time, according to the purpose. One might guess that small-angle collisions are not very important in creating complications, so that the "transport collision time" τ_1, gives the right measure.

In special circumstances this might be different. For example, in a magnetic field the electron motion in the plane perpendicular to the field is quantized because the motion consists of closed orbits. A deflection by quite a small angle might be sufficient to throw the electron from one such orbit to a neighbouring one, and hence it might be conjectured that in the magnetoresistance problem $\hbar/\tau$, not $\hbar/\tau_1$, measures the strength of the coupling.

XIV. OMISSION OF OFF-DIAGONAL TERMS. RIGOROUS DERIVATIONS

We again meet the problem of how to justify an equation in terms of probabilities, i.e. not involving phase relations, or off-diagonal elements of the density matrix.

There have been many approaches to this problem, the best known being those by Van Hove (1955, 1956) and Prigogine (1962) (see also the review by Chester, 1963).

It is not possible to give an account of such methods within the scope of these lectures, but I shall give some brief indication of the ideas of Van Hove.

Van Hove starts by considering the time evolution operator

$$e^{-iHt/\hbar} \qquad (52)$$

where H contains the Hamiltonian of independent particles and also their interaction, but not any external field. In order to keep track of the orders of magnitude it is convenient to define the interactions with a parameter λ :

$$H = H_0 + \lambda V \quad . \qquad (53)$$

Perturbation theory then amounts to an expansion in powers of λ.

If a direct expansion is terminated after a few powers of λ, it will be a reasonable approximation only for very short times. To study the approach to equilibrium, one has to go at least to times of the order of the collision time, which is proportional to λ^{-2}. (We again ignore the possibility that the individual interaction might be too strong to be described in first-order Born approximation.)

This suggests that in evaluating (53) one should treat λ as small, and t as large, the quantity $\lambda^2 t$ being finite. One includes all powers of $\lambda^2 t$, but neglects $\lambda^m t^n$ with $n < 2m$ (terms with $n > 2m$ do not occur).

The collection of the relevant terms is assisted by the "diagonal singularity" property of all physically interesting operators. By this one means that the off-diagonal elements are, in the limit of an infinite system, negligible compared to the diagonal ones. This is reminiscent of the result used by Greenwood in the scattering from random centres, where it was a consequence of the stochastic uniformity of the system.

Evidently the system considered by Van Hove must be statistically uniform, otherwise it would be difficult to go to the limit of infinite size. This accounts for the diagonality in momentum, or wave vector.

I am not clear whether Van Hove's argument is adequately represented by this remark. In a Bloch problem electric states are described by wave vector and band label. Should we expect matrix elements to be small if they are diagonal in wave vector but non-diagonal in the band label? In many cases such inter-band terms belong to large energy differences and are probably small on that account. But there are cases of degenerate bands, in which two states in different bands with the same wave vector can have nearly the same energy.

Mr. G. Kirczenow has convinced me that the diagonality may not be confined to the cases resulting from uniformity and I must leave this question open.

Using these concepts, Van Hove can show that the sum of the relevant terms leads to a quantity whose time-dependence is given by the Boltzmann equation. However, since λ has been treated as small, this is, of course, still dominated by the limitation that the collisions are not too frequent, as we have discussed before.

XV. ILLUSTRATIONS. LATTICE THERMAL CONDUCTIVITY

I shall now turn to specific applications. So far we have discussed mainly general principles. I want to look at particular examples both to clarify the general methods by illustration, and also to make a little propaganda for the Boltzmann equation. Of course this is not universally applicable, and we know it does not work for dense systems. But in its domain of applicability it has advantages compared to more sophisticated methods, as we shall see.

We have already discussed a particularly simple example, the electric conductivity, in the case of scattering by imperfections or other fixed scattering centres. For not too great a density of scatterers this can be treated by the Boltzmann equation, and I then pointed out that one has to take care with the concept of a collision time. The uncritical use of a collision time or mean free path, is, in fact, less sophisticated than the Boltzmann equation.

Let us now consider an example of mutual collisions, and I shall choose for this the thermal conductivity of non-metallic crystals. I have some affection for this problem, on which I wrote my Ph.D. thesis in 1929 (Peierls, 1929, 1955), and even after this long time there remain points worth further discussion.

This problem has an interesting history, because a great many plausible attempts were, in fact, incorrect. Of these I shall mention that of Debye. He realized that the heat was carried by lattice waves (today we talk of "phonons") and that their scattering, without which the conductivity would be infinite, was due to the forces not being exactly harmonic. To estimate this scattering he used an elegant and simple idea, typical of Debye's approach to physics.

The main effect of the anharmonicity is to make the elastic constants, and hence the sound velocity, density dependent. The presence of lattice vibrations causes fluctuations in the density, and hence of the wave propagation, and this results in scattering, just as the fluctuations of the refractive index for light cause light scattering. Debye estimated in this way the thermal conductivity by analogy with the case of light and found that (at least in the classical region $T > \Theta_{Debye}$) the conductivity K at temperature T was proportional to T^{-1}, in apparent agreement with experiment. The empirical situation was not as clear as one might wish because at that time the only relevant experiments were those of Eucken in about 1908, measuring

κ for a few solids at room temperature and in liquid air. Even today the high-temperature behaviour of κ is not known very accurately.

In spite of its appealing simplicity, Debye's approach is incorrect. The reason is that it treats the density fluctuations as static. This is a justified approximation for light, because the fluctuations, which are due to the thermal vibration of the lattice, move with the velocity of sound, which is very slow compared to light. But this approximation is not good for the scattering of the lattice waves themselves. As a result, Debye's argument would predict, for example, a finite thermal conductivity for an anharmonic elastic continuum, whereas, as we shall see, its conductivity would be infinite.

Pauli realized that a more detailed theory was needed, and did some preliminary calculations on which he reported at a conference. The published abstract of this talk is, I believe, Pauli's only wrong published result. However, he evidently was not satisfied and asked me to study the problem further.

XVI. CONTENT OF BOLTZMANN EQUATION

In a crystal, the potential energy U is a function of the atomic displacements u, and has the form

$$U = \text{const.} + \alpha u^2 + \beta u^3 + \ldots \qquad (54)$$

This is symbolic, since there are many atoms, and each displacement has three components. There is no linear term, since $u = 0$ represents the equilibrium, by definition. The usual theory of harmonic lattice vibrations is obtained by including only the quadratic term, and the cubic term can be treated as a perturbation and will then cause transitions. Since u is a linear combination of phonon amplitudes, each of which has matrix elements increasing or decreasing a phonon number by 1, u^3 contains terms in which two phonons are destroyed and one created, or vice versa,

$$\underline{k}_1 \text{ and } \underline{k}_2 \rightleftarrows \underline{k}_3 \qquad (55)$$

There are also matrix elements corresponding to the creation or destruction of three phonons, but such a transition could never conserve energy.

The Boltzmann equation contains only energy-conserving transitions for which (since the phonon energy is $\hbar\omega$)

$$\omega_1 + \omega_2 = \omega_3 \tag{56}$$

In addition, there is wave vector conservation, which is due to the translation symmetry of the problem. In a continuous medium, invariant under infinitesimal translations, the sum of all wave vectors would be conserved like the sum of the momenta in the collision of free particles. A lattice is invariant only under translations by a multiple of the basic lattice vectors, and therefore the conservation law is

$$\underline{k}_1 + \underline{k}_2 = \underline{k}_3 + \underline{K} \tag{57}$$

where $\underline{K}$ is any reciprocal lattice vector.

If $\underline{K} \neq 0$, we speak of an Umklapp process, a rather ugly German term which I used for this, and which has become accepted. To illustrate it, consider a linear chain of spacing a, where K is zero or $\pm 2\pi/a$. Two waves travelling to the right as shown, k_1 and k_2

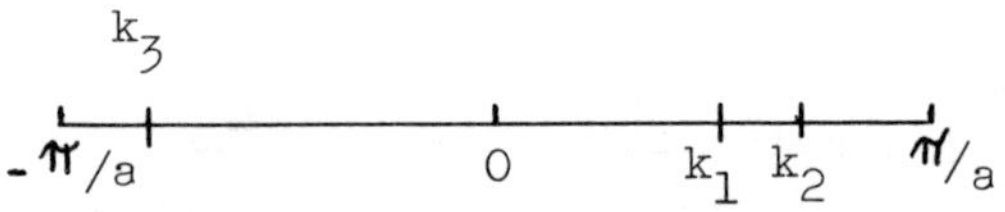

may be such that $k_1 + k_2 > \pi/a$

and therefore lies outside the defining interval, so that the result of the interaction would be $k_3 = k_1 + k_2 - \frac{2\pi}{a}$, a wave travelling to the left. We can think of an Umklapp process as an ordinary interaction of lattice waves together with a Bragg scattering.

The collision term in the Boltzmann equation will contain a factor

$$\left\{(N_1+1)(N_2+1)N_3 - N_1N_2(N_3+1)\right\} \tag{58}$$

where $N_1 \equiv N(k_1)$ is the number of phonons of wave vector k_1, etc. The first term comes from the transition from right to left in (55), the second from the inverse.

Near equilibrium we can write

$$N(k) = N^0(\omega_k) + g(\underline{k}) \frac{\partial N^0(\omega_k)}{\partial \omega} \tag{59}$$

where N^o is the Planck distribution and the factor $\frac{\partial N^o}{\partial \omega}$ has been included in the definition for convenience. Inserting in (58) and keeping only linear terms in g, we find, apart from factors,

$$(N_1 + 1)(N_2 + 1)N_3(g_3 - g_1 - g_2) \tag{60}$$

This vanishes, and therefore the phonon distribution is stationary, if $g(k) \propto \omega(k)$, since for all allowed transitions (60) then vanishes by (56). This arises from energy conservation and means that a change in the phonon temperature will not be removed by the collisions.

XVII. EXPONENTIAL BEHAVIOUR

If there were no Umklapp processes, i.e. if $\underline{K}$ were always zero in (57), then there would be another stationary distribution if $g(\underline{k}) \propto \underline{k}$. This would represent a phonon drift, which would evidently carry a heat transport, without the need for a temperature gradient, hence the thermal conductivity would still be infinite. This would be the position in a continuous medium.

In a real crystal, Umklapp processes will destroy such a drift, but they are rare at low temperatures since they require the presence of at least one short-wave phonon, and their number decreases exponentially. We thus expect at low temperatures an increase in the thermal conductivity of the form

$$K \sim e^{\gamma \Theta / T} \tag{61}$$

where Θ is the Debye temperature, and γ a numerical factor less than unity.

This result shows that a discussion in terms of collision times, or phonon lifetimes, can be very misleading. The natural way of defining a phonon lifetime would be to assume one phonon added to the equilibrium distribution, and to watch it decay. In that case no problem of a drift arises, and the lifetime will not grow exponentially at low temperatures.

Although the rise indicated by (61) was predicted in 1929, it was found only in 1951 by Berman (1951). There were two reasons why it was hard to find. Firstly, we expected it to appear at temperatures somewhat lower than Θ. In fact, Debye's definition of Θ is such that the completely classical situation is reached immediately above Θ, but the extreme quantum limit is reached only at temperatures of $\frac{1}{10}\Theta$ or so, as one can see from the specific heat curve.

To show the effect, one has therefore to go to very low temperatures, which means using very pure crystals, since otherwise impurities could mask the phonon interaction.

The second reason is embarrassing for the theoreticians. It was not immediately realized that to be pure for the present purpose a crystal must also be isotopically pure. Most crystals contain a random mixture of isotopes, whose different masses affect the kinematics of lattice waves, and lead to scattering. This was pointed out by Pomeranchuk (1943) but in wartime conditions his paper was at first overlooked. In his experiments, Berman noticed that only some materials showed the exponential rise, and these were the ones containing only one dominant isotope for each constituent.

The same need for Umklapp processes arises in the electric conductivity of pure metals. In his theory, Bloch derived the well-known T^5 law for the resistance by assuming the phonons always to remain in equilibrium. This is justified if the relaxation of the phonon distribution is faster than that of the electrons, which is true at high temperatures. For equilibrium it is, however, essential that the wave vector conservation be broken by Umklapp processes in electron-phonon, or phonon-phonon interactions (or by impurity scattering). In many metals it seems certain that ultimately Umklapp processes will become rare and there should be a law of the type of (61), though with a smaller γ.

This behaviour was expected, in particular for the alkalis, since 1930, I was interested to hear here from Professor Gitterman that there exist recent experiments, which I had overlooked, which show this effect in the alkalis. After all nature seems to behave as it ought to.

XVIII. THE USE OF MODERN METHODS

I have set out the phonon conductivity problem in such detail because it is a good example of the logical (if not always mathematical) simplicity of the Boltzmann equation.

A more modern approach, using the techniques of many-body theory, was tried by J. Ranninger (1965). He found a solution by which the thermal conductivity of the lattice went to infinity at $T = 0$, which is satisfactory. However it did not behave exponentially like (61) but only as a power of T. Indeed, it would have given a finite conductivity even without Umklapp processes.

In view of this manifestly incorrect result, Ranninger re-examined his derivation and found a very subtle way in which his previous result had to be corrected. It is doubtful if the correct way of handling the ingenious and sophisticated approach would ever have been found without the knowledge of the correct answer from the Boltzmann equation.

Another example concerns the use of modern techniques to calculate the electric conductivity of metals with impurity resistance, which we have already discussed several times. I refer to work by Edwards (1958) and by Chester and Thellung (1959), which was done in the same spirit, though with different techniques.

Chester and Thellung start from the Kubo formula, which is not easy to evaluate directly in the case of a constant field, as we have seen. The important quantity to evaluate in that formula is

$$j(t)\; j(0) \tag{62}$$

where $j(t)$ is the electron current at time t. This can also be expressed as

$$U(t)j(0)\; U(-t)j(0) \tag{63}$$

where $U(t)$ is the time evolution operator for the full Hamiltonian, including the scattering potential, but not the electric field. $U(t)$ is now expanded in powers of the scattering potential and the leading terms collected, following the principles set out by van Hove.

In applying van Hove's ideas to an expression of the form (63), it is natural to pay attention to the terms which would contribute to the expectation value of $U(t)$. If only these terms were included, one would obtain for the conductivity the result $e^2 n \tau / m$, with $1/\tau$ being the total collision rate. As we saw earlier, this is not correct, because the conductivity depends on τ_1, the transport collision time. This knowledge helped the authors to look for the source of the difference, and they found that it was essential to combine the expansion of $U(t)$ and $U(-t)$; there are terms in each which vanish when averaged separately but whose product is non-zero.

In the isotropic case this immediately led to the extra $\cos \Theta$ term in the definition of the transport collision rate, and hence to the right answer.

For the anisotropic case we know that the transport collision time is defined by an integral equation which, in general, does not have an explicit solution. In cases in which the Boltzmann equation is justified, any other correct evaluation of the conductivity must therefore lead to the same integral equation, or its equivalent.

Chester and Thellung found that, in the anisotropic case, the terms which must be taken into account from an infinite series of powers not of a number, but of an operator. The way to sum the series is first to diagonalize the operator. The equation defining the eigenvalues of the operator turns out to be identical with the equation defining the eigenvalues of the Boltzmann collision operator, and from this the equivalence follows fairly directly.

These examples show the power of the Boltzmann equation when it is appropriate. It is certainly not as general as the Kubo formula, but difficulties arise in the evaluation of the latter. If one uses very abstract methods for its evaluation, one may lose sight of the physical content of the approximations made. This applies particularly to the use of Green's function techniques.

These lead, in principle, to an infinite set of coupled equations, which cannot be solved directly. It is customary to simplify these by some kind of "decoupling" in which one assumes the higher-order functions expressible in terms of products of the lower-order ones. It is usually very hard to understand the nature of the error made in this step, and it is hard to check the reliability of the approximation, unless one already knows the answer from simpler arguments - in which case the more abstract method is not really necessary.

XIX CONCLUSIONS

Our discussion has been confined to a very simple class of transport problems. Apart from the exclusion of dense or strongly coupled systems, we have excluded the possibility of long-range interactions, as in plasmas and other dense systems, all cases of spatial inhomogeneity. Nevertheless the limited range of problems gave us an opportunity of examining some general basic principles, which are applicable much more generally than the particular problems considered.

REFERENCES

BERMAN, R., 1951, Proc. Roy. Soc. A, 208, 90.

CHESTER, G.V. and THELLUNG, A., 1959, Proc. Phys. Soc. 73, 745.

CHESTER, G.V., 1963, Reports on Progress in Physics 26, 411.

EDWARDS, S.F., 1958, Phil. Mag. 3, 1020.

GREENWOOD, D.A., 1958, Proc. Phys. Soc. 71, 585.

KUBO, R., 1956, Canad. J. Phys. 34, 1274.
1958, Lectures in Theoretical Physics, Boulder 1, 120.

PAULI, W., 1928, Festschrift zum 60.Geburtstage Arnold Sommerfelds (Leipzig, Hirzel), p. 30.

PEIERLS, R.E., 1929, Ann. der Physik, 3, 1055.
1955, Quantum Theory of Solids (Oxford, Clarendon Press).

POMERANCHUK, 1943, J. of Physics U.S.S.R. 7, 197.

PRIGOGINE, I., 1962, Non-Equilibrium Statistical Mechanics (New York; Interscience).

RANNINGER, J., 1965, Phys. Rev. 140A, 2031.

VAN HOVE, L., 1955, Physica 21, 517 and 901.
1956, Physica 22, 343.

VAN WIERINGEN, J.S., 1954, Proc. Phys. Soc. A, 67, 206.

ENTROPY, DYNAMICS AND SCATTERING THEORY

I. Prigogine and F. Mayné

Faculté des Sciences
Université Libre de Bruxelles
Bruxelles, Belgium

I. INTRODUCTION
II. THE MAC KEAN MODEL
III. IRREVERSIBILITY AS A SYMMETRY BREAKING
IV. STAR UNITARY TRANSFORMATION
V. CONSTRUCTION OF THE Λ - TRANSFORMATION
VI. POTENTIAL SCATTERING
VII. CONCLUDING REMARKS
REFERENCES

I. INTRODUCTION

Time reversal invariance has always been associated with the basic laws of dynamics. The discovery of irreversible processes satisfying quite different evolution equations (Fick's law, heat conduction ...) at the beginning of the 19th century came therefore as a great surprise.

The existence of irreversible processes is summarized in the second law or principle of Thermodynamics. This principle states that there exists a function, the entropy, whose time variation can be split into two parts

$$d\,S = d_e\,S + d_i\,S \qquad (1.1)$$

such that $d_i\,S$, the entropy production inside the system is non-negative. Therefore, when the system is isolated ($d_e S = 0$), the entropy never decreases.

But irreversible processes appear not only in connection with classical phenomena such as heat conduction but also at a much more fundamental level, at the very core of quantum mechanics, in the measurement process, or in elementary particle physics, through the existence of unstable states.

For a very long time, the second law was mainly used to describe the final equilibrium state, corresponding to maximum entropy or minimum free energy. However, the emphasis has been shifting towards non equilibrium processes, to a situation more and more away from equilibrium, because one of the unexpected features which has been noticed in the last years is that deviation from equilibrium may be a source of "non-Boltzmannian type of order". - This is certainly a very important aspect for the understanding of many manifestations of nature around us.

Therefore, one has to understand more clearly what entropy and irreversibility mean from the point of view of classical or quantum mechanics.

We shall first discuss Boltzmann's interpretation of Entropy [1].

Boltzmann's approach is based essentially on the recognition that we are dealing with very complicated dynamical systems. For this reason he felt free to replace the dynamical description by a stochastic process, through the use of a kinetic equation.

From this kinetic equation, one can define an $\mathcal{H}$-quantity and one can prove from this equation that $\mathcal{H}$ can only decrease. Boltzmann was therefore led to identify his $\mathcal{H}$-quantity with entropy.

Numerical experiments [2][3] can be performed, showing indeed the decrease of the $\mathcal{H}$-quantity as predicted by Boltzmann.

However, more sophisticated experiments can also be done where at some instant, all velocities are reversed. In that case the Boltzmann $\mathcal{H}$-quantity first goes back to its initial value. (Loschmidt paradox [4]).

This behavior appears as very unsatisfactory from the point of view of thermodynamics. If this was true, it would mean that over a macroscopic period, there would be a decrease of entropy corresponding to an antithermodynamic behavior. If the entropy produced during one period could indeed be suppressed during a later period, then the very definition of irreversible processes would become questionable.

Also, if there would be thermodynamic as well as non-thermodynamic types of behavior, could we then speak at all of a second law of thermodynamics?

Boltzmann recognized that when we reverse the velocities, we cannot expect $\mathcal{H}$ to decrease, the reason being that, when we reverse the velocities we introduce correlations between the particles and the hypothesis of molecular chaos is no longer valid. This then would lead to an antithermodynamic behavior.

However, molecular chaos is only a special initial condition. Therefore, if irreversibility will be based on this initial condition, we would again be in trouble to understand the generality of the second law.

To show in more detail the difficulty involved in Boltzmann's interpretation of irreversibility, we shall use for illustration a simple model, the Mac Kean Model.

II. THE MAC KEAN MODEL

The Mac Kean model [5][6][7][8] consists of a system of n particles each of which can have only the velocity +1 or -1, and with a very simple law of collision. When two particles with velocities

e_1 (± 1) and e_2 (± 1) collide, they emerge with velocities e_1^*, e_2^* respectively. With probability $1/2$, the final velocities are

$$e_1^* = e_1 \quad , \quad e_2^* = e_1 e_2$$

or

$$e_1^* = e_1 e_2 \quad , \quad e_2^* = e_2 \tag{2.1}$$

The different possibilities are

$$+1,+1 \begin{cases} +1,+1 \\ +1,+1 \end{cases} \qquad -1,-1 \begin{cases} -1,+1 \\ +1,-1 \end{cases}$$

$$+1,-1 \begin{cases} +1,-1 \\ -1,-1 \end{cases} \qquad -1,+1 \begin{cases} -1,+1 \\ -1,+1 \end{cases} \tag{2.2}$$

The corresponding master equation for the N distribution function takes the form

$$\frac{\partial \rho(e_1, e_2, \dots e_N, t)}{\partial t} = \frac{1}{N} \sum_{1 \le i < j \le N} \{ \rho(e_1, \dots e_i, \dots e_i e_j, \dots e_N, t)$$

$$+ \rho(e_1, \dots e_i e_j, \dots e_j, \dots e_N, t) - 2 \rho(e_1 \dots e_i \dots e_j, \dots e_N, t) \} \tag{2.3}$$

One can look immediately for factorized distribution functions corresponding to molecular chaos:

$$\rho(\{e_i\}, t) = \prod_{i=1}^{N} f(e_i, t) \tag{2.4}$$

with

$$f(+1, t) + f(-1, t) = 1 \tag{2.5}$$

Replacing in the master equation (2.3) ρ by (2.4), one obtains immediately a non-linear equation for the one-particle distribution function

$$\frac{d f(e)}{dt} = f(e) f(+1) + f(-e) f(-1) - f(e) \tag{2.6}$$

which can be written using (2.5) as

$$\frac{df_+}{dt} = 1 - 3 f_+ + 2 f_+^2 \tag{2.7}$$

in which $f_+ \equiv f(+1, t)$

The remarkable feature of this model is that this non-linear equation can be solved immediately. One obtains

$$f_+(t) = \frac{1}{2} + \frac{\Lambda}{2} \frac{e^{-t}}{1 - \Lambda(1 - e^{-t})} \quad ; \quad \Lambda = 2 f_+(t=0) - 1 \tag{2.8}$$

One can now introduce Boltzmann's $\mathcal{H}$ quantity

$$\mathcal{H}_B = \int dv \, f \log f$$

which in this case takes the form:

$$\begin{aligned} \mathcal{H}_B &= f_+ \log f_+ + f_- \log f_- \\ &= f_+ \log f_+ + (1 - f_+) \log (1 - f_+) \end{aligned} \tag{2.9}$$

and verify that it can only decrease in time

$$\frac{d\mathcal{H}_B}{dt} = (1 - f_+)(1 - 2 f_+) \log \frac{f_+}{1 - f_+} \leq 0 \tag{2.10}$$

The equilibrium condition is the extremum of $\mathcal{H}_B$: $f_+ = 1/2$. (One has to discard the solution $f_+ = 1$ corresponding to a trivial invariant of motion due to the collision law which does not alter two particles both having the velocity $+1$).

One can now discuss the evolution of the Mac Kean model without introducing a priori the factorization condition (2.4). (see F. Henin [7][8] for a detailed analysis). The evolution of the system can be described in terms of the moments

$$\langle e_1^{\alpha_1} e_2^{\alpha_2} \dots e_n^{\alpha_n} \rangle \tag{2.11}$$

and as

$$\begin{aligned} e_i^{\alpha_i} &= 1 \qquad \alpha_i \text{ even} \\ e_i^{\alpha_i} &= e_i \qquad \alpha_i \text{ odd} \end{aligned} \tag{2.12}$$

the only independent moments are

$$x_p = \langle e_1 e_2 \dots e_p \rangle \tag{2.13}$$

Using the master equation (2.3) one obtains easily the equation describing the evolution of the moments

$$\frac{dx_p}{dt} = \frac{p(p-1)}{N} x_{p-1} - \frac{p(N-1)}{N} x_p + \frac{p(N-p)}{N} x_{p+1} \tag{2.14}$$

for $1 \leq p \leq N$

If one looks for moments such that p is much smaller than N $(p \ll N)$ one obtains a simplified form

$$\frac{dx_p}{dt} = -p(x_p - x_{p+1}) \tag{2.15}$$

which admits the factorized solution

$$x_p = \langle e_i \rangle^p \tag{2.16}$$

with

$$\frac{d\langle e_1 \rangle}{dt} = -\langle e_1 \rangle (1 - \langle e_1 \rangle) \tag{2.17}$$

But this is a very special case corresponding to particular initial conditions. It simply shows that, given a factorized initial condition, it propagates during the evolution.

However, one can immediately consider situations in which Boltzmann's entropy has even a wrong qualitative behavior. For illustration consider equation (2.15) for the one particle moment

$$\frac{dx_1}{dt} = -(x_1 - x_2)$$

an initial condition $x_2(0) = x_1^2(0)$ would correspond to the Boltzmann situation, but if one chooses $|x_2| > |x_1|$ with opposite sign, one obtains an "antithermodynamic behavior". Boltzmann's $\mathcal{H}$ quantity will first increase! This does not mean at all that the system is not going to equilibrium. In fact, Mac Kean's model corresponds to a Poisson process and it can be proved by studying the spectrum of equation (2.14) that the distribution always goes to equilibrium whatever the initial conditions.

One can even construct a Liapounoff function such as

$$\sum_{\{e_i\}} \rho^2(e_1, \dots e_N, t) \tag{2.18}$$

which in terms of the moments is given by [7] [8]

$$\sum_{\{e_i\}} \rho^2 = \frac{1}{2^N} \sum_{p=0}^{N} \frac{N!}{p!(N-p)!} x_p^2 > 0 \tag{2.19}$$

This function decreases in time whatever the initial conditions

$$\frac{d}{dt}\sum_{\{e_i\}} \rho^2 = -(N-1)\sum_{p=0}^{N-2}\frac{(N-2)!}{p!(N-2-p)!}\left(x_{p+1}-x_{p+2}\right)^2 \leq 0 \qquad (2.20)$$

Equilibrium is reached when

$x_1 = x_2 = \dots = x_n$ for which (2.20) vanishes. This shows that irreversibility is not related to molecular chaos but has a much deeper origin. Also entropy in general is not linked to the one particle distribution, but in general has to be linked to the Liapounoff function (2.18) which is related to the complete dynamics of the system. Only in very special situations can one reduce entropy to a function of the one particle distribution function. Only near equilibrium, can one express it in terms of macroscopic quantities and then the second law takes its phenomenological form. But in general the second law is a <u>theorem in "dynamics"</u> and not in phenomenological Physics.

The consideration of Mac Kean's model shows clearly (see [7] [8]) that if one starts from a non-factorized initial condition, in the course of evolution the correlations die out, the longest correlation time being contained in the linearized Boltzmann equation. Before reaching equilibrium, the system evolves through the linearized Boltzmann equation. But in general the system goes to equilibrium without satisfying at any time the non linearized Boltzmann equation. All these results can be generalized to other soluble models, such as the Kac model (Henin (to appear)).

The empirical success of Boltzmann's equation, the physical character of the assumption on which it is based, show that it constitutes an important step into the right direction. However, Boltzmann's arguments contain a lot of plausible assumptions which have to be made more explicit and related to dynamics.

In the following we shall present a method to construct a Liapounoff function playing the role of Boltzmann's $\mathcal{H}$ quantity, and which can be constructed for a wide class of dynamical systems.

Our main problem will be to understand how cross sections (or related quantities) can at all be introduced into the equations describing the time evolution of dynamic systems. It has to be kept in mind that both in classical and quantum mechanics we start usually with a hamiltonian description. The question is then what is the relation between a dynamic description in terms of the hamiltonian and the dynamic description in terms of cross sections or more generally in terms of physical "processes". We shall see that we begin to be able to give an answer to this fascinating question.

III. IRREVERSIBILITY AS A SYMMETRY BREAKING

Our starting point is the Liouville equation (see for instance [9])

$$i\partial_t \rho = L\rho \tag{3.1}$$

where

$$L\rho = \begin{cases} -i\{H,\rho\} & \text{Poisson Bracket} \\ [H,\rho]_- & \text{Commutator} \end{cases} \tag{3.2}$$

L is a hermitian superoperator (in the quantum case, it acts on the space of density operators) (see [10] [11])

$$L = L^+ \tag{3.3}$$

The most fundamental property of the Liouville equation is its <u>" Lt - invariance"</u>, that is, it remains invariant under the simultaneous changes

$$\begin{aligned} L &\to -L \\ t &\to -t \end{aligned} \tag{3.4}$$

This kind of property does not exist in macroscopic equations for thermodynamic quantities. For instance, in the heat conduction equation

$$\frac{\partial T}{\partial t} = \kappa \frac{\partial^2 T}{\partial x^2} \tag{3.5}$$

the change $t \to -t$ leads to a completely different equation as it has no meaning to reverse the sign of the heat conductivity κ which is a positive number.

The Fourier equation describing the evolution to uniform temperature in the future is changed by the operation $t \to -t$ into an "Anti-Fourier equation" leading to uniform temperature in the past.

Another example is the evolution equation for free particles

$$\frac{\partial \rho}{\partial t} + \underline{v}\frac{\partial \rho}{\partial \underline{x}} = 0 \tag{3.6}$$

which is Lt -invariant ($L \to -L$ corresponds here to $v \to -v$).
But if one adds a collision term

$$\frac{\partial \rho}{\partial t} + \underline{v}\,\frac{\partial \rho}{\partial \underline{x}} = -\,\frac{\rho - \rho^{eq}}{\tau} \qquad (3.7)$$

the equation is no more invariant as the collision term is not affected by the change ($t \to -t,\ v \to -v$).

The Lt-invariance does not prevent, for instance, mixing but this kind of process is described by laws symmetrical in time which do not permit the introduction of a Liapounoff function which could then be related to entropy.

The recent development of ergodic theory has greatly increased our understanding of the dynamic conditions to be satisfied to obtain "mixing". However, for the reason we have just mentioned, the ergodic approach has till now been unable to come even near to the problem we are discussing here: the microscopic meaning of entropy. On the other hand, the method we shall discuss now, permits to formulate this problem in general and to solve it rigorously in simple cases.

Let us start with the formal solution of the Liouville equation (3.1) which can be written as

$$\rho(t) = e^{-iLt}\rho(0) = \frac{1}{2i\pi}\int_C dz\, e^{-izt}\,\frac{1}{z-L}\,\rho(0) \qquad (3.8)$$

in terms of the resolvent $(z-L)^{-1}$ of the Liouville operator, and of the contour C.

In the case of the initial value problem, the contour C^+ has to be traced in the upper half plane corresponding to the complex z variable. For a final value problem, on the contrary, the contour, must be taken below the real axis [9][12]

The question is to know if the different choices of contour generate different solutions. If the solutions are the same, we shall be in the "reversible" case. This is the situation when the singularities of the resolvent are isolated poles on the real axis.

However, if continuous parts appear in the spectrum, we'll have to take into account analytic continuation. In that case the solutions will in general be different and we may expect "thermodynamic behavior".

To discuss the resolvent, it is convenient to introduce orthogonal hermitian projection superoperators P, Q such that

$$P^2 = P \quad ; \quad Q^2 = Q \qquad PQ = QP = 0$$

$$P + Q = 1 \tag{3.9}$$

fixing the language in which the results will be formulated. Generally P is chosen as projecting onto the diagonal elements in the representation in which some model Hamiltonian H_0 is diagonal

$$\langle m|P\rho|n\rangle = \langle m|\rho|m\rangle \delta_{m,n}$$

For this reason $P\rho$ is called "vacuum of correlations"

Let us also introduce a set of auxiliary irreducible operators, the collision operator

$$\psi(z) = PLQ \frac{1}{z - QLQ} QLP \tag{3.10}$$

the creation (of correlations) operator

$$\mathcal{C}(z) = \frac{1}{z - QLQ} QLP \tag{3.11}$$

the destruction operator

$$\mathcal{D}(z) = PLQ \frac{1}{z - QLQ} \tag{3.12}$$

and the propagation of correlations operator

$$\mathcal{P}(z) = \frac{1}{z - QLQ} \tag{3.13}$$

the most important of which, the collision operator, $\psi(z)$ corresponds to a transition from the vacuum of correlations to the vacuum of correlations through a dynamical evolution in the correlation space.

One can then express the resolvent $(z - L)^{-1}$ in terms of these operators. The result is [13][14][15]

$$\frac{1}{z - L} = \{P + \mathcal{C}(z)\} \frac{1}{z - \psi(z) - PLP} \{P + \mathcal{D}(z)\} + \mathcal{P}(z) \tag{3.14}$$

In general this expression is quite complicated to analyse, and for the sake of illustration, we shall consider a simple case, coming back later to the general situation. When the relaxation time is much larger than the duration of a collision, the asymptotic limit (for large t) can be described in terms of the limit $z \to 0$. [9] This leads for the evolution of the diagonal elements ρ_0 of the density operator to the expression

$$i \frac{\partial}{\partial t} \rho_0 = \psi(0) \rho_0 \tag{3.15}$$

with

$$\psi(0) = \lim_{z \to +i\epsilon} \left\{ PLQ \frac{1}{z - QLQ} QLP \right\} \tag{3.16}$$

(P has been chosen such that $PLP = 0$)
In the limit of large systems, the sum over the intermediate states Q involves an integration. Then one can formally write

$$-i\psi(0) = -\pi PLQ\, \delta(QLQ)\, QLP + i PLQ \frac{1}{QLQ} QLP \tag{3.17}$$

the second term being understood as a principal part.

When the first term in $\delta(QLQ)$ exists, we obtain a new type of behavior. Indeed if we now perform the change $L \to -L$, this term does not change sign.

$\psi(0)$ which is the operator describing the evolution of ρ_0 (3.15) contains an even part and an odd part in L. We have a breaking of the symmetry Lt.

Furthermore we have two other very important properties which appear at this point.

First the even part of the collision operator $-i\psi(0)$ has a well defined sign

$$-i\overset{e}{\psi}(0) < 0$$

This is exactly the property necessary to ensure that the system will go to equilibrium. The even part acts as a kind of "friction". This friction leads then in turn to the validity of the second law without any appeal to probabilistic or stochastic processes. On the contrary the relation with probability theory emerges, as a consequence of dynamics. Indeed, in the simplest cases (see [16]) it may be understood in terms of Markoff processes.

In conclusion this theory gives the symmetry breaking and this symmetry breaking induces the right sign to obtain the second law of thermodynamics as well as the properties necessary to go from a dynamic description to a description in terms of probabilistic processes.

IV. STAR UNITARY TRANSFORMATION

The striking difference between the Liouville equation and equations as Boltzmann, Fokker-Planck equations lies in the appearance in the collision operator of an even part in L. Before we discuss the origin of the "symmetry breaking" more in detail, let us make some preliminary remarks. Let us write the Liouville equation expliciting the different components

$$i\partial_t \rho_o = L_{oo}\rho_o + L_{oc}\rho_c \tag{4.1a}$$

$$i\partial_t \rho_c = L_{co}\rho_o + L_{cc}\rho_c \tag{4.1b}$$

in which $L_{oo} = PLP$, $L_{oc} = PLQ$, $L_{co} = QLP$ ...

In an arbitrary representation, one cannot talk of well defined units. For instance, starting from an initial condition in which there would be no correlation, no relation between the units($\rho_c = 0$), after some time the correlations because of (4.16) will appear.

Now if one goes to the representation in which the Hamiltonian is diagonal, then the Liouville equation takes the form

$$i\partial_t \rho_o = 0 \tag{4.2a}$$

$$i\partial_t \rho_c = L_{cc}\rho_c \tag{4.2b}$$

which is certainly not the answer we are looking for. The units are now distributed once and for all on the different levels and remain there. However what we want to obtain is a description of the dynamic evolution in terms of physical processes such as decay of excited states or collisions between the particles. Clearly the units obtained through diagonalization of the Hamiltonian, which are by definition not interacting, are distinct from the units one observes, and participate in the various physical processes.

Now any canonical transformation will lead to a form either of type (4.1), either of type (4.2).

Having exhausted the possibilities of canonical transformation, we have to try non-canonical transformation which would lead to a description of the following type

$$i\partial_t \rho_0 = \Phi \rho_0 \tag{4.3}$$

where the Φ operator would be dissipative. The problem is to know if such non-canonical transformations exist and which type of transformation is needed. The remarkable point is that if such transformation exists, the physical nature of the problem requires a very special class of non-unitary transformation, the class of star-unitary transformation [10][11].

The star-hermitian conjugation $*$ is defined as the combination of the hermitian conjugation and the L-inversion, (we shall denote) introduced previously

$$\mathcal{U}^*(L) = \mathcal{U}^+(-L) = \mathcal{U}^{+\prime}(L) \tag{4.4}$$

This conjugation is fundamental and it can be verified immediatly that the various operators which have been considered ($iL, -i\psi(0)$) are star hermitian. They are such that

$$\mathcal{U}^* = \mathcal{U} \tag{4.5}$$

In general, there are different realisations for star hermitian operators; either they are even in L and hermitian, either they are odd in L and anti-hermitian, either contain both parts.

Having defined star-hermitian conjugation one can introduce now star-unitary transformation $\mathcal{U}$ that is, such that

$$\mathcal{U}^+(-L) = \mathcal{U}(L)^{-1} \tag{4.6}$$

The importance of star-unitary operators stems from the fact that one can show on physical grounds they are the only generalisation of unitary operators which leave invariant the average values of observables.

Indeed in the representation obtained by an L-dependent transformation [10] [11].

$$\begin{aligned} {}^{\Phi}\rho &= \Lambda^{-1}\rho \\ {}^{\Phi}A &= \Lambda^{-1'}\rho \end{aligned} \qquad (4.7)$$

Note that because of the equivalence between Schrödinger and Heisenberg picture, if ${}^{\Phi}\rho$ is defined through the Λ transformation, ${}^{\Phi}A$ should be defined through Λ') one obtains immediately

$$\begin{aligned} \langle A \rangle &= \mathrm{Tr}\{ {}^{\Phi}A^{+}\, {}^{\Phi}\rho \} = \mathrm{Tr}\{(\Lambda^{-1'}A)^{+}(\Lambda^{-1}\rho)\} \\ &= \mathrm{Tr}\{A^{+}\Lambda^{-1*}\Lambda^{-1}\rho\} \\ &= \mathrm{Tr}\{A^{+}\rho\} \end{aligned} \qquad (4.8)$$

using star-unitarity conditions.

Furthermore the star-unitary transformation preserves the star-hermitian character of operators but not the hermitian or anti-hermitian properties as does the unitary transformation. As a consequence in the transform of the Liouville equation

$$i\partial_t\, {}^{\Phi}\rho = \Phi\, {}^{\Phi}\rho \qquad (4.9)$$

with

$$\Phi = \Lambda^{-1} L \Lambda \qquad (4.10)$$

$i\Phi$ is star hermitian but has no longer the anti-hermitian property of iL. It will contain in general two parts, a hermitian part even in L and an odd anti-hermitian part.

We still need another important property which will be assumed for the moment and proved later (§5) to be a consequence of the construction of the Λ transformation: the even part $i\Phi^{e}$ is non negative

$$i\overset{e}{\phi} \geq 0 \tag{4.11}$$

We can then construct a Liapounoff function [11] which is the following quadratic functional of the density operator in the <u>physical representation</u> (see (2.18))

$$\Omega = \mathrm{Tr}\{\, {}^{p}\rho^{+}\, {}^{p}\rho \,\} \tag{4.12}$$

which can only decrease in time

$$\begin{aligned} \frac{d\Omega}{dt} &= \mathrm{Tr}\{(-i\phi\, {}^{p}\rho)^{+}\, {}^{p}\rho\} + \mathrm{Tr}\{{}^{p}\rho^{+}(-i\phi\, {}^{p}\rho)\} \\ &= -2\,\mathrm{Tr}\{{}^{p}\rho^{+}(i\overset{e}{\phi})\, {}^{p}\rho\} \leq 0 \end{aligned} \tag{4.13}$$

The system will evolve to equilibrium until Ω takes its minimum value compatible with the normalisation of ρ. It is easy to show that thermodynamical equilibrium corresponds in the physical represenation to the situation where all the quantum states have the same probabilities and random phases.

The remarkable property (4.13) of Ω gives the possibility of a dynamical interpretation of entropy which does not present the difficulties of Boltzmann's, in which for instance the Loschmidt paradox disappears.

Indeed if one considers an experiment involving an inversion of all the velocities, the Boltzmann $\mathcal{H}$ quantity would increase instead of decreasing at some stage of the evolution.

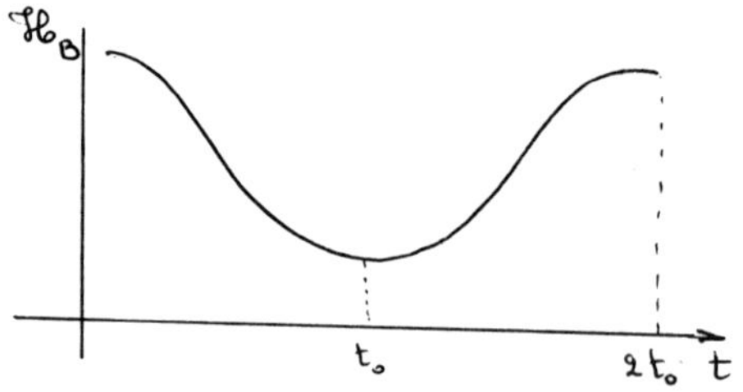

[11] On the contrary, the Ω function will be at any time decreasing

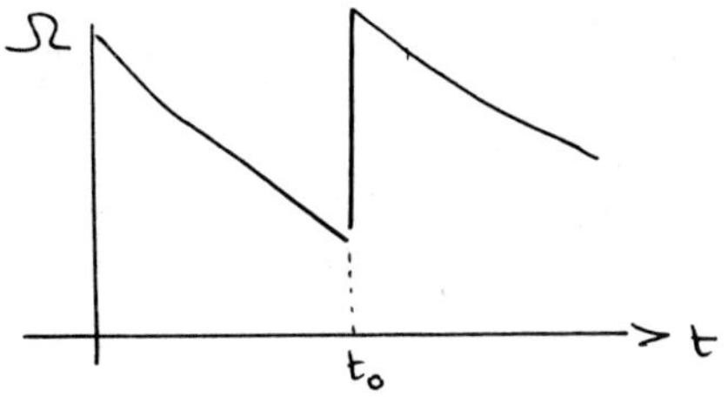

However at time t_o when one inverses the velocities, it takes account of the fact that the system at that moment is not isolated. One has to introduce external devices which return the velocities and that requires some entropy flow to bring the system into a new highly organised state having long range correlations. Afterwards, the system being isolated again, the Ω quantities decrease.

V. CONSTRUCTION OF THE Λ-TRANSFORMATION

The construction of the star-unitary transformation Λ involves three steps [10],[17] (see also [11], [18], [19]).

We construct first a special solution of the Liouville equation.

Out of this special solution one builds a projector which in general is a star-hermitian projection operator.

From the star-hermitian projector, using Kato theory [20], one constructs the Λ transformation. Starting from the Liouville equation (4.1)

$$i\partial_t \rho_o = L_{oo}\rho_o + L_{oc}\rho_c \tag{5.1a}$$

$$i\partial_t \rho_c = L_{co}\rho_o + L_{cc}\rho_c \tag{5.1b}$$

one looks for a special solution $\tilde{\rho}$ satisfying the condition that the diagonal elements obey a separate equation and that the non-diagonal elements are functionals of the diagonal ones:

$$i\partial_t \tilde{\rho}_o = \Theta \tilde{\rho}_o \tag{5.2a}$$

$$i\partial_t \tilde{\rho}_c = C\Theta\tilde{\rho}_o \quad ; \quad \tilde{\rho}_c = C\tilde{\rho}_o \tag{5.2b}$$

The conditions for $\tilde{\rho}_o, \tilde{\rho}_c$ to be a special solution of Liouville equation is that the Θ operator satisfies the non-linear integral equation

$$\Theta = L_{oo} - iL_{oc}\int_0^\infty d\tau\, e^{-iL_{cc}\tau} L_{co}\, e^{i\Theta\tau} \tag{5.3a}$$

and that C is related to Θ by

$$iC = \int_0^\infty d\tau\, e^{-iL_{cc}\tau} L_{co}\, e^{i\Theta\tau} \tag{5.3b}$$

These integral equations imply the following relations between C and Θ

$$\Theta = L_{oo} + L_{oc}C \tag{5.4}$$

$$C\Theta = L_{co} + L_{cc}C \tag{5.5}$$

and combining these two equations, the non-linear equation for C

$$CL_{oo} + CL_{oc}C = L_{co} + L_{cc}C \tag{5.6}$$

It is simple to see that introducing (5.4) and (5.5) into (5.2):

$$i\partial_t \tilde{\rho}_o = L_{oo}\tilde{\rho}_o + L_{oc}\tilde{\rho}_c$$

$$i \partial_t \tilde{\rho}_c = L_{co} \tilde{\rho}_o + L_{cc} \tilde{\rho}_c$$

one simply recovers the initial Liouville equation. This means that we have constructed an exact, special solution of Liouville equation.

Let us introduce now star-hermitian conjugate operators of C and

$$\mathcal{D} = C^* \tag{5.7}$$

$$\eta = -\Theta^* \tag{5.8}$$

for which we have the relations similar to (5.4-6)

$$\eta = L_{oo} + \mathcal{D} L_{co} \tag{5.9}$$

$$\eta \mathcal{D} = L_{oc} + \mathcal{D} L_{cc} \tag{5.10}$$

$$L_{oo} \mathcal{D} + \mathcal{D} L_{co} \mathcal{D} = L_{oc} + \mathcal{D} L_{cc} \tag{5.11}$$

As a consequence of these definitions and of the relations obtained, one has the following commutation relations:

$$(P + \mathcal{D}C)\Theta = \eta(P + \mathcal{D}C) \tag{5.12}$$

$$(P + \mathcal{D})L = \eta(P + \mathcal{D}) \tag{5.13}$$

$$L(P + C) = (P + C)\Theta \tag{5.14}$$

The last two expressions being particularly interesting as they permit to prove that our special solution generates a projector.

Comparing the formal solution of

$$(P+D)\, i \frac{\partial \rho}{\partial t} = (P+D) L \rho = \eta (P+D) \rho \tag{5.15}$$

that is

$$(P+D)\, \rho(t) = e^{-i\eta t} (P+D)\, \rho(0) \tag{5.16}$$

with the special solution

$$\tilde{\rho}_0(t) = e^{-i\Theta t} \tilde{\rho}_0(0) \tag{5.17}$$

one can easily make the connection between (5.16) and (5.17) if one multiplies (5.17) at the left by

$$A^{-1} = P + DC \tag{5.18}$$

to obtain

$$A^{-1} \tilde{\rho}_0(t) = A^{-1} e^{-i\Theta t} \tilde{\rho}_0(0) = e^{-i\eta t} A^{-1} \tilde{\rho}_0(0) \tag{5.19}$$

So one can make the identification

$$\tilde{\rho}_0(0) = A\, (P+D)\, \rho(0) \tag{5.20}$$

$$\tilde{\rho}_0(t) = A\, (P+D)\, \rho(t) \tag{5.21}$$

showing that in the course of time these equations keep their form. Considering also the non-diagonal part $\tilde{\rho}_c$ (5.2b) and the evolution equation (5.17) one obtains for $\tilde{\rho}(t)$

$$\begin{aligned} \tilde{\rho}(t) &= (P+C)\, e^{-i\Theta t} A\, (P+D)\, \rho(0) \\ &\equiv \Sigma(t)\, \rho(0) \end{aligned} \tag{5.22}$$

for any $t \geq 0$, which is a special solution for positive time, with the semi-group property

$$\Sigma(t_1 + t_2) = \Sigma(t_1)\,\Sigma(t_2) \qquad (5.23)$$

for $t_1, t_2 \geq 0$

The limit of the $\Sigma(t)$ operator when t goes to zero by positive value, is the operator Π

$$\Pi \equiv (P + C)\, A\, (P + D) \qquad (5.24)$$

which has obviously the property of a projection operator as

$$\begin{aligned} \Pi^2 &= (P + C)\ A\ (P + D)\ (P + C)\ A\ (P + D) \\ &= (P + C)\ A\ (P + DC)\ A\ (P + D) \\ &= (P + C)\ A\ (P + D) = \Pi \end{aligned}$$

so that

$$\Pi^2 = \Pi \qquad (5.25)$$

and furthermore by construction Π is star-hermitian

$$\Pi^* = \Pi \qquad (5.26)$$

The most remarkable property of the Π projection operator is that it commutes with the Liouville operator L

$$\Pi L = L \Pi \qquad (5.27)$$

The demonstration follows from the different commutation relations (5.12-14).

$$\begin{aligned} \Pi L &= (P + C)\ A\ (P + D)\ L \\ &= (P + C)\ A\,\eta\,(P + D) \\ &= (P + C)\,\Theta\,A\ (P + D) \\ &= L\ (P + C)\ A\ (P + D) \\ &= L\,\Pi \end{aligned}$$

This property has a deep physical meaning related to the fact that because of it, $\Pi\rho$ satisfies a separate equation of evolution. Π defines a subdynamics.

It is to be noticed that we have been considering only positive times. The same procedure can also be applied for negative time ($t < 0$) and in that case one would have obtained a projector $\Pi' = \Pi(-L)$ such that

$$\Pi' L = L \Pi' \tag{5.28}$$

which should be used in connection with final value problems while (5.27) has to be applied for initial value problems. The description of time has been brought into the mathematical structure of the space in which we are working.

As far as the relationship with spectral theory is concerned, the eigenvalue problem:

$$\Theta u_n = z_n u_n \tag{5.29}$$

when one compares the integral equation (5.3a) Θ has to satisfy and the formal expression of $\psi(z)$ as given by (3.10), one can immediately see that it is equivalent to

$$\psi(z_n) u_n = z_n u_n \tag{5.30}$$

or in a more compact form to the resolution of the non-linear problem [21]:

$$z - \psi(z) = 0 \tag{5.31}$$

$\psi(z)$ being a non-hermitian operator, there is no guarantee that this eigenvalue problem has necessarily a solution. However if it possesses a solution, one can make the following classification. Either the eigenvalue is situated on the real axis and in that case it will give a hermitian operator Π

$$\Pi = \Pi^{+} \tag{5.32}$$

Either its solution is in the lower half-plane giving a star-hermitian projector

$$\Pi = \Pi^{*} \tag{5.33}$$

The last step proceeds with the construction of the star-unitary transformation Λ with the use of Kato's theory [20].

Kato's construction permits us to make a similitude between two projection operators. Suppose a given P and a family of projection operators $\Pi(\lambda)$ depending on a parameter λ (which may be considered as the strength of the interaction:

$$H = PH + \lambda QH$$

with the condition

$$\Pi(0) = P \tag{5.34}$$

then

$$\Pi(\lambda) = \Lambda(\lambda) P \Lambda^{-1}(\lambda)$$

(5.35)

will give us a unitary operator if Λ is hermitian

$$\Pi = \Pi^{+} \quad \Longrightarrow \quad \Lambda^{+} = \Lambda^{-1} \tag{5.36}$$

and a star-unitary operator Λ for star-hermitian Π

$$\Pi = \Pi^{*} \quad \Longrightarrow \quad \Lambda^{*} = \Lambda^{-1} \tag{5.37}$$

The main theorem of Kato [20] states that you can generate the Λ transformation through the following differential equation

$$\frac{d\Lambda}{d\lambda} = Z(\lambda)\Lambda(\lambda) \tag{5.38}$$

where $Z(\lambda)$ is given by

$$Z(\lambda) = \left[\frac{\partial \Pi}{\partial \lambda}, \Pi\right]_{-} \tag{5.39}$$

So that once we have Π, and the condition (5.34) one integrates the differential equation to obtain Λ.

If Λ is written as

$$\Lambda = C\chi \tag{5.40}$$

the operator χ has to satisfy a differential equation which corresponds to the one derived by Mandel and Turner [22],[23] before the relation between our approach and Kato's theory was understood.

It is very easy to show that the main requirements imposed on the transformation are satisfied. The dissipativity condition which requires that the operator

$$\Phi = \Lambda^{-1} L \Lambda \tag{5.41}$$

and in particular

$$\Phi_0 = P\Phi P = \chi_0 \Theta \chi_0^{-1} \tag{5.42}$$

with

$$\chi_0 = P\Lambda P \tag{5.43}$$

has a part with a <u>definite sign</u> , comes simply from the fact that being related by a similitude to Θ corresponding to the eigenvalues of $z - \psi(z)$, it has a negative imaginary part (Im $z_n < 0$), that is, one considers contributions from singularities in the lower half plane.

Another property of the kinetic operator is its <u>bloc. diagonalisation</u>, which is by construction. As $\Pi L = L\Pi$ (5.27), one has immediately

$$\begin{aligned} \Lambda^{-1}(\Pi L - L\Pi)\Lambda &= \Lambda^{-1}\left[\Lambda P \Lambda^{-1} L - L \Lambda P \Lambda^{-1}\right]\Lambda \\ &= P\Phi - \Phi P = 0 \end{aligned} \tag{5.44}$$

It is also important to note that Kato's theorem gives the essential physical meaning to which of the roots of the dispersion relation $z - \psi(z)$ we have to consider. The dispersion equation has in general many roots but if we want to solve Kato's equation we have to consider the boundary condition (5.34). That means that one has to take as the root the root which, when λ goes to zero, comes to the real axis (it is not necessarily the closest to the real axis).

The contribution from the other roots are not lost, they are distributed on the other subspaces.

In fact the version presented here is somewhat simplified. One should have considered a complete set of projection operators

$$\sum P_\nu = 1 \quad ; \quad P_\nu^2 = P_\nu \quad ; \quad P_\nu P_\mu = 0 \tag{5.45}$$

and continue them into a complete set of star-hermitian projection operators

$$\sum \Pi_\nu = 1 \quad ; \quad \Pi_\nu^2 = \Pi_\nu \qquad \Pi_\nu \Pi_\mu = 0$$
$$\Pi_\nu = \Pi_\nu^* \tag{5.76}$$

This would involve the solution of an infinite number of dispersion equations each of them satisfying the Kato continuity condition. This procedure would have given us a sufficient number of roots to represent any arbitrary initial value problem.

VI. POTENTIAL SCATTERING

This chapter is devoted to the illustration of the general formulation of the star-unitary transformation on the example of scattering theory.

The derivation of a cross section in potential scattering appears as the simplest example which exhibits already most of the features for which our approach is of interest and therefore very suitable for such an illustration.

Furthermore collision theory is quite an important subject as much of our information on the interaction between the particles comes from scattering experiments and also, it was through the use of the invariants of collision that Boltzmann was able to establish his $\mathcal{H}$ theorem. From that time, the treatment of dissipative processes has always been in terms of "physical processes", such as scattering, spontaneous and induced emission of light, ... Outstanding examples are the papers by Planck on the black body radiation [24] and by Einstein on the interaction between matter and radiation [25].

We know how to calculate, at least in simple cases, the scattering cross sections from the interaction in terms of the t matrix (or S-matrix). Here however we are emphasizing a different problem: in which sense is scattering describing a temporal process? how to go to a time behaviour of a dynamical system in which the collisions are the generators of evolution? Usually the starting point of scattering theory is the Schrödinger equation

$$i\partial_t \psi(r) = H\psi(r) \tag{6.1}$$

where H is a sum of a kinetic part H_0 and of an interaction part $\mathcal{V}$ (assumed to be a short range potential)

$$H = H_0 + \mathcal{V} \tag{6.2}$$

The solution of Schrödinger equation is analysed in terms of, for instance, the outgoing eigenfunctions of H

$$H|\psi_\alpha^+\rangle = E_\alpha|\psi_\alpha^+\rangle \tag{6.3}$$

solutions of the Lippman-Schwinger integral equation

$$|\psi_\alpha^+\rangle = |\alpha\rangle + \frac{1}{i\epsilon - H_0 + E_\alpha}\mathcal{V}|\psi_\alpha^+\rangle \tag{6.4}$$

the $|\alpha\rangle$ being a complete set of eigenfunctions of H_0 with the same eigenvalues as the $|\psi_\alpha^+\rangle$)

$$H_0|\alpha\rangle = E_\alpha|\alpha\rangle \tag{6.5}$$

In terms of the t-matrix defined as

$$\frac{1}{z-H}\mathcal{V} = \frac{1}{z-H_0}t(z) \tag{6.6}$$

the $|\psi_\alpha^+\rangle$ can be written also as

$$|\psi_\alpha^+\rangle = |\alpha\rangle + \frac{1}{i\epsilon - H_0 + E_\alpha} t(E_\alpha + i\epsilon)|\alpha\rangle \qquad (6.7)$$

Expressing the scattered part of $\psi(t)$ in terms of $|\psi_\alpha^+\rangle$, one can calculate the probability of finding a given free state in the scattered part, the time derivative of which in the limits of long time and infinite beam gives the cross section. As we mentioned previously, we are interested in a different problem. We want to describe scattering as a temporal process, in which the collisions are the generators of the evolution.

In the first place, let's emphasise that in a strict sense, cross sections are not reducible to an Hilbert space concept. A temporal description of scattering is only possible in the space of density operators (we have called superspace). To illustrate this point, let us consider the time evolution of the density operator ρ as given by the Liouville-von Neumann equation (3.1).

Its formal solution may be written in the interaction representation as

$$\rho(t) = U(t,0)\,\rho(0)\,U(0,t) \qquad (6.8)$$

where the unitary operator $U(t,t')$ defined through

$$U(t,t') = e^{iH_0 t}\, e^{-iH(t-t')}\, e^{-iH_0 t'} \qquad (6.9)$$

satisfies the integral equation

$$U(t,t') = 1 + \frac{\lambda}{i}\int_{t'}^{t} d\tau\, H'(\tau)\, U(\tau,t') \qquad (6.10)$$

with $$H'(\tau) = e^{iH_0 t}\, H\, e^{-iH_0 t}$$

To make our point clearer, let us consider the lowest order contribution (in the coupling parameter λ) to the scattering from a state $|\alpha\rangle$ to a state $|\beta\rangle$:

$$\rho_{\alpha\alpha}(t) = U_{\alpha\beta}(t,0)\, \rho_{\beta\beta}(0)\, U_{\beta\alpha}(0,t)$$

This involves the integral

$$\int_0^t d\tau\, e^{i(E_\beta - E_\alpha)\tau} \int_0^t d\tau'\, e^{-i(E_\beta - E_\alpha)\tau'}$$

The asymptotic limit of this expression for large t is

$$t\,\delta(E_\alpha - E_\beta)$$

If we had been working with the wave function and taken the asymptotic limit on the wave function before performing the product, we would have obtained the divergent result

$$[\delta(E_\alpha - E_\beta)]^2$$

As it has been noticed by Van Hove [26] and others[27] this result has a simple physical interpretation: the contribution arises from events which are quasi-instantaneous on the long time scale corresponding to the interval $0 - t$

$$\frac{\tau - \tau'}{t} \to 0$$

In other words the t factor comes from "constructive interferences" between the two wave function spaces involved in the time evolution of the density operator. Furthermore, along with the long time limit, in the usual theory, one has in order to obtain a non-vanishing transition rate, to perform other limiting procedures as the infinite beam limit which has also to be performed on the square of the amplitude. Therefore, scattering introduces well defined phase relations and as a result the scattered beam can no longer be described in the space of wave functions [28]. One can now take two attitudes: the basic description is in terms of wave functions satisfying the Schrödinger equation (or classically in terms of trajectories). The result corresponds then to some asymptotic approximation (large t, large wave packet) of the real description, performed on the level of the probabilities (square of the amplitudes).

But one can go one step further and argue that the exact motion of the wave packet (or the classical trajectories) are of no physical interest in the experimental situation we are dealing with.

In this case we have to reformulate dynamics in such a way that cross sections become the generators of evolution. As scattering introduces interference effect, such a description will be associated with the space of density operators.

One ultimate goal will be to obtain an equation of the form

$$\partial_t P_\alpha(t) = \sum_\beta \sigma_{\alpha\beta} P_\beta(t)$$

(which is the definition of cross sections) as a direct, exact consequence of the Liouville equation.

We shall treat here the problem of potential scattering. We shall consider a system of entities specified by some quantum number k, enclosed in a quantization box of volume L^3. Their energies are given by ω_k $(0 < \mu \leq \omega_k \leq \infty)$ the spectrum of which becomes continuous in the limit of large volume.

The free Hamiltonian of this system is taken as

$$H_0 = \sum_k \omega_k |k\rangle\langle k| \tag{6.11}$$

$|k\rangle$ being a complete set of normalised eigenfunctions corresponding to the eigenvalues k. The interaction is assumed to have the form

$$H^I = \sum_{kk'} f_{kk'} |k\rangle\langle k'| \tag{6.12}$$

in which $f_{kk'}$ is of order L^{-3} and because of the hermiticity of the Hamiltonian is such that

$$f_{kk'} = f^*_{k'k} \tag{6.13}$$

All the quantities appearing in the formalism are derived from the resolvant of the Liouville operator $(z - L)^{-1}$ the matrix elements of which can be expressed as a convolution of the matrix elements of the resolvant of the Hamiltonian

$$\langle k_1 k_2 | \frac{1}{z - L} | k_3 k_4 \rangle = -\frac{1}{2i\pi} \int_{C'} dz' \langle k_1 | \frac{1}{z' - H} | k_3 \rangle \times \langle k_4 | \frac{1}{z' - z - H} | k_2 \rangle \tag{6.14}$$

where the contour C' runs from right to left in the above half plane, below z $(\operatorname{Im} z' < \operatorname{Im} z)$

Using the identity

$$\frac{1}{z-H} = \frac{1}{z-H_0} + \frac{1}{z-H_0} H^I \frac{1}{z-H} \tag{6.15}$$

and the definition of the t-operator

$$t(z) \frac{1}{z-H_0} = H^I \frac{1}{z-H} \tag{6.16}$$

the resolvant of H can be written as

$$\frac{1}{z-H} = \frac{1}{z-H_0} + \frac{1}{z-H_0} t(z) \frac{1}{z-H_0} \tag{6.17}$$

Its matrix elements are

$$\langle k_1 | \frac{1}{z-H} | k_2 \rangle = \frac{1}{z-\omega_{k_1}} \delta_{k_1 k_2} + \frac{1}{z-\omega_{k_1}} t_{k_1 k_2}(z) \frac{1}{z-\omega_{k_2}} \tag{6.18}$$

in terms of the t-matrix elements which are solutions of the integral equations

$$t_{kk'}(z) = f_{kk'} + \sum f_{kk''} \frac{1}{z-\omega_{k''}} t_{k''k'}(z)$$

$$t_{kk'}(z) = f_{kk'} + \sum t_{kk''}(z) \frac{1}{z-\omega_{k''}} f_{k''k'} \tag{6.19}$$

One obtains for (6.17):

$$\langle k_1 k_2 | \frac{1}{z - L} | k_3 k_4 \rangle = - \frac{1}{2i\pi} \int_{c'} dz' \, \frac{1}{z' - \omega_{k_1}}$$

$$\times \frac{1}{z' - z - \omega_{k_4}} \left[\delta_{k_1 k_3} + t_{k_1 k_3}(z') \frac{1}{z' - \omega_{k_3}} \right]$$

$$\times \left[\delta_{k_4 k_2} + t_{k_4 k_2}(z' - z) \frac{1}{z' - z - \omega_{k_2}} \right] \qquad (6.20).$$

At this point a remark should be made about the procedure and the approximations which we will use in the following.

As already emphasised, the limit to the continuous spectrum is essential as it introduces the symmetry breaking, the distinction between the retarded and advanced solutions.

As it is usually done, we shall delay as long as possible the limiting procedure $(L^3 \to \infty)$ but we shall keep in all expressions only the dominant term in the volume.

That is, if in an expression, one has terms independant of the volume, one shall neglect terms in L^{-3}. If the leading term is in L^{-3}, one has to drop contributions of order L^{-6}. For illustration, if in the expression (6.20) for the matrix elements of the resolvant, one has $k_1 = k_3$, $t_{k_1 k_3}$, which is of order L^{-3}, has to be neglected with respect to the first term $\delta_{k_1 k_3}$. But, if the expression appears in a summation over k_3 both terms have to be kept as they are giving a contribution of order unity in the volume.

$\overset{\circ}{\Pi}$ Subdynamics

In this problem one can easily construct the different subdynamics. In the case of discrete spectrum, the decomposition into subdynamics is a quite trivial problem, as the contour can be easily split into a sum of contours around the singularities of the resolvant which are isolated, each of these singularities given the contribution to a subdynamic.

However, when one goes to the limit of large volume , these singularities coalesce into a cut and the decomposition in subdynamics when the limit is taken becomes a quite delicate procedure. Nevertheless this decomposition is still possible through the use of proper analytic continuation procedures. Here we shall be mainly interested in the $\overset{\circ}{\Pi}$ subdynamic which arises from the contribution of the contour γ_0 encircling the $z = 0$ singularity.

The element of the $\overset{\circ}{\Pi}$ projector are given by

$$\langle k_1 k_2 | \overset{\circ}{\Pi} | k_3 k_4 \rangle \equiv \langle k_1 k_2 | CAD | k_3 k_4 \rangle$$

$$= \frac{1}{2i\pi} \int_{\gamma_0} dz \, \langle k_1 k_2 | \frac{1}{z - L} | k_3 k_4 \rangle \tag{6.21}$$

where γ_0 is a contour around $z = 0$. From the above expression, one obtains

$$A_{kk,kk} \equiv \langle kk | \overset{\circ}{\Pi} | kk \rangle$$

$$= -\frac{1}{2i\pi} \int_{\gamma_0} dz \, \frac{1}{2i\pi} \int_{c'} dz' \, \frac{1}{z' - \omega_k} \, \frac{1}{z' - z - \omega_k}$$

$$= 1 \tag{6.22}$$

in which the terms of order L^{-3} have been neglected with respect to the term 1.

For $A_{kk,k'k'}$ which can be written as

$$A_{kk,k'k'} = -\frac{1}{2i\pi}\int_{\gamma_0} dz \frac{1}{2i\pi}\int dz' \frac{1}{z^2} t_{kk'}(z')$$

$$t_{k'k}(z'-z)\left(\frac{1}{z'-z-\omega_k} - \frac{1}{z'-\omega_k}\right)\left(\frac{1}{z'-z-\omega_{k'}} - \frac{1}{z'-\omega_{k'}}\right)$$

one has to evaluate the residue of the integrand at the double pole $z = 0$:

$$A_{kk,k'k'} = \frac{1}{2i\pi}\lim_{z\to+i\epsilon}\frac{\partial}{\partial z}\int_{-\infty}^{+\infty} dx\left(\frac{1}{x-z-\omega_k} - \frac{1}{x+i\epsilon-\omega_k}\right)$$

$$\times\left(\frac{1}{x-z-\omega_{k'}} - \frac{1}{x+i\epsilon-\omega_{k'}}\right) t_{kk'}(x+i\epsilon)\, t_{k'k}(x-z)$$

To obtain this expression, one has taken into account the fact that $\mathrm{Im}\, z > \mathrm{Im}\, z'$ and that at the limit $z \to +i\epsilon$, the C' contour goes to the real axis from above.

It is now an easy matter to perform the derivation with respect to z and to take the limit $z \to +i\epsilon$. The result is

$$A_{kk,k'k'} = \frac{1}{(\omega_{k'}-\omega_k-i\epsilon)^2} t_{kk'}(\omega_{k'}+i\epsilon)\, t_{k'k}(\omega_{k'}-i\epsilon)$$

$$+ \frac{1}{(\omega_k-\omega_{k'}-i\epsilon)^2} t_{kk'}(\omega_k+i\epsilon)\, t_{k'k}(\omega_k-i\epsilon)$$

$$+ 2i\pi\,\delta(\omega_k-\omega_{k'})\, t_{kk'}(\omega_k+i\epsilon)\left[\frac{\partial}{\partial z} t_{k'k}(\omega_k-z)\right]_{z=i\epsilon}$$

(6.23)

The other elements of $\overset{\circ}{\Pi}$ subspace are simple to evaluate. The creation fragments are obtained from

$$(CA)_{k_1k_2,kk} \equiv \overset{\circ}{\Pi}_{k_1k_2,kk} \tag{6.24}$$

For illustration, let us again derive in detail the expression for $C_{k'k,kk}$

$$C_{k'k,kk} = -\frac{1}{2i\pi}\int_{\gamma_0} dz \, \frac{1}{2i\pi}\int_{\mathcal{C}'} dz' \, \frac{1}{z'-\omega_{k'}} \, t_{k'k}(z')$$

$$\times \frac{1}{z}\left(\frac{1}{z'-z-\omega_k} - \frac{1}{z'-\omega_k}\right)$$

$$= \lim_{z\to i\epsilon} \frac{1}{2i\pi}\int_{-\infty}^{+\infty} dx \left(\frac{1}{x-z-\omega_k} - \frac{1}{x+i\epsilon-\omega_k}\right)$$

$$t_{k'k}(x+i\epsilon) \; \frac{1}{x+i\epsilon-\omega_{k'}}$$

so that one obtains

$$C_{k'k,kk} = \frac{t_{k'k}(\omega_k+i\epsilon)}{i\epsilon+\omega_k-\omega_{k'}} \tag{6.25}$$

and similarly, one gets the following results

$$C_{kk',kk} = -\frac{t_{kk'}(\omega_k-i\epsilon)}{i\epsilon-\omega_k+\omega_{k'}} \tag{6.26}$$

$$C_{k'k'',kk} = -\frac{t_{k'k}(\omega_k+i\epsilon)}{i\epsilon+\omega_k-\omega_{k'}} \; \frac{t_{kk''}(\omega_k-i\epsilon)}{i\epsilon-\omega_k+\omega_{k''}} \tag{6.27}$$

The destruction fragments are evaluated from the relation

$$(AD)_{kk,k_1k_2} = \overset{\circ}{\pi}_{kk\,k_1k_2} \tag{6.28}$$

and are given by

$$D_{kk,k'k} = \frac{t_{kk'}(\omega_k + i\epsilon)}{i\epsilon + \omega_k - \omega_{k'}} \tag{6.29}$$

$$D_{kk,kk'} = - \frac{t_{k'k}(\omega_k - i\epsilon)}{i\epsilon - \omega_k + \omega_{k'}} \tag{6.30}$$

$$D_{kk\,k'k''} = - \frac{t_{kk'}(\omega_k + i\epsilon)}{i\epsilon + \omega_k - \omega_{k'}} \frac{t_{k''k}(\omega_k - i\epsilon)}{i\epsilon + \omega_{k''} - \omega_k} \tag{6.31}$$

We can now evaluate the important collision operator Θ solution of $z - \psi(z) = 0$, the matrix elements of which are given by

$$\Theta_{kk,k'k'} = \sum_{k_1k_2} \langle kk | L | k_1k_2 \rangle C_{k_1k_2\,k'k'} \tag{6.32}$$

One obtains immediatly

$$\Theta_{kk,k'k'} = 2i\pi\, \delta(\omega_k - \omega_{k'})\, t_{kk'}(\omega_k + i\epsilon) \times t_{k'k}(\omega_k - i\epsilon). \tag{6.33}$$

and for the kk,kk element:

$$\Theta_{kk,kk} = 2i\, \mathrm{Im}\, t_{kk}(\omega_k + i\epsilon). \tag{6.34}$$

If one wishes to go the the $\overset{\circ}{\pi}$ component of the physical representation, one has still to introduce the dressing operator χ.

It is easy to show that the dominant contribution in the volume is

$$\chi_{kk,kk} = 1 \tag{6.35}$$

and

$$\chi_{kk,k'k'} = O\left(\frac{1}{L^6}\right).$$

which we have not explicited here as it would not be needed.

Very important properties are related to the so called star conjugation. If one replaces L by -L in the elements of the resolvant (6.14), one obtains

$$\langle k_1 k_2 | \frac{1}{z+L} | k_3 k_4 \rangle = \frac{1}{2i\pi} \int_{C'} dz' \langle k_1 | \frac{1}{z'+H} | k_3 \rangle$$

$$\times \langle k_4 | \frac{1}{z'-z+H} | k_2 \rangle \tag{6.36}$$

in terms of the resolvant $(z + H)^{-1}$ of the Hamiltonian which can be derived from the following relation:

$$t(-z) \frac{1}{z+H_0} = H^I \frac{1}{z+H} \tag{6.37}$$

as

$$\frac{1}{z+H} = \frac{1}{z+H_0} - \frac{1}{z+H_0} t(-z) \frac{1}{z+H_0} \tag{6.38}$$

With this expression, one can evaluate in the same way as above the C′, A′, D′ with the result

$$C' = D^+ \tag{6.39}$$

$$D' = C^+$$

The transformed collision operator is given by

$$-i\Theta'_{k'k',kk} = -i\Theta_{kk,k'k'} \tag{6.40}$$

and with the use of the relation

$$\{t_{k\ell}(z)\}^{c.c} = t_{\ell k}(z^*) \tag{6.41}$$

one can verify the properties of the collision operator derived in the formalism.

$(-i\theta)$ is a real, star-hermitian and adjoint symmetrical operator, and its even part in L is hermitian and definitely positive

$$(-i\overset{e}{\theta})_{kk,k'k'} \geq 0 \tag{6.42}$$

In order to have all the elements of the tranformation, one should also study the correlation subspaces. This can easily be done following the procedure used for $\overset{o}{\pi}$ subspace and using proper analytic continuation. The details can be found in [29]. The Λ transformation which defines the physical representation is given by

$$\Lambda = C\chi \tag{6.43}$$

The most important property of this transformation is its star unitarity. This property is quite immediate to verify and in fact reduces to the property of unitarity of the S matrix:

$$t_{kk_1}(\omega_k + i\epsilon) - t_{kk_1}(\omega_{k_1} - i\epsilon)$$

$$- \sum_{k'} t_{kk'}(\omega_{k'} - i\epsilon)\, t_{k'k_1}(\omega_{k'} + i\epsilon)$$

$$\times \left\{ \frac{1}{i\epsilon + \omega_k - \omega_{k'}} + \frac{1}{i\epsilon + \omega_{k'} - \omega_{k_1}} \right\} = 0 \tag{6.44}$$

and

$$t_{kk_1}(\omega_k + i\epsilon) - t_{kk_1}(\omega_{k_1} - i\epsilon)$$

$$- \sum_{k'} t_{kk'}(\omega_k + i\epsilon)\, t_{k'k_1}(\omega_{k_1} - i\epsilon)$$

$$\times \left\{ \frac{1}{i\epsilon + \omega_k - \omega_{k'}} + \frac{1}{i\epsilon + \omega_{k'} - \omega_{k_1}} \right\} = 0. \tag{6.45}$$

And from the star unitarity of the Λ, it is easy to prove the projection properties of the Π subspaces. Indeed

$$\mathbb{1} = \Lambda\Lambda^{*} = C\chi\chi^{*}D = CAD = \sum_{i} \overset{i}{\Pi} \qquad (6.46)$$

is nothing else than the completeness relation and

$$\mathbb{1} = \Lambda^{*}\Lambda = \chi^{*}DC\chi \qquad (6.47)$$

is easily related to the idempotence and orthogonality relations of the Π's .

A point has still to be emphasised : Λ is <u>not</u> a unitary transformation

$$\Lambda\Lambda^{+} \neq 1 \neq \Lambda^{+}\Lambda \qquad (6.48)$$

which can be checked immediately on particular examples. The non unitary of the Λ transformation is the property which in fact enables us to introduce in a very direct way the concept of cross-sections.

The physical or causal representation is defined as

$$^{\phi}\rho = \Lambda^{*}\rho$$

and in this representation the evolution is described by a kinetic equation

$$i\partial_t \, ^{p}\rho = \Phi \, ^{p}\rho$$

with the collision operator given by

$$\Phi = \Lambda^{*} L \Lambda$$

the $\overset{o}{\Pi}$ elements of which are the cross sections given previously.

If we had taken the factorizable unitary tranformation induced by the Møller wave operator

$$M = \{ U^{(-)} \times U^{(-)+} \}$$

we should have obtained a vanishing cross section as for the diagonal elements

$$M^{+} L M = 0$$

as can be verified immediately

VII. CONCLUDING REMARKS

We have shown that for an important class of dynamical systems it was possible through a non-canonical transformation to go from a Hamiltonian description of dynamics to a description in terms of processes which incorporates explicitly the dissipativity. For this class of systems one has two basically different formulations of dynamics which can be summarized in the following chart:

	Space time description	Processes
basic elements	forces	cross section life-times
collision	limiting case of dynamical processes	elementary processes
generator	of motion: Hamiltonian	of evolution: cross section life times
description	Space time (world lines) "nothing begins nothing ends"	non local in space time
mathematical properties	Group	Semi group.

Now there exists systems for which only one of these descriptions holds. On one hand, there are simple non-dissipative systems. In that case, our transformation leads to the usual static description corresponding to non-interacting units obtained by diagonalization. On the other hand, chemical kinetics is a good example of a system without a Hamiltonian description but where the description in terms of elastic and inelastic collisions is the most appropriate.

Also in high energy problems the existence of a Hamiltonian can by no means be considered as granted.

The epistemological consequences of the existence of this new tranformation theory have been discussed elsewhere [30]. Let us only mention here that it gives a precise mathematical meaning to the concept of complementarity as introduced by Bohr and Rosenfeld. It is indeed the description in terms of processes which is essential when we want to use a material system as a measurement apparatus as we want then to relate its temporal evolution to the various physical processes which are going on in the system. This description is "complementary" in the Bohr-Rosenfeld sense to the description in terms of the motions of the various particles which form this material system to which it is related by our star-unitary transformation. In these lectures we have emphasised the physical aspects of our problem: the relation between dynamics and irreversibility. It should be obvious that this leads to new and fascinating problems in Mathematics. The central part is the discussion of possible "equivalencies" between hermitian and dissipative operators of motion. This problem is the natural generalisation of the problem of equivalencies between hermitian operators which is at the core of transformation theory both in classical and quantum mechanics.

ACKNOWLEDGEMENTS

We thank all members of our group,especially F. Henin, C. George, A.Grecos for elaboration of the ideas summarized in these lectures. The results described here are due to the continuous effort of groups in Brussels and Austin. We also want to express our deep appreciation to our colleague and friend L. Rosenfeld. His untimely death has deprived us of his inspiration. His enthusiasm and his constructive criticisms have influenced our work in a decisive fashion.

REFERENCES

1. L. BOLTZMANN, Wien. Ber., 66, 275 (1872).
2. B.J. ALDER and T.E. WAINWRIGHT, J. Chem. Phys. 33, 1434 (1960).
3. A. BELLEMANS and J. ORBAN. Phys. Lett. 24A, 620 (1967).
4. J. LOSCHMIDT, Wien. Ber., 73, 139 (1876).
5. H.P. MacKEAN Jr., J. of Comb. Theory. 2, 358, (1967).
6. M. KAC, in the Boltzmann Equation, Theory and Applications, ed. E.G.D. Cohen and W. Thirring (Springer Verlag, Wien, New York, 1973) pp 379-400.
7. F. HENIN and I. PRIGOGINE, PNAS (to appear).
8. F. HENIN, Physica (to appear).
9. I. PRIGOGINE, Non Equilibrium Statistical Mechanics (Wiley, Interscience, London, New York, 1962)
10. C. GEORGE, I. PRIGOGINE and L. ROSENFELD, Koningl. Dansk Vid. Mat-Fys. Medd. 38, 12, (1972).
11. I. PRIGOGINE, C. GEORGE, F. HENIN and L. ROSENFELD, Chemica Scripta 4, 5, (1973).
12. R. BALESCU, Statistical Mechanics of Charged Particles, Wiley Interscience, New York, 1963.
13. M. BAUS, Acad. Roy. Belg. Bull. Cl. 53, 1291, 1332, 1352 (1967).
14. L. LANZ and L.A. LUGIATO, Physica 44, 532 (1969).
15. A. GRECOS, Physica 51, 50 (1970).
16. M. De HAAN and F. HENIN. Physica 67, 197 (1973).
17. I. PRIGOGINE and A. GRECOS. PNAS to appear 1974.
18. L. ROSENFELD in "Irreversibility in the Many-Body problem, Sitges International School of Physics 1972. Plenum Press, New York, 1972.
19. R. BALESCU In Irreversibility in the Many-body problem, Sitges International School of Physics 1972. Plenum Press, New York.
20. T. KATO, Perturbation Theory for Linear Operators, Springer Verlag, 1966.
21. B. SZ. NAGY and C FOIAS, Harmonic Analysis of Operators on Hilbert Space, North Holland Publishing Co. 1970.
22. P. MANDEL, Physica 50, 77 (1970).
23. J.W. TURNER, Physica 51, 351 (1971).
24. M. PLANCK, Ver.d. D. Physik. Ges. 2, 237 (1900).
25. A. EINSTEIN, Ver. d. D. Physik Ges. 18, 316(1916); 19, 82(1917).
26. L. VAN HOVE, Physica 21, 517 (1955).
27. J. PRIGOGINE, Superfluidite et Equation de Transport Quantique I.I.S.N. Bruxelles 1960.
28. F. MAYNE and I. PRIGOGINE, Physica 63, 1, 1973.
29. F. MAYNE, I. PRIGOGINE, A. GRECOS, C. GEORGE (To appear).
30. I. PRIGOGINE, in Connaissance Scientifique et Philosophie, Colloque du Bicentenaire de l'Académie Royale de Belgigue 1973 (in the press)

RESPONSE, RELAXATION AND FLUCTUATION

Ryogo Kubo

Department of Physics, University of Tokyo
Tokyo, Japan

I. INTRODUCTION

II. CLASSICAL BROWNIAN MOTION AND ITS GENERALIZATIONS

III. RICE'S METHOD (HARMONIC ANALYSIS)

IV. DIRECT INTEGRATION, PATH INTEGRAL REPRESENTATION

V. STOCHASTIC LIOUVILLE EQUATION

VI. RETARDED FRICTION, FLUCTUATION-DISSIPATION THEOREMS

VII. FORCE CORRELATIONS

VIII. SOME EXAMPLES

IX. SOME COMMENTS

X. DAMPING-THEORETICAL METHOD

XI. CONCLUDING REMARKS

NOTES AND REFERENCES

I. INTRODUCTION

The real world is extremely complex. Almost everything occurring there is a many-body process in which an enormous number of atoms, molecules, living cells, individuals etc. are involved. There are different levels of observation and description, from the uttermost microscopic level on the one extreme to the most macroscopic level on the other. The principal aim of physics is to bridge successive levels, completing logical connection from the micro- to the macro-level. As is very well known, the microscopic dynamics gradually loses its preponderance giving way to the laws of probability as one climbs up the staircase, although the conservation laws invariably keep their predominant role. It is thus quite understandable that macroscopic physical laws can be so general as to apply to a large family of physical systems, irrespective of the nature of their constituent elements. Also a considerable part of the laws of statistical physics can be applied to non-physical systems, thanks to the great generality of logics of probability.

Shifting the level of observation or description from one to another more macroscopic, means a reduction in the number of variables describing the system, that is, a coarse-graining of the space of state-variables, which is usually accompanied by a coarse-graining of time if we are concerned with a process occurring in space and time. Eliminated degrees of freedom are not, however, totally ignored but they act on the remaining variables as sources of noise, imparting therby a stochastic nature to the process projected by the coarse-graining. So that we have a hierarchy of stochastic processes, each member constituting a projection or a contraction of the foregoing one. This means a certain restriction on the nature of these stochastic processes. That is to say, we are not quite free in choosing a model at a level of this hierarchy, as the model should satisfy certain conditions. We are not ready to give a complete answer to the question of what these conditions really are. However at least we know an example, the Einstein relation, or the fluctuation-dissipation theorem as its generalization. The fluctuation-dissipation theorem asserts the existence of an internal relationship between fluctuations and the average behaviour of a system and is well established for systems near equilibrium. How far this theorem can be generalized to systems far from equilibrium or to non-physical systems is an open question of great interest.

Physicists would like, whenever it is possible, to start from the first principles, that is from the ultimate microscopic level, and working via elaborate many-body theoretical techniques, reach some result; the result is more pleasing and more satisfactory if it is simple enough to appeal to our physical intuition. Here we shall not work that way but will take a phenomenological view focusing our attention on the process of coarse-graining in the sense mentioned above. This is along the line which I myself have been pursuing, but I have not completely cleared the way yet, so that the present lectures remain as a sort of preliminary approach for a more systematic program.

II. CLASSICAL BROWNIAN MOTION AND ITS GENERALIZATIONS

From our viewpoint, the classical theory of Brownian motion to statistical physics, is like the theory of ideal gases to the traditional statistical mechanics. It is an idealization of the stochastic processes exhibited by a many-body system when observed by means of a small number of projected variables. Here the whole complexity is represented only by the motion of the Brownian particle, on which random forces,due to the thermal motion of the bath molecules, are acting. As such a typical example of statistical physics, the classical Brownian motion can be treated in several different ways affording thus illuminating examples of different approaches to more general problems of statistical physics. Also, the idealization makes the problem so simple and so transparent that it provides us with a standard problem. To this we return for reference when necessary, and from this we seek for ways of generalization.

Let us start from the Langevin equation of motion of a Brownian particle of mass m,

$$m\dot{u} = F(t) = -m\gamma u + R(t) \tag{2.1}$$

where u is the velocity and F is the force at the instant t. F is the total resultant of impulses given to the Brownian particle by the surrounding molecules. It is divided into two parts: the frictional force, $-m\gamma u$, and the random force $R(t)$. The idealization of the classical theory consists in the following assumptions:

A. the equation of motion is linear in the random source $R(t)$; that is to say the friction is proportional to the velocity u with the friction constant $m\gamma$.

B. The random force $R(t)$ has a white noise spectrum; that is, its correlation time τ_c is zero.

C. $R(t)$ is a stationary Gaussian process.

The assumption A seems the simplest possible assumption adequate for a Brownian particle, but it is related, as we shall see later, to the assumption B, which is a very natural one if the Brownian particle is very much heavier than the bath molecules. The Brownian particle would change its state of motion only after experiencing a great number of collisions with bath molecules. Namely the time scale τ_r of the Brownian particle is very much larger than that of bath molecules,

$$\tau_c \ll \tau_r \tag{2.2}$$

which is idealized to B. The assumption C is also justified by the same reason because $R(t)$ is then essentially an accumulated effect of a large number of more or less uncorrelated collisions for which some sort of the central limit theorem of probability theory would work. The Gaussian assumption C together with the linearity assumption A nicely reproduces the Maxwellian law for the velocity distribution in equilibrium.

These assumptions make the problem well-defined and easily tractable. It should, however, be noticed at the same time that the assumption B introduces a difficulty. Mathematically this means that the process $u(t)$ derived from the basic process by eq. (2.1) is everywhere non-differentiable as is very well-known since the classical work of Wiener, so that the stochastic equation (2.1) should be treated properly with great care. This can be done with the use of Wiener integrals and Ito's stochastic differentiation and integration procedures [7] . On the other hand, most physicists prefer more familiar ways of calculation to such unusual ones. That is, we regard a purely white noise spectrum or a delta-type correlation function as a limit of less singular spectra or correlation functions. This should be allowed for most purposes, with due caution. Throughout these lectures we adopt this attitude. Thus the correlation function of $R(t)$ is assumed by B to be

$$\langle R(t_1)R(t_2)\rangle = 2\pi I_R \,\delta(|t_1-t_2|) \tag{2.3}$$

where I_R is the power spectrum of $R(t)$. This is considered as a limit of the Wiener-Khintchin theorem

$$\langle R(t_1)R(t_2)\rangle = \int_{-\infty}^{\infty} I_R(\omega)e^{i\omega(t_1-t_2)}d\omega \tag{2.4}$$

where we may assume, for example,

$$\langle R(t_1)R(t_2)\rangle = \frac{\pi I_R}{\tau_c} e^{-|t_1-t_2|/\tau_c} \tag{2.5}$$

or

$$I_R(\omega) = 2\pi I_R \frac{1}{1+\omega^2\tau_c^2} . \tag{2.6}$$

In the limit of $\tau_c \to 0$, this becomes a white spectrum and the correlation function (2.5) is regarded as a delta-function (2.3).

The ideal Brownian motion is like the ideal gas in statistical thermodynamics. We would like to generalize it to less ideal models in order to simulate more realistic processes in nature. Assumptions A to C may be replaced by:

A′ the observed variable is non-linear in the random force,

B′ the random force $R(t)$ is a non-white noise, that is, its correlation time τ_c is finite, no longer very short compared with the time scale τ_r of the observable,

C′ the random force $R(t)$ is no longer Gaussian.

Negation of each idealization means an enormous range of possibility and there are innumerable combinations of generalizations. Although most of them are in fact beyond our capacity for mathematical analysis, still we can choose some of them as standards useful for elucidating the principles of statistical physics.

We shall begin in the following with the classical ideal Brownian motion. We review first several methods of solving eq. (2.1), that is of finding the process $u(t)$ as derived from the given underlying process $R(t)$. This part will be supplementary to the classical review article of Wang and Uhlenbeck [8] and will be an introduction to the theories of non-ideal Brownian motion, which will be discussed in the later lectures.

III. RICE'S METHOD (HARMONIC ANALYSIS) [8]

Taking the advantage of the linearity of eq. (2.1), we perform harmonic analysis of $u(t)$ and $R(t)$ over a given time interval $0 < t < T$ as follows:

$$u(t) = \sum_{n=-\infty}^{\infty} u(\omega_n) e^{i\omega_n t}$$

$$R(t) = \sum_{n=-\infty}^{\infty} R(\omega_n) e^{i\omega_n t}$$

choosing $\omega_n = 2\pi n/T$ with integers n from $-\infty$ to ∞. Equation (2.1) then gives

$$u(\omega_n) = \frac{1}{m} \frac{R(\omega_n)}{i\omega_n + \gamma} \tag{3.1}$$

which yields at once

$$I_u(\omega) = \frac{1}{m} \frac{I_R(\omega)}{\omega^2 + \gamma^2} \tag{3.2}$$

for the power spectrum of $u(t)$. The correlation function of $u(t)$ is then

$$\langle u(t_1) u(t_2) \rangle = \frac{\pi I_R}{m^2 \gamma} e^{-\gamma |t_1 - t_2|} \tag{3.3}$$

and in particular

$$\langle u^2 \rangle = \frac{\pi I_R}{m^2 \gamma} = \frac{kT}{m} \tag{3.4}$$

if we assume

$$m\gamma kT = \pi I_R = \int_0^{\infty} \langle R(t_1) R(t_1 + t) \rangle \, dt \tag{3.5}$$

Since a linear combination of Gaussian processes is also Gaussian, the assumption C imparts a Gaussian property to $u(t)$, which assures the equilibrium Maxwellian distribution. The assumption B together with the fact that eq. (2.1) is first order in the time derivative implies the Markoffian nature of the process $u(t)$. The Doob theorem [8] tells us that a Gaussian-Markoffian process ought to have an exponential decay in the correlation function as in fact is shown by eq. (3.3).

The two-time distribution $W_2(u_1t_1, u_2t_2)$ of $u(t)$ is characterized by the variance $\langle u(t_1)^2\rangle = \langle u(t_2)^2\rangle$ and $\langle u(t_1)u(t_2)\rangle$. The transition probability (or the Green function) $P(u_1t_1 | u_2t_2)$ from u_1 at t_1 to u_2 at t_2 is then obtained from the relation

$$W_2(u_1t_1, u_2t_2) = W_1(u_1t_1)P(u_1t_1|u_2t_2) \tag{3.6}$$

where W_1 is the (equilibrium) one-time distribution function. Thus we have

$$P(u_1t_1 | u_2t_2) = \left[\frac{m}{2\pi kT}\right]^{\frac{1}{2}}(1-e^{-2\gamma(t_2-t_1)})^{-\frac{1}{2}} \exp\left[-\frac{m}{2kT}\frac{\{u_2-u_1 e^{-\gamma(t_2-t_1)}\}^2}{1-e^{-2\gamma(t_2-t_1)}}\right]. \tag{3.7}$$

Since $u(t)$ is Gaussian and Markoffian this completes the solution of Eq. (2.1).

The transition probability (3.7) is the fundamental solution of the Fokker-Planck equation

$$\frac{\partial}{\partial t} P = \frac{\partial}{\partial u}(\gamma u + D_u \frac{\partial}{\partial u})P \tag{3.8}$$

with the diffusion constant D_u in the velocity space defined by

$$D_u = \frac{\gamma kT}{m} = \frac{\pi I_R}{m^2} . \tag{3.9}$$

IV. DIRECT INTEGRATION AND PATH INTEGRAL REPRESENTATION

Interpreting eq. (2.1) as an ordinary differential equation satisfied by an arbitrary sample of the stochastic process in question, we integrate this equation in the usual way. The solution is

$$u(t) = u(t_0)e^{-\gamma(t-t_0)} + \int_{t_0}^{t} dt' e^{-\gamma(t-t')} R(t')/m \qquad t > t_0 \tag{4.1}$$

if the initial value of u is given as $u(t_0)$. Taking the average of $u(t)$ over the whole ensemble of samples we obtain

$$\langle u(t)\rangle = u(t_0)e^{-\gamma(t-t_0)} \tag{4.2}$$

and

$$\langle (u(t)- u_0 e^{-\gamma(t-t_0)})^2\rangle = \int_{t_0}^{t} dt_1 \int_{t_0}^{t} dt_2 e^{-\gamma(t-t_1)-\gamma(t-t_2)} \frac{\langle R(t_1)R(t_2)\rangle}{m^2}$$

which becomes with the use of eq. (2.3)

$$\langle (u(t)-\langle u(t)\rangle)^2\rangle = \frac{\pi I_R}{m^2\gamma}(1-e^{-2\gamma(t-t_0)}) \tag{4.3}$$

if the initial distribution of u is sharp.

As mentioned earlier, $u(t)$ in eq. (4.1) is linear in $R(t')$ $(t_0 < t' < t)$ so that it is Gaussian, and its probability distribution is characterized by $\langle u(t)\rangle$ and $\langle (u(t) - \langle u(t)\rangle)^2\rangle$ as given by eqs. (4.2) and (4.3). It is the transition probability $P(u_0 t_0 | ut)$, eq. (3.7).

One comment seems appropriate at this point. The present method has an advantage over Rice's method; namely that it is applicable even when the derived process does not have a power spectrum of finite intensity. If γ in eq. (2.1) is zero, the power spectrum (3.2) ceases to exist and then Rice's method is useless. Consider the pure Wiener process of diffusion

$$\dot{x} = u(t) \tag{4.4}$$

where the velocity $u(t)$ (which here replaces $R(t)$ in eq. (2.1)) is a pure white noise. Equations (4.2) and (4.3) now read as

$$\langle x(t)\rangle = x(t_0) = x_0$$

$$\langle (x(t)-x_0)^2\rangle = 2\pi I_u (t-t_0)$$

where I_u is the power intensity of the white noise $u(t)$. Defining the diffusion constant by

$$D = \pi I_u \tag{4.5}$$

we get

$$P(x_0t_0|xt) = [4\pi D(t-t_0)]^{-\frac{1}{2}} \exp\left[-\frac{(x-x_0)^2}{4D(t-t_0)}\right] . \tag{4.6}$$

Should Rice's method be used, the same result can be obtained by taking the limit of $\gamma\to 0$ and writing x for u and D for $2kT\gamma/m$ in the final result Eq. (3.7).

In Eq. (4.4) the velocity $u(t)$ should more properly be considered as the random process driven by a random force, namely by Eq. (2.1),

$$m\ddot{x} = -m\gamma\dot{x} + R(t) \tag{4.7}$$

Rice's method cannot be used directly for solving this equation (it can be used if the particle is harmonically bound to the origin, and the Brownian motion can be treated as the limit of a vanishing elastic force). But Eq. (4.1) can be again integrated to

$$x(t)-x(t_0) = \frac{u_0}{\gamma}(1-e^{-\gamma(t-t_0)}) + \int_{t_0}^{t} dt_1 \int_{t_0}^{t_1} dt_2 e^{-\gamma(t_1-t_2)} R(t_2)/m. \tag{4.8}$$

Since $x(t)$ is linear in $R(t')$ $(t_0<t'<t)$, the process $x(t)$ is Gaussian so that its average and variance define the distribution. The second term on the r.h.s. of Eq. (4.8) is written as

$$\int_{t_0}^{t} dt_2 \int_{t_2}^{t} dt_1 e^{-\gamma(t_1-t_2)} R(t_2)/m = \int_{t_0}^{t} dt' \frac{1-e^{-\gamma(t-t')}}{\gamma} \frac{R(t')}{m}$$

If the equilibrium is assumed for the initial distribution of u_0, the average displacement vanishes and the variance of displacement in the time interval (t_0,t) is easily calculated with the use of Eqs. (2.3) and (3.4) to be

$$\langle(x(t)-x(t_0))^2\rangle = \frac{\langle u^2\rangle}{\gamma}\left[t-t_0-\frac{1}{\gamma}\{1-e^{-\gamma(t-t_0)}\}\right] \tag{4.9}$$

which gives the transition probability

$$P(0,0|xt) = [4\pi D\{t-(1-e^{-\gamma t})/\gamma\}]^{-1/2}$$
$$\exp\left[-\frac{x^2}{4D\{t-(1-e^{-\gamma t})/\gamma\}}\right] \tag{4.10}$$

where the diffusion constant is defined by

$$D = \frac{\langle u^2\rangle}{\gamma} = \frac{kT}{m\gamma} \quad \text{(Einstein relation)} \tag{4.11}$$

It should be noticed that the transition probability (4.10) does not satisfy the Chapman-Kolomogorov equation, which means that the process x(t) is non-Markoffian. This is because Eq. (4.7) is not first order but second order in the time-derivatives of x. If we consider the variables x and u together, the Brownian motion in phase space is a two-dimensional Markoffian process, but the projection onto the displacement-coordinate is not by itself Markoffian. It is, however, very important to recognize the fact that it recovers a Markoffian nature if the space and time are coarse-grained. Namely, if the time t is much longer than the correlation time τ_c of velocity u, that is, $t \gg \tau_c = 1/\gamma$, the terms, $\exp(-\gamma t)$, may safely be ignored in Eq.(4.10). Then this is reduced to the expression (4.6), which is of course the fundamental solution of the diffusion equation

$$\frac{\partial P}{\partial t} = D\frac{\partial^2}{\partial x^2}P. \tag{4.12}$$

The coarse-graining in space is also implied here because we have to sacrifice a more detailed description of the probabilities of small displacements over the distance of the mean free path;

$$\Delta x \gg l = \langle u^2\rangle^{1/2}\tau_c$$

Another method of direct integration is worth noting here. The probability of realizing a sample path of u(t) as determined by Eq.(2.1) or Eq. (4.1) is equal to that of realizing a path of R(t). Choosing $t_j = t_0 + j\Delta t$ ($j = 1,2, \ldots n$, $t_n \equiv t$), we ask for the probability $P(R_1, R_2 \ldots R_n)$ of realizing R(t) as R_1 at t_1, R_2 at t_2, ..., and R_n at $t_n = t$. Because R(t) is Gaussian and white-noise, this is easily seen to be given by

$$P(R_1, \ldots, R_n) = \left[\frac{\Delta t}{4\pi^2 I_R}\right]^{\frac{n}{2}} \exp\left[-\frac{\Delta t}{4\pi I_R}(R_1^2 + \ldots + R_n^2)\right]$$

(4.13)

by calculating the characteristic function of P as

$$Q(\xi_1 \ldots \xi_n) = \langle \exp i(\xi_1 R_1 + \ldots + \xi_n R_n) \rangle$$

$$= \exp\left[-\frac{1}{2}\sum_j^n \sum_l^n \xi_j \xi_l \langle R_j R_l \rangle\right] \longrightarrow$$

$$\longrightarrow \exp\left[-\frac{1}{2(\Delta t)^2}\int_{t_0}^t dt_1 \int_{t_0}^t dt_2 \xi(t_1)\xi(t_2)\langle R(t_1)R(t_2)\rangle\right]$$

$$= \exp\left[-\frac{1}{(\Delta t)^2}\int_{t_0}^t \xi(t')^2 dt' \cdot 2\pi I_R\right]$$

$$= \exp\left[-\frac{2\pi I_R}{\Delta t}(\xi_1^2 + \ldots + \xi_n^2)\right]$$

In this calculation we assumed a very large n. The exponent on the r.h.s. of Eq. (4.13) becomes, with the use of Eq. (3.9),

$$-\frac{1}{4\pi I_R}\int_{t_0}^t R^2(t')dt' = -\frac{1}{4D_u}\int_{t_0}^t (\dot{u}(s) + \gamma u(s))^2 ds$$

(4.14)

in this limit. Thus, the transition probability (3.7) can be expressed as

$$P(u_0 t_0 | ut) = \int d\mathcal{D}[u(s)] \exp\left\{\int_{t_0}^t \mathcal{L}(u(s), \dot{u}(s))ds\right\}$$

(4.15)

with the Langrangian

$$\mathcal{L} = -\frac{1}{4D_u}(\dot{u}(s) + \gamma u(s))^2$$

$$= -\frac{m}{4\gamma kT}(\dot{u}(s) + \gamma u(s))^2$$

(4.16)

where the paths $\{u(s)\}$ cover all possible paths with $u(t_0) = u_0$ and $u(t) = u$, for each of which the action integral gives the weight.

(The measure of paths is defined by Eq. (4.13)). The path integral formula for the case $\gamma = 0$ was first obtained by Wiener, so that path integrals of this sort may be called Wiener integrals. The corresponding path integral expression for quantal systems is quite analogous to this, but it is called the Feynman integral.

Onsager and Machlup [9] formulated their fluctuation theory as a classical Brownian motion and used the same expression as Eq. (4.15).

V. STOCHASTIC LIOUVILLE EQUATION

The Langevin equation (2.1) determines a sample path u(t) for each given sample of R(t). We consider now an ensemble of Brownian particles at each time point and denote the density of representative points in u at time t by a distribution function f(u,t). This satisfies the Liouville equation

$$\frac{\partial}{\partial t} f(u,t) = - \frac{\partial}{\partial u} \dot{u} f \tag{5.1}$$

or

$$\frac{\partial}{\partial t} f(u,t) = \frac{\partial}{\partial u} (\gamma u - \frac{1}{m} R(t) f \tag{5.2}$$

by Eq. (2.1). The Lionville equation may be written more concisely as

$$\frac{\partial f}{\partial t} = \Omega(t) f \tag{5.3}$$

where $\Omega(t)$ denotes the "Liouville" operator on the r.h.s. of Eq. (5.2). Since the random force R(t) is a stochastic process, the operator Ω is also stochastic. In this sense Eq. (5.2) or (5.3) will be called a <u>stochastic Liouville equation</u>, which is a Schroedinger picture corresponding to the Heisenberg picture in the Langevin equation of motion. A formal solution of Eq. (5.3) can be written as

$$f(u,t) = \left[\exp \int_{t_0}^{t} \Omega(t')dt' \right] \cdot f(u,t_0) \tag{5.4}$$

for a given sample of R(t). In particular if the initial ensemble is given by

$$f(u,t_0) = \delta(u-u_0) \tag{5.5}$$

f(u,t) is also a delta-function, $\delta(u-u(t))$, the final point being determined by Eq. (1.2) for a given sample of R(t). Taking the average of f over all possible samples we then obtain the transition probability,

$$\begin{aligned} P(u_0t_0|ut) &= \langle f(u,t)\rangle \\ &= (u|\langle \exp \int_{t_0}^{t} \Omega(t')dt' \rangle|u_0) \end{aligned} \tag{5.6}$$

where the bracket notation, like that of Dirac's, means the kernel of the integral representation of the operator inside. Generally, the operator $\Omega(t)$ may not be commutable for different times and the co-responding exponential operator should be interpreted as an ordered exponential. However this precaution is not needed for Eq. (5.2) and the average of the exponential operator in Eq. (5.6) is easily obtained by using the cumulant theorem noticing that cumulants higher than of the second order identically vanish for a Gaussian process;

$$\begin{aligned} &P(u_0t_0|ut) \\ &= \exp\left[\int_{t_0}^{t} dt_1 \frac{\partial}{\partial u}\gamma u + \frac{1}{2m^2}\int_{t_0}^{t} dt_1 \int_{t_0}^{t} dt_2 \frac{\partial^2}{\partial u^2}\langle R(t_1)\, R(t_2)\rangle\right]\cdot\delta(u-u_0) \\ &= \exp\left[(t-t_0)\left\{\frac{\partial}{\partial u}\gamma u + \frac{\partial^2}{\partial u^2} D_u\right\}\right]\delta(u-u_0) \end{aligned} \tag{5.7}$$

where

$$D_u = \frac{1}{m^2}\int_0^{\infty}\langle R(t_1)R(t_1 + t)\rangle\, dt = \frac{\pi I_R}{m^2} \tag{5.8}$$

is the diffusion coefficient in the velocity space. Equation (5.7) means that the transition probability $P(u_0t_0|u,t)$ is the fundamental solution (Green function) of Eq. (3.8). The relation (3.9) ensures that the equilibrium distribution P_e satisfying the equation

$$(\gamma u + D_u \frac{\partial}{\partial u})\, P_e = 0$$

is Maxwellian with the temperature T.

It should be noticed that Eq. (3.8) is the Fokker-Planck equation for the process u(t). The Markoffian property comes from the white noise assumption B for R(t). The Fokker-Planck property, that is the diffusion-type nature of the process, is due to the Gaussian assumption C. A simpler example is the process (4.4), for which the same procedure immediately leads to the diffusion equation (4.12). Another example is a Brownian motion in a potential field V(x), for which we write the Langevin equation,

$$\frac{dx}{dt} = u, \qquad \dot{u} = -\frac{1}{m}\frac{\partial V}{\partial x} - \gamma u + \frac{1}{m}R(t) \tag{5.9}$$

and the stochastic Liouville equation,

$$\frac{\partial}{\partial t}f(x,u,t) = \left[-\frac{\partial}{\partial x}u + \frac{\partial}{\partial u}\left(\frac{1}{m}\frac{\partial V}{\partial x} + \gamma u - \frac{1}{m}R(t)\right)\right]f(x,u,t).$$

The corresponding Fokker-Planck equation is the Kramers equation

$$\frac{\partial}{\partial t}P(x,u,t) = \left[-\frac{\partial}{\partial x}u + \frac{\partial}{m\partial u}\frac{\partial V}{\partial x} + \frac{\partial}{\partial u}\gamma u + D_u\frac{\partial^2}{\partial u^2}\right]P(x,u,t) \tag{5.10}$$

VI. RETARDED FRICTION, FLUCTUATION-DISSIPATION THEOREMS [4,5]

Now we turn to non-ideal Brownian motions. The first thing we should do seems to be to remove the restriction B, the white noise assumption for the random force, because a frequency-dependent impedence is quite common in nature, which, as we shall soon see, is inseparably related to a non-white noise.

Consider the response of a Brownian system to an external force. The Langevin equation (2.1) may be replaced by

$$m\dot{u} = -m\gamma u + R(t) + K(t) \tag{6.1}$$

in the presence of an external driving force K(t). If K is periodic with a frequency ω, the response will be on the average

$$\langle u(t) \rangle = \mathrm{Re}\, \mu[\omega] K_0 \, e^{i\omega t} \tag{6.2}$$

with

$$\mu[\omega] = \frac{1}{m} \, \frac{1}{i\omega + \gamma} \tag{6.3}$$

which is the mobility, or more generally the admittance, of the Brownian system. In many cases, the admittance is not so simple but can be a more complex function of ω. Thus it should be generalized to

$$\mu[\omega] = \frac{1}{m} \, \frac{1}{i\omega + \gamma(\omega)} \tag{6.4}$$

with a frequency-dependent friction. Such examples are found in many mechanical as well as electrical systems, e.g., for a Brownian particle, if its mass is no longer much larger than that of a bath molecule, the condition (2.2) will not hold and the assumption B has to be discarded. This leads to a retarded friction, or a friction with a memory effect, in the Langevin equation. Thus we write

$$m\dot{u} = - m \int_{-\infty}^{t} \gamma(t-t') \, u(t')dt' + R_{\infty}(t) + K(t) \tag{6.5}$$

instead of Eq. (6.1). The average response $\langle u(t) \rangle$ to a periodic K is of the form Eq. (6.2), with $\mu[\omega]$ given by Eq. (6.4) in which the friction $\gamma[\omega]$ is defined as the Fourier-Laplace transform of the retarded friction kernel, i.e.

$$\gamma[\omega] = \int_{0}^{\infty} \gamma(t) e^{-i\omega t} dt \tag{6.6}$$

Being a friction in the Brownian motion, $\gamma[\omega]$ may be assumed to satisfy the following conditions [6];

i) $\mathrm{Re}\, \gamma[\omega] > 0$ (6.7)

ii) $\lim_{\omega \to \infty} |\gamma[\omega]| = \text{finite} < \infty$ (6.8)

We consider the Brownian motion in the absence of the external field (K = 0). If it is aged in a bath at an equilibrium temperature T, it should obey the fluctuation-dissipation theorem (F-D theorem), which can be expressed in the following two forms

$$\text{(i)} \quad \mu[\omega] = \frac{1}{kT}\int_0^\infty dt \langle u(t_0)u(t_0+t)\rangle\, e^{-i\omega t} \tag{6.9}$$

$$\text{(ii)} \quad \gamma[\omega] = \frac{1}{mkT}\int_0^\infty dt \langle R_\infty(t_0)R_\infty(t_0+t)\rangle\, e^{-i\omega t} \tag{6.10}$$

These may be called the first and the second F-D theorems. Since Eq. (6.5) (with K=0) is linear, Rice's method can be used to obtain the power spectrum of u as

$$I_u(\omega) = \frac{1}{m^2}\frac{I_R(\omega)}{|i\omega + \gamma(\omega)|^2} \tag{6.11}$$

The relation (6.10) gives

$$I_R(\omega) = \frac{m\ k\ T}{\pi}\,\mathrm{Re}\,\gamma(\omega) \tag{6.12}$$

by the Wiener-Khintchin Theorem, which gives in turn

$$\langle u(t_0)u(t_0+t)\rangle = \frac{kT}{2\pi m}\int_{-\infty}^{+\infty}\left(\frac{1}{i\omega+\gamma(\omega)} + \frac{1}{-i\omega+\gamma^*(\omega)}\right)e^{i\omega t}d\omega$$

$$= \frac{kT}{2\pi m}\int_{-\infty}^{+\infty}\frac{e^{i\omega t}}{i\omega+\gamma(\omega)}\,d\omega \tag{6.13}$$

As the function $(-i\omega + \gamma^*(\omega))^{-1}$ does not have any pole in the upper half plane of complex ω because of the condition (6.7) so that the integral reduces to the latter expression. In particular, this is evaluated for $t = 0^+$ by the residue around $\omega = \infty$ with the use of (6.8) as

$$\lim_{t\to 0^+}\langle u(t_0)u(t_0 + t)\rangle = \langle u^2\rangle = kT/_m \tag{6.14}$$

which is the equipartition law. If $R_\infty(t)$ is assumed to be Gaussian, this assures the equilibrium Maxwellian distribution of u. Equation (6.13) means also the relation (6.9) for the response to an external force.

Thus we see that the Langevin equation (6.5) with the condition (6.7) and (6.8) does properly represent a generalized Brownian motion of a light particle as long as non-linear effects are still disregarded.

Equation (6.13) implies that this may as well be represented by

$$m\dot{u}(t) = -m \int_{t_0}^{t} \gamma(t-t')u(t')dt' + R(t) \quad , \quad t > t_0 \tag{6.15}$$

for K=0. The process u(t) is stationary in the sense that the initial time t_0 is chosen arbitrarily. The random force R(t) is not quite the same as R(t) in Eq. (6.5) but should satisfy the conditions

$$\langle u(t_0)R(t)\rangle = 0 \qquad\qquad t > t_0 \tag{6.16}$$

and

$$\langle R(t_0)R(t_0+t)\rangle = \langle R(t_1)R(t_1+t)\rangle = mkT\gamma(t) \tag{6.17}$$

With Eq. (6.16), we obtain from Eq. (6.15)

$$\langle u(t_0)\dot{u}(t_0+t)\rangle = -\int_0^t \gamma(t-t')\,\langle u(t_0)u(t_0+t')\rangle\,dt'$$

which is Laplace-transformed to result in

$$\int_0^\infty \langle u(t_0)u(t_0+t)\rangle\, e^{-i\omega t}dt = \frac{\langle u^2\rangle}{i\omega + \gamma[\omega]} \tag{6.18}$$

or equivalently in Eq. (6.13). The correlation function of R,(6.17) is calculated from Eq. (6.15) using (6.16) and (6.18).

The first F-D theorem reduces to the Einstein relation (4.11) for the classical Brownian motion at zero-frequency (Eqs. (3.3), (3.4) and (6.3)). The second F-D theorem corresponds to the Nyquist theorem of thermal noise and generalizes the relation (3.5). The linear response theory provides us with a general proof of the F-D theorem from a statistical-mechanical viewpoint. The diffusion constant D can be defined by

$$\begin{aligned} D &= \lim_{t\to\infty}\langle(x(t) - x(0))^2\rangle/2t \\ &= \lim_{\infty} \frac{1}{2t}\int_0^t dt_1 \int_0^t dt_2 \langle u(t_1)u(t_2)\rangle \\ &= \int_0^\infty \langle u(t_0)u(t_0+t)\rangle dt \end{aligned} \tag{6.19}$$

With Eqs. (3.3) and (3.4) this gives the Einstein relation (4.11) or

$$D = \mu kT \tag{6.20}$$

for a classical Brownian particle. Thus the first F-D theorem is regarded as a generalization of the Einstein relation. The second F-D theorem corresponds to the Nyquist theorem and generalizes the relation (3.5).

The linear response theory [11, 12] proves the first F-D theorem on the basis of statistical mechanics and gives the second F-D theorem as a collorary. In this sense the generalized Langevin equation (6.5) can be regarded as a stochastic representation of the linear response theory. To this point we shall come back later to discuss the problem from a damping-theoretical point of view as developed by Mori. We emphasize here that the white noise assumption B in the classical theory is thus necessarily related to a friction without retardation. It is thus logical to regard it as a limit $\tau_c \to 0$ such as that in Eq. (2.5).

The above treatment can be generalized to Brownian motions with more than one variable. A generalized Langevin equation may be written in the form

$$\dot{X}(t) = i\Omega X - \int_{t_0}^{t} \Gamma(t-t')X(t')dt' + R(t), \quad t > t_0 \tag{6.21}$$

where X, is a column vector with components $X_1, \ldots X_n$, Ω a constant retardation and R is the random force. If an external force is present, this should be added on the r.h.s.. If the system is quantal rather than classical, the definition of correlation functions should be generalized to the canonical correlations [5,10], which are defined, for example,

$$\langle A(t_1);B(t_2)\rangle = \frac{1}{\beta}\int_0^{\beta} d\lambda \, \mathrm{Tr}\, \rho_e e^{\lambda \mathcal{H}} A(t_1) e^{-\lambda \mathcal{H}} B(t_2) \tag{6.22}$$

where

$$\rho_e = e^{-\beta \mathcal{H}} / \mathrm{Tr}\, e^{-\beta \mathcal{H}}$$

is the equilibrium canonical density matrix with the Hamiltonian $\mathcal{H}$ at a temperature $T=1/k\beta$ and $A(t)$ and $B(t)$ are the Heisenberg operators corresponding to the relevant quantities A and B. The canonical correlations have the proper symmetry and positive definiteness. For the vectors X and R in Eq. (6.21) the corresponding row vectors $\tilde{X}$ and $\tilde{R}$ are defined and the variables are normalized for convenience as

$$\langle X_j(t);\ X_k(t)\rangle = \delta_{jk}$$

or

$$\langle X(t);\ \tilde{X}(t)\rangle = \mathbb{1} \tag{6.23}$$

and

$$i\Omega = \langle \dot{X}(t);\ \tilde{X}(t)\rangle = -\langle X(t);\ \dot{\tilde{X}}(t)\rangle \tag{6.24}$$

The random force is assumed to satisfy the conditions

$$\langle R(t)\rangle = 0 \text{ and } \langle R(t);\ \tilde{X}(t_0)\rangle = 0 \qquad (t > t_0) \tag{6.25}$$

The F-D theorems are now written as

$$\langle X(t)\rangle = \mathrm{Re}\,\Lambda(\omega)K_0 e^{i\omega t} \tag{6.26}$$

$$\Lambda(\omega) = \int_0^\infty \langle X(t_0+t);\ \hat{X}(t_0)\rangle e^{-i\omega t}dt = \frac{1}{i(\omega-\Omega) + \Gamma(\omega)} \tag{6.27}$$

$$\Gamma(\omega) = \int_0^\infty \Gamma(t)e^{-i\omega t}dt = \int_0^\infty \langle R(t_0+t);\ \hat{R}(t_0)\rangle e^{-i\omega t}dt \tag{6.28}$$

The simplest example is the Brownian motion of a harmonic oscillator for which X is composed of the coordinate x and the momentum p. A more complex example is a linearized Boltzmann equation for the one-particle distribution function $f_1(p,x)$ generalized to the form

$$\frac{\partial}{\partial t} f_1(p,x,t) = i\mathcal{L}_0 f_1 - \int^t \Gamma(pxt|p'x't') f_1(p',x',t')dt' + R \tag{6.29}$$

where $i\mathcal{L}_0 f_1$ is the drift term and $-\Gamma$ is the linearized collision operator. The noise R is regarded as the source of fluctuation of f_1 around its average and should be related to the collision operator $-\Gamma$ if Eq. (6.29) describes a system near thermal equilibrium.

The generalized Langevin equation (6.21) or (6.15) is sometimes called the Mori formalism [13], which derives this form from the basic Liouville equation by a damping-theoretical (projection operator) method. We shall come back to this later.

It should also be noted, that the generalized Langevin equation discussed here does not necessarily assume a Gaussian property of the random force. It is considered as a representation of fluctuations near thermal equilibrium and it correctly describes the correlation functions. More information is obtained only when the stochastic nature of the random force is defined precisely: This means at the same time that the Langevin equation has a great generality.

VII. FORCE CORRELATIONS [5]

In Eq. (6.5) or (6.15) the force from bath molecules is divided into the systematic and the random part. The correlation function of the total force

$$F = m\dot{u} \tag{7.1}$$

is easily found from (6.18). In its Fourier-Laplace transform it is given by

$$\frac{1}{m^2\langle u^2\rangle}\int_0^\infty \langle F(t_0+t)\,F(t_0)\rangle\, e^{-i\omega t}dt \equiv \gamma_t[\omega] = \frac{i\omega\,\gamma[\omega]}{i\omega + \gamma[\omega]} \tag{7.2}$$

which is written as

$$\gamma[\omega] = \frac{i\omega\gamma_t[\omega]}{i\omega - \gamma_t[\omega]}, \quad \text{or} \quad \frac{1}{\gamma_t[\omega]} - \frac{1}{\gamma[\omega]} = \frac{1}{i\omega} \tag{7.3}$$

Correspondingly, if we write Eq. (6.21) as

$$\dot{X}(t) = i\Omega X + F(t) \tag{7.4}$$

Eq. (7.2) is generalized to

$$\int_0^\infty \langle F(t_0 + t);\, \tilde{F}(t_0)\rangle\, e^{-i\omega t} dt \doteqdot \Gamma_t(\omega) = \Gamma \frac{1}{i(\omega-\Omega) + \Gamma}\; i(\omega-\Omega) \tag{7.5}$$

or equivalently

$$\Gamma_t[\omega] = i(\omega-\Omega) \frac{1}{i(\omega-\Omega) + \Gamma(\omega)} \Gamma(\omega) \tag{7.6}$$

and Eq. (7.3) to

$$\frac{1}{\Gamma_t[\omega]} = \frac{1}{\Gamma[\omega]} + \frac{1}{i(\omega-\Omega)} \tag{7.7}$$

Equation (7.2) means $\gamma_t[0] = 0$ or

$$\int_0^\infty \langle F(t_0 + t);\, F(t_0)\rangle\, dt = 0 \tag{7.8}$$

namely that the time integral of the total force over an infinite time interval should identically vanish, in contrast to

$$\int_0^\infty \langle R(t_0 + t);\, R(t_0)\rangle\, dt = mkT\gamma[0] \neq 0 \tag{7.9}$$

More generally, if the system has proper frequencies, the power spectrum of the force F vanishes at these frequencies as is seen by Eq. (7.5). For the example of a harmonic oscillator, Eq. (7.5) reduces to

$$\frac{1}{mkT} \int_0^\infty \langle F(t_0 + t)\, F(t_0)\rangle\, e^{-i\omega t} dt = \frac{(\omega_0^2 - \omega^2)\gamma[\omega]}{\omega_0^2 - \omega^2 + i\omega\gamma[\omega]} \tag{7.10}$$

Another simple example may be instructive in order to see the difference of F(t) and R(t). A light Brownian particle is supposed to experience elastic collisions with very heavy scatterers, randomly distributed in space. A simple model of this random motion is usually described by the transport equation,

$$\frac{\partial}{\partial t} f(u,t) = -\frac{1}{\tau} f(u,t) + \frac{1}{\tau}\int \Phi(u',u)\, f(u',t)du' \tag{7.11}$$

for the velocity distribution function f(u,t). The second term on the r.h.s. represents transitions in the velocity, Φ being the scattering probabilities at each collision. The velocity correlation function is easily calculated for this model to yield

$$\langle u(t_0 + t)\, u(t_0)\rangle = \frac{\langle u_x^2\rangle}{i\omega + \gamma}\mathbb{1} \tag{7.12}$$

where

$$\gamma = \frac{1}{\tau}(1-\alpha), \qquad \alpha = \overline{\cos\theta} \tag{7.13}$$

θ being the deflection angle in each collision. In this model the duration time τ_c of collision is assumed to be infinitesimally short in comparison with the mean free time, so that the random force R(t) is a sequence of irregular pulses forming a white noise, but generally non-Gaussian. The real force F is also such a sequence of pulses, which however are correlated. If $\mathbf{p}_0, \mathbf{p}_1, \mathbf{p}_2$ ..... is a sequence of the momentum values,

$$\int_{t_j-0}^{t_j+0} \mathbb{F}(t)dt = \mathbf{p}_j - \mathbf{p}_{j-1} \equiv \Delta\mathbf{p}_j$$

is the impulse given by the j-th collision occuring at time t_j. Then we have

$$\mathbb{F}_t[\omega] \equiv \int_0^t \mathbb{F}(t')\, e^{-i\omega t'} dt' = e^{-i\omega t_1}\Delta\mathbf{p}_1 + \dots + e^{-i\omega t_n}\Delta\mathbf{p}_n \tag{7.14}$$

if n collisions take place in the time interval (0,t). The correlation function of $\mathbb{F}(t)$ can be obtained from the formula,

$$\int_0^\infty \langle \mathbb{F}_t[\omega]\mathbb{F}_t[-\omega]\rangle\, e^{-st}dt$$
$$= \frac{1}{s^2}\int_0^\infty e^{-st}\left\{ e^{-i\omega t}\langle \mathbb{F}(t_0+t)\mathbb{F}(t_0)\rangle + e^{i\omega t}\langle \mathbb{F}(t_0)\mathbb{F}(t_0+t)\rangle\right\} \tag{7.15}$$

The successive impulses are in fact correlated as one finds by

$$\langle \Delta \mathbb{P}_1^{\,2}\rangle = \langle(\mathbb{P}_1 - \mathbb{P}_0)^2\rangle = 2\langle \mathbb{P}\mathbb{P}\rangle(1-\alpha)$$
$$\langle \Delta\mathbb{P}_j\, \Delta\mathbb{P}_1\rangle = \langle(\mathbb{P}_j - \mathbb{P}_{j-1})(\mathbb{P}_1 - \mathbb{P}_0)\rangle$$
$$= -\langle \mathbb{P}\mathbb{P}\rangle\alpha^{j-2}(1-\alpha)^2 \quad , \quad j \geqslant 2 \tag{7.16}$$

so that one gets

$$\int_0^\infty \langle \mathbb{F}(t_0 + t)\,\mathbb{F}(t_0)\rangle\, e^{-i\omega t}dt = \langle \mathbb{P}\mathbb{P}\rangle\,\frac{i\omega\gamma}{i\omega+\gamma} \tag{7.17}$$

corresponding to

$$\langle \mathbb{F}(t_0 + t)\,\mathbb{F}(t_0)\rangle = \gamma\langle \mathbb{P}\mathbb{P}\rangle\left\{2\delta(t) - \gamma e^{-\gamma|t|}\right\} \tag{7.18}$$

whereas

$$\langle \mathbb{R}(t_0 + t)\,\mathbb{R}(t_0)\rangle = 2\,\gamma\langle \mathbb{P}\mathbb{P}\rangle\,\delta(t) \tag{7.19}$$

This shows that the correlation effects a negative tail to cancel out the integration over the initial time interval of the order τ_c. In other words the friction formula (7.9) may be approximated by

$$\gamma\cdot\mathbb{1} \sim \frac{1}{mkT}\int_0^{\tau'}\langle \mathbb{F}(t_0 + t)\mathbb{F}(t_0)\rangle dt \tag{7.20}$$

if a suitable time τ' ($\tau_c < \tau' < \tau$) can be so chosen that the integral attains a plateau. This is the well-known Kirkwood assumption and Eq. (7.20) was his formula for the friction coefficient of a Brownian particle [14]. It should be kept in mind that this approximation essentially means a nearly white noise of the random force and fails badly if a non-white behaviour is dominant.

VIII. SOME EXAMPLES

As an interesting example of the foregoing treatment, we shall consider first of all Brownian fluctuations of the electric polarization of a spherical particle of a metal [15]. It consists of free electrons in the background medium with a dielectric constant ε_0. If a uniform electric field $\mathbf{E}_0$ with a frequency ω is applied from outside, which is vacuum, the electric field inside the particle will be

$$\mathbf{E} = \mathbf{E}_0' + \mathbf{E}' \tag{8.1}$$

where

$$\mathbf{E}_0' = \frac{3}{\varepsilon_0 + 2} \mathbf{E}_0 \tag{8.2}$$

is the applied field modified by the dielectric constant ε_0 and

$$\mathbf{E}' = - \frac{4\pi}{\varepsilon_0 + 2} \mathbf{P} \tag{8.3}$$

is the field produced by electronic charges

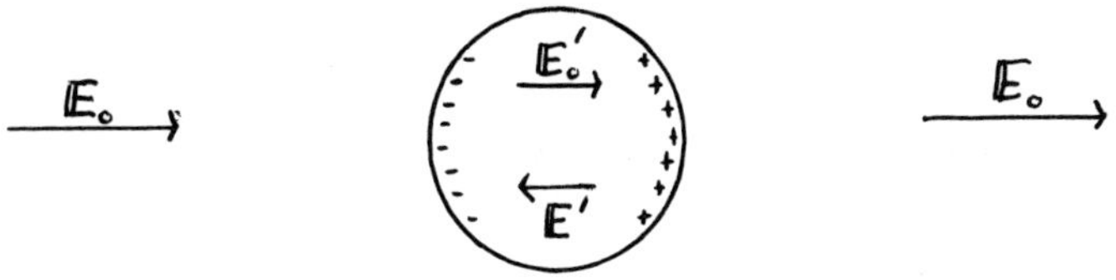

Fig.1.

accumulated on the surface, which correspond to a uniform polarization $\mathbf{P}$. The total electric moment of the sphere is

$$\mathbf{M} = \Omega \mathbf{P} \quad , \tag{8.4}$$

where $\Omega = 4\pi a^3/3$ is the volume of the particle with radius a. The applied field interacts with the particle and the energy of interaction is

$$\mathcal{H}'_{ext} = - \mathbf{M} \cdot \mathbf{E}'_0 \tag{8.5}$$

We consider such a uniform polarization in a given direction and its Brownian fluctuation. The current J associated with the polarization is defined by

$$\dot{M}(t) = J(t) \tag{8.6}$$

The current responding to an external periodic field is written as

$$\begin{aligned} \langle J(t) \rangle &= \mathrm{Re}\, \sigma^e[\omega] \mathbf{E}'_0 \, e^{i\omega t} \\ &= \mathrm{Re}\, \sigma^e[\omega] \frac{3}{\varepsilon_0 + 2} \mathbf{E}_0 \, e^{i\omega t} \, . \end{aligned} \tag{8.7}$$

The linear response theory gives the admittance coefficient $\sigma^e[\omega]$ as

$$\sigma^e[\omega] = \beta \int_0^\infty dt \; e^{-i\omega t} \langle J(0); J(t) \rangle / \Omega \tag{8.8}$$

where the current correlation function represents the Brownian fluctuation in the absence of external forces.

We observe now that the Brownian current (8.6) consists of two parts, the systematic and the random currents. The former is the current induced by the self-field E' caused by the polarization. This may be expressed as

$$\begin{aligned} J_s(t) &= \Omega \int_0^t \sigma(t-t')E'(t')dt' \\ &= - \int_0^t 4\pi\bar{\sigma}(t-t')M(t')dt' \end{aligned} \tag{8.9}$$

where $\sigma(t)$ is the retarded kernel of the complex conductivity $\sigma[\omega]$ and $\bar{\sigma}$ is given by

$$\bar{\sigma}(\omega) = \int_0^\infty \bar{\sigma}(t) e^{-i\omega t} dt = \frac{\sigma[\omega]}{\varepsilon_0 + 2} \tag{8.10}$$

as is seen from Eqs. (8.3) and (8.4). Equation (8.6) is written as

$$\dot{M}(t) = J_s(t) + J'(t) \tag{8.11}$$

and is regarded as the Langevin equation for the Brownian motion. The second F-D theorem (6.10) gives

$$\frac{4\pi\sigma(\omega)}{\varepsilon_0 + 2} = \frac{1}{\langle M;M \rangle} \int_0^\infty \langle J'(t) J'(0) \rangle e^{-i\omega t} dt \tag{8.12}$$

Now remember that the Debye-Kirkwood-Fröhlich formula [16] for static susceptibility is

$$\frac{\partial M}{\partial \mathbb{E}_0'} = \beta \langle M;M \rangle \tag{8.13}$$

(which is the static limit of the first F-D theorem [11]). If $\mathbf{E}_0$ is static, the self-field $\mathbb{E}'$ should cancel $\mathbb{E}_0'$ to make $\mathbb{E}$ equal to zero in Eq.(8.1); that is to say

$$\beta \langle M;M \rangle = \frac{\Omega(\varepsilon_0 + 2)}{4\pi}$$

Therefore Eq. (8.12) gives

$$\sigma[\omega] = \beta \int_0^\infty dt\ e^{-i\omega t} \langle J'(t); J'(0) \rangle / \Omega \tag{8.14}$$

which is the well known conductivity formula. Note that the random current $J'(t)$ is not exactly equal to the actual current in the metallic particle; the self-field is ignored in the random force driving the current. The conductivity $\sigma[\omega]$, is the <u>intrinsic</u> conductivity of the metal, whereas $\sigma^e[\omega]$, Eq.(8.8) is <u>extrinsic</u> and vanishes for zero frequency as it should be. These two are related to each other by the general formula Eq. (7.3).

Equation (8.14) can be transformed into

$$\sigma[\omega] = \frac{\omega_p^2}{i\omega} - \frac{1}{\omega^2}\int_0^\infty dt\, e^{-i\omega t} \langle \dot{J}'(t);\dot{J}'(0)\rangle/\Omega \tag{8.15}$$

by partially integrating twice. Here ω_p is the plasma frequency [17]. This expresses the conductivity in terms of the force correlation (not including the self-field). This formula is particularly useful for computing optical or high-frequency response of conduction electrons because the force comes from electron-phonon, electron-impurity, or boundary scattering and these interactions can be treated by straightforward perturbations.

We can of course treat an infinite system of particles. Assuming that it is spatially uniform, fluctuations and responses are Fourier-analysed. For example, consider the k-component of density $n_k(t)$ and write the equation of continuity as

$$\dot{n}_k(t) = -\, i\mathbf{k}\mathbf{j}_k(t) = -\, i\mathbf{k}\mathbf{j}_k^s(t) - i\mathbf{k}\mathbf{j}_k'(t) \tag{8.16}$$

where $\mathbf{j}_k$ stands for the total current which is composed of the systematic part and the random part. The systematic current is driven by the self-field produced by the density fluctuations. In order to see what the self-field should be, consider a spring at an elongation Δx, under an external force F, which is balanced with the elastic force, $-\alpha\Delta x$. In the same way, if the density response to an external potential ϕ_k^e is defined by

$$\langle n_k(t)\rangle = -\ \mathrm{Re}\ \chi[k,\omega]\phi_k^e\, e^{i\omega t} \tag{8.17}$$

the self-field ϕ_k^s is defined by

$$\phi_k^s = n_k/\chi[k,0]. \tag{8.18}$$

In the presence of an external potential ϕ_k^e, the effective local field is the resultant of ϕ_k^e and the averaged self-field; namely

$$\phi_k^*(t) = \phi_k^e(t) + \frac{\langle n_k(t)\rangle}{\chi[k,0]} = \phi_k^e(t) - \frac{\chi[k,\omega]}{\chi[k,0]}\,\phi_k^e(t)$$

or (8.19)

$$\phi_k^*(t) = \frac{1}{\varepsilon[k,\omega]}\,\phi_k^e(t)$$

with the shielding factor $\varepsilon[k,\omega]$ defined by

$$\varepsilon\frac{1}{\varepsilon[k,\omega]} = 1 - \frac{\chi[k,\omega]}{\chi[k,0]}$$

(8.20)

The current response can be written in two ways, as

$$\begin{aligned}\langle j_k(t)\rangle &= \mathrm{Re}\,\mu^e[k,\omega]\quad ik\phi_k^e \\ &= \mathrm{Re}\,\mu^*[k,\omega]\;\; ik\,\phi_k^*\end{aligned} \tag{8.21}$$

in terms of the external (extrinsic) mobility μ^e or of the local (intrinsic) mobility μ^*. These two mobilities are related to each other by

$$\mu^e[k,\omega] = \frac{\mu^*[k,\omega]}{\varepsilon[k,\omega]}$$

(8.22)

Now we see that the systematic current j_k^s in Eq. (8.16) is given by

$$j_k^s(t) = \int_0^t \mu^*(k,t-t')\,ik\phi_k^s(t')\,dt'$$

(8.23)

or

$$-ikj_k^s(t) = -\int_0^t \gamma_k(t-t')\;n_k(t')\,dt'$$

(8.24)

with $\gamma_k(t)$ corresponding to the Fourier-Laplace image

$$\gamma_k[\omega] = \frac{\mathbb{k}\, \mu^*[\mathbb{k},\omega]\mathbb{k}}{\chi[k,0]} \tag{8.25}$$

considering Eq. (8.18). Equation (8.16) is a generalized Langevin equation of the same form as Eq. (6.15), provided that j_k' is in fact random satisfying the condition (6.16). The density correlation function is obtained from this equation giving

$$\begin{aligned} \Lambda[k,\omega] &= \int_0^\infty dt\, e^{-i\omega t}\langle n_k(t);\, n_{-k}(0)\rangle \\ &= \frac{\langle n_k(0);\, n_{-k}(0)\rangle}{i\omega + \gamma_k[\omega]} = \frac{\chi[k,0]}{\beta} \quad \frac{1}{i\omega + \gamma_k[\omega]} \end{aligned} \tag{8.26}$$

The linear response theory gives

$$\begin{aligned} \chi[k,\omega] &= \int_0^\infty dt\, e^{-i\omega t}\langle (n_{-k}(0),\, n_k(t))\rangle \\ &= \chi[k,0] - i\omega\beta\Lambda[k,\omega] \end{aligned} \tag{8.27}$$

as an F-D theorem, which reduces to

$$\chi[k,0] = \beta\langle n_k(0);\, n_{-k}(0)\rangle \tag{8.28}$$

for the static susceptibility. This last relation is used in Eq. (8.26). In Eq. (8.27) the round bracket means Poisson (quantal or classical) bracket.

The F-D theorems can also be written here as

$$\mu^e[k,\omega] = \beta\int_0^\infty dt\, e^{-i\omega t}\langle \mathbb{j}_k(t);\, \mathbb{j}_{-k}(0)\rangle \tag{8.29}$$

and

$$\mu^*[k,\omega] = \beta\int_0^\infty dt\, e^{-i\omega t}\langle \mathbb{j}_k'(t);\, \mathbb{j}_{-k}'(0)\rangle \tag{8.30}$$

The continuity equation (8.16) means that

$$i\omega\,\chi[k,\omega] - \mathbb{k}\mu^{e}[k,\omega]\,\mathbb{k} = 0$$

(8.31)

Equation (7.2) is now written as

$$\mu^{e}[k,\omega] = \frac{\mu^{*}[k,\omega]}{1+\dfrac{\mathbb{k}\;\mu^{*}[k,\omega]\;\mathbb{k}}{i\,\omega\chi[k,\,0]}} \qquad (8.32)$$

which is inverted into

$$\mu^{*}[k,\omega] = \frac{\mu^{e}\;[k,\omega]}{1-\dfrac{\mathbb{k}\mu^{e}\;[\,k,\omega]\;\;\mathbb{k}}{i\,\omega\chi[k,\,0]}}$$

(8.33)

The last equation is identical with (8.22) by the relation (8.31). Equations (8.32) and (8.33) obviously correspond to Eqs. (7.2) and (7.3). With the use of Eqs. (8.31) and (8.32), the shielding factor $\varepsilon[k,\omega]$ is expressed as

$$\varepsilon[k,\omega] = 1+\frac{\gamma_k[\omega]}{i\omega} = 1+\frac{\mathbb{k}\;\mu^{*}[k,\omega]\;\mathbb{k}}{\omega\chi[k,0]} \quad .$$

(8.34)

This is a useful formula, because the local mobility μ^{*} may be approximated in some simple way.

If the system consists of a single species of particles and no force field is acting to prevent the conservation of total linear momentum, the local mobility μ^{*} in Eq. (8.21) has a pole of first order in ω. The corresponding pole of $\Lambda[k,\omega]$, Eq. (8.26), gives the dispersion relation of the longitudinal sound wave. On the other hand, if the momentum conservation law does not hold, μ^{*} remains finite at $\omega=0$. Then the pole of $\Lambda[k,\omega]$ gives the diffusive mode. The generalized diffusion coefficient may be defined by

$$D[k,\omega] = \mu^{*}[k,\omega]/\chi[k,0]$$

(8.35)

where the static density susceptibility $\chi[k,0]$ satisfies the thermodynamic relation

$$\lim_{k\to 0} \chi[k,0] = \left(\frac{\partial n}{\partial \zeta}\right)_T = n^2 \kappa \tag{8.36}$$

where ζ is the chemical potential and κ the isothermal compressibility. Equation (8.34) together with (8.35) is a generalization of the Einstein relation. Equation (8.18) means that the self-field for small k's can be identified with the change of the chemical potential induced by density variation.

For charged particles, the self-field is mainly due to coulomb interactions. For small wave numbers, $\chi[k,0]^{-1}$ is dominated by

$$\chi[k,0]^{-1} \sim \frac{4\pi e^2}{\varepsilon_0 k^2} \tag{8.37}$$

where ε_0 is the dielectric constant of the background medium for the charged particles. If only this is retained, the shielding factor ε is the dielectric constant. Equation (8.20) shows that $1/\varepsilon$ can be expressed directly in terms of the density response function (8.27). This formula was given by Nozierés and Pines [18]. By Eqs. (8.31) and (8.29), the density response functions can be replaced by the current correlation function (8.29), and also can be expressed, by Eq (8.32) in terms of the correlation function of the random current Eq. (8.30). The commonly used RPA (random phase approximation) simply replaces this by the current correlation of non-interacting particles. Then one can easily calculate the static correlations, $\langle n_k \, n_{-k} \rangle$ by inverting the expression (8.26) and the free energy of the interacting particles using the equation

$$2\frac{\partial}{\partial v_k} \frac{F}{L^3} = \langle n_{-k} n_k \rangle = \frac{1}{\pi} \int_{-\infty}^{\infty} d\omega \frac{h\chi[k,0]}{1-e^{-\beta h \omega}} \; \mathrm{Im} \frac{1}{\varepsilon[k,\omega]} \tag{8.38}$$

where v_k is the k-th Fourier component of the two-particle interaction potential. For the jelly-model of electrons this approximation gives the well-known high density approximation first obtained by Gellman and Brueckner. Similar procedure is possible for other systems, for example, for magnetic systems, and leads to the RPA results. There are a number of such examples, although they are often expressed in somewhat different contexts. As an interesting example somewhat beyond the usual RPA level of approximation, a recent work of Moriya and Kawabata [19] may be worth noting; they treat the problem of spin fluctuations and the itinerant electron ferromagnetism.

IX. SOME COMMENTS

Before leaving the subject of retarded frictions, a few points may be commented rather briefly.

i) If a system is described as a Markoffian process by a set of stochastic variables, the whole set is said to be complete. If we concentrate our attention only on a subset of the variables, the process is no longer Markoffian. Such a projection usually recovers the Markoffian property by coarse-graining of time scale. This we have seen for the example of a Brownian motion (section IV). It should be noted, however, that this recovery is not always realizable. In some cases, a non-Markoffian nature persists in long time scales. Somewhat ironically, a realistic Brownian particle should exhibit this singular behaviour. More delicate and complex examples are fluctuations of order parameters in the very neighbourhood of a second order phase transition. The latter problem is very interesting but is beyond the scope of this lecture.

In the classical theory of Brownian motion, the friction is usually assumed to be the Stokes resistance; namely

$$m\gamma u = 6\pi a\eta u \equiv \zeta u \tag{9.1}$$

where a is the radius of the Brownian particle and η is the viscosity of the fluid. This is not quite right, because for a non-steady motion one should apply the Boussinesq equation

$$F(t) = -\zeta u(t) - \tfrac{1}{2}m_0\dot{u}(t) - \alpha\pi^{-\frac{1}{2}}\int_0^t dt'\,(t-t')^{-\frac{1}{2}}\,\dot{u}(t') \tag{9.2}$$

for the systematic force assuming the fluid incompressible and the stick condition at the particle-fluid boundary. Here $m_0 = 4\pi a^3\rho_0/3$ is the mass of the fluid displaced by the spherical particle and

$$\alpha = \zeta a \nu^{-1/2} \qquad (\nu = \eta/\rho_0)$$

Equation (9.2) shows that the retarded friction kernel (6.6) is

$$m\gamma[\omega] = \zeta + \alpha (i\omega) + \tfrac{1}{2} m_0 i\omega \tag{9.3}$$

and that the auto-correlation function (6.13) behaves as

$$\langle u(t_0)u(t_0+t)\rangle \simeq \frac{2}{3} \frac{kT}{\rho_0} (4\pi\nu|t|)^{-3/2} \tag{9.4}$$

As was pointed out by Lorentz a long time ago, the ideal Brownian motion theory is justified only with the assumption that the density inside the particle is far larger than that of the surrounding fluid (which is a more strict statement than that the Brownian particle is heavy). This is not at all surprising but has attracted much attention rather recently since the computer experiment by Alder and Wainright [20] showed this explicitly. The F-D theorem [11] implies that the auto-correlation function of the velocity is equal to the average velocity of the Brownian particle after a time t if a unit impulse was exercised on it at t=0. The imparted momentum diffuses among the fluid molecules within a volume proportional to $t^{3/2}$ giving the particle an average momentum of the order of $t^{-3/2}$.

In this problem, the set of stochastic variables can be completed, besides those of the Brownian particle, by adding the velocity field v(r,t) of the fluid. The noise source is present in the fluid as a random stress tensor, which is assumed to be white-noise both in spatial and temporal Fourier components and to assure the basic Markoffian property of the whole process. The projection onto the Brownian particle results in Eq.(9.4). The Boussinesq friction is simply derived from the linearized hydrodynamic equation, but it should be related to the random force R(t) on the particle by the F-D theorem. An explicit proof of this has recently been given by Hauge and Martin-Löf [21].

ii) The electrical conductivity $\sigma[\omega]$ is given in terms of a current correlation function. As we have seen already the consideration of coulomb interaction presents a problem. In Eqs. (8.14) and (8.30), coulomb interactions as the self-field should be eliminated. In fact, most practical calculations are carried in that way; usually, electric carriers, say electrons, are assumed to be scattered by phonons or imperfections. There still remains a difficult problem of renormalization or dressing of particles and interactions, which has been a challenge to a great number of many-body theorists in the past years. I shall not go into this problem but will comment on a point which seems to have caused some confusion.

For a system of conduction electrons, Eq.(8.14) may be written as

$$\sigma[\omega] = \beta \frac{\langle J;J\rangle/\Omega}{i\omega + \gamma[\omega]} = \frac{e^2 n}{m} \frac{1}{i\omega + \gamma[\omega]} \tag{9.5}$$

Here J/Ω is the electronic current density, for which the generalized equipartition law holds irrespective of interactions and statistics

$$\langle J;J\rangle = \frac{e^2 n}{m} kT \cdot \Omega \tag{9.6}$$

corresponding to the conductivity sum-rule

$$\int_{-\infty}^{\infty} \sigma[\omega]\, d\omega = \frac{ne^2}{m} . \tag{9.7}$$

Equation (9.5) gives

$$\rho[\omega] = \frac{m}{e^2 n}(i\omega + \gamma[\omega]) \tag{9.8}$$

and in particular

$$\rho[0] = \frac{m}{e^2 n}\gamma[0] = \frac{1}{e^2 nkT}\int_0^{\infty} \langle R(t);R(0)\rangle dt \tag{9.9}$$

where

$$R(t) = \dot{J}'(t) \tag{9.10}$$

is the random force acting on the electrons. Equation (9.9) is often interpreted as a direct expression of the zero-frequency resistance in terms of force correlation [22]. It is useful, but seems to have been sometimes misunderstood. It is exact as long as the random force is <u>properly</u> defined. In the literature the random force is often confused with the total force F; namely sometimes it is taken as

$$R(t) \sim F(t) = \dot{J}(t) = \frac{1}{i\hbar}\left[J(t), \mathcal{H}\right] = \frac{1}{i\hbar}\left[J(t), \mathcal{H}'(t)\right] \tag{9.11}$$

where $\mathcal{H}$ is the total hamiltonian and $\mathcal{H}'$ is the perturbation noncommutable with the current. When this is done, Eq.(9.9) is no longer exact although it can be a useful approximation if another dishonesty is committed at the same time. This we have seen already in section VII for the example of Kirkwood's equation for resistance (7.20). In many cases, one does not even care whether the plateau of the time integral really exists, but one carries out the integration over infinite times obtaining a useful, realistic answer. The trick is the use of lower order perturbational calculation, say the Born-approximation, which does not give the correct time-dependence of the correlation function. The effectiveness of this approximation further depends on another condition, that is a narrow spectrum of relaxation frequencies. For example, consider non-degenerate electrons in a semiconductor scattered by static impurities. For a group of electrons with an infinitesimal spread of energy, $\gamma[0]$ yields, with the prescription described in the above, a reasonable approximation for the relaxation frequency, or the inverse relaxation time. The conductivity is determined by the averaged relaxation time, whereas the approximation, Eq. (9.9),corresponds to an average of relaxation frequencies unless the electrons are grouped for different energies,and so it can be a bad approximation. On the other hand, the method can be much better if the electrons are degenerate. In fact it is a simple calculation leading to the Grueneisen formula for a metal [23], because the dispersion of relaxation times is rather small. The approximation is essentially equivalent to calculating the average relaxation frequencies from a Boltzmann-Bloch equation, or to using a variational solution of this equation, in which the collision kernel is obtained by a Born-approximation of electron-phonon scattering processes. The same approximation is often useful in the calculation of the effects of configurational disorder: say random spins or random distribution of atoms or impurities [24]. But it should be kept in mind that such a calculation remains an approximation, that can be good only when certain conditions are satisfied.

iii) The random force in the Langevin equation , eq. (6.15), can also be regarded as a random process driven by a noise source lying at a more microscopic level. Its Langevin equation may be written as (denoting now R as R_1)

$$\dot{R}_1(t) = -\int_{t_o}^{t} dt' \, \gamma_2(t-t') \, R_1(t') + R_2(t) \quad , \; t > t_o \tag{9.12}$$

with the same conditions for the new random force $R_2(t)$, namely,

$$\langle R_2(t) \rangle = 0, \qquad \langle R_1(t_o) \, R_2(t) \rangle = 0 \quad , \; t > t_o \quad .$$

Then the correlation function of $R_1(t)$ is given in the form

$$\int_0^\infty dt \, e^{-i\omega t} \langle R_1(t_o+t) \, R_1(t_o) \rangle = \frac{\langle R_1^2 \rangle}{i\omega + \gamma_2[\omega]} \tag{9.13}$$

where $\gamma_2[\omega]$ is the Fourier-Laplace image of $\gamma_2(t)$. In the same way as for eq. (6.15), the retardation kernel $\gamma_2(t)$, or its Fourier-Laplace image, is related to the random force $R_2(t)$ by

$$\gamma_2[\omega] = \frac{1}{\langle R_1^2 \rangle} \int_0^\infty \langle R_2(t_o+t) \, R_2(t_o) \rangle \, e^{-i\omega t} dt \tag{9.14}$$

Thus the admittance $\mu[\omega]$, eq. (6.9), is now written as

$$\mu[\omega] = \frac{1}{m} \; \frac{1}{i\omega + \dfrac{\langle R_1^2 \rangle}{m^2 \langle u^2 \rangle} \; \dfrac{1}{i\omega + \gamma_2[\omega]}} \tag{9.15}$$

Repeating this process again, we have an expression of the form

$$\mu[\omega] = \cfrac{\Delta_0^2}{i\omega + \cfrac{\Delta_1^2}{i\omega + \cfrac{\Delta_2^2}{i\omega + \gamma_3[\omega]}}} \tag{9.16}$$

where $\gamma_3[\omega]$ is related to the correlation function of the third order random force $R_3(t)$ by

$$\gamma_3[\omega] = \frac{1}{\langle R_2^2 \rangle} \int_0^\infty \langle R_3(t_0 + t)\, R_3(t_0) \rangle\, e^{-i\omega t}\, dt \tag{9.17}$$

By repeating the same procedure infinitely many times, we can write the expression (9.16) as a continued fraction, which was first introduced by Mori [25] and was carefully examined by Dupuis [26] .

The expression (9.16) has been used often as a theoretical approximation for calculating density response of liquids, spin response of magnetic systems and other things. It has been also used as a method of semi-empirical analysis of experimental data of such responses obtained by various kinds of spectroscopic experiments including inelastic scattering of neutrons. The parameters Δ_1, Δ_2, etc., are related to the moments of the spectrum, or the time-derivatives of the corresponding response function (or the correlation function), evaluated at time t=0 [2, 5, 11]; for example,

$$\Delta_1^2 = m_2 \qquad \Delta_2^2 = \left[m_4 - (m_2)^2\right]/m_2$$

$$\Delta_3^2 = \left[m_6\, m_2 - (m_4)^2\right] / \left[m_2(m_4 - m_2^2)\right] \quad \text{etc.} \tag{9.18}$$

where the m_{2n}'s are even moments of the spectrum and are equilibrium averages of the squares of higher derivatives of the force acting on the mode of our particular interest. Each of the functions $\gamma_n[\omega]$ appearing in the continued fraction is a correlation function of the corresponding random force of a certain order, and will satisfy the conditions (6.7)

The simplest use of eq. (9.16) would be to assume $\gamma_3[\omega]$ to be a constant, Δ_3 over the frequency range of observation. If ω is scaled by Δ_3 , the r.h.s. of eq. (9.16) depends only on the parameters

$$a = \Delta_1/\Delta_3 \quad \text{and} \quad b = \Delta_2/\Delta_3 \tag{9.19}$$

besides the scaled frequency ω/Δ_3. Another choice often used is the assumption,

$$\frac{1}{i\omega + \gamma_3[\omega]} = \int_0^\infty dt\, e^{-i\omega t - \frac{\Delta_3^2}{2} t^2} \quad . \tag{9.20}$$

The essential difference between these two choices is that all the moments are convergent for the second one, whereas the moments higher than the 6-th cease to exist for the first one. Nevertheless, general features of the spectrum are nearly the same for the two choices. Depending on the values of the parameters a and b, eq. (9.19), the spectrum shows one, two or three peaks. Figs. 2 and 3 are due to a recent work of Tomita and Makishima. Fig. 2 corresponds to the assumption γ_3 = const and Fig. 3 to the assumption of eq. (9.20), and I, II, and III indicate the domains of the parameters where the spectrum is singly, doubly or triply peaked. The peak around $\omega = 0$ is often called the central diffusion peak, and the side peaks are associated with damped oscillatory modes [27].

As the Brownian variables, one could also choose a set of suitable variables and apply eq. (6.21). For many variable cases, the expression (6.27) can be extended to a continued fraction form as a matrix. This will be sometimes more useful and more direct in the sense that we can select, from the beginning, some particular modes of collective motion, which are in many cases related to conserved quantities of the system. On the other hand, a simple continued fraction, of the form (9.16), is expected to work well if the introduction of higher order random forces automatically chooses the relevant modes. It should also be kept in mind, however, that a formal expansion of an admittance function in a continued fraction expression does not necessarily mean that each of the $\gamma_n[\omega]$ appearing in the expression is connected with a meaningful physical quantity; it can be only a mathematical convenience.

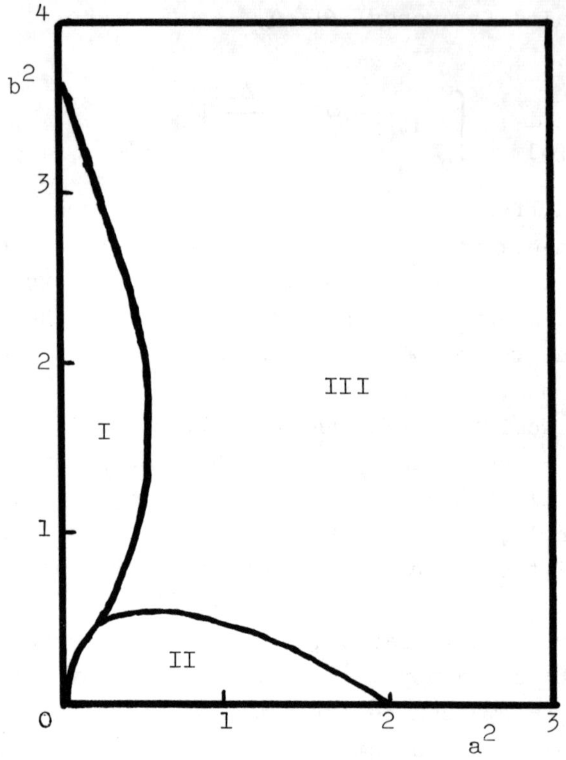

Fig. 2.

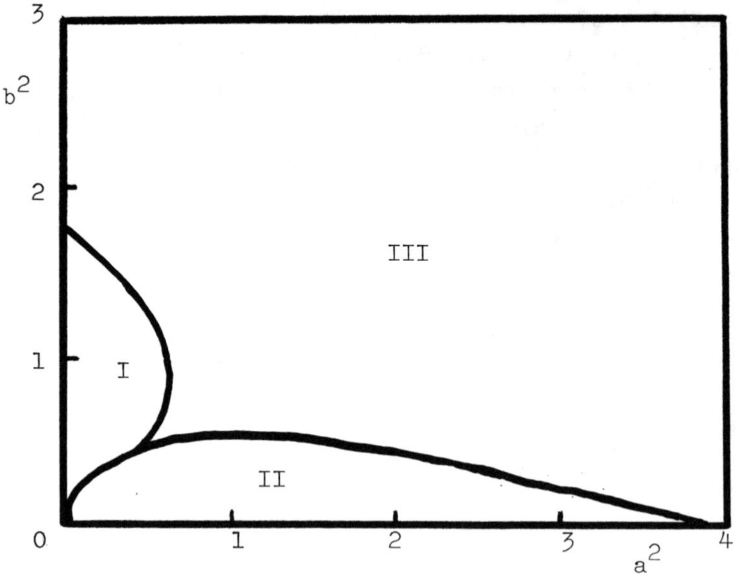

Fig. 3.

X. DAMPING-THEORETICAL METHOD [28]

As was stressed in the introduction, shifting the level of observation to a more macroscopic level means a reduction in the number of variables throwing away the irrelevant ones. A mathematical formalism for this reduction has been known for a long time as the damping theory. It has been extended greatly in various contexts. In this school we have already heard about some of the ambitious, newer attempts from Prof. Prigogine. Since our phenomenological approach has to be founded on a more microscopic approach, this formalism is very important for our purpose. It is not possible and perhaps unnecessary to discuss this here in great detail, but it seems desirable to sketch some aspects rather briefly.

We consider a Markoffian equation,

$$\frac{\partial}{\partial t} f\,(q_1 \dots q_n, t) = \Gamma f \tag{10.1}$$

where the variables $(q_1 \dots q_n)$ are a complete set and Γ is the evolution operator (which can be a Liouville operator $i\mathcal{L}$). Let us divide the set into two subsets ,

$$q' = (q_1 \dots q_m) \quad , \qquad \text{and} \qquad q'' = (q_{m+1} \dots q_n) \tag{10.2}$$

and suppose that q" is going to be <u>hidden</u>. The projected process consisting of q' is described in terms of a projection

$$g = \mathcal{P} f \tag{10.3}$$

where the function g is essentially dependent on the variables q'. More explicitly, it will be of the form

$$\mathcal{P} f = g\,(q', q'', t) = \bar{g}\,(q', t)\ \varphi_o\,(q'') \tag{10.4}$$

where $\varphi_o(q'')$ is a given function of q" independent of t. Thus the projection (10.3) will be written as

$$g = \varphi_o\,(q'') \int \psi_o(q'')\, d\, q''\ f\,(q', q'', t) \tag{10.5}$$

where $\psi_o(q'')$ is also a given function satisfying the condition

$$\int \psi_o(q'') \, \varphi_o \, (q'') \, dq'' = 1 \tag{10.6}$$

From eq. (10.5) we have

$$\bar{g} \, (q',t) = \int \psi_o(q'') \, dq'' \, f(q', q'', t) \tag{10.7}$$

and the condition (10.6) guarantees

$$\mathcal{P}^2 = \mathcal{P} \tag{10.8}$$

for this projection procedure.

The evolution equation (10.1) is separated into two equations,

$$\frac{\partial}{\partial t} \mathcal{P} f = \mathcal{P} \Gamma \mathcal{P} f + \mathcal{P} \Gamma Q f \tag{10.9}$$

$$\frac{\partial}{\partial t} Qf = Q \Gamma Q \, f + Q \Gamma \mathcal{P} f \tag{10.10}$$

where

$$Q = 1 - \mathcal{P} \tag{10.11}$$

is also a projection operator. For a given initial condition

$$f(q,t\;) \equiv f_o = \mathcal{P} f_o + Q \, f_o \quad , \tag{10.12}$$

eq. (10.10) is integrated to

$$Qf = \int_{t_o}^{t} e^{(t-\tau)Q\Gamma} \, Q \, \Gamma \, \mathcal{P} \, f(\tau) \, d\tau + e^{(t-t_o)Q\Gamma} \, Qf_o \tag{10.13}$$

and is inserted into eq. (10.9) to give

$$\frac{\partial}{\partial t} \mathcal{P} f = \mathcal{P} \Gamma \mathcal{P} f + \mathcal{P} \Gamma \int_{t_o}^{t} d\tau \;\; e^{(t-\tau)Q\Gamma} \, Q \Gamma \mathcal{P} f \; (\tau)$$

$$+ \quad \mathcal{P} \Gamma \, e^{(t-t_o)Q\Gamma} Q \, f_o \qquad (t > t_o) \tag{10.14}$$

This is very well known. The projected process is no longer Markoffian, but carries a memory of the initial state through the third term of the r.h.s., and it is governed by the retarded (non-Markoffian) evolution operator in the second term. Equation (10.14) may written in a somewhat more transparent form

$$\left[s - \mathcal{P}\Gamma - \mathcal{P}\Gamma \frac{1}{s - Q\Gamma} Q\Gamma \right] \mathcal{P} F = \mathcal{P} f_o + \mathcal{P}\Gamma \frac{1}{s - Q\Gamma} \, Q \, f_o \tag{10.15}$$

if the Laplace transform

$$F\,(q\,,\;s) = \int_0^{\infty} e^{-st}\; f\,(q,\;t)\;dt$$

is introduced. Equation (10.15) is nothing more than what has been called the damping theory in quantum mechanics, where $\mathcal{P}$ usually means a projection to the diagonal elements in the unperturbed states of the system. Very often the initial state f_0 is chosen to make $Q\,f_0$ vanish. With such an initial condition eq. (10.14) or (10.15) gives the evolution law of $\mathcal{P}f$ or $\bar{g}$, eq. (10.4). This will be reduced to

$$\frac{\partial}{\partial t}\,\bar{g} = \bar{\Gamma}\,\bar{g} \qquad , \tag{10.16}$$

if we replace the r.h.s. of eq. (10.14) by

$$\phi_0\,\bar{\Gamma}\,\bar{g} = \mathcal{P}\Gamma\mathcal{P}f + \mathcal{P}\Gamma\int_{-\infty}^{t} d\tau\; e^{(t-\tau)Q\Gamma}\; Q\Gamma\mathcal{P}f \;, \tag{10.17}$$

assuming that the time t is coarse-grained with a scale τ_r, which is much longer than τ_c scaling the rate of decrease of the retardation kernel in the integrand.

A simple example of such projection is a Brownian motion, in a potential field, eq. (5.13), which is written as

$$\frac{\partial}{\partial t}\,f\,(x,u,t) = (\,\Gamma_0 + \Gamma_1)f \tag{10.18}$$

with

$$\Gamma_0 = \gamma\,\frac{\partial}{\partial u}\,(u + \frac{1}{m\,\beta}\;\frac{\partial}{\partial u}\,) \quad , \tag{10.19}$$

$$\Gamma_1 = -\,u\,\frac{\partial}{\partial x} + \frac{1}{m}\;\frac{\partial V}{\partial x}\;\frac{\partial}{\partial u} \tag{10.20}$$

The projection $\mathcal{P}$, (10.4), is defined by

$$\mathcal{P}f\,(x,\;u,\;t) = \phi_0\,(u)\int_{-\infty}^{\infty} du\;f\,(x,\;u,\;t) \tag{10.21}$$

where $\phi_0(u)$ is the normalized Maxwellian distribution,

$$\phi_0\,(u) = (\,\frac{2\,\pi}{m\,\beta})^{-1/2}\;\exp\,(\,-\,\frac{\beta}{2}\,m\,u^2\,)$$

It is easily seen, that from the definitions (10.19) to (10.21),

$$\Gamma_o \mathcal{P} = \mathcal{P}\Gamma_o = 0 \quad \text{and} \quad \mathcal{P}\Gamma_1\mathcal{P} = 0 \ . \tag{10.22}$$

These make the first term on the r.h.s. of eq. (10.14) vanish and reduce Γ , in the second term, to Γ_1. The third term is absent if the Maxwellian distribution is initially assumed. The above observation simplifies the equation considerably. In order to obtain the familiar equation of diffusion in the potential field V, we require the conditions

$$\ell \left| \frac{1}{\bar{g}} \frac{\partial \bar{g}}{\partial x} \right| \ll 1 \text{ and } \ell \left| \frac{1}{V} \frac{\partial V}{\partial x} \right| \ll 1 \tag{10.23}$$

$$(\ell = \langle u^2 \rangle^{1/2} / \gamma) \quad ,$$

which mean that the distribution and the potential are slowly varying over distances of the order of the mean free path ℓ , and that the operator Γ_1 can be treated as a perturbation relative to Γ_o. Thus we approximate the operator exp $(t-\tau)\, Q\Gamma$ by exp $(t-\tau)\,\Gamma_o$ to obtain

$$\mathcal{P}\Gamma_1 \int_{t_o}^{t} d\tau \; e^{(t-\tau)\Gamma_o} \; Q \; \Gamma \; \mathcal{P} f_o \,(\tau) \tag{10.24}$$

for the integral operator in eq. (10.14). But we further notice that

$$\mathcal{P}\Gamma_1\mathcal{P} \propto \phi_o(u) \int_{-\infty}^{\infty} du \Big(- u \frac{\partial}{\partial x} + \frac{1}{m} \frac{\partial V}{\partial x} \frac{\partial}{\partial u} \Big) \phi_o(u) = 0$$

which saves Q in eq. (10.24). Noticing that

$$\Gamma_o \, u \, \phi_o(u) = - \gamma \, u \, \phi_o \,(u)$$

we write the integral operator as

$$- \int_{-\infty}^{\infty} du \; \Big(-u \frac{\partial}{\partial x} + \frac{1}{m} \frac{\partial V}{\partial x} \frac{\partial}{\partial u} \Big) \; u \, \phi_o(u) \int_{t_o}^{t} d\tau \; e^{-\gamma(t-\tau)} \Big(\frac{\partial}{\partial x} + \beta \frac{\partial V}{\partial x} \Big) \bar{g} \; (x,t)$$

By integration over u, the term with $\partial/\partial u$ on the left vanishes. Assuming a slow change of $g(\bar{x},t)$ in t, namely

$$\Delta t \gg \tau_c = 1/\gamma \quad ,$$

this finally becomes

$$\frac{\partial}{\partial x} \frac{\langle u^2 \rangle}{\gamma} \Big(\frac{\partial}{\partial x} + \beta \frac{\partial V}{\partial x} \Big) \; \bar{g} \; (x, \; t) \tag{10.25}$$

which is the diffusion operator in the presence of the potential V. Now the velocity u is hidden and the Brownian motion is observed only with respect to the spatial part. The Markoffian property of diffusion is recovered by coarse-graining in space and time, provided that the potential is sufficiently slowly changing.

The derivation of the so-called master equation, from a classical or quantal Liouville equation, can be made in the same way. The operator Γ is now a Liouville operator $i\mathcal{L}$, which is divided into $i\mathcal{L}_0$ and $i\mathcal{L}_1$. If the projection $\mathcal{P}$ is properly defined to satisfy the conditions

$$i\mathcal{L}_0\mathcal{P} = \mathcal{P}i\mathcal{L}_0 = 0$$

eq. (10.15) takes the form

$$\left[s - \mathcal{P}i\mathcal{L}_1\mathcal{P} - \mathcal{P}i\mathcal{L}_1 \frac{1}{s-Qi\mathcal{L}} Qi\mathcal{L}_1\right]\mathcal{P}f = \mathcal{P}f_0 \tag{10.26}$$

with the initial condition $Q\,f_0 = 0$. If the operator $Q\,i\mathcal{L}$ in the denominator on the l.h.s. is simply replaced by $i\mathcal{L}_0$, and s is sent to 0^+ in the same denominator, this becomes a master equation, in which the transition probability is calculated by the golden rule. The conditions to justify this approximation are rather well known. From my own point of view, this is very closely related to the phenomena of motional narrowing, for which I have no more time in these lectures, and refer you to some of my previous articles [3, 10].

As was shown by Mori [13], the damping theory can be used to give a microscopic basis of the Langevin equation. Instead of applying it to a distribution function, we now apply it to dynamical equations. For the sake of clarity, we consider a classical system. A dynamical quantity A_t is a function of time because it is a phase function, namely

$$A_t = A(p_t, q_t) = A(p,q,t) \tag{10.27}$$

In the last expression, (p, q) is the <u>initial</u> phase and (p_t, q_t) is its image after time t, as determined by the Hamiltonian equation of motion. For A_t, (10.27), we can write the equation of motion

$$\frac{\partial A_t}{\partial t} = -\,i\mathcal{L}\,A_t \tag{10.28}$$

with the Liouville operator (note the sign). Now we define the <u>projection</u> $\mathcal{P}$ of any phase function g (p,q) by

$$\mathcal{P}g = A\,(A, g)\,/\,(A, A) \tag{10.29}$$

where the bracket is defined by

$$(f,g) = \iint dp\, dq\; C\, e^{-\beta H(p,q)} f(p,q)\, g(p,q)$$

with the normalized canonical distribution $C \exp(-\beta H)$.
Because of the stationary nature of the canonical distribution, we have generally

$$(\dot{g}, g) = (g, \dot{g}) = 0, \qquad (g_t, h_t) = (g, h) \tag{10.30}$$

The damping formula (10.14) can be applied to eq. (10.28) with the projection $\mathcal{P}$ defined by eq. (10.29). Equations (10.30) give

$$\mathcal{P}(-i\mathcal{L}A) = \mathcal{P}\dot{A} = 0$$

$$\mathcal{P}(-i\mathcal{L})g = A\,(A, \dot{g})/(A, A) = -A\,(\dot{A}, g)/(A, A) \tag{10.31}$$

The damping equations (10.13) and (10.14) are now written as

$$\frac{d}{dt} A_t = -\int_{t_0}^{t} \gamma(t-\tau)\, A_\tau\, d\tau + R_t \quad , \qquad t > t_0 \tag{10.32}$$

with the random force defined by

$$R_t = e^{-i(t-t_0)Qi\mathcal{L}}\, \dot{A} \quad , \tag{10.33}$$

and the retarded function

$$\gamma(t) = (R_0, R_t) \,/\, (A, A) \quad . \tag{10.34}$$

Since $\mathcal{P}R_t = 0$, eq. (10.32) projected by $\mathcal{P}$ gives eq. (10.14), or equivalently, in this case

$$\int_0^\infty (A_{t_0}, A_{t_0+t})\, e^{-i\omega t} dt = \frac{(A, A)}{i\omega + \gamma[\omega]} \tag{10.35}$$

The part $Q\,A_t$ is obtained from eq. (10.32), as a convolution of R_t and the correlation function (10.35). This is equivalent to eq. (10.13).

Equation (10.32) was first obtained by Mori [13] and is the generalized Langevin equation, which we assumed for our phenomenological treatments. Equation (10.34) is equivalent to eq. (6.17), and represents an F-D theorem of the second kind.

It should be noted that the damping-theory is a formalism, which allows a very wide range of applicability. For example, if the projection is defined so as to project the distribution function of a many-particle system to a product of one-particle distribution functions, this formalism can be used for deriving the Boltzmann equation. However, a careful examination is required in order to see under what conditions the reduced equation can really be used.

XI. CONCLUDING REMARKS

Since I have used up my time, I have to finish now. As the reader may have noticed, in the introduction, I wished to include in these lectures some other points to generalize the Brownian motion theory. Let me make a few very brief comments on some of these points

i) A stochastic equation of motion can be non-linear; for example,

$$\dot{x}(t) = v\,(x) + R\,(x,\, t) \tag{11.1}$$

where $R(x, t)$ is a random noise. If it is a white noise, then the process x is Markoffian. To derive such a Markoffian equation, it is most convenient to use the method of stochastic equation as discussed in section V. If, furthermore, $R(x, t)$ is Gaussian, the Markoffian equation is reduced to a Fokker-Planck equation, which is a diffusion-type equation. Generally, we do not really require the stationarity of the process, so that the drift term and the diffusion term may explicitly depend on time. They can also be non-linear functions of x. These are two ways of interpreting a stochastic differential equation of the type, eq. (11.1), in the case where $R(x, t)$ has a singular nature of this kind. This point was briefly mentioned in the introduction, and the reader is referred to, for example, a paper by Mortensen [7]. Further references should also be made to a series of papers by Lax [29].

As is very well known, extensive work has been done on applications of Brownian motion theory to laser problems, that is to quantum aspects of Brownian motion of laser modes coupled with atoms.

The theory now extends to much wider categories of problems. The reader is referred to a review paper by H. Haken [30].

ii) A particularly important example of non-linear Brownian systems is an oscillator on which the random noise exerts not only amplitude modulations but also frequency modulations. Numerous examples are found in physics, in resonance problems. The concept of motional narrowing played a very important role in NMR, ESR and other kinds of spectroscopy. From the author's point of view, it gives us a very nice example, which helps understand the physical and the logical structures of stochastic processes [3] . The narrowing means that the correlation time of R is very much shorter than the relevant time constants of x , and so it means a Markoffian character for the process x . Transport equations, such as Boltzmann equations, require some sort of narrowing conditions, either as weak perturbations or as localized perturbations. Such narrowing conditions may not be satisfied in practice. In such cases we have retardation effects and memory effects, which can generally be non-linear effects.

iii) Relatively simple examples of such un-narrowed perturbations are stochastic theories of spectral lines. Equation (11.1) may be thought of as a Schrödinger equation, or a Liouville equation for the density matrix, which contains a random perturbation from the environment. If the stochastic nature of this perturbation is simple enough, the response or relaxation of the system can be explicitly treated. An easy case arises if the random perturbation is basically Markoffian. Assuming simple but typical types of such perturbations, the line shape problem has been discussed by the author [31] .

iv) A Gaussian property of a stochastic process is attributed to a certain kind of central limit theorem. In this respect, we could ask the question of whether a Gaussian nature may in fact be proved for macrovariables, which consist of a great number of small contributions from a numerous constituent units of the system. I would have liked to have time to discuss this problem, but here can only refer to our recent work [32] . In thermal equilibrium, the distributions of extensive thermodynamic quantities are usually Gaussian, the variances being also extensive, except in the very neighborhood of a phase transition where the distributions become non-Gaussian, which is now very well recognized as being related to the critical singularities. In non-equilibrium situations, it can be shown that the probability distribution of an extensive variable is generally Gaussian if the size of the system is very large, except in the neighborhood of a certain critical condition. Such an extensive variable can be regarded as a Gaussian process not necessarily stationary. For example, large deviations from equilibrium, relaxation from far from equilibrium, and response to a large external force, behave, in general, non-linearly. The evolution of the average behavior and the fluctuations may be described by a Fokker-Planck equation, which is generalized, in the sense that the coefficients are time-dependent, and that the equation, by itself, is incomplete unless it is supplemented by other evolution equations to determine the time-dependence of the coefficients. This was first discussed by van Kampen [33].

NOTES AND REFERENCES

1) These lectures are, in a sense, a continuation of three previous summer school lecture series (Ref. 2, 3, 4) given by the present author, and of a review report (Ref. 5) of the same author. Some of the references made previously will be omitted. A great part of this set of lectures is based on the author's lectures at the University of Tokyo and also on the chapters in Statistical Physics (Ref. 6).

2) R. KUBO, in Lectures in Theoretical Physics, vol. 1, ed. W. Brittin,(Interscience), New York, 1959, p.120.

3) R. KUBO, in Fluctuation, Relaxation and Resonance in Magnetic Systems, ed. D. ter Haar, (Oliver and Boyd) Edinburgh, 1962, p. 23.

4) R. KUBO, in Tokyo Summer Lectures in Theoretical Physics, 1965 Part I, Many-Body Theory, ed. R. Kubo,(Shokabo) Tokyo and (Benjamin) New York.

5) R. KUBO, Rep. on Progress in Physics, Vol. 29, Part I, (1966) 225.

6) R. KUBO, in Tokei Butsurigaku (Statistical Physics) ed. R. Kubo and M. Toda, Chaper 5 (Brownian Motion), Chapter 6 (Physical Processes as Stochastic Processes),(Iwanami Pub.) Tokyo, 1973, (in Japanese).

7) Physicists may get some idea of such mathematical problems from R.E. MORTENSEN, J. Stat. Phys. 1, 271 (1969).

8) M.C. WANG and G.E. UHLENBECK, Rev. Mod. Phys. 17, 327 (1945). Also see N. Wax (ed.) Selected Papers on Noise and Stochastic Processes,(Dover Pub.), New York, 1954.

9) L. ONSAGER and S. MACHLUP, Phys. Rev. 91, 1505 (1953). S. MACHLUP and L. ONSAGER, Phys. Rev. 91, 1512 (1953).

10) R. KUBO, J. Math. Phys. 4, 174 (1963).

11) R. KUBO, J. Phys. Soc. Japan, 12, 570 (1957).

12) R. KUBO in Statistical Mechanics of Equilibrium and Non-Equilibrium, ed. J. Meixner, (North Holland), Amsterdam 1965, p. 80.

13) H. MORI, Prog. Theor. Phys. Kyoto 33, 423 (1965)

14) J. KIRKWOOD, J. Chem. Phys. 14, 180 (1946)

15) A. KAWABATA and R. KUBO, J. Phys. Soc. Japan 21, 1765 (1966) R. KUBO in Cooperative Phenomena, ed. H. Haken and M. Wagner, (Springer-Verlag), 1973, p. 140.

16) H. FROHLICH, Theory of Dielectrics, (Oxford Clarendon Press), (1949).

17) Here we have used the generalized equipartition law

$$\langle J' ; J' \rangle = \langle J ; J \rangle = Ne^2kT/m$$

where N is the total number of electrons. This is seen from the definition of the canonical correlation (6.22) (see also eq.(9.6))

18) P. NOZIERES and D. PINES, Nuovo Cimento X 9, 470 (1958). See also articles in Many Body Problems, ed. D. Pines, (Benjamin) New York, (1962).

19) T. MORIYA and A. KAWABATA, J. Phys. Soc. Japan, 34, 639 (1973), 35, 669 (1973).

20) B.J. ALDER and T.E. WAINWRIGHT, Phys. Rev. A1, 18 (1970). Since this discovery of Alder and Wainwright, there has been a great accumulation of theoretical papers, but so far only one experimental paper has appeared claiming observation of the long tail (Y.W. KIM and J.E. MATTA, Phys. Rev. Letters, (1973)). For other references see the lectures by P. Mazur in this volume.

21) E.H. HAUGE and A. MARTIN-LÖF, J. Stat. Phys. 7, 259 (1973). P. MAZUR and D. BEDEAUX, to appear in Physica.

22) There are quite a few papers on this subject. Here we only refer to: S.F. EDWARDS, Proc. Phys, Soc. 86, 977 (1965). J.S. ROUSSEAU, J.C. STODDART and N.H. MARCH, J. Phys. C 5, L 175 (1972). V.M. KENKRE and M. DRESDEN, Phys. Rev. A6, 769 (1972). W.G. CHAMBERS, J. Phys. C 6, 2586 (1973). P.N. ARGYRES and J.L. SIGEL, Phys. Rev. Letters, 31 1397 (1973)

23) This can be easily seen in the following way: We write the Boltzmann-Bloch equation in the form

$$\Gamma[f] = E\frac{\partial f_0}{\partial p}$$

and assume $f=f_0(1+\psi)$, linearizing the collision operator to get

$$f_0\tilde{\Gamma}\psi = E\frac{\partial f_0}{\partial p}.$$

The conductivity is expressed in the form

$$(p,\ f_0\psi) = (p,\ f_0\tilde{\Gamma}^{-1}\frac{1}{f_0}\frac{\partial f_0}{\partial p})$$

which is a sort of average of relaxation frequencies. An approximation to the above expression would be

$$(p\ f_0\psi) = \frac{(p,\ \frac{\partial f_0}{\partial p})^2}{(p,\ f_0\tilde{\Gamma}\ f_0^{-1}\frac{\partial f_0}{\partial p})}$$

This leads to the Grueneisen formula for metallic electrons interacting with phonons in equilibrium (Kubo, 1943).

24) For example, I. MANNARI, Prog. Theor. Phys. 26, 51 (1961). This sort of approximation, for the conductivity change due to disorder, has the advantage that it expresses the resistance in terms of space-time correlations of the disorder. The spirit of the calculation is common to other kinds of relaxation processes (see Ref. 2).

25) H. MORI, Prog. Theor. Phys. 33, 424 (1965).

26) M. DUPUIS, Prog. Theor. Phys. 37, 502 (1967).

27) There are a number of papers on this subject. The expression (9.16) with a constant γ_3 is called the three-pole approximation. An obvious generalization of eq. (9.16) is to include natural frequencies of corresponding modes in the form

$$\gamma_n[\omega] = \frac{\Delta_n^2}{i(\omega-\Omega_n)+\gamma_{n+1}[\omega]}$$

by starting from two conjugate Brownian modes $(x_1,\ x_2)$. For

more details, the reader is referred to the following articles: W. MARSHALL and S.W. LOVESEY, Theory of Thermal Neutron Scattering, (Oxford University Press), 1971; K. TOMITA and H. TOMITA, Prog. Theor. Phys. 45, 1407 (1971); H. TOMITA and H. MAKISHIMA, Prog. Theor. Phys. 48, 1133 (1972); K. TOMITA and H. MAKISHIMA, Prep. (1974); S.W. LOVESEY and R.A. MESERVE, Phys. Rev. Letters 28, 614 (1972), J. Phys. C 6, 79 (1973).
It should perhaps be noted that the so-called central peak problem, which called much attention recently, seems somewhat deeper than that appears in the three-pole approximation or its simple modifications.

28) Excellent treatments of the damping theory are found in text books on quantum mechanics such as HEITLER's Quantum Theory of Radiation and MESSIAH's Quantum Mechanics. Applications to statistical mechanics have been made by many authors. Early works are: S. NAKAJIMA, Prog. Theor. Phys. 20, 948 (1958) (also R. KUBO, as quoted in this paper); R. ZWANZIG, J. Chem. Phys. 33, 1338 (1960), also in Lectures in Theoretical Physics, Vol. III, ed. W.E. Brittin, B.W. Downs and J. Downs (Interscience Pub.), New York, 1961.

29) M. LAX, Rev. Mod. Phys. 32, 625 (1960), 38, 359 (1966), 38, 541 (1966).

30) H. HAKEN, Rev. Mod. Phys. to appear.

31) R. KUBO, in Stochastic Processes in Chemical Physics, ed. Shuler, (John Wiley and Sons, Inc.), 1969, p. 101.

32) R. KUBO, K. MATSUO and K. KITAHARA, J. Statist. Phys. 9, 51 (1973).

33) N.G. VAN KAMPEN, Can. J. Phys. 39, 551 (1961), and in Fluctuation Phenomena in Solids, ed. R.E. Burgers, (Academic Press), New York, 1965.

FLUCTUATING HYDRODYNAMICS AND RENORMALIZATION OF SUSCEPTIBILITIES AND TRANSPORT COEFFICIENTS

P. Mazur

Instituut Lorentz
University of Leiden, Netherlands

I. GENERAL INTRODUCTION

1.1. Fluctuating hydrodynamics and renormalization of transport coefficients

1.2. Stochastic differential equations

II. ON THE CRITICAL BEHAVIOUR OF THE DIELECTRIC CONSTANT FOR A NON-POLAR FLUID

2.1. Introduction

2.2. Formal Theory

2.3. Expansion of the dielectric tensor in terms of density-density correlation functions

2.4. The extinction coefficient

2.5. The static, non-dispersive dielectric tensor

2.6. The real dispersive part of the dielectric tensor and the velocity of light

III. RENORMALIZATION OF THE DIFFUSION COEFFICIENT IN A FLUCTUATING FLUID

3.1. Derivation of a general formula for the renormalized diffusions coefficient

3.2. An expansion of the renormalized coefficient in correlation functions

3.3. The long time behaviour of the diffusion memory kernel, and the $\bar{k},\omega$ dependent renormalized diffusion constant

3.4. Expansion of the renormalized diffusion coefficient in terms of the renormalized diffusion propagator

REFERENCES

I. GENERAL INTRODUCTION *

1.1 Fluctuating hydrodynamics and renormalization of transport coefficients

In recent years relaxation phenomena in non-linear systems have been studied extensively. Two aspects of this problem have in particular been investigated. On the one hand the renormalization of transport coefficients in the critical region has been discussed within the framework of the mode-mode coupling theory by Kawasaki and others [4,5]. On the other hand slow decaying contributions to time correlation functions and memory kernels, as first found in a computer experiment by Alder and Wainwright [6], were studied using either hydrodynamic, kinetic or related arguments [6-13]. As has been pointed out, for example, by Kawasaki [14] and Zwanzig [15] both aspects are intimately related.

In these lectures we shall discuss the renormalization of certain transport-coefficients or susceptibilities consistently within the framework of fluctuating hydrodynamics.** To this end we extend, so to say, the Landau-Lifshitz theory of linear fluctuating hydrodynamics [16], to include the non-linear (fluctuating) terms which occur in the hydrodynamic equations. A theory of this type is particularly well suited to study phenomena which occur at small wave vectors and long times when hydrodynamic considerations are applicable. We shall restrict ourselves to situations in which the non-linearity of the fluctuating equations considered has a very simple structure.

We shall on the one hand consider the behaviour of the dielectric constant in a fluctuating non-polar fluid and show that the deviations from the validity of the Clausius-Mossotti formula may be interpreted, within our framework, as being due to a renormalization of the dielectric constant as a result of density fluctuations. Explicit formulae can then be obtained to describe the behaviour of the frequency and wave vector dependent dielectric tensor in the critical region.

---

* The subject matter of these lectures is essentially contained in a series of papers by D. Bedeaux and the author [1-3].

** A similar point of view was taken by Giterman and Gorodetskii [31].

We shall furthermore study the diffusion of tagged particles through an incompressible fluctuating fluid, starting from a non-linear generalized Langevin equation for the density of the tagged particles (the non-linear Landau-Lifshitz equation for the diffusion problem). We show that, averaging this equation over the fluctuations a linear macroscopic diffusion equation is obtained with a renormalized wave vector and frequency dependent diffusion coefficient, for which we obtain a closed expression. An analysis of this expression leads amongst other things to the well known $t^{-3/2}$ asymptotic behaviour for the diffusion memory kernel. We are able to discuss in detail the question whether the expression for the asymptotic behaviour of the memory kernel should contain the renormalized diffusion coefficient, as argued by several authors [4-8], rather than the bare one.

We will first discuss briefly a class of stochastic differential equations in their relation to the theory presented.

1.2 Stochastic differential equations

The theory which we will present is essentially based on the consideration of stochastic differential equations of a certain class. Basically these stochastic differential equations may be of the form

$$L^* x(t) = K(t) . \tag{1.1}$$

Here $x(t)$ is a random process, $K(t)$ some "external generalized force" (a given function of time t) and L^* a stochastic differential operator of the form

$$L^* = \frac{\partial}{\partial t} + \gamma_0 + \gamma_1 y(t) \tag{1.2}$$

where γ_0 and γ_1 are constants, and $y(t)$ is a second stationary random process, with zero mean

$$\langle y(t) \rangle = 0, \tag{1.3}$$

and given stochastic properties.

The formal solution of eq. (1.1) is

$$x(t) = L^{*-1} K(t) = G^* K(t) \tag{1.4}$$

with L^{*-1} the inverse of L^* and $G^* = L^{*-1}$ the random Green function operator of the problem. If one averages eq. (1.4) one obtains

$$\langle x(t)\rangle = \langle L^{*-1}\rangle K(t) = \langle G^*\rangle K(t) = G\,K(t) \tag{1.5}$$

where G is the propagator for the averaged equation. Defining

$$L = G^{-1} \tag{1.6}$$

we obtain for $\langle x(t)\rangle$ the equation

$$L\langle x\rangle = K(t) \tag{1.7}$$

with

$$L = \langle L^{*-1}\rangle^{-1} \tag{1.8}$$

Let us now define an operator γ' through

$$L \equiv \frac{\partial}{\partial t} + \gamma_0 + \gamma' = L_0 + \gamma' \tag{1.9}$$

so that

$$L = L_0(1 + L_0^{-1}\gamma') = L_0(1 + G_0\gamma') \tag{1.10}$$

with

$$G_0 = L_0^{-1} \tag{1.11}$$

On the other hand

$$L^* = L_0\left(1 + \gamma_1 G_0 y(t)\right) \tag{1.12}$$

so that

$$\langle L^{*-1}\rangle^{-1} = L_0\langle(1 + \gamma_1 G_0 y)^{-1}\rangle^{-1} \tag{1.13}$$

If we then substitute eqs. (1.9) and (1.13) into eq. (1.8) we obtain

$$\gamma' = -L_0 + L_0 \left\langle (1+\gamma_1 G_0 y)^{-1} \right\rangle^{-1}$$

$$= \left\langle -L_0 (1+\gamma_1 G_0 y)^{-1} + L_0 \right\rangle \left\langle (1+\gamma_1 G_0 y)^{-1} \right\rangle^{-1}$$

$$= \gamma_1 \left\langle y (1+\gamma_1 G_0 y)^{-1} \right\rangle \left\langle (1+\gamma_1 G_0 y)^{-1} \right\rangle^{-1}. \tag{1.14}$$

Equation (1.14) defines the macroscopic operator γ' in terms of correlation functions of $y(t)$. If we were to call γ_0 the bare kinetic coefficient and $\gamma_0 + \gamma_1 y(t)$ the random kinetic coefficient, then $\gamma_0 + \gamma'$ represents, so to say, the macroscopic "renormalized" kinetic coefficient since, according to eqs. (1.7) and (1.9), $\langle x \rangle$ obeys the equation

$$\frac{\partial \langle x(t) \rangle}{\partial t} = -(\gamma_0 + \gamma')\langle x(t) \rangle + K(t). \tag{1.15}$$

Note that γ' is still an operator, which turns out to be a convulution operator in time such that eq. (1.15) has the form

$$\frac{\partial \langle x(t) \rangle}{\partial t} = -\gamma_0 \langle x(t) \rangle - \int_{-\infty}^{t} \gamma'(t-t') \langle x(t) \rangle dt' + K(t').$$

In the next chapters we shall consider specific problems described by stochastic differential equations, more or less of the type of eq. (1.1). We will consider as stated in Section 1.1 the dielectric properties of a fluctuating fluid, and the diffusion of tagged particles through a fluctuating fluid. We shall evaluate the corresponding "renormalized kinetic coefficient", i.e. the dielectric tensor, and the diffusion memory kernel, to some order of approximation. Stochastic differential equations of the type of eq. (1.1) have recently been studied by van Kampen [17]. We refer to his papers for further references.

II. ON THE CRITICAL BEHAVIOUR OF THE DIELECTRIC CONSTANT FOR A NON-POLAR FLUID

2.1 Introduction

In the theory of the propagation of light through a fluid two aspects have received special attention. On the one hand one has studied the scattering of light from an incident beam. The phenomenological theory of this phenomenon goes back to the work of Smoluchowski [18] and Einstein [19]. Molecular theories of light scattering have subsequently been given by, among others, Yvon [20] and Zimm [21] and also by Fixmann [22], who was able to justify Einstein's result. On the other hand the propagation of the transmitted beam has been studied by calculation of the refractive index of the system. Theories for the refractive index have been given by, among others, Yvon [20], Hoek [23], Rosenfeld [24], Mazur and Mandel [25]. All these theories start essentially from a molecular basis. The theory of the refractive index is related to the theory of light scattering to the extent that the imaginary part of the refractive index, in the absence of true absorption, is directly related to the total intensity of the scattered light.

All molecular theories both of light scattering and of the refractive index suffer from the drawback that they are based on series expansions in the polarizability per unit volume and the calculation of the first few terms of the series. Strictly speaking they are therefore limited to systems where the refractive index is not appreciably different from unity. This makes their application questionable, for instance, in the critical region. The phenomenological theory does not suffer from this drawback.

We shall develop here a theory of the dielectric constant in a fluctuating fluid, which yields an expression for the macroscopic wave vector and frequency dependent dielectric tensor in terms of density fluctuation correlation functions. This tensor describes the response of the system to an arbitrary electromagnetic field. Its transverse part, with k and ω satisfying the usual dispersion relation, yields the index of refraction of the medium in terms of these correlation functions. The theory is not restricted to small values of the polarizability per unit volume and is therefore well suited to the study of the behaviour of the dielectric tensor in the critical region. In particular we shall study the dispersion of the velocity of light in the medium when it is close to the critical point. For small values of the polarizability per unit of volume the results can be shown to be in agreement with those of previous molecular theories.

2.2 Formal theory

The Maxwell equations in a fluctuating fluid are

$$\operatorname{curl} \vec{E} = -\frac{\partial \vec{B}}{\partial t}, \quad \operatorname{div} \vec{D} = 0, \quad \operatorname{curl} \vec{H} = \frac{\partial \vec{D}}{\partial t}, \quad \operatorname{div} \vec{B} = 0, \tag{2.1}$$

where $\vec{E}(\vec{r},t)$ and $\vec{B}(\vec{r},t)$ are the fluctuating electric and magnetic fields respectively, and $\vec{D}(\vec{r},t)$ and $\vec{H}(\vec{r},t)$ the fluctuating inductions; units are chosen in such a way that the velocity of light is unity in vacuum. Neglecting the magnetic properties we have

$$\vec{H} = \vec{B}, \quad \vec{D} = \vec{E} + \vec{P} \tag{2.2}$$

with $\vec{P}$ the fluctuating polarization.

We define Fourier transforms with respect to $\vec{r}$ and t of a field f by

$$f(\vec{k},\omega) = \int d\vec{r}\, dt\, e^{-i(\vec{k}\cdot\vec{r}-\omega t)} f(\vec{r},t). \tag{2.3}$$

From equations (2.1) and (2.2) we then obtain the vector wave equation

$$(\vec{k}\vec{k} - k^2 + \omega^2)\vec{E}(\vec{k},\omega) = -\omega^2 \vec{P}(\vec{k},\omega) \tag{2.4}$$

The general retarded solution of this equation is

$$\vec{E}(\vec{k},\omega) = \vec{E}^{\circ}(\vec{k},\omega) - \vec{\vec{F}}(\vec{k},\omega)\cdot\vec{P}(\vec{k},\omega) \tag{2.5}$$

where $\vec{E}^{\circ}$ is a solution of the homogeneous equation and is therefore the incident field in vacuum. In the presence of externally controlled sources $\vec{E}^{\circ}$ contains also the vacuum fields generated by these sources. The retarded propagator of the electromagnetic field in vacuum $\vec{\vec{F}}$

is given by the diagonal elements in $\vec{k},\omega$ representation

$$\overleftrightarrow{F}(\vec{k},\omega) = \left[\vec{k}\vec{k} - k^2 + (\omega+i0)^2\right]^{-1}\omega^2 = \left[k^2 - (\omega+i0)^2\right]^{-1}(\vec{k}\vec{k} - \omega^2) \tag{2.6}$$
$$= \left[\tfrac{1}{2} i\pi k^{-1}\left\{\delta(k-\omega) - \delta(k+\omega)\right\} + \mathcal{P}\frac{1}{k^2-\omega^2}\right](\vec{k}\vec{k} - \omega^2)$$

where 0 represents an infinitesimally small positive number.

From eq. (2.4) or (2.5) it can be seen that the electric field will fluctuate because of the fluctuations in the polarization of the dielectric. These in turn will be caused for a non-polar fluid by fluctuations in the density $\rho(\vec{r},t)$ and in other thermodynamic variables. Averaging over these fluctuations, denoted by $\langle\ldots\rangle$, yields the macroscopic quantities. In order to come to a closed description we must relate the polarization to the electric field. In linear optics $\vec{P}(\vec{r},t)$ is given by

$$\vec{P}(\vec{r},t) = \chi^*(\rho(\vec{r},t))\vec{E}(\vec{r},t) \tag{2.7}$$

where $\chi^*(\rho(\vec{r},t))$, the fluctuating susceptibility, is related to the density by the Clausius-Mossotti formula

$$\chi^*(\rho) = \alpha_0\rho\left(1 - \tfrac{1}{3}\alpha_0\rho\right)^{-1} \tag{2.8}$$

where α_0 is a constant frequency independent molecular polarizability. This restricts the applicability of the theory to the transmission of fields at non-resonant frequencies. The validity of eq. (2.8) is assumed here for density fields which vary sufficiently slowly in time and space. Eq. (2.8) can be justified on a molecular basis [1]. Deviations from eq. (2.8) occur if ρ varies appreciably over molecular distances. Now upon substitution of eq. (2.8) into eq. (2.5) one obtains

$$\vec{E} = \vec{E}^0 - \overleftrightarrow{F}\cdot\chi^*\vec{E} \tag{2.9}$$

with the formal solution

$$\vec{E} = (1 + \overleftrightarrow{F}\cdot\chi^*)^{-1}\cdot\vec{E}^0 \tag{2.10}$$

In both eqs. (2.9) and (2.10), χ^* and $\vec{E}$ have to be interpreted as operators.

According to eq. (2.8), χ^* is diagonal in the $\vec{r},t$ representation. It is therefore a convolution operator in the $\vec{k},\omega$ representation. The propagator $\overleftrightarrow{F}$ is diagonal in the $\vec{k},\omega$ representation (cf. eq. (2.6)) and a convolution operator in the $\vec{r},t$ representation.

The macroscopic electric field is obtained if one averages eq. (2.10),

$$\langle \vec{E} \rangle = \langle (1+\overleftrightarrow{F}\chi^*)^{-1} \rangle \cdot \vec{E}^{\circ} \tag{2.11}$$

On the other hand, substituting eq. (2.10) into eq. (2.8), and using also eq. (2.11), yields for the average polarization

$$\begin{aligned} \langle \vec{P} \rangle &= \langle \chi^* \vec{E} \rangle = \langle \chi^* (1+\overleftrightarrow{F}\chi^*)^{-1} \rangle \cdot \vec{E}^{\circ} \\ &= \langle \chi^* (1+\overleftrightarrow{F}\chi^*)^{-1} \rangle \cdot \langle (1+\overleftrightarrow{F}\chi^*)^{-1} \rangle^{-1} \cdot \langle \vec{E} \rangle \end{aligned} \tag{2.12}$$

This equation defines the macroscopic dielectric susceptibility tensor $\overleftrightarrow{\chi}$ and dielectric tensor $\overleftrightarrow{\varepsilon}$

$$\overleftrightarrow{\varepsilon} - \mathbf{1} = \overleftrightarrow{\chi} = \langle \chi^* (1+\overleftrightarrow{F}\chi^*)^{-1} \rangle \cdot \langle (1+\overleftrightarrow{F}\chi^*)^{-1} \rangle^{-1}. \tag{2.13}$$

It will be the starting point of our further discussion.

In the absence of fluctuations it follows from the above equation that the macroscopic dielectric constant is given by the Clausius-Mossotti formula

$$\varepsilon_0 - \mathbf{1} = \chi_0 = \alpha_0 \rho_0 \left(1 - \tfrac{1}{3}\alpha_0 \rho_0\right)^{-1}, \quad \rho_0 = \langle \rho \rangle \tag{2.14}$$

We will now investigate in more detail the deviations from the Clausius-Mossotti formula, caused by the density fluctuations. We define the propagator in the absence of fluctuations

$$\begin{aligned} \overleftrightarrow{F}_{\varepsilon_0} &= (1+\overleftrightarrow{F}\chi_0)^{-1}\overleftrightarrow{F}(\vec{k},\omega) = \left[\vec{k}\vec{k} - k^2 + (\omega+i0)^2\varepsilon_0\right]^{-1}\omega^2 \\ &= \varepsilon_0^{-1}\left[k^2 - (\omega+i0)^2\varepsilon_0\right]^{-1}(\vec{k}\vec{k} - \omega^2\varepsilon_0) \end{aligned} \tag{2.15}$$

and write

$$\chi^* = \chi_0 + \Delta\chi^* \tag{2.16}$$

If we then substitute eq. (2.16) into eq. (2.13), we find that

$$\varepsilon = \varepsilon_0 + \langle \Delta\chi^* (1 + \overset{\rightleftarrows}{F}_{\varepsilon_0} \Delta\chi^*)^{-1} \rangle \cdot \langle (1 + \overset{\rightleftarrows}{F}_{\varepsilon_0} \Delta\chi^*)^{-1} \rangle \, . \tag{2.17}$$

Equation (2.17), which is equivalent to equation (2.13), may be used to obtain expansions for $\overset{\rightleftarrows}{\varepsilon}$ in terms of density-density correlation functions. We note that the "renormalized" propagator $\overset{\rightleftarrows}{F}_{\varepsilon_0}$ accounts for the fact that the fluctuations in χ^* interact via the medium rather than through vacuum. Each term in the expansion of eq. (2.17) corresponds to a partial resummation of the terms in the expansion of eq. (2.13) in powers of χ^* . The formal transformation from eq. (2.13) to eq. (2.17) performs these resummations to all orders. It should also be mentioned that one can easily convince oneself, using the fact that the fluid is translationally invariant in space and time, that $\overset{\rightleftarrows}{\varepsilon}$ is diagonal in the $\vec{k}, \omega$ representation, as of course it should be.

Before proceeding with the analysis of the influence of density fluctuations on the behaviour of the dielectric constant, we shall derive two further identities.

Define the "cut-out" propagator:

$$\overset{\rightleftarrows}{H} = \overset{\rightleftarrows}{F} - \frac{1}{3} \tag{2.18}$$

and the Clausius-Mossotti function:

$$\overset{\rightleftarrows}{\gamma} = 3\,(\overset{\rightleftarrows}{\varepsilon} - 1)(\overset{\rightleftarrows}{\varepsilon} + 2)^{-1} \tag{2.19}$$

or

$$\overset{\rightleftarrows}{\varepsilon} - 1 = \overset{\rightleftarrows}{\gamma} \cdot (1 - \tfrac{1}{3}\overset{\rightleftarrows}{\gamma})^{-1} \tag{2.20}$$

With these definitions and eq. (2.8), one finds after some straight forward transformations from eq. (2.13) that

$$\overset{\rightleftarrows}{\gamma} = \langle \rho\alpha_0 (1 + \overset{\rightleftarrows}{H} \rho\alpha_0)^{-1} \rangle \cdot \langle (1 + \overset{\rightleftarrows}{H} \rho\alpha_0)^{-1} \rangle \, , \tag{2.21}$$

an expression for the Clausius-Mossotti function in terms of the density correlation functions.

In the absence of fluctuations, eq. (2.21) leads to the Clausius-Mossotti formula in its usual form. An even more useful identity is found if we write

$$\rho = \rho_0 + \Delta\rho \ , \quad \rho_0 = \langle \rho \rangle \tag{2.22}$$

Substitution of eq. (2.22) into eq. (2.21) leads to

$$\overset{\rightrightarrows}{\gamma} = \rho_0 \alpha_0 + \langle \alpha_0 \Delta\rho (1 + \overset{\rightrightarrows}{K} \alpha_0 \Delta\rho)^{-1} \rangle \cdot \langle (1 + \overset{\rightrightarrows}{K} \alpha_0 \Delta\rho)^{-1} \rangle \tag{2.23}$$

with the renormalized propagator

$$\begin{aligned} \overset{\rightrightarrows}{K} &\equiv (1 + \overset{\rightrightarrows}{H} \alpha_0 \rho_0)^{-1} \overset{\rightrightarrows}{H} \\ &= \left[1 + (\overset{\rightrightarrows}{F} - \tfrac{1}{3}) \alpha_0 \rho_0 \right]^{-1} \cdot (\overset{\rightrightarrows}{F} - \tfrac{1}{3}) \ . \end{aligned} \tag{2.24}$$

If one uses the relations (cf. eq. (2.14) and (2.15)

$$\frac{\varepsilon_0 - 1}{\varepsilon_0 + 2} = \frac{1}{3} \rho_0 \alpha_0 \tag{2.25}$$

$$\begin{aligned} \overset{\rightrightarrows}{F}_{\varepsilon_0} &= \left[1 + \overset{\rightrightarrows}{F} (\varepsilon_0 - 1) \right]^{-1} \cdot \overset{\rightrightarrows}{F} \\ \overset{\rightrightarrows}{F} &= \left[1 - \overset{\rightrightarrows}{F}_{\varepsilon_0} (\varepsilon_0 - 1) \right]^{-1} \cdot \overset{\rightrightarrows}{F}_{\varepsilon_0} \end{aligned} \tag{2.26}$$

one finds from eq. (2.24) a relation between the propagator $\overset{\rightrightarrows}{K}$ and the propagator $\overset{\rightrightarrows}{F}_{\varepsilon_0}$ occurring in eq. (2.17)

$$\overset{\rightrightarrows}{K} = \left(\frac{\varepsilon_0 + 2}{3} \right)^2 \left(\overset{\rightrightarrows}{F}_{\varepsilon_0} - \frac{1}{\varepsilon_0 + 2} \right) \ . \tag{2.27}$$

The general expression eq. (2.23) relates the corrections to the Clausius-Mossotti formula directly to density fluctuation correlation functions. The propagator $\overset{\rightrightarrows}{K}$ accounts for the fact that these density fluctuations interact via the medium. As we see from eq. (2.27), $\overset{\rightrightarrows}{K}$ is essentially the propagator in the medium with dielectric constant ε_0, modified by Lorentz corrections.

2.3 Expansion of the dielectric tensor in terms of density-density correlation functions

We shall now study eq. (2.33) in more detail. Expanding the right hand side in powers of $\alpha_0 \Delta\rho$, we obtain

$$\vec{\vec{\Gamma}}(\vec{k},\omega) = \alpha_0 \rho_0 \left\{ 1 - (2\pi)^{-4} \alpha_0 \rho_0 \int_{k'<k_0} S_2(\vec{k}-\vec{k}', \omega-\omega') \vec{\vec{K}}(\vec{k}',\omega')\, d\vec{k}' d\omega' + \text{higher order terms} \right\} \tag{2.28}$$

where the density-density correlation function S_2 is defined as

$$S_2(\vec{r}_2-\vec{r}_1; t_2-t_1) = \rho_0^{-2} \langle \Delta\rho(\vec{r}_1,t_1) \Delta\rho(\vec{r}_2,t_2) \rangle \tag{2.29}$$

and where the integral in (2.28) is cut off at some inverse molecular distance. The introduction of the cut off in eq. (2.28) is related to the fact that eq.((2.8) does not hold any more if ρ varies appreciably over molecular distances. In a molecular version of the present theory, in which corrections to eq. (2.8) are taken into account, it is not necessary to introduce an explicit cut off [1]. The Fourier transform of S_2 is defined in the usual manner. If one also expands $\vec{\vec{K}}$ in $\alpha_0 \rho_0$, one obtains a series expansion in powers of $\alpha_0 \rho_0$, which is in fact closely related to the series obtained by previous authors from molecular theories. Such a series will rapidly converge if $\alpha_0 \rho_0 \ll 1$ or if $\|\vec{\vec{\varepsilon}} - 1\| \ll 1$, i.e. for sufficiently low densities. However, the expansion in eq. (2.28) will also rapidly converge if $\alpha_0 \rho_0$ is of order unity, but the integrals involving the correlation functions become progressively smaller, i.e. if the density fluctuations are sufficiently small, so that $\|\vec{\vec{\varepsilon}} - \varepsilon_0\| \ll 1$. Such an expansion is therefore well suited to study the behaviour of the dielectric tensor in the critical region.

In the next sections we will calculate the correction to the Clausius-Mossotti function originating from the density-density function S_2, assuming therefore that under most circumstances this is the dominant contribution. Using (2.25), eq. (2.28) yields for $\vec{\vec{\varepsilon}}$:

$$\overleftrightarrow{\varepsilon}(\vec{k},\omega) \simeq \varepsilon_0 - (\varepsilon_0-1)^2(2\pi)^{-4}\int S_2(\vec{k}-\vec{k}',\omega-\omega')\overleftrightarrow{K}(\vec{k}',\omega')\,d\vec{k}'d\omega'$$

$$= \varepsilon_0 - (\varepsilon_0-1)^2\left(\frac{\varepsilon_0+2}{3}\right)^2(2\pi)^{-4}\int S_2(\vec{k}-\vec{k}',\omega-\omega')\left[\overleftrightarrow{F}_{\varepsilon_0}(\vec{k}',\omega') - \frac{1}{\varepsilon_0+2}\right]d\vec{k}'\,d\omega' \tag{2.30}$$

where we have also used eq. (2.27).

At this point it should be mentioned that eq. (2.30) can also be obtained from the general formula (2.17), if one expands the r.h.s. consistently up to powers quadratic in $\Delta\rho$.

The correction to the dielectric tensor in eq. (2.30) has the form of a mode-mode coupling expression, encountered in the now familiar theories for the renormalization of transport coefficients. Here the coupling is between the electromagnetic mode and the modes governing the behaviour of S_2 (hydrodynamic modes). In the language of the mode-mode coupling theories we may call ε_0 the "bare" dielectric constant and $\overleftrightarrow{\varepsilon}$ the renormalized dielectric tensor.

Up to corrections of order $(v/c)^2$ (v is the thermal velocity in the fluid) we may neglect the motion of the fluid, so that we can use the static approximation for the correlation function S_2

$$S_2(\vec{k}-\vec{k}',\omega-\omega') = 2\pi\, S_2(\vec{k}-\vec{k}')\,\delta(\omega-\omega') \tag{2.31}$$

Equation (2.30) then becomes

$$\overleftrightarrow{\varepsilon} = \varepsilon_0 - (\varepsilon_0-1)^2\left(\frac{\varepsilon_0+2}{3}\right)^2(2\pi)^{-3}\int S_2(\vec{k}-\vec{k}')\left[\overleftrightarrow{F}_{\varepsilon_0}(\vec{k}',\omega) - \frac{1}{\varepsilon_0+2}\right]d\vec{k}' \tag{2.32}$$

For isotropic systems we may write

$$\overleftrightarrow{\varepsilon}(\vec{k},\omega) = \varepsilon_{tr}(\vec{k},\omega)(1-\vec{k}\vec{k}/k^2) + \varepsilon_\ell(\vec{k},\omega)\,\vec{k}\vec{k}/k^2, \tag{2.33}$$

where the transverse and longitudinal dielectric constants are given by

$$\varepsilon_{tr}(\vec{k},\omega) = \varepsilon_0 - (2\pi)^{-3}(\varepsilon_0-1)^2\left(\frac{\varepsilon_0+2}{3}\right)^2\int S_2(|\vec{k}-\vec{k}'|)\,\vec{u}\cdot\left[\overleftrightarrow{F}_{\varepsilon_0}(\vec{k}',\omega) - \frac{1}{\varepsilon_0+2}\right]\cdot\vec{u}\,d\vec{k}' \tag{2.34}$$

with $\vec{u}$ a unit vector orthogonal to $\vec{k}$, $\vec{u}\cdot\vec{k}=0$,

$$\varepsilon_l(\vec{k},\omega) = \varepsilon_0-(\varepsilon_0-1)^2\left(\frac{\varepsilon_0+2}{3}\right)^2(2\pi)^{-3}\int S_2(|\vec{k}-\vec{k}'|)\frac{\vec{k}}{k}\cdot\left[\vec{\vec{F}}_{\varepsilon_0}(\vec{k}',\omega)-\frac{1}{\varepsilon_0+2}\right]\cdot\frac{\vec{k}}{k}\,d\vec{k}' \qquad (2.35)$$

For k in the hydrodynamic region the density-density correlation function has, in good approximation the Ornstein-Zernike* [26,27] form

$$S_2^H(k) = k_B T \kappa\,(1+k^2\xi^2)^{-1} \qquad (2.36)$$

where k_B is Boltzmann's constant, T the temperature, κ the isothermal compressibility and ξ the hydrodynamic correlation length.

For larger values of k one has to add terms describing the molecular structure at small intermolecular distances. We therefore write in general

$$S_2(k) = S_2^H(k) + S_2^M(k)\,. \qquad (2.37)$$

We shall assume that S_2^M is a sufficiently slowly varying function of k/k_0 and becomes small for hydrodynamic values of k.

In the following sections we shall evaluate the various contributions to $\varepsilon(k,\omega)$ by substituting eq. (2.37) into eq. (2.32).

2.4 The extinction coefficient

We shall first compute the imaginary part of dielectric tensor in the approximation given in eq. (2.31). This quantity is directly related to the extinction of light in the fluid. Substituting eq. (2.15) into eqs. (2.34) and (2.35), gives after integration over k' and with $\vec{\Omega} = \vec{k}'/k'$,

$$\mathrm{Im}\,\varepsilon_{tr}(\vec{k},\omega) = \frac{\omega^3}{32\pi^2}\,n_0\,(\varepsilon_0-1)^2\left(\frac{\varepsilon_0+2}{3}\right)^2\int d\vec{\Omega}\,S_2(|\vec{k}-\omega n_0\vec{\Omega}|)(1+\cos^2\theta) \qquad (2.38)$$

---

* The calculations of sections 4-6 can, in principle, also be made by assuming that the Ornstein-Zernike form is slightly modified:

$$S_2^H = k_B T \kappa\,(1+k^2\xi^2)^{-1+\eta/2}.$$

where θ is the angle between $\vec{k}$ and $\vec{\Omega}$, and also

$$\mathrm{Im}\,\varepsilon_\ell(k,\omega) = \frac{\omega^3}{16\pi^2}\, n_0\,(\varepsilon_0-1)^2\left(\frac{\varepsilon_0+2}{3}\right)^2 \int d\vec{\Omega}\; S_2(|\vec{k}-\omega n_0\vec{\Omega}|)\sin^2\theta \qquad (2.39)$$

$n_0 \equiv \varepsilon_0^{1/2}$ is the "bare" index of refraction.

The extinction coefficient is defined as

$$\alpha(\omega) \equiv 2\omega\,\mathrm{Im}\,n(\omega) \qquad (2.40)$$

with

$$n(\omega) = \left(\varepsilon_{tr}(k(\omega),\omega)\right)^{1/2}, \qquad (2.41)$$

where $k(\omega)$ is the solution of the dispersion relation

$$k^2 = \omega^2\,\varepsilon_{tr}(k,\omega)\,. \qquad (2.42)$$

To lowest order,

$$k(\omega) = n_0\,\omega \qquad (2.43)$$

so that we have to first order from eqs. (2.40) – (2.42) together with eq. (2.38) that

$$\alpha(\omega) = \frac{\omega^4}{32\pi^2}\,(\varepsilon_0-1)^2\left(\frac{\varepsilon_0+2}{3}\right)^2 \int d\vec{\Omega}\; S_2(n_0\omega|\vec{\Omega}_0-\vec{\Omega}|)(1+\cos^2\theta), \qquad (2.44)$$

where $\vec{\Omega}_0 = \vec{k}/n_0\omega$.

In formulae (2.38), (2.39) and (2.44) we can substitute the Ornstein-Zernike form for S_2, since only hydrodynamic values of the argument occur. The integrations can then be performed. This yields in particular for $\alpha(\omega)$

$$\alpha(\omega) = -\,\frac{\omega^4}{32\pi\,\varepsilon_0}\,(\varepsilon_0-1)^2\left(\frac{\varepsilon_0+2}{3}\right)^2 \frac{k_B T k}{\omega^2\xi^2}\,\alpha'(\xi\omega n_0) \qquad (2.45)$$

with

$$\alpha'(\xi\omega n_0) = \frac{1+u^2}{u^2} + \left[\left(\frac{1+u^2}{2u^2}\right)^2 + 1\right] \ln\left(\frac{1-u^2}{1+3u^2}\right), \tag{2.46}$$

$$u \equiv \xi\omega n_0 \left(1+\xi^2\omega^2 n_0^2\right)^{-1/2} \tag{2.47}$$

Furthermore,

$$\alpha'(\xi\omega n_0) \simeq -\frac{16}{3} n_0^2 \omega^2 \xi^2 \quad \text{if} \quad \omega^2\xi^2 << 1 \tag{2.48}$$

$$\alpha'(\xi\omega n_0) \simeq -4\left[\ln(2\omega\xi n_0) - 1/2\right] \tag{2.49}$$

For the explicit results for $\mathrm{Im}\,\varepsilon_{tr}(k,\omega)$ and $\mathrm{Im}\,\varepsilon_{\ell}(k,\omega)$, the reader is referred to [1].

The expression for the extinction coefficient contains the factor $((\varepsilon_0+2)/3)^2$, which also appears in the expressions obtained from the phenomenological theory of light scattering. The first molecular derivation of this factor was given by Fixman [22]. For $\alpha_0\rho_0 << 1$, eq. (2.45) reduces to the expression derived by Rosenfeld in that limit. For $\alpha_0\rho_0$ of order unity Rosenfeld's theory does not strictly apply.

We also note that if the system is in the critical region, but not extremely close to the critical point, eq. (2.48) applies and $\alpha(\omega)$ appears to diverge as k, which diverges as ξ^2. However, extremely close to the critical point, the behaviour is described by eq. (2.49), and $\alpha(\omega)$ is logarithmically divergent. Furthermore the extinction coefficient close to the critical point behaves roughly as ω^2 rather than ω^4. This is the phenomenon of critical opalescence.

2.5 The static, non-dispersive dielectric tensor

Let us now study $\overleftrightarrow{\varepsilon}(\vec{k},\omega)$, for $\omega = 0$ and $\vec{k} = 0$. We have in this case from eq. (2.32), using also eqs. (2.15) and (2.37)

$$\overleftrightarrow{\varepsilon}(0,0) = \varepsilon_0 + \varepsilon^M + \varepsilon^H , \tag{2.50}$$

where

$$\varepsilon^M = (27\pi^2\varepsilon_0)^{-1}(\varepsilon_0-1)^3(\varepsilon_0+2)\int_0^{k_0} S_2^M(k)\,k^2\,dk \tag{2.51}$$

$$\varepsilon^H = (27\pi^2\varepsilon_0)^{-1}(\varepsilon_0-1)^3(\varepsilon_0+2)\,k_BTk\,\xi^{-3}\int_0^{k_0\xi}(1+x^2)^{-1}x^2\,dx \tag{2.52}$$

We cannot evaluate the contribution $\Delta\varepsilon^M$ explicitly. It does not depend critically, however, on the thermodynamic variables. On the other hand the integral in (2.52) is quite elementary if $k_0\xi \gg 1$, and one obtains

$$\varepsilon^H = (27\pi^2\varepsilon_0)^{-1}(\varepsilon_0-1)^3(\varepsilon_0+2)\,k_BTk\,\xi^{-3}\left(k_0\xi - \frac{\pi}{2}\right) , \quad k_0\xi \gg 1 . \tag{2.53}$$

If $k_0\xi \ll 1$, which is typically the case if the system is sufficiently dilute, eq. (2.52) gives

$$\varepsilon^H = (81\pi^2\varepsilon_0)^{-1}(\varepsilon_0-1)^3(\varepsilon_0+2)\,k_BTk\,k_0^3 \tag{2.54}$$

As is seen from eq. (2.52), the contribution $\Delta\varepsilon^H$ to the static dielectric constant exhibits a marked critical behaviour. In the next section it will become apparent that this term must be taken into account when discussing the critical behaviour of the velocity of light.

2.6 The real dispersive part of the dielectric tensor and the velocity of light

If we define the dispersive part of $\overleftrightarrow{\varepsilon}(\vec{k},\omega)$ as

$$\Delta\overleftrightarrow{\varepsilon}(\vec{k},\omega) = \overleftrightarrow{\varepsilon}(\vec{k},\omega) - \overleftrightarrow{\varepsilon}(0,0) \tag{2.55}$$

we obtain from eq. (2.32), using eqs. (2.15) and (2.37),

$$\mathrm{Re}\,\Delta\overleftrightarrow{\varepsilon}(\vec{k},\omega) = \varepsilon_0^{-1}(6\pi)^{-3}(\varepsilon_0-1)^2(\varepsilon_0+2)^2 k_B T k\xi^{-3}\,\overleftrightarrow{I}(\vec{k}\xi,\omega n_0) \tag{2.56}$$

where

$$\overleftrightarrow{I}(\vec{x},y) = \int_{x'<\xi k_0} d\vec{x}'\,(1+|\vec{x}-\vec{x}'|^2)^{-1}\left[\left(1-3\frac{\vec{x}'\vec{x}'}{x'^2}\right)+3y^2(x'^2-y^2)^{-1}\left(1-\frac{\vec{x}'\vec{x}'}{x'^2}\right)\right] \tag{2.57}$$

In eq. (2.57) we have neglected the contribution to the integral arising from the molecular part S_2^M of S_2. This implies that we have neglected terms of order $(\omega/k_0)^2$ and $(k/k_0)^2$, which is certainly permissible for optical frequencies and wave numbers. If furthermore $k_0\xi >> 1$, the integrations in eq. (2.57) may be performed and analytic expressions for the real part of $\Delta\overleftrightarrow{\varepsilon}$ can be obtained. We again refer to [1] for these expressions. Here we shall only discuss explicitly the velocity of light, which is defined as

$$c(\omega) \equiv \left[\mathrm{Re}\,n(\omega)\right]^{-1} \tag{2.58}$$

where $n(\omega)$ is given by (2.41) with (2.42). To first order in the corrections, we can again use eq. (2.43) so that

$$\begin{aligned} c(\omega) &= \left\{\mathrm{Re}\left[\varepsilon_{tr}(n_0\omega,\omega)\right]^{1/2}\right\}^{-1} \\ &= n_0^{-1}\left\{1-\tfrac{1}{2}n_0^{-2}\varepsilon^M-\tfrac{1}{2}n_0^{-2}\varepsilon^H-\mathrm{Re}\,\Delta\varepsilon_{tr}(n_0\omega,\omega)\right\} \end{aligned} \tag{2.59}$$

With the result of the integrations in eq. (2.57), and using also eq. (2.53), one then finds that

$$c(\omega) = n_0^{-1} - \frac{1}{2} n_0^{-3} \varepsilon^M - \frac{n_0^{-5}}{54\pi^2} (\varepsilon_0 - 1)^3 (\varepsilon_0 + 2)(k_0 \xi - \pi/2) k_B T k \xi^{-3}$$

$$- \frac{1}{32\pi} \varepsilon_0^{-2} (\varepsilon_0 - 1)^2 \left(\frac{\varepsilon_0 + 2}{3}\right)^2 k_B T k \xi^{-2} \omega \Xi(\xi \omega n_0), \tag{2.60}$$

where

$$\Xi(\xi \omega n_0) = \frac{1}{4u^4} (5u^4 + 2u^2 + 1) \arcsin u - \frac{1}{4u^3} (5u^2 + 3)(1 - u^2)^{1/2} +$$

$$+ \left[1 + \left(\frac{1 + u^2}{2u^2}\right)^2\right] \arctan\left(\frac{u(1 - u^2)^{1/2}}{1 + u^2}\right) \tag{2.61}$$

with u given by eq. (2.47).

Furthermore

$$\Xi(\xi \omega n_0) = \frac{44}{15} \xi \omega n_0, \quad \xi \omega n_0 << 1 \tag{2.62}$$

$$\Xi(\xi \omega n_0) = \pi, \quad \xi \omega n_0 >> 1. \tag{2.63}$$

Equation (2.60) describes the behaviour of $c(\omega)$ in the critical region: not too close to the critical point $c(\omega)$ appears to diverge like ξ. Close to the critical point eq. (2.63) applies and shows that $c(\omega)$ remains finite. If one expands eq. (2.60) in powers of $\alpha_0 \rho_0$ and only retains quadratic terms, one obtains an expression identical with the expression obtained in that order for $c(\omega)$ by Larsen, Mountain and Zwanzig [28] from a molecular theory. Inspection of expression (2.60) shows, however, that the critical non-dispersive term in c, which is of order $\alpha_0^3 \rho_0^3$ can have the same magnitude, not too close to the critical point, as the dispersive contribution which is of order $\alpha_0^2 \rho_0^2$.

To conclude we may remark that the theory developed in this chapter for the dielectric tensor, on the basis of a theoretical phenomenological-fluctuation approach, enables one to give a detailed description of the behaviour of the dielectric tensor in the critical region, and applies even when the refractive index is appreciably different from unity. The theory has been generalized to the description of binary fluids (liquids) [29].

III. RENORMALIZATION OF THE DIFFUSION COEFFICIENT IN A FLUCTUATING FLUID

3.1 Derivation of a general formula for the renormalized diffusion coefficient

We consider a density distribution $n(\vec{r},t)$ of tagged particles in an incompressible fluid. It is assumed that the solution of the tagged particles is sufficiently dilute so that the fluctuations of the fluid are those of the pure fluid in equilibrium. The density $n(\vec{r},t)$ satisfies the conservation law

$$\frac{\partial}{\partial t} n(\vec{r},t) = -\operatorname{div} \vec{j}(\vec{r},t) \tag{3.1}$$

where $j(\vec{r},t)$ is the current density of the tagged particles. In the hydrodynamic regime this current is assumed to be given by

$$\vec{j}(\vec{r},t) = \vec{v}(\vec{r},t) n(\vec{r},t) - D_0 \operatorname{grad} n(\vec{r},t) + \vec{j}_R(\vec{r},t) \tag{3.2}$$

where $\vec{v}(\vec{r},t)$ is the fluctuating velocity field of the fluid. The first term on the right hand side of eq. (3.2) represents the convective part of the current. The second term represents the diffusive part of the current with respect to the moving fluid, with a "bare" diffusion coefficient D_0 which is assumed to be constant. The last term represents the random current which is assumed to have the property that

$$\langle \vec{j}_R(\vec{r},t) \rangle_{fl} = 0 \tag{3.3}$$

where the average is taken for given values of the hydrodynamic fluid field $\vec{v}(\vec{r},t)$.

Substituting eq. (3.2) into eq. (3.1), we find that $n(\vec{r},t)$ obeys the non-linear Langevin-equation

$$\left(\frac{\partial}{\partial t} - D_0\nabla^2\right) n(\vec{r},t) = -\,\mathrm{div}\ \vec{v}\, n(\vec{r},t) - \mathrm{div}\ \vec{j}_R(\vec{r},t)$$

(3.4)

where $\vec{v}$ is a vector operator. Even though this equation is linear in the density of the tagged particles the coupling to the fluid is non-linear.

The Fourier transform of a field $f(\vec{r},t)$ is defined as

$$f(\vec{k},\omega) \equiv \int e^{-i(\vec{k}\cdot\vec{r}-\omega t)} f(\vec{r},t)\, d\vec{r}\, dt \tag{3.5}$$

In the $\vec{k},\omega$ representation eq. (3.4) becomes

$$(i\omega - D_0 k^2)\, n(\vec{k},\omega) = i\vec{k}\cdot\vec{v}\, n(\vec{k},\omega) + i\vec{k}\cdot\vec{j}_R(\vec{k},\omega) \tag{3.6}$$

where we note that the operator $\vec{v}$ in this representation has matrix elements

$$\vec{v}(\vec{k},\omega|\vec{k}',\omega') = (2\pi)^{-4}\, \vec{v}(\vec{k}-\vec{k}', \omega-\omega') \tag{3.7}$$

Eq. (3.6) has the formal solution:

$$n(\vec{k},\omega) = n_0(\vec{k},\omega) - i\, G_0 \vec{k}\cdot\vec{v}\, n(\vec{k},\omega) - i\, G_0 \vec{k}\cdot\vec{j}_R(\vec{k},\omega), \tag{3.8}$$

where $n_0(\vec{k},\omega)$ is a solution of the homogeneous equation and therefore a non-fluctuating quantity. G_0 is the bare diffusion propagator (in the absence of fluctuations)

$$G_0(\vec{k},\omega) \equiv -(i\omega - D_0 k^2)^{-1} \tag{3.9}$$

The formal solution can be rewritten as

$$n(\vec{k},\omega) = \left[1 + i\, G_0 \vec{k}\cdot\vec{v}\right]^{-1} \left\{n_0(\vec{k},\omega) - i\, G_0 \vec{k}\cdot\vec{j}_R(\vec{k},\omega)\right\} \tag{3.10}$$

This yields for the current of the tagged particles after some algebra

$$\vec{j}(\vec{k},\omega) = (\vec{v} - iD_0\vec{k})\left[1 + iG_0\vec{k}\cdot\vec{v}\right]^{-1} n_0(\vec{k},\omega) - i\omega\left[1 + iG_0\vec{k}\cdot\vec{v}\right]^{-1} G_0\vec{j}_R(\vec{k},\omega) \tag{3.11}$$

Averaging eqs. (3.10) and (3.11) first over the random current and then over the fluctuations of the fluid we find for the mean particle and current density of the tagged particles using also eq. (3.3)

$$\langle n(\vec{k},\omega)\rangle = \left\langle\left[1 + iG_0\vec{k}\cdot\vec{v}\right]^{-1}\right\rangle n_0(\vec{k},\omega) \tag{3.12}$$

$$\langle \vec{j}(\vec{k},\omega)\rangle = \left\langle(\vec{v} - iD_0\vec{k})\left[1 + iG_0\vec{k}\cdot\vec{v}\right]^{-1}\right\rangle n_0(\vec{k},\omega) \tag{3.13}$$

where we have introduced the shorthand notation

$$\langle\ldots\rangle = \langle\langle\ldots\rangle_{fl}\rangle \tag{3.14}$$

Eliminating n_0, the mean current can be written as

$$\langle \vec{j}(\vec{k},\omega)\rangle = \left\{-iD_0\vec{k} + \left\langle\vec{v}\left[1 + iG_0\vec{k}\cdot\vec{v}\right]^{-1}\right\rangle\left\langle\left[1 + iG_0\vec{k}\cdot\vec{v}\right]^{-1}\right\rangle^{-1}\right\} \times \langle n(\vec{k},\omega)\rangle \tag{3.15}$$

This relation defines the "renormalized" diffusion coefficient

$$D = D_0 + k^{-2}\left\langle i\vec{k}\cdot\vec{v}\left[1 + iG_0\vec{k}\cdot\vec{v}\right]^{-1}\right\rangle\left\langle\left[1 + iG_0\vec{k}\cdot\vec{v}\right]^{-1}\right\rangle^{-1} \tag{3.16}$$

D is an operator which is, using translational invariance and stationarity of the fluid, diagonal in the $\vec{k},\omega$ representation. From eqs. (3.1), (3.15) and (3.16) it then follows that the mean tagged particle density satisfies the diffusion equation with the renormalized diffusion coefficient

$$\left(i\omega - D(\vec{k},\omega)k^2\right)\langle n(\vec{k},\omega)\rangle = 0 \,. \tag{3.17}$$

The macroscopic diffusion propagator is therefore diagonal in the $\vec{k},\omega$ representation and is given by

$$G(\vec{k},\omega) \equiv -\left(i\omega - k^2 D(\vec{k},\omega)\right)^{-1} \tag{3.18}$$

In subsequent sections we will evaluate D by expanding the right hand side of eq. (3.16) in fluctuation correlation functions of the fluid. Note also that

$$G = \langle G^* \rangle \tag{3.19}$$

where G^* is the fluctuating diffusion propagator

$$G^* \equiv -(i\omega - D_0 k^2 - i\vec{k}\cdot\vec{v})^{-1} = \left[1 + i\,G_0\,\vec{k}\cdot\vec{v}\right]^{-1} G_0 \tag{3.20}$$

Eq. (3.19) can be obtained from eq. (3.16) with some straightforward algebra. See in this connection section 1.2.

3.2 An expansion of the renormalized transport coefficient in correlation functions

In this section we will expand the renormalized transport coefficient in terms of the correlation functions of the fluid. Expanding eq. (3.16) in powers of $\vec{v}$ yields

$$\begin{aligned} D = D_0 &+ k^{-2}\langle \vec{k}\cdot\vec{v}\,G_0\,\vec{k}\cdot\vec{v}\rangle - ik^2\langle \vec{k}\cdot\vec{v}\,G_0\,\vec{k}\cdot\vec{v}\,G_0\,\vec{k}\cdot\vec{v}\rangle \\ &- k^{-2}\langle \vec{k}\cdot\vec{v}\,G_0\,\vec{k}\cdot\vec{v}\,G_0\,\vec{k}\cdot\vec{v}\,G_0\,\vec{k}\cdot\vec{v}\rangle \\ &+ k^{-2}\langle \vec{k}\cdot\vec{v}\,G_0\,\vec{k}\cdot\vec{v}\rangle\,G_0\,\langle \vec{k}\cdot\vec{v}\,G_0\,\vec{k}\cdot\vec{v}\rangle + \text{higher order terms} \end{aligned} \tag{3.21}$$

To second order in $\vec{v}$ one has

$$D = D_0 + k^{-2}\langle \vec{k}\cdot\vec{v}\,G_0\,\vec{k}\cdot\vec{v}\rangle = D_0 + D_1 \tag{3.22}$$

where all contributions are diagonal in $\vec{k},\omega$ representation due to translational invariance and stationarity of the fluid. We note that $\vec{v}$ in eq. (3.22) should be interpreted as an operator which is diagonal in the $\vec{r},t$ representation and a convolution operator in the $\vec{k},\omega$ representation (c.f. (3.7)). The diagonal elements of D_1

can now be written as

$$D_1(\vec{k},\omega) = k^{-2}(2\pi)^{-4}\int d\vec{k}'d\omega'\,\vec{k}\cdot\overleftrightarrow{S}^{vv}(\vec{k}',\omega')\cdot(\vec{k}-\vec{k}')\,G_0(\vec{k}-\vec{k}',\omega-\omega') \tag{3.23}$$

where the correlation function of $\vec{v}$ is given by

$$\langle \vec{v}(\vec{k},\omega)\vec{v}^*(\vec{k}',\omega')\rangle = (2\pi)^4\,\overleftrightarrow{S}^{vv}(\vec{k},\omega)\,\delta(\vec{k}-\vec{k}')\,\delta(\omega-\omega') \tag{3.24}$$

For an isotropic incompressible fluid we furthermore have

$$\overleftrightarrow{S}(\vec{k},\omega) = \left(1-\frac{\vec{k}\vec{k}}{k^2}\right)S^{vv}_{tr}(\vec{k},\omega) \tag{3.25}$$

Substituting eq. (3.25) into eq. (3.6) we therefore have

$$D_1(\vec{k},\omega) = k^{-2}(2\pi)^{-4}\int d\vec{k}'d\omega'\,(k^2-k'^{-2}(\vec{k}'\cdot\vec{k})^2)\,S^{vv}_{tr}(\vec{k}',\omega')\,G_0(\vec{k}-\vec{k}',\omega-\omega') \tag{3.26}$$

or in $\vec{k},t$ representation

$$D_1(\vec{k},t) = k^{-2}(2\pi)^{-3}\int d\vec{k}'\,(k^2-k'^{-2}(\vec{k}'\cdot\vec{k})^2)\,S^{vv}_{tr}(\vec{k}',t)\,G_0(\vec{k}-\vec{k}',t) \tag{3.27}$$

This is the memory kernel for the diffusion process.

The coefficient $D_1(\vec{k},\omega)$ as given in eq. (3.26) will in general diverge because we use a continuum description of the system. One therefore has to impose a cut off k_0 in k space where k_0 is of the order of an inverse molecular length. The coefficient $D_1(\vec{k},t)$ does not suffer from this drawback, however the answer will only be reliable for sufficiently small k and sufficiently large t. As a final remark we note that $G_0(\vec{k},t)$ is according to eq. (3.9) given by

$$G_0(\vec{k},t) = -\frac{1}{2\pi}\int d\omega\, e^{-i\omega t}(i\omega-D_0k^2)^{-1} = \begin{cases} 0 & \text{if } t<0 \\ e^{-k^2D_0t} & \text{if } t\geq 0 \end{cases} \tag{3.28}$$

and therefore satisfies a causality condition. Consequently $D_1(k,t)$ satisfies a similar causality condition.

3.3 The long time behaviour of the diffusion memory kernel, and the $\vec{k}, \omega$ dependent renormalized diffusion constant

In this section we will first evaluate the memory kernel $D_1(k,t)$ as given in eq. (3.27). For this purpose we shall first use the fluid correlation function, as found from linearized hydrodynamics with constant coefficients. We will come back to the more general case at the end of this section. The correlation function S_{tr}^{vv} for hydrodynamic values of k and t is then given by [15]

$$S_{tr}^{vv}(k,t) = \rho_0^{-1} k_B T_0 \exp(-\nu k^2 |t|) \tag{3.29}$$

where k_B is Boltzmann's constant and ν the kinematic viscosity.

Substitution of eqs. (3.28) and (3.29) into eq. (3.27) yields

$$D_1(k,t) = \begin{cases} 0 & \text{if } t<0 \\ \frac{1}{2}\rho_0^{-1} k_B T_0 (2\pi)^{-2} \int_{-\infty}^{\infty} dk' \int_{-1}^{1} d\xi\, k'^2 (1-\xi^2) \exp\left[-t\nu k'^2 - tD_0(k'^2+k^2-2kk'\xi)\right] & \text{if } t>0 \end{cases} \tag{3.30}$$

Performing the integrals one obtains for hydrodynamic times and wave vectors

$$D_1(k,t) = \frac{2}{3}\rho_0^{-1} k_B T_0 \left[4\pi t(\nu + D_0)\right]^{-3/2} f\left(\frac{k^2 + D_0^2}{\nu + D_0}\right) \exp\left(-\frac{k^2 D_0 \nu t}{\nu + D_0}\right) \quad \text{if } t>0 \tag{3.31}$$

where

$$f(\alpha) = \frac{3}{2}\alpha^{-1}\left[1 + \frac{i}{2}\sqrt{\pi}\,\alpha^{-1/2} e^{-\alpha} \operatorname{erf}(i\sqrt{\alpha})\right] \tag{3.32}$$

The notation erf indicates the error function. Furthermore $f(\alpha)$ is positive and real for positive values of α. For small values of α

$$f(\alpha) = 1 - \frac{2}{5}\alpha + O(\alpha^2) \tag{3.33}$$

For large values of α

$$f(\alpha) = \frac{3}{4}\alpha^{-2}\left[1+2\alpha\right] + O(\alpha^{-3}) \tag{3.34}$$

For $k = 0$ one finds the usual form

$$D_1(0,t) = \frac{2}{3}\rho_0^{-1} k_B T_0 \left[4\pi t(\nu + D_0)\right]^{-3/2} \quad , \; t > 0 \tag{3.35}$$

For k larger than zero one can distinguish two regions in which the time dependence is essentially different

$$D_1(k,t) = \begin{cases} \frac{2}{3}\rho_0^{-1} k_B T_0 \left[4\pi t(\nu + D_0)\right]^{-3/2} & \text{if } \; 0 \leq \frac{k^2 t D_0}{D_0+\nu} \max\{\nu, D_0\} << 1 \\ \frac{3}{4\pi}\frac{k^2 D_0^2}{(D_0+\nu)^2}\rho_0^{-1} k_B T\left[4\pi t(\nu + D_0)\right]^{-1/2} \exp\left(-\frac{k^2 \nu D_0 t}{\nu + D_0}\right) & \text{if } \; \frac{k^2 t D_0^2}{D_0+\nu} >> 1 \end{cases} \tag{3.36}$$

In the general case when the transport coefficients of the fluid are wave vector and frequency dependent, eq. (3.29) is still valid asymptotically for large times using the values of the transport coefficients in the hydrodynamic limit

$$\left(\lim_{\omega\to 0}\lim_{k\to 0}\right)$$

These transport coefficients may then also be interpreted as the renormalized coefficients of the fluid. All the results derived in this section for the long time tails are therefore still valid asymptotically in this order of the expansion (3.21). In a subsequent section we will discuss the behaviour of higher order terms in eq. (3.21). These contain, it turns out, next to faster decaying terms, contributions to the asymptotic $t^{-3/2}$ behaviour. It should also be mentioned that the $t^{-7/4}$ contribution to the long time behaviour of $D(0,t)$ found by Pomeau [8] and Ernst and Dorfmann [10] would follow if one takes into account a term proportional to $k^{\frac{1}{2}}$ in the viscosity of the fluid, as found by these authors.

We shall now calculate the diffusion coefficient $D(k,\omega)$ as given in eq. (3.26) by choosing a suitable molecular cut off wave vector k_0. The correlation function in the (k,ω) representation is

$$S_{tr}^{vv}(k,\omega) = 2\,\frac{k_B T_0}{\rho_0}\,\frac{\nu k^2}{\omega^2 + \nu^2 k^4} \quad , \tag{3.37}$$

the Fourier transform of eq. (3.30). Substitution of eqs. (3.9) and (3.37) into eq. (3.26) yields

$$D_1(k,\omega) = \frac{k_B T_0}{4\pi^3 \rho_0} \int_0^{k_0} dk' \int_{-1}^{1} d\xi \int_{-\infty}^{\infty} d\omega' \, k'^2 (\xi^2 - 1) \frac{\nu k'^2}{\omega'^2 + \nu^2 k'^4} \left[i(\omega - \omega') - D_0 (k^2 + k'^2 - 2kk'\xi) \right]^{-1} \tag{3.38}$$

If one evaluates the integral in eq. (3.38) one obtains

$$D_1(k,\omega) = \frac{k_B T_0}{3\pi^2 (D_0 + \nu) \rho_0} \left\{ k_0 - \frac{3}{8} \pi (D_0 + \nu)^{-1/2} (D_0 k^2 - i\omega)^{1/2} f(r) \right\} \tag{3.39}$$

with

$$r \equiv \left(\frac{D_0}{D_0 + \nu} \right)^{1/2} \left(1 - \frac{i\omega}{D_0 k^2} \right)^{-1/2} \text{ and } f(r) \equiv \left(1 - \frac{1}{2} r^2\right) \sqrt{1 - r^2} + \frac{1}{2} r^{-3} \arcsin r \tag{3.40}$$

For small values of $|r|$, which is the case if $D_0 k^2 << |\omega|$, one has

$$f(r) = \frac{4}{3} - \frac{2}{5} r^2 + \mathcal{O}(r^4) \tag{3.41}$$

so that

$$D_1(0,\omega) = \frac{k_B T_0}{3\pi^2 (D_0 + \nu) \rho_0} \left\{ k_0 - \frac{1}{2} \pi (D_0 + \nu)^{-1/2} (-i\omega)^{1/2} \right\} \tag{3.42}$$

Furthermore one finds that

$$D_1(k,0) = \frac{k_B T_0}{3\pi^2 (D_0 + \nu) \rho_0} \left[k_0 - \frac{3}{8} \pi k \left(\frac{D_0}{D_0 + \nu} \right)^{1/2} f\left(\left\{ \frac{D_0}{D_0 + \nu} \right\}^{1/2} \right) \right] \tag{3.43}$$

In the zero frequency and wave vector limit we find

$$D_1(0,0) = \lim_{\omega \to 0} \lim_{k \to 0} D_1(k,\omega) = \lim_{k \to 0} \lim_{\omega \to 0} D_1(k,\omega) = \frac{k_B T_0 k_0}{3\pi^2 (D_0 + \nu) \rho_0} \tag{3.44}$$

If the tagged particle is sufficiently large so that $D_0 = 0$, eq. (3.44) gives, with $\eta = \nu \rho_0$ the viscosity,

$$D(0,0) = \frac{k_B T_0 k_0}{3\pi^2 \eta} \tag{3.45}$$

This is the Stokes-Einstein law if one chooses the cut off equal to

$$k_0 = \pi(\text{diameter})^{-1} = \tfrac{1}{2}\pi R^{-1} \qquad (3.46)$$

where R is the radius of the tagged particles. See for a proper derivation of this result ref. 30 .

3.4 Expansion of the renormalized diffusion coefficient in terms of the renormalized diffusion propagator

If one investigates the behaviour of the higher order terms in the expansion (3.21) one finds that they also contain contributions to the $t^{-3/2}$ long time behaviour, next to faster decaying terms. Some of these contributions can easily be resummed and then lead for $k = 0$ to an asymptotic time behaviour

$$D^r(k=0,t) = \frac{2}{3}\rho_0^{-1} k_B T_0 \left[4\pi t(\nu + D_0 + D_1(k=0,\omega=0))\right]^{-3/2} \qquad (3.47)$$

where D^r is the resummed part of D and where $D_1(k=0, \omega=0)$ is given by the second order approximation to $D(k,\omega)$, eq. (3.44).

Formula (3.41) suggests that summation over all the contributions to the $t^{-3/2}$ behaviour leads to the asymptotic formula

$$D(k=0,t) = \frac{2}{3}\rho_0^{-1} k_B T_0 \left[4\pi t(\nu + D(k=0,\omega=0))\right]^{-3/2} \qquad (3.48)$$

containing the full rather than the bare diffusion coefficient. Eq. (3.48) could be obtained by extensive partial resummation as is in fact done by Kawasaki [4] in the context of the mode-mode coupling theory. We shall obtain this result here by an alternative method, without the use of resummation techniques.

In order to proceed with our analysis we rewrite the full diffusion propagator eq. (3.18) in the form

$$G = -(i\omega - k^2 D)^{-1} = G_0 - G_0 k^2 (D - D_0) G = \left[1 + G_0 k^2 (D - D_0)\right]^{-1} G_0 \qquad (3.49)$$

With the help of this equality one can rewrite after some straight forward algebra the general formula (3.16) for the diffusion coefficient in the form

$$D-D_0 = k^{-2}\left\langle i\vec{k}\cdot\vec{v}\left[1-G\left\{k^2(D-D_0)-i\vec{k}\cdot\vec{v}\right\}\right]^{-1}\right\rangle\cdot\left\langle\left[1-G\left\{k^2(D-D_0)-i\vec{k}\cdot\vec{v}\right\}\right]^{-1}\right\rangle^{-1} \quad (3.50)$$

The right hand side can be expanded in powers of correlation functions of $\vec{v}$ and of $(D-D_0)k^2$. The quantity $(D-D_0)k^2$ can then be eliminated from the r.h.s. by iteration. The result is an expansion for $(D-D_0)k^2$ in terms of correlation functions of $\vec{v}$, but always contains the full diffusion propagator rather than the bare one. Up to second order one finds

$$D-D_0 = +k^{-2}\langle\vec{k}\cdot\vec{v}\, G\, \vec{k}\cdot\vec{v}\rangle \quad (3.51)$$

This expression which has again the well known mode-mode coupling form gives, for the asymptotic long time behaviour, after an analysis similar to the one in section 3.3 the result (3.48).

If one analyses the contributions of fourth order in $\vec{v}$ (assuming $\vec{v}$ to be a Gaussian process for hydrodynamic k,ω, so that odd orders in $\vec{v}$ vanish) one finds that these decay for $k = 0$ as t^{-2}. The sixth order contributions, which have also been analysed, decay as $t^{-5/2}\ln t$ For details of this analysis we refer to reference [3].

To summarize, it would therefore seem that an expansion starting from formula (3.50), in terms of the fluid correlation functions and the renormalized diffusion propagator, is indeed a systematic one for the long time behaviour and that eq. (3.48) is asymptotically correct to all orders.

REFERENCES

1) Bedeaux, D. and Mazur, P., Physica 67 (1973) 23.
2) Bedeaux, D. and Mazur, P., Physica (1974), to be published.
3) Mazur, P. and Bedeaux, D., Physica (1974), to be published.
4) Kawasaki, K., Ann. Phys. 61 (1970) 1 and references therein.
5) Kadanoff, L.P. and Swift, J., Phys. Rev. 166 (1968) 89.
6) Alder, B.J. and Wainwright, T.E., Phys. Rev. Lett. 18 (1967) 988, phys. Rev. A1 (1970) 18.
7) Ernst. M.H., Hauge, E.H. and van Leeuwen, J.M.J., Phys. Rev. Lett. 25 (1970) 1254; Phys. Rev. A4 (1971) 2055.
8) Pomeau, Y., Phys. Rev. A5 (1972) 2569; A7 (1973) 1134.
9) Dormfman, J.R. and Cohen, E.G.D., Phys. Rev. Lett. 25 (1970) 1257; Phys. Rev. A6 (1972) 776.
10) Ernst, M.H. and Dorfman, J.R., Phys. Lett. 38A (1972) 269; Physica 61 (1972) 157.
11) Résibois, P., preprint.
12) Zwanzig, R. and Bixon, M., Phys. Rev. A2 (1970) 2005.
13) Keyes, T. and Oppenheim, I., Phys. Rev.A7 (1973) 1384.
14) Kawasaki, K., Phys. Lett. 32A (1970) 379; 34A (1971) 12.
15) Zwanzig, R., Statistical Mechanics, Proc. Sixth I.U.P.A.P. Conf. on Statistical Mechanics (The University of Chicago Press 1972).
16) Landau, L. and Lifshitz, E., Fluid Mechanics Pergamon Press New York (1959).
17) Van Kampen, N.G., preprints.
18) Smoluchowski, M., Ann. Physik 25 (1908) 205.
19) Einstein, A., Ann. Physik 33 (1910) 1275.
20) Yvon, J., Recherches sur la Théorie cinétique des liquides, Hermann et Cie. (Paris 1937).
21) Zimm, B.H., J. chem. Phys. 13 (1945) 141.
22) Fixmann, M., J. chem Phys. 23 (1955) 2074.
23) Hoek, H., Thesis (Leiden, 1939).
24) Rosenfeld, L., Theory of Electrons, North-Holland Publ. Comp. (Amsterdam, 1951).
25) Mazur, P. and Mandel, M., Physica 22 (1956) 299.
26) Zernike, F., Proc. Acad. Sci. Amsterdam 18 (1916) 1520.
27) Landau, L.D. and Lifshitz, E.M., Statistical Physics, Pergamon Press (London-Paris, 1958).
28) Larsen, S.V., Mountain, R.D. and Zwanzig, R., J. chem. Phys. 42 (1965) 2187.
29) Kim, Shoon K. and Mazur, P., Physica 71 (1974) 579.
30) Bedeaux, D. and Mazur, P., to be published.
31) Giterman, M.Sh., and Gorodetskii, E.E., JETP 29 (1969) 347, 30 (1970) 348.

IRREVERSIBILITY OF THE TRANSPORT EQUATIONS

J. Biel

Facultad de Ciencias
Universidad de Valencia
Valencia, Spain

I. INTRODUCTION

II. GENERAL REMARKS ON IRREVERSIBILITY

2. Statistical description of a system
3. Natural motion in phase space
4. Entropy and information
5. Coarse-grained probability density
6. Generalised η-theorem

III. THE IRREVERSIBILITY OF THE BOLTZMANN TRANSPORT EQUATION

7. The Boltzmann equation
8. Boltzmann's H-theorem
9. The BBGKY-hierarchy
10. The thermodynamic limit
11. A new hierarchy
12. Introduction of the irreversibility
13. Binary collisions
14. Molecular chaos

IV. THE IRREVERSIBILITY OF OTHER EQUATIONS

15. The Prigogine-Brout equation
16. The Langevin and Fokker-Planck equations
17. The Choh-Uhlenbeck equation
18. Final remarks

REFERENCES

I. INTRODUCTION

It is well-known that the official, common language for physics is not English but "broken English". You have already all realised that the language which I speak is, instead, "awfully-broken English". I apologize for this and I hope that, nevertheless, you will understand me.

The aim of my talks here will be to remind you of a few fundamental ideas on the special character of the transport equations, i.e., the equations governing the irreversible behaviour of macroscopic systems. I will not give an account here of the results obtained recently by the Brussels group, which can be regarded as a rigorous mechanical theory of irreversibility. These will be presented to you by Professors Prigogine and Mayné. Instead, I will deal with the ideas underlying the various derivations of transport equations as they have appeared in the literature during the last fifteen years. I hope that my talks will make the comparison of these approaches to irreversibility easier for you.

The question of whether the world is actually reversible or not, is irrelevant here. Irreversibility may be seen as an objective property of Nature, and thus one could think, for instance, that some ten thousand millions years ago a creation process was begun and that since that time the world has been evolving irreversibly towards a so-called "Ω-point". On the other hand, irreversibility may also be considered as a "human illusion" according to von Smoluchowski. The important point for us is that, whenever we observe macroscopic systems, the results which we find are the same as if the systems were behaving irreversibly. This irreversible behaviour is expressed by means of equations which are said to be "irreversible". These are for instance the phenomenological equations.

In thermodynamics, irreversibility is introduced for an isolated system through the inequality

$$dS/dt > 0 \ . \tag{1.1}$$

This inequality means that every real process which takes place in an isolated system increases its entropy. According to Gibbs, entropy is defined in statistical mechanics as

$$S = -k\,\overline{\ln \varrho} \qquad (1.2)$$

where k is Boltzmann's constant and ϱ is the probability density in phase space associated with the system. The bar denotes the corresponding ensemble average. Since the ensemble probability density is in general time-dependent, the entropy defined in formula (1.2) seems also to be time-dependent. However, the probability density satisfies the Liouville theorem

$$d\varrho/dt = 0 \qquad (1.3)$$

and with the use of (1.3) it is very easy to prove that the definition (1.2) implies

$$dS/dt = 0 \,. \qquad (1.4)$$

Expressions (1.1) and (1.4) are in contradiction because the latter means that it is possible for an isolated system not in equilibrium to evolve keeping its entropy constant. Indeed, to obtain equation (1.1) we have not made use of any steady state condition for ϱ. The contradiction between (1.1) and (1.4) arises as a consequence of the use of Liouville's theorem (1.3).

Let us now review briefly the methods by means of which statistical mechanics is able to describe macroscopic systems from microscopic laws. This review will aid us in understanding the reasons behind the contradiction involved in the statistical definition of entropy and, moreover, will give us the procedure for removing the difficulties. We will see that the mechanical evolution involves no irreversibility at all. In order to obtain irreversible equations it will be necessary to interrupt the dynamical evolution at every instant of time.

After that, we will deal with the most famous transport equation, the Boltzmann equation, and we will try to investigate how irreversibility is involved in it. We will review one of the methods (due to Van Hove) for introducing irreversibility into reversible equations and we will see how this method is applied to derive some irreversible equations, namely: the Prigogine-Brout, the Langevin, the Fokker-Planck and the Choh-Uhlenbeck equations.

Perhaps some if not all of the topics of my talks are already familiar to most of you. However I think that it is sometimes necessary to review the things which we know from a unified point of view. And it is now a good time to do this because Professors Prigogine and Mayné will deal with the same problems from a different viewpoint. In any case, I am sure that you will be amused when you hear me present matters well known to you in a way that will indeed be worse than if you had done it yourselves.

II. GENERAL REMARKS ON IRREVERSIBILITY

2. Statistical description of a system

Perhaps the most characteristic feature of the microscopic and macroscopic descriptions of a large system is the amount of information that each one of them gives us about the system. Indeed, the exact microscopic description involves an enormously large number of parameters(of the order of the number of internal degrees of freedom of the system); on the other hand, for a macroscopic description one only needs to specify the values of a small number of variables (such as volume, energy, etc.). We say that by means of the macroscopic description one specifies macrostates of the system and we give the name of microstates to the different states described microscopically. So it is clear that for a particular macrostate of the system there are many microstates compatible with it. In other words, by passing from the microscopic description to the macroscopic one, one loses information.

The Gibbs' ensemble method for treating statistical mechanics consists in considering for each macrostate of the system an ensemble of equivalent independent systems, all of them in the same macrostate but realising each one of them through a different microstate among those compatible with the given macrostate. This ensemble is represented in the phase space of the system as a cloud of representative points and is characterised by the so-called density in phase space, which for a given point of phase space at a given instant of time is proportional to the number of representative points around the given point in phase space at that time. Thus one can define the probability of a microstate at a given time as being proportional to the density in phase space for the phase point corresponding to the given microstate. In order to associate with each macrostate of the system a density in phase space it is necessary to introduce an "a priori" statistical postulate. The most natural one is that the relative probabilities of finding the system in specified regions of the phase space are proportional to the volume of these regions if they represent equally well the information which we have about the system.

Let us now emphasize that the statistical description, to which we are here referring, reflects the lack of microscopic information which we have about our system. Indeed, instead of specifying our system by means of its microstate (as would be the case in the mechanical or microscopic description), we specify only the probability of a microstate; that is, we deal with a set of many points in a small region of phase space instead of with only one point.

3. Natural motion in phase space

As a consequence of the equations of motion, each one of the representative points in phase space is moving throughout the accessible region of this space. This motion will be referred to as the natural motion. Therefore the number of representative points in a fixed region of phase space will in general change with time.

Since representative points can neither be created nor annihilated in the course of motion, their flow in phase space must be conservative, that is

$$\frac{\partial \rho}{\partial t} = -\operatorname{div}(\vec{v}\rho) = -\vec{v}\cdot \operatorname{grad}\rho - \rho \operatorname{div}\vec{v}. \tag{3.1}$$

Here

$$\rho = \rho(t) = \rho(\vec{r}^N, \vec{p}^N; t) = \rho(\vec{r}_1, \ldots, \vec{r}_N, \vec{p}_1, \ldots, \vec{p}_N; t) \tag{3.2}$$

is the probability density in phase space, which depends on the dynamical variables of the system composed of N point-particles as well as on the time; and

$$\vec{v} = (\dot{\vec{r}}^N, \dot{\vec{p}}^N), \ \operatorname{div}\vec{v} = \sum_{j=1}^{N}\left(\frac{\partial}{\partial \vec{r}_j}\cdot\dot{\vec{r}}_j + \frac{\partial}{\partial \vec{p}_j}\cdot\dot{\vec{p}}_j\right), \operatorname{grad}\rho = \sum_{j=1}^{N}\left(\frac{\partial \rho}{\partial \vec{r}_j} + \frac{\partial \rho}{\partial \vec{p}_j}\right). \tag{3.3}$$

Now, using Hamilton's equations

$$\dot{\vec{r}}_j = \frac{\partial H}{\partial \vec{p}_j}, \quad \dot{\vec{p}}_j = -\frac{\partial H}{\partial \vec{r}_j} \quad (j = 1, \ldots, N) \tag{3.4}$$

where $H = H(\vec{r}^N, \vec{p}^N)$ is the Hamiltonian function of the system, we get

$$\operatorname{div}\vec{v} = 0 \tag{3.5}$$

and therefore

$$\frac{d\rho}{dt} = \frac{\partial \rho}{\partial t} + \sum_{j=1}^{N}\left(\frac{\partial \rho}{\partial \vec{r}_j}\cdot\dot{\vec{r}}_j + \frac{\partial \rho}{\partial \vec{p}_j}\cdot\dot{\vec{p}}_j\right) = \frac{\partial \rho}{\partial t} + \vec{v}\cdot\operatorname{grad}\rho = 0. \tag{3.6}$$

This equation means that the density in phase space (proportional to ρ) remains constant around a moving point; that is, the probability density is an integral of the motion. This is one of the forms of Liouville's theorem, which states that the cloud of representative points moves like an incompressible fluid. By introducing Liouville's operator

$$L = -i\vec{v}\cdot\text{grad} = i\sum_{j=1}^{N}\left(\frac{\partial H}{\partial \vec{r}_j}\cdot\frac{\partial}{\partial \vec{p}_j} - \frac{\partial H}{\partial \vec{p}_j}\cdot\frac{\partial}{\partial \vec{r}_j}\right) \tag{3.7}$$

equation (3.6) becomes

$$\frac{\partial \rho}{\partial t} = -iL\rho \tag{3.8}$$

which is known as Liouville's equation and provides us with the time variation of the probability density for a fixed point in phase space. It is important to emphasize that this variation of $\rho(t)$ is entirely due to the natural motion of the representative points, that is to a mechanical motion. It is for this reason that one can say that equations (3.4) and (3.8) are equivalent, since both equations describe the natural motion of the system.

Hamilton's equations must be used to find the microstate of the system at a time t if its microstate at time $t=0$ is known; whereas the Liouville equation must be used to find the probability at time t of a given microstate if its probability at time $t=0$ is known. In both cases the same dynamical problem must be solved.

Natural motion may also be seen as a set of transformations which for any time t map the phase space onto itself. These transformations are induced by the equations of motion (3.4) and have the property (equivalent to Liouville's theorem) that they conserve volume, that is, the volume of a region $d\Gamma_0 = d\vec{r}_0^N d\vec{p}_0^N$ of phase space is equal to the volume of the transformed region $d\Gamma_t = d\vec{r}_t^N d\vec{p}_t^N$. This property is easily proved by writing

$$d\Gamma_0 = J\, d\Gamma_t \tag{3.9}$$

where J, the jacobian of the transformation, is equal to unity by virtue of Hamilton's equations.

Since the density in phase space is proportional to the probability density, the number of representative points at time t_0 in a given infinitesimal region of volume $d\Gamma_0$ around $(\vec{r}_0^N, \vec{p}_0^N)$ is proportional to $\rho(\vec{r}_0^N, \vec{p}_0^N; t_0)d\Gamma_0$. Now $\rho(\vec{r}_t^N, \vec{p}_t^N; t)d\Gamma_t$ is proportional to the number of representative points in the transformed region, which at time t contains the representative points of the original region.

Hence

$$\rho(\vec{r}^N_0, \vec{p}^N_0; t_0)d\Gamma_0 = \rho(\vec{r}^N_t, \vec{p}^N_t; t)d\Gamma_t \ . \tag{3.10}$$

Since the volumes $d\Gamma_0$ and $d\Gamma_t$ are equal, one returns to the preceding form (3.6) of Liouville's theorem: $d\rho/dt = 0$. From this latter derivation of Liouville's theorem it is also evident that the equation for the variation of the probability density is equivalent to Hamilton's equations.

Formally the solution of Liouville's equation (3.8) can be written as

$$\rho(\vec{r}^N, \vec{p}^N; t) = e^{-itL} \rho(\vec{r}^N, \vec{p}^N; 0) . \tag{3.11}$$

Here it is quite easy to see that $\rho(t)$ involves the same lack of microscopic information as $\rho(0)$. Indeed the operator e^{-itL} with the aid of which one passes from $\rho(0)$ to $\rho(t)$ is a mechanical one (see eq. (3.7)) and no loss of information is involved.

Therefore, the time variation of ρ as expressed by equation (3.8) implies no loss of information at all. The initial information which we have about our system is preserved during its natural motion.

4. Entropy and information

We have said that when one passes from the microscopic description to the macroscopic one, one loses information. Let us now look for a quantity measuring the lack of information which we have about our system when its state is described in terms of ρ. Using information theory it is possible to show that the lack of information is given by

$$I = -k \int_\Gamma \rho \ln \rho \, d\Gamma \ . \tag{4.1}$$

For systems in equilibrium this expression is identical to the statistical entropy if k is Boltzmann's constant. Indeed, for an isolated system in equilibrium, ρ is uniform throughout the accessible region Ω of phase space as a consequence of the postulate of equal a priori probabilities. Then: $\rho = 1/W$, where $W = \int_\Omega d\Gamma$ is a measure of the number of microstates in the region Ω. Thus

$$I = k \int_\Omega \frac{1}{W} \ln W \, d\Gamma = k \ln W . \tag{4.2}$$

The identification of I with the entropy can be done directly since Boltzmann's definition of the entropy in a microcanonical ensemble (which is that corresponding to the isolated system in equilibrium) is

$$S = k \ln W . \tag{4.3}$$

We may also see from expression (4.1) that, among the various ensembles which we could choose for representing an isolated system in equilibrium, the microcanonical ensemble gives us the least amount of microscopic information about the system.

To do this we calculate the maximum of (4.1) for the variable ρ using the condition

$$\int_\Omega \rho \, d\Gamma = 1 . \tag{4.4}$$

Thus:

$$\delta I = -k \int_\Omega (1 + \ln\rho)\, \delta\rho \, d\Gamma = -k \int_\Omega (1 + \alpha + \ln\rho)\, \delta\rho \, d\Gamma = 0$$

since from (4.4)

$$\alpha \int_\Omega \delta\rho \, d\Gamma = 0 . \tag{4.5}$$

Hence, on Ω one has:

$$\rho = e^{-(\alpha+1)} = \left[\int_\Omega d\Gamma \right]^{-1} = \frac{1}{W} \tag{4.6}$$

which is the probability density of the microcanonical ensemble. Therefore expression (4.3) gives the least amount of information about the isolated system in equilibrium.

Let us now assume that our system is not in equilibrium, but represented by a time-dependent probability density in phase space $\rho(\vec{r}^N, \vec{p}^N; t)$. The information which we have about such a system at time t_0 is given by

$$I = -k \int_\Gamma \rho(\vec{r}_0^N, \vec{p}_0^N; t_0) \ln \rho(\vec{r}_0^N, \vec{p}_0^N : t_0)\, d\Gamma_0 . \tag{4.7}$$

Using Liouville's theorem,

$$\rho(\vec{r}_0^N, \vec{p}_0^N; t_0) = \rho(\vec{r}_t^N, \vec{p}_t^N ; t), \tag{4.8}$$

we also have

$$I = -k\int_{\Gamma} \rho(\vec{r}^{\,N}_t, \vec{p}^{\,N}_t ; t) \ln \rho(\vec{r}^{\,N}_t, \vec{p}^{\,N}_t ; t) d\Gamma_t . \tag{4.9}$$

We have written here $d\Gamma_t$ instead of $d\Gamma_0$, because in doing so the integral, which is taken over the whole phase space, does not change its value. We now change the variables of integration and obtain

$$I = -k\int_{\Gamma} \rho(\vec{r}^{\,N}_0, \vec{p}^{\,N}_0 ; t) \ln \rho(\vec{r}^{\,N}_0, \vec{p}^{\,N}_0 ; t) d\Gamma_0 . \tag{4.10}$$

Comparison of (4.7) and (4.10) shows that the information does not change with time as a consequence of the dynamical evolution of the system. This result is the same as what we obtained by looking into the formal solution of Liouville's equation (3.11).

In view of this result we conclude that the entropy of a system as defined by (4.1) cannot change during the evolution. This contradicts the laws of macroscopic systems. However, Boltzmann's definition of entropy (4.3) agrees with the Second Principle of Thermodynamics and gives correct results for systems in equilibrium.

It is clear that we might study the irreversible behaviour of a system from two different viewpoints. Consider, for instance, an isolated system in equilibrium with an internal constraint. With this system we associate a microcanonical ensemble, and we calculate the value of W and - by means of the definition (4.3) - the entropy of the system. We then remove the internal constraint and, after some time, the system reaches a new state of equilibrium. We associate now another microcanonical ensemble with the system and calculate the new value of W and the entropy in the new state of equilibrium by means of (4.3). It is well-known that the entropy of the new state is larger than the initial one:

$$S_f > S_i . \tag{4.11}$$

The reason for this is that the number of microstates compatible with the final macroscopic situation is much larger than the number of microstates compatible with the initial macrostate. Considering expression (4.2), we conclude that there is more information in the initial situation than in the final one. We have lost information when we have changed the microcanonical ensemble.

We may also try to solve the same problem in a different way. Consider, as before, the isolated system in equilibrium with an internal constraint. Then the constraint is removed and a non-equilibrium ensemble, characterised by a time dependent probability density $\rho(t)$, is associated with the system. From what we have said until now, we know that our information about the system will not change with time. Therefore, the entropy will remain constant and we cannot describe the irreversible process which actually takes place in the isolated system by means of the variation of the density $\rho(t)$.

Irreversibility is therefore associated with a loss of information but this loss of information is not the one implied by the mechanical statistical description. The lack of microscopic information in the statistical description is entirely contained in $\rho(t)$; given by (4.1) and conserved during the natural motion of the system. To obtain irreversibility it is necessary to introduce an additional lack of information.

In the study of the irreversible behaviour of the isolated system which we did by means of the two microcanonical ensembles, the additional lack of information was introduced when we changed the representative equilibrium ensemble. On the other hand, the study of the same problem by means of a non-stationary ensemble has shown that the introduction of an additional lack of information is something that cannot be done by the system itself during its mechanical evolution.

Let us consider the problem once again in a slightly different way by assuming that our isolated system is in a given microstate m_0 at time t=0. At a later time, t, its microstate will be m_t . Consider now the set $\{m_0\}$ of all microstates compatible with the initial macrostate M_0 . It is evident that at time t the set $\{m_0\}$ will be transformed into the set $\{m_t\}$. Is the set $\{m_t\}$ the same as the set of all microstates compatible with the macrostate M_t of the system at time t ? Of course not, because the number $W(M_t)$ of microstates compatible with M_t must be larger than the number $W(M_0)$ of microstates compatible with M_0 . The number of microstates in the set $\{m_t\}$ is however equal to the number of microstates in $\{m_0\}$. Since the mechanical evolution cannot increase the number of microstates and, on the other hand, the information depends here on the number of possible microstates, one must conclude that the mechanical evolution never can imply irreversibility (loss of information).

5. Coarse-grained probability densities

The loss of information implied in the change from the microscopic description to the statistical is enough to derive the macroscopic equilibrium properties of large systems but, on the other hand, is not enough to give an account of irreversibility. The reason for this failure of the statistical description presented above is that the probability density in phase space (which replaces in this description the mechanical microstate of the system) is still too fine a quantity because its evolution is given by Liouville's equation which is a purely mechanical equation. We may think, therefore, that we could solve our difficulties if we describe our system by means of a less fine probability density - one which does not satisfy Liouville's equation.

To define this probability density we divide the phase space into fixed small cells and average the original probability density (which we shall refer to as the fine-grained probability density) over each one of the cells. Let γ_i be one of these cells and $\mu(\gamma_i)$ its volume:

$$\mu(\gamma_i) = \int_{\gamma_i} d\Gamma \, . \tag{5.1}$$

The average of $\varrho(t)$ on this cell is given by

$$P_i(t) = \frac{1}{\mu(\gamma_i)} \int_{\gamma_i} \varrho(t) \, d\Gamma \tag{5.2}$$

and will be referred to as coarse-grained probability density. We assume that the only information which we have about our system is represented by the coarse-grained probability density and therefore that the fine-grained probability density which would give us much more information, is unknown.

It is clear that

$$\int_{\gamma_i} P_i(t) \, d\Gamma = P_i(t)\mu(\gamma_i) = \int_{\gamma_i} \varrho(t) \, d\Gamma \tag{5.3}$$

because $P_i(t)$ is uniform in the cell γ_i . Moreover, if $\varrho(t)$ is normalized to unity

$$\sum_i P_i(t)\mu(\gamma_i) = \int_{\Gamma} \varrho(t) \, d\Gamma = 1 \, . \tag{5.4}$$

The summations over the cells can be replaced by integrations over the whole phase space in the form

$$\sum_i P_i(t)\mu(\gamma_i) = \int_\Gamma P(t)\,d\Gamma = 1 \tag{5.5}$$

where $P(t) = P(\vec{r}^N, \vec{p}^N; t)$ is proportional to the probability that the representative point of the system is in the cell $\gamma_i = \Delta\vec{r}^N \Delta\vec{p}^N$ at time t. If we now define the entropy

$$S = -k \int_\Gamma P(t) \ln P(t)\,d\Gamma, \tag{5.6}$$

in the same way as in (4.1), we may prove as we did for (4.1) that for an isolated system in equilibrium the expression for $P(t)$ which maximizes (5.6) is $P = 1/W$ on the accessible region of phase space, that is, the microcanonical ensemble. Actually, we may describe all <u>equilibrium</u> systems by means of $P(t)$ as well as by means of $\rho(t)$.

On the other hand, it is evident that $P_i(t)$ <u>does not verify the Liouville equation</u> since

$$\frac{\partial P_i(t)}{\partial t} = \frac{1}{\mu(\gamma_i)} \int_{\gamma_i} \frac{\partial \rho(t)}{\partial t}\,d\Gamma = \frac{1}{\mu(\gamma_i)} \int_{\gamma_i} (-iL\rho(t))\,d\Gamma \neq -iL P_i(t). \tag{5.7}$$

Since we do not now have the same mechanical evolution for $P_i(t)$ as we had for $\rho(t)$, we may hope that the entropy defined by (5.6) will be a time-dependent function which increases with time.

6. <u>Generalized η-theorem</u>

To study the time dependence of the above defined entropy, let us introduce the quantity

$$\eta(t) = \sum_i \mu(\gamma_i) P_i(t) \ln P_i(t). \tag{6.1}$$

The function $\eta(t)$ depends essentially on the size and character of the cells introduced in our description of the system. Now, since $P_i(t)$ is uniform within γ_i,

$$\eta(t) = \sum_i \int_{\gamma_i} P(t) \ln P(t)\,d\Gamma = \int_\Gamma P(t) \ln P(t)\,d\Gamma \tag{6.2}$$

and we see that the above defined entropy is

$$S = -k\,\eta(t). \tag{6.3}$$

A different form for the η-function can be obtained by using the definition (5.2) of the coarse-grained probability densities:

$$\eta(t) = \sum_i \left(\int_{\gamma_i} \rho(t)\, d\Gamma \right) \ln P_i(t) = \sum_i \int_{\gamma_i} \rho(t) \ln P_i(t)\, d\Gamma = \int_\Gamma \rho(t) \ln P(t)\, d\Gamma = \overline{\ln P(t)} \tag{6.4}$$

where we have again used the fact that $P_i(t)$ is uniform within γ_i and denoted with the bar the ensemble average.

Let us now assume that at time $t = t_0$ we have information about our system which allows us to assign a value $P_i(t_0)$ to the probability density of each cell γ_i. Since this is the only information which we have, we construct an ensemble of systems in such a manner that the representative points are uniformly distributed in each cell. The fine-grained probability density $\rho(t_0)$ characteristic of this ensemble is therefore equal to $P(t_0)$:

$$P(t_0) = \rho(t_0). \tag{6.5}$$

The value of the η-function at $t = t_0$ is thus

$$\eta(t_0) = \int_\Gamma P(t_0) \ln P(t_0)\, d\Gamma = \int_\Gamma \rho(t_0) \ln \rho(t_0)\, d\Gamma. \tag{6.6}$$

If $\rho(t_0)$ is the probability density corresponding to a stationary ensemble, $\partial\rho/\partial t = 0$ and ρ is constant in time within each cell γ_i. Therefore $P(t) = \rho(t)$ for all t and the description of the corresponding equilibrium system may be made equally well in terms of P or of ρ. However, if $\partial\rho/\partial t \neq 0$, the number of representative points within each fixed region of phase space changes with time and in general $P(t) \neq \rho(t)$. For this latter case let us choose an instant of time $t_1 > t_0$. Considering (6.4) the value of the η-function at t_1 is

$$\eta(t_1) = \int_\Gamma \rho(t_1) \ln P(t_1)\, d\Gamma. \tag{6.7}$$

Now since from Liouville's theorem we have

$$\frac{d}{dt} \int_\Gamma \rho(t) \ln \rho(t)\, d\Gamma = 0, \tag{6.8}$$

the difference between (6.6) and (6.7) may be written as

$$\eta(t_0)-\eta(t_1)=\int_\Gamma\left[\rho(t_1)\ln\rho(t_1)-\rho(t_1)\ln P(t_1)\right]d\Gamma$$

$$=\int_\Gamma\left[\rho(t_1)\ln\rho(t_1)-\rho(t_1)\ln P(t_1)-\rho(t_1)+P(t_1)\right]d\Gamma, \tag{6.9}$$

where the last equality follows from (5.3). The integrand Q of this expression is not negative because its derivative with respect to ρ,

$$\frac{\partial Q}{\partial\rho}=\ln\frac{\rho(t_1)}{P(t_1)}, \tag{6.10}$$

is positive if $\rho(t_1)>P(t_1)$, negative if $\rho(t_1)<P(t_1)$ and zero if $\rho(t_1)=P(t_1)$. Therefore, Q is minimal if $\rho(t_1)=P(t_1)$. But in that case, Q is also zero. Hence it will be positive if $\rho(t_1)\neq P(t_1)$. We conclude that if $\rho(t)$ is different from $P(t)$ we always have

$$\eta(t_0)>\eta(t)\quad\text{for}\quad t>t_0. \tag{6.11}$$

This result is very interesting since it tells us that the value of the η-function is larger at the initial time t_0 than at some later time t_1. As this function is proportional to the information, the inequality just derived allows us to conclude that in passing from t_0 to t_1 some information has been lost.

One might think that this η-theorem is an explanation of irreversibility, because it seems to predict a monotonic decreasing of η with time, i.e., in view of (6.3), a monotonic increasing of entropy. However, this is not exactly the case. To see this, we firstly analyse how the information has been lost in passing from t_0 to t_1.

We are describing our system by means of the coarse-grained probability densities $P(t)$. The only information which we have about the system is provided by $P(t)$, i.e., the average of $\rho(t)$ over each cell γ_i. In the case which we have considered our information is contained in $P(t_0)$ at t_0 and in $P(t_1)$ at t_1. At t_0 we have constructed an ensemble of systems according to our information and thus the description provided by this ensemble must be exactly the same as that provided by $P(t_0)$, so $P(t_0)=\rho(t_0)$. In this case, i.e., at time t_0 we know therefore $\rho(t_0)$. As time goes on,

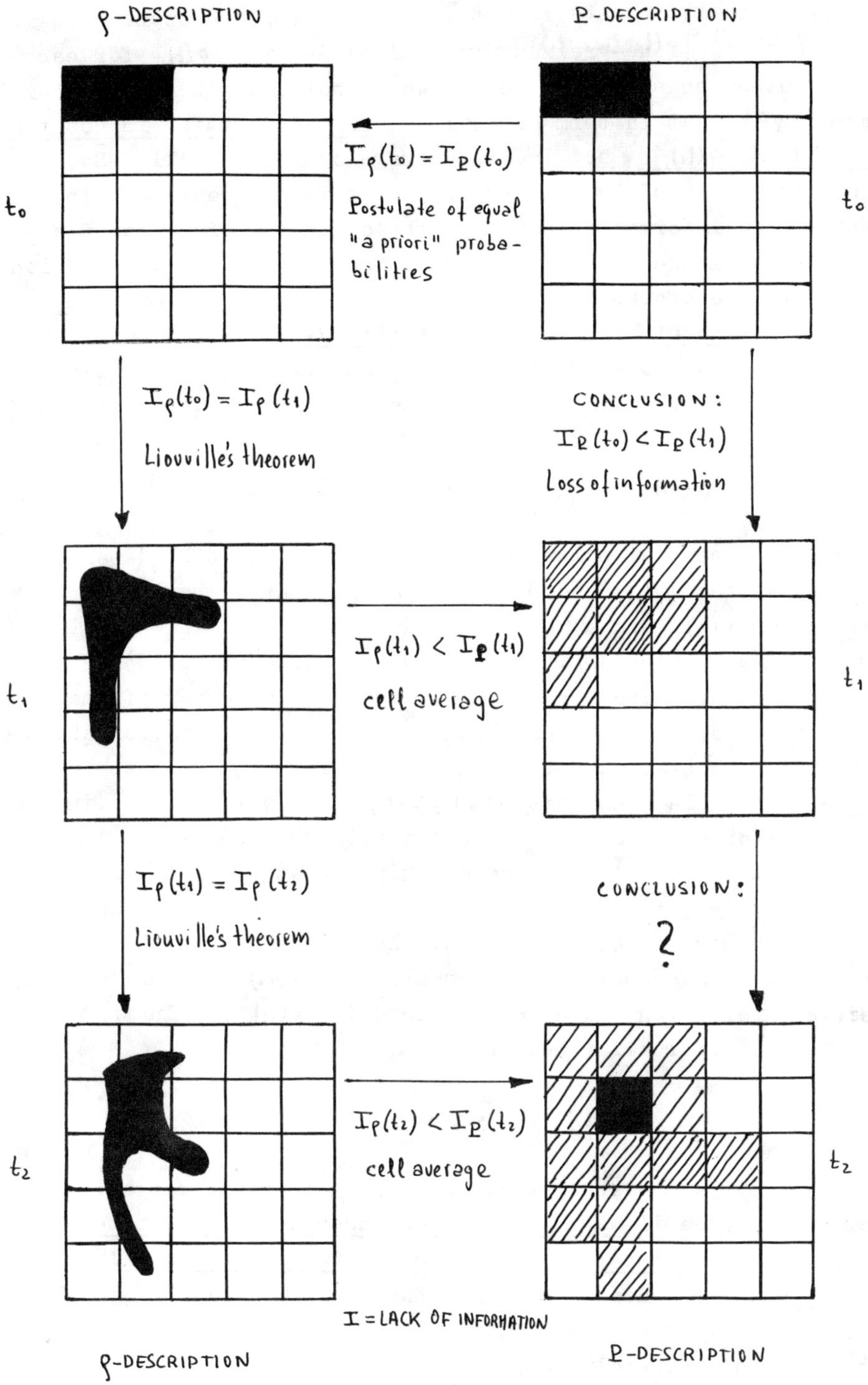
ρ-DESCRIPTION
P-DESCRIPTION
$I_\rho(t_0) = I_P(t_0)$
Postulate of equal "a priori" probabilities
t_0
t_0
$I_\rho(t_0) = I_\rho(t_1)$
Liouville's theorem
CONCLUSION:
$I_P(t_0) < I_P(t_1)$
Loss of information
$I_\rho(t_1) < I_P(t_1)$
cell average
t_1
t_1
$I_\rho(t_1) = I_\rho(t_2)$
Liouville's theorem
CONCLUSION:
?
$I_\rho(t_2) < I_P(t_2)$
cell average
t_2
t_2
I = LACK OF INFORMATION
ρ-DESCRIPTION
P-DESCRIPTION

the representative points move throughout the phase space changing the value of the fine-grained probability density $\rho(t)$ for each fixed point of phase space. Since we do not describe the system in terms of $\rho(t)$ but in terms of $P(t)$, our information is not that contained in $\rho(t_1)$ but in its average over each cell. So $P(t_1) \neq \rho(t_1)$ and the information contained in our description is less than the information given by the ensemble. But from Liouville's theorem the latter is exactly the same as the information at time t_0; therefore, in the coarse-grained description there is less information at time t_1 than at time t_0.

Let us note, however, that we cannot rigorously establish an inequality such as (6.11) for two times t_1 and t_2 different from t_0:

$$\eta(t_1) \overset{?}{>} \eta(t_2) \quad \text{for} \quad t_2 > t_1 . \tag{6.12}$$

Indeed, the essential point for the derivation of (6.11) was the equality between $\rho(t_0)$ and $P(t_0)$ and at present we have not a similar equality for $\rho(t_1)$ and $P(t_1)$.

If we want to establish an inequality such as (6.12) we must "forget" our earlier description at time t_0 by means of the ensemble of systems, and, by assuming that all possible information about our system is now contained not in $\rho(t_1)$ but in $P(t_1)$, construct a different ensemble such that $\rho'(t_1) = P(t_1)$. So we are now in the same situation as before and we can obtain the inequality (6.12). It is important to "forget" the description at time t_0, because if we fail to do so, the whole possible information at t_1 would not be contained in $P(t_1)$ because we would have in addition the information that at t_0 the system was represented by $\rho(t_0)$. The totality of possible information would be contained in $\rho(t_1)$. In this manner, we could prove for all times that

$$d\eta(t)/dt < 0 . \tag{6.13}$$

Note however that this behaviour of the η-function can only be rigorously proved if we introduce new ensembles at all times. This has the effect of interrupting at every time the dynamical evolution of the probability density. This change of ensembles corresponds to the change of microcanonical ensembles in the equilibrium statistical description of the evolution of an isolated system (see inequality (4.11)).

I will now present two different methods for <u>interrupting at every time</u> the dynamical evolution of a mechanical system and so introducing a loss of information, that is, irreversibility.

III. THE IRREVERSIBILITY OF THE BOLTZMANN TRANSPORT EQUATION

7. The Boltzmann equation

I will deal firstly with some ideas underlying the <u>traditional method</u> for deriving the Boltzmann transport equation

$$\frac{\partial f}{\partial t} + \vec{v}\cdot\frac{\partial f}{\partial \vec{r}} + \vec{a}\cdot\frac{\partial f}{\partial \vec{v}} = \iiint (f'f'_1 - ff_1)\, g\, b\, db\, d\varepsilon\, d\vec{v}_1 \,. \tag{7.1}$$

The distribution function $f(\vec{r},\vec{v},t)$ is defined as

$$f(\vec{r},\vec{v};t) = \frac{n(t)}{\delta\vec{r}\,\delta\vec{v}} \tag{7.2}$$

i.e., the ratio of the number $n(t)$ of particles which at time t are in a small region of the 6-dimensional μ-space around a point $(\vec{r},\vec{v})$ to the volume $\delta\vec{r}\,\delta\vec{v}$ of the region. If the total number of particles in the system is N, one has

$$\int_\mu f(\vec{r},\vec{v};t)\, d\vec{r}\, d\vec{v} = N\,. \tag{7.3}$$

The quantity $f(\vec{r},\vec{v};t)\frac{d\vec{r}\,d\vec{v}}{N}$ may be considered as the probability that a particle is at the point with coordinates $(\vec{r},\vec{v})$ in μ-space. If there are no collisions between the particles, then

$$f(\vec{r}+\vec{v}\,\delta t,\vec{v}+\vec{a}\,\delta t\,;t+\delta t) = f(\vec{r},\vec{v};t)\,; \tag{7.4}$$

but if collisions are possible, then there is a non-vanishing difference between these values of f :

$$\left(\frac{\partial f}{\partial t}\right)_C \delta t = f(\vec{r}+\vec{v}\,\delta t,\vec{v}+\vec{a}\,\delta t\,;t+\delta t) - f(\vec{r},\vec{v};t)\,. \tag{7.5}$$

In order to calculate this difference one may consider a particle within a small region of volume $\delta\vec{r}$ about $\vec{r}$ and with velocity in

$\delta\vec{v}$ about $\vec{v}$. In the same region one assumes that there are particles with velocities in $\delta\vec{v}_1$ about $\vec{v}_1$ which constitute the beam incident on the first particle. Then one calculates the number of particles of velocity $\vec{v}$ which by collision change their velocities. In the same way one can calculate the increase in the number of particles with velocity in $\delta\vec{v}$ about $\vec{v}$ due to collisions, and the difference provides the rate of change of the number of particles with velocity in $\delta\vec{v}$ about $\vec{v}$ due to collisions. Thus:

$$\left(\frac{\partial f}{\partial t}\right)_C = \iiint (f'f'_1 - ff_1)\, g\, b\, db\, d\varepsilon\, d\vec{v}_1 , \qquad (7.6)$$

where f, f_1 and f', f'_1 are respectively the distribution functions of the particles after and before the collisions, $g = |\vec{v}_1 - \vec{v}| = |\vec{v}'_1 - \vec{v}'|$ the modulus of their relative velocities, b the impact parameter, and ε the scattering angle. To obtain the expression (7.6) one must make use of a very important assumption, namely that within the integral of this expression the probability of having in $\delta\vec{r}$ about $\vec{r}$ two particles with velocities between $\vec{v}$ and $\vec{v}+\delta\vec{v}$ and between $\vec{v}_1$ and $\vec{v}_1+\delta\vec{v}_1$ respectively, is proportional to

$$\left(f(\vec{r},\vec{v},t)\,\delta\vec{r}\,\delta\vec{v}\right)\left(f(\vec{r},\vec{v}_1;t)\,\delta\vec{r}\,\delta\vec{v}_1\right) . \qquad (7.7)$$

This assumption is known as the <u>molecular chaos hypothesis</u> and states that the probability of having at the same time one particle about $(\vec{r},\vec{v})$ and another one about $(\vec{r},\vec{v}_1)$ in μ-space can be replaced in the collision integral by the product of the probabilities of having separately each particle in these regions of μ-space. Since the probability of the composite event is assumed to be equal to the product corresponding to the two simple events, the hypothesis of molecular chaos states that <u>both events can be considered as statistically independent</u>. That is, the correlations between the velocities of the particles before the collision do not contribute to the value of $(\partial f/\partial t)_C$. The effect of this hypothesis is the same as that of <u>assuming that the velocities of the particles are uncorrelated before the collision</u>. It is clear however, that during the evolution of the gas, collisions occur among the particles and therefore, the velocity of a given particle depends on its previous history. Consequently the velocities before the following collisions must be correlated.

The hypothesis of molecular chaos is then not a consequence of the dynamics of collisions: it seems rather to be incompatible with the laws of mechanics. Moreover, one cannot easily understand why this hypothesis has to be verified in the mechanical system. However, the molecular chaos will appear as a fundamental factor in irreversibility.

8. Boltzmann's H-theorem

One of the reasons which make the Boltzmann equation so important in the theory of irreversibility is that its solutions verify the so-called H-theorem. This theorem was stated earlier than the generalised φ-theorem and may be formulated as follows.

Let us define a quantity

$$H(t) = \iint_{\mu} f(t) \ln f(t) \, d\vec{r} \, d\vec{v} . \qquad (8.1)$$

Then, if $f(t)$ is a solution of the Boltzmann equation, one has

$$\frac{\partial H(t)}{\partial t} = \iint_{\mu} \frac{\partial f(t)}{\partial t} \left[1 + \ln f(t) \right] d\vec{r} \, d\vec{v}$$

$$= \frac{1}{4} \iiiint\!\!\int (f'f'_1 - ff_1 + \ln ff_1 - \ln f'f'_1) g \, b \, db \, d\varepsilon \, d\vec{v}_1 \, d\vec{v} \, d\vec{r} , \qquad (8.2)$$

and since the integrand can never be positive, one finds

$$\frac{dH(t)}{dt} \leq 0 . \qquad (8.3)$$

This result means that the Boltzmann equation is irreversible in the sense that, for any initial condition, one has

$$\lim_{t \to \infty} f(\vec{r}, \vec{v}, t) = f^{MB}(\vec{r}, \vec{v}) , \qquad (8.4)$$

where $f^{MB}(\vec{r}, \vec{v})$ is the Maxwell-Boltzmann distribution function. Indeed $H(t)$ is a decreasing function of time and constant only when $f'f'_1 = ff_1$, i.e., when f is the Maxwell-Boltzmann distribution function. Since for this equilibrium function the value of H is minimal, the H-theorem implies an irreversible evolution of any initial distribution function towards the Maxwell-Boltzmann distribution.

This irreversible behaviour is incompatible with the laws of mechanics. Indeed, if for the distribution function of a particular conservative system the H-theorem is valid - and one consequently concludes that the system evolves towards a particular equilibrium state - we may imagine a state exactly the same but where the velocities of the particles are reversed. It is then clear - from the mechanical point of view - that from this latter "equilibrium state" the system would "irreversibly" evolve towards the initial "non-equilibrium" state.

It is important to note, however, that the H-function (8.1) has only the property (8.3) of being a decreasing function of time if the distribution function through which it is defined is a solution of the Boltzmann equation. Now, the solutions of the Boltzmann equation do not evolve in time through mechanical equations but, instead, through the Boltzmann equation which is not an entirely mechanical one, since in its derivation one deals not simply with mechanical collisions - one has to introduce the molecular chaos hypothesis which is not mechanical. Since this is the only non-mechanical hypothesis, we must conclude that it is responsible for the irreversible behaviour of systems described by means of Boltzmann's equation.

Therefore, if we want to study the conditions for irreversibility, we must analyse the meaning of the hypothesis of molecular chaos. To this end I will present a different derivation of the Boltzmann equation by means of which we "derive" the molecular chaos. This derivation will be "as mechanical as possible", but we shall see that in order to obtain irreversible equations it is necessary to introduce some hypotheses. The character of these specific hypotheses, whether mechanical or not, is, in my opinion a question of language. I prefer to say that they are not mechanical. They are, however, very plausible. On the contrary the hypothesis of molecular chaos is clearly a non-mechanical assumption.

9. The BBGKY hierarchy

We describe our system (a gas) by means of a statistical ensemble which is not in equilibrium and which is defined by a probability density in phase space (6N-dimensional if the system consists of N point particles without constraints) which depends explicitly on time. The probability density satisfies Liouville's equation

$$\frac{\partial}{\partial t} \rho_N(\vec{r}^N, \vec{p}^N; t) = -i L_N \rho_N(\vec{r}^N, \vec{p}^N; t). \tag{9.1}$$

It is important to distinguish between the Liouville equation (i.e. the evolution equation of $\rho_N(t)$) and the equation expressing the dynamical evolution of a function representing a microscopic quantity and not depending explicitly on time:

$$\frac{d}{dt} A(\vec{r}_t^N, \vec{p}_t^N) = i L_N A(\vec{r}_t^N, \vec{p}_t^N). \tag{9.2}$$

Here, $\vec{r}_t^N$ and $\vec{p}_t^N$ represent the positions and momenta of the N particles at time t. If at t=0 they are respectively $\vec{r}_0^N$ and $\vec{p}_0^N$,

$$\begin{aligned} \vec{r}_t &= \vec{R}(\vec{r}_0^N, \vec{p}_0^N; t) = G(\vec{r}_0^N, \vec{p}_0^N; t)\vec{r}_0 \\ \vec{p}_t &= \vec{P}(\vec{r}_0^N, \vec{p}_0^N; t) = G(\vec{r}_0^N, \vec{p}_0^N; t)\vec{p}_0, \end{aligned} \tag{9.3}$$

where $G(\vec{r}_0^N, \vec{p}_0^N; t)$ is the <u>evolution operator</u> of the system. If the system is conservative (i.e. its Hamiltonian function is independent of time) $G(t)$ has the form

$$G(t) = e^{itL_N}. \tag{9.4}$$

Therefore

$$\vec{r}_t = e^{itL_N}\vec{r}_0 \quad , \qquad \vec{p}_t = e^{itL_N}\vec{p}_0 \tag{9.5}$$

The formal integration of (9.2) gives

$$A(\vec{r}_t^N, \vec{p}_t^N) = e^{itL_N} A(\vec{r}_0^N, \vec{p}_0^N) \tag{9.6}$$

or, using (9.5)

$$e^{itL_N} A(\vec{r}_0^N, \vec{p}_0^N) = A\left(e^{itL_N}\vec{r}_0^N, e^{itL_N}\vec{p}_0^N\right). \tag{9.7}$$

Applying this property to the probability density, we also have

$$\rho(\vec{r}_t^N, \vec{p}_t^N; \tau) = e^{itL_N} \rho(\vec{r}_0^N, \vec{p}_0^N; \tau) \tag{9.8}$$

which is similar to (9.6), but in (9.8) τ only acts as a constant parameter. On the other hand, by formal integration of the Liouville equation (9.2) one obtains

$$\rho(\vec{r}^N, \vec{p}^N; t) = e^{-itL_N} \rho(\vec{r}^N, \vec{p}^N; 0). \tag{9.9}$$

Making $\tau = t$ in (9.8) and $\vec{r} = \vec{r}_0$ and $\vec{p} = \vec{p}_0$ in (9.9), one finds

$$\rho(\vec{r}^N_t, \vec{p}^N_t; t) = e^{itL_N} \rho(\vec{r}^N_0, \vec{p}^N_0; t) = e^{itL_N} e^{-itL_N} \rho(\vec{r}^N_0, \vec{p}^N_0; 0) = \rho(\vec{r}^N_0, \vec{p}^N_0; 0) \tag{9.10}$$

which is the expression of Liouville's theorem. It is important to note that (9.6) and (9.8) are valid for any phase function depending, or not explicitly depending, on time. But (9.9) and (9.10) are only valid for the probability density in phase space because it is only to this function that Liouville's theorem is applicable.

We assume that the particles of our system interact only with each other and through a repulsive finite-range potential $\varphi_{ij} = \varphi(|\vec{r}_i - \vec{r}_j|)$. The force between two particles is

$$-\frac{\partial \varphi(|\vec{r}_i - \vec{r}_j|)}{\partial \vec{r}_i} \neq 0 \qquad \text{only if} \qquad |\vec{r}_i - \vec{r}_j| \leq \lambda \tag{9.11}$$

λ being the range of the interactions. The Liouville operator is then

$$L_N = -i \sum_{j=1}^{N} \frac{\vec{p}_j}{2m} \cdot \frac{\partial}{\partial \vec{r}_j} + i \frac{1}{2} \sum_{j=1}^{N} \sum_{\substack{k=1 \\ j \neq k}}^{N} \frac{\partial \varphi_{jk}}{\partial \vec{r}_j} \cdot \left(\frac{\partial}{\partial \vec{p}_j} - \frac{\partial}{\partial \vec{p}_k} \right) . \tag{9.12}$$

Solving equation (9.1) is equivalent to solving the equations of motion of the system, a very complex problem. Indeed, to do this it would be necessary to know $\rho_N(0)$, which involves very detailed information about the microscopic state of the system. This information is neither necessary nor helpful. It is not necessary because the transport coefficients (and, in general, every important property of the system) can be expressed in terms of the phases of a very small number n of particles ($n \approx 1, 2$). And, on the other hand, it is not convenient to have too much information at one's disposal because it is the lack of microscopic information that gives the irreversibility. The reason for this is that irreversibility is characteristic of the evolution of the macrostates and not of the microstates.

We need, therefore, to reduce the information contained in the probability density. To this end, we introduce the reduced distribution functions $F_n(t)$, through integrations of $\rho_N(t)$:

$$F_n(t) = F_n(\vec{r}^n, \vec{p}^n; t) = V^n \iint_{\Gamma} d\vec{r}^{(N-n)} d\vec{p}^{(N-n)} \rho_N(\vec{r}^N, \vec{p}^N; t), \tag{9.13}$$

where V is the volume occupied by the system and the integrations with respect to the $\vec{r}$'s are taken over V. Here we have used the notation

$$\vec{r}^{\,n}=(\vec{r}_1,\dots,\vec{r}_n),\ \vec{r}^{\,(N-n)}=(\vec{r}_{n+1},\dots,\vec{r}_N),\ \vec{p}^{\,n}=(\vec{p}_1,\dots,\vec{p}_n),\ \vec{p}^{\,(N-n)}=(\vec{p}_{n+1},\dots,\vec{p}_N). \tag{9.14}$$

From the normalization of $\rho_N(t)$

$$\iint_\Gamma d\vec{r}^{\,N} d\vec{p}^{\,N} \rho_N(\vec{r}^{\,N},\vec{p}^{\,N};t) = 1 \tag{9.15}$$

we deduce the normalization of $F_n(t)$:

$$\iint_\Gamma d\vec{r}^{\,n} d\vec{p}^{\,n} F_n(\vec{r}^{\,n},\vec{p}^{\,n};t) = V^N \tag{9.16}$$

It suffices now to integrate formally the Liouville equation (9.1) for particles $n+1$, $n+2$, ..., N to obtain

$$\frac{\partial F_n(t)}{\partial t} = -i L_n F_n(t) - i\,\frac{N-n}{V}\,\mathcal{L}_{n,n+1} F_{n+1}(t), \tag{9.17}$$

where

$$\mathcal{L}_{n,n+1} = i\sum_{j=1}^{n}\iint_\Gamma d\vec{r}_{n+1} d\vec{p}_{n+1}\,\frac{\partial\varphi(|\vec{r}_j-\vec{r}_{n+1}|)}{\partial\vec{r}_j}\cdot\frac{\partial}{\partial\vec{p}_j} \tag{9.18}$$

and L_n is the Liouville operator for an isolated system of n particles (that is, the expression (9.12) replacing N by n). Equation (9.17) cannot be solved for $F_n(t)$ without knowing $F_{n+1}(t)$, which is not known without solving a similar equation, etc. Therefore, in order to determine $F_n(t)$ it would be necessary to solve the whole set of coupled equations of type (9.17), known as BBGKY-hierarchy (Bogolubov, Born, Green, Kirkwood, Yvon). The exact solution of the hierarchy is therefore equivalent to that of the Liouville equation. It may seem then that there is no advantage in writing Liouville's equation (9.1) in the form of the hierarchy (9.17), because we have not lost information. However, the hierarchy is better suited than the Liouville equation to the removal of the undesired information as we shall now see.

10. Thermodynamic limit

The average particle density of a system consisting of N particles in a volume V is D = N/V. If the system is finite (i.e. if it has a finite number of particles within a finite volume and the energy of the system is finite) Poincaré's theorem is applicable and, consequently, for almost every initial state the system evolves in such a manner that after a finite time it will be in a microscopic state as close as desired to the initial one. In other words, according to Poincaré's theorem a finite system evolves by cycles of finite duration. The duration of a cycle is called recurrence time and when N is large it is of the order of e^N.

Clearly, by virtue of this theorem, a finite system cannot be irreversible in the sense that, from a given initial non-equilibrium state, it would evolve towards an equilibrium state. According to Poincaré's theorem, when the system reaches the equilibrium state, it continues evolving until it again reaches the initial state. This is actually so, but the macroscopic systems are so large that the recurrence time is enormously large (of the order of $10^{10^{10}}$) and the system never reaches the initial non-equilibrium state starting from the "equilibrium" state. For studying a macroscopic system we have to introduce in our equations the condition that both N and V are very large. In this manner, the recurrences due to Poincaré's cycles do not appear. Actually, N is of the order of Avagadro's number, 10^{23} and V is very large when compared to the volume of each molecule.

The condition that the system is large is introduced through the so-called thermodynamic limit:

$$N \to \infty \ , \quad V \to \infty \ , \quad N/V = D \text{ finite}. \tag{10.1}$$

In the thermodynamic limit D is kept finite to avoid introducing in the equations governing the system other than those effects due exclusively to the large number of particles and the large size of the system. So, when introducing the thermodynamic limit, the exact equations derived from the Liouville equation are valid for a system which has the same properties as those described by (10.1) but which is infinitely large.

One of the differences between the large and the small systems will be in the role played by the surface effects due to the walls of the container. These may be relevant in the case of a small system; but if the system is large and the surface effects are proportional to the area of the walls, they will be negligible when compared to the volume effects, as the ratio of surface to volume of a "normal system" tends to zero for an infinitely large system. For this reason, since the surface contribution is zero in the thermodynamic limit, it is possible to choose arbitrary boundary conditions. In statistical mechanics it is common to deal with periodic conditions. For instance, integrating by parts and using periodic boundary conditions at the limits of the phase integrations, we have

$$\iint_{\Gamma} d\vec{r}^N d\vec{p}^N A(\vec{r}^N, \vec{p}^N) L_N B(\vec{r}^N, \vec{p}^N) = - \iint_{\Gamma} d\vec{r}^N d\vec{p}^N B(\vec{r}^N, \vec{p}^N) L_N A(\vec{r}^N, \vec{p}^N) \tag{10.2}$$

for any phase functions A and B. This property is very often used, as we have done already in deriving the BBGKY-hierarchy.

Another remarkable difference between a large and a small system, in the case when the particle interactions are of finite range, appears in integrations of type

$$\frac{1}{V} \int d\vec{r} \, \frac{\partial \varphi(|\vec{r}|)}{\partial \vec{r}} \, A(\vec{r}^N). \tag{10.3}$$

In general, the $\vec{r}$-integrations are extended to the whole volume V of the system. However, because of (9.11), in expressions such as (10.3), the integration is restricted to a sphere of radius λ which is finite in the thermodynamic limit. Since all phase functions are summable in phase space, the expression (10.3) is zero in the thermodynamic limit.

By introducing the thermodynamic limit into exact expressions we obtain expressions "exact in the thermodynamic limit", but these will not be exact for a macroscopic system because it is not infinitely large. In other words: by means of the thermodynamic limit we introduce an approximation in the study of a macroscopic system, consisting in the elimination of some terms, i.e. a loss of information.

As we have already said, this loss of information is responsible for a kind of irreversibility, that is, for the removal of Poincare's cycles. However, this is not the only kind of irreversibility present in macroscopic systems. In fact, an infinitely large mechanical system does not evolve by cycles but this does not imply the validity of an H-theorem. So that in order to obtain irreversibility it is necessary to introduce more assumptions.

11. A new hierarchy

The solution of the BBGKY-hierarchy provides the reduced distribution functions $F_n(\vec{r}^n, \vec{p}^n; t)$ defined in (9.13). We will obtain now a hierarchy for the momentum distribution functions

$$f_n(\vec{p}^n; t) = \frac{1}{V^n} \int d\vec{r}^n \, F_n(\vec{r}^n, \vec{p}^n; t) \tag{11.1}$$

which are proportional to the probabilities of finding the particles 1, 2, ..., n with momenta between $\vec{p}^n$ and $\vec{p}^n + d\vec{p}^n$ independently of their positions. To this end, we use the so-called projection operators. We define

$$P_n \ldots = \frac{1}{V^n} \int d\vec{r}^n \ldots \tag{11.2}$$

which is a short notation for saying that P_n acts on any phase function $A(\vec{r}^N, \vec{p}^N; \tau)$ as follows:

$$P_n A(\vec{r}^N, \vec{p}^N; \tau) = \frac{1}{V^n} \int d\vec{r}^n \, A(\vec{r}^N, \vec{p}^N; \tau). \tag{11.3}$$

In particular, according to (11.1), we have

$$f_n(\vec{p}^n; t) = P_n F_n(\vec{r}^n, \vec{p}^n; t). \tag{11.4}$$

The name of projection operator takes its origin from the fact that P_n is an idempotent operator:

$$P_n^2 \ldots = P_n P_n \ldots = P_n \ldots \tag{11.5}$$

Moreover, if we define the operator $(1 - P_n)$ by

$$(1 - P_n) A = A - P_n A, \tag{11.6}$$

we see that it is also a projection operator:

$$(1 - P_n)^2 = 1 - 2P_n + P_n^2 = 1 - 2P_n + P_n = (1 - P_n). \tag{11.7}$$

Also, P_n and $(1 - P_n)$ are orthogonal:

$$P_n(1-P_n) = (1-P_n)P_n = 0, \tag{11.8}$$

and, in addition:

$$P_n + (1-P_n) = 1. \tag{11.9}$$

If we project the BBGKY-hierarchy (9.17) with P_n (that is, multiplying both sides by P_n) we get in the thermodynamic limit

$$\frac{\partial f_n(t)}{\partial t} = -i \lim_{V\to\infty} P_n \mathcal{L}_{n,n+1} h_{n+1}(t), \tag{11.10}$$

where

$$h_n(\vec{r}^n, \vec{p}^n; t) = F_n(\vec{r}^n, \vec{p}^n; t) - f_n(\vec{p}^n; t) = (1-P_n) F_n, \tag{11.11}$$

where we have considered the set of n particles to be far from the walls of the container, so that

$$\mathcal{L}_{n,n+1} f_{n+1} = 0. \tag{11.12}$$

In the same manner, by projecting the BBGKY-hierarchy with $(1-P_n)$, we have, in the thermodynamic limit

$$\frac{\partial h_n(t)}{\partial t} = -iL_n f_n(t) - iL_n h_n(t) - iD \lim_{V\to\infty} (1-P_n)\mathcal{L}_{n,n+1} h_{n+1}(t). \tag{11.13}$$

The set of equations (11.10) and (11.13) is equivalent to the BBGKY hierarchy (9.17) in the thermodynamic limit, because $F_n = f_n + h_n$. We can now formally solve for $h_n(t)$ in equation (11.13) and taking the solution for n+1 replace it into equation (11.10). We obtain

$$\begin{aligned} \frac{\partial f_n(t)}{\partial t} = D \int_0^t \frac{\partial K_{n,n+1}(\tau)}{\partial \tau} f_{n+1}(t-\tau)\, d\tau + D\, K_{n,n+1}(t)\, h_{n+1}(0) \\ + D^2 \int_0^t J_{n,n+1,n+2}(\tau)\, h_{n+2}(t-\tau)\, d\tau, \end{aligned} \tag{11.14}$$

where we have defined the operators

$$K_{n,n+1}(t) = -i \lim_{V\to\infty} P_n \mathcal{L}_{n,n+1} e^{-itL_{n+1}}, \tag{11.15}$$

$$J_{n,n+1,n+2}(t) = -i \lim_{V\to\infty} K_{n,n+1}(t)(1-P_{n+1})\mathcal{L}_{n+1,n+2} \tag{11.16}$$

We will not use equation (11.13) any more but only equation (11.14). Since we will not deal with both (11.13) and (11.14), we will lose information about the behaviour of the system. This lack of information is due neither to an approximation nor to an assumption. It is merely due to the fact that we are not interested in knowing the whole reduced distribution function F_n but only a part f_n of it. Therefore, this lack of information does not imply irreversibility. Equation (11.14) is exact in the thermodynamic limit. Let us now neglect its last term which, for finite times, is of the order of D^2. This term is studied in detail elsewhere. For the moment it is enough to note that, since we are interested in dilute gases, $D \ll 1$, and we may hope that the contribution of the last term in equation (14), if it is not zero, will be very small. The equation

$$\frac{\partial f_n(t)}{\partial t} = D\int_0^t \frac{\partial K_{n,n+1}(\tau)}{\partial \tau} f_{n+1}(t-\tau)\,d\tau + D\,K_{n,n+1}(t)\,h_{n+1}(0) \tag{11.17}$$

is actually a hierarchy very similar to the BBGKY-hierarchy. The most important differences between the two hierarchies is that (11.17) gives the evolution of the momentum distribution functions instead of the reduced distribution functions. Another difference is that (11.17) contains a term which depends on the initial conditions of the system.

According to the prescriptions of the statistical mechanics method which we are developing, we are free to choose the initial conditions to be imposed on the system. We may prepare the system at a given time and then we must allow it to evolve mechanically from the initial state.

We might assume that the system had no correlations at all at t=0, but it is enough to assume that the correlations have a finite range at t=0 and that the system is initially spatially uniform. We write

$$F_n(\vec{r}^n,\vec{p}^n;0) = \prod_{j=1}^{n} F_1(\vec{r}_j,\vec{p}_j;0)\left[1 - g_n(\vec{r}^n,\vec{p}^n;0)\right], \tag{11.18}$$

which defines the correlation function $g_n(0)$ at the initial time, and our assumption implies that

$$g_n(\vec{r}^n, \vec{p}^n; 0) = 0 \quad \text{if} \quad |\vec{r}_j - \vec{r}_k| > \varkappa \quad \text{for all pairs } (j,k). \tag{11.19}$$

An extra assumption (which is not necessary but which we make for the sake of simplicity) is that $g_n(0)$ is independent of the momenta. Now, if the system is initially spatially uniform, $F_n(0)$ is invariant under a spatial translation and, therefore, $F_1(0)$ must be independent of $\vec{r}$. Then, according to (11.1) $F_1(0) = f_1(0)$. Instead of (11.18) we may therefore write

$$F_n(\vec{r}^n, \vec{p}^n; 0) = \prod_{j=1}^{n} f_1(\vec{p}_j; 0) \left[1 + g_n(\vec{r}^n; 0) \right] \tag{11.20}$$

and by using (11.1) again,

$$f_n(\vec{p}^n; 0) = \prod_{j=1}^{n} f_1(\vec{p}_j; 0) \left[1 + \frac{1}{V^n} \int d\vec{r}^n \, g_n(\vec{r}^n; 0) \right]. \tag{11.21}$$

In the thermodynamic limit this equation reduces to

$$f_n(\vec{p}^n; 0) = \prod_{j=1}^{n} f_1(\vec{p}_j; 0) \tag{11.22}$$

as, according to (11.19), the integrations in (11.21) corresponding to the relative coordinates of the particles, must be taken over the sphere of finite radius $\varkappa$ and then the last term vanishes in the thermodynamics limit. Our initial conditions imply, therefore, that the momenta of the particles are not correlated at t = 0. In other words: $f_n(0)$ factorizes in the form (11.22). From the definition (9.13) of $F_n(t)$ and the definition (11.20) of $g_n(0)$, it can be obtained that

$$g_n(\vec{r}^n; 0) = \frac{1}{V} \int d\vec{r}_{n+1} \, g_{n+1}(\vec{r}^{n+1}; 0) \tag{11.23}$$

and from the definition of $h_n(t)$ and (11.20) and (11.22),

$$h_n(\vec{r}^n, \vec{p}^n; 0) = f_n(\vec{p}^n; 0) \, g_n(\vec{r}^n; 0). \tag{11.24}$$

Therefore, the hierarchy (11.17) may be written as

$$\frac{\partial f_n(t)}{\partial t} = D\int_0^t \frac{\partial K_{n,n+1}(\tau)}{\partial \tau} f_{n+1}(t-\tau)\,d\tau + D\,K_{n,n+1}(t) f_{n+1}(0)\, g_{n+1}(0)$$

$$= D\sum_{j=1}^{n}\left\{\int_0^t \frac{\partial K^{(j,n+1)}(\tau)}{\partial \tau} f_{n+1}(t-\tau)\,d\tau + K^{(j,n+1)}(t) f_{n+1}(0)\, g_2(\vec{r}_j - \vec{r}_{n+1};0)\right\} \tag{11.25}$$

since it is possible to show that in the thermodynamic limit

$$K_{n,n+1}(t) = \sum_{j=1}^{n} K^{(j,n+1)}(t) \lim_{V\to\infty} \frac{1}{V^{n-1}} \int\cdots\int d\vec{r}_1 \ldots d\vec{r}_{j-1}\, d\vec{r}_{j+1} \ldots d\vec{r}_n\,, \tag{11.26}$$

where

$$K^{(j,n+1)}(t) = \lim_{V\to\infty} \frac{1}{V} \int d\vec{p}_{n+1} \int d\vec{r}_{n+1} \int d\vec{r}_j \frac{\partial \varphi(|\vec{r}_j - \vec{r}_{n+1}|)}{\partial \vec{r}_j} \cdot \frac{\partial}{\partial \vec{p}_j} e^{-itL^{(j,n+1)}} \tag{11.27}$$

and where $L^{(j,n+1)}$ is the Liouville operator corresponding to an isolated system consisting only of particles j and $n+1$.

12. Introduction of the irreversibility

Let us introduce new time variables defined by

$$s = Dt\,, \quad \sigma = D\tau \tag{12.1}$$

and let

$$\tilde{f}_n(s) = f_n(s/D) = f_n(t) \tag{12.2}$$

be the momentum distribution function $f_n(t)$ in the new time scale, where we use s instead of t. The hierarchy (11.25)now reads

$$\frac{\partial \tilde{f}_n(s)}{\partial s} = \sum_{j=1}^{n}\left\{\int_0^s \frac{\partial K^{(j,n+1)}(\sigma/D)}{\partial \sigma} \tilde{f}_{n+1}(s-\sigma)\,d\sigma + K^{(j,n+1)}(s/D)\tilde{f}_{n+1}(0)\, g_2(\vec{r}_j - \vec{r}_{n+1};0)\right\}$$

$$= \sum_{j=1}^{n}\left\{K^{(j,n+1)}(s/D)\tilde{f}_{n+1}(0) - \int_0^s K^{(j,n+1)}(\sigma/D) \frac{\partial \tilde{f}_{n+1}(s-\sigma)}{\partial \sigma}\,d\sigma\right.$$

$$\left. + K^{(j,n+1)}(s/D)\tilde{f}_{n+1}(0)\, g_2(\vec{r}_j - \vec{r}_{n+1};0)\right\}, \tag{12.3}$$

where we have made a partial integration and used the fact that

$$K^{(j,n+1)}(0)\, f_{n+1} = 0 . \tag{12.4}$$

We emphasize that this is exactly the same hierarchy as (11.25). To derive the Boltzmann equation we must now introduce a new factor of irreversibility. Let us see why.

When we considered the thermodynamic limit we removed from our system the possibility of cyclic evolution as required by Poincaré's theorem for finite systems. The thermodynamic limit is then an important factor for the irreversibility of the macroscopic systems. But it is not enough to ensure that the system will have an irreversible behaviour. Indeed, if we change s into $-s$, σ into $-\sigma$ and $\vec{p}^N$ into $-\vec{p}^N$, we see that the functions $\hat{f}_n(\vec{p}^n;s) = \tilde{f}_n(-\vec{p}^n;-s)$ verify the same equation as the $\tilde{f}_n(\vec{p}^n;s)$. That means that the equation (12.3) is reversible, that is, <u>its solutions cannot verify a η-theorem</u>. In fact, we may reverse the velocities of the system and obtain a reversed evolution with an increase of the η-function. Both evolutions, the direct and the reversed one, would be possible, because the functions $\tilde{f}_n(\vec{p}^n;s)$ and $\hat{f}_n(\vec{p}^n;s)$ are solutions of the same equation (12.3) if we choose the initial conditions in such a manner that $\tilde{f}_n(\vec{p}^n;0) = \hat{f}_n(\vec{p}^n;0)$, i.e. as an even function of the momenta.

On the other hand, we are interested in a <u>dilute gas</u> in which only the binary collisions are relevant. We have to consider therefore the limit $D \to 0$. In equation (11.25) we may see that, <u>for finite times t, the limit $D \to 0$ would imply that $f_n(\vec{p}^n;t)$ remains constant</u>. Actually $f_n(t)$ changes only through the collisions of the particles. Therefore, to have a change in the value of $f_n(t)$ a time interval of the order of the <u>average duration between two successive collisions is necessary</u>.

Thus, t must be of the order of t_c $(t/t_c \sim 1)$ and, on the other hand, one has that t_c is of the order of the inverse of D, because for smaller densities the mean free path of the particles and hence t_c are larger. Therefore, to see the evolution of $f_n(\vec{p}^n;t)$ when $D \to 0$, it is necessary to allow that at the same time $t \to \infty$, in such a manner that $Dt \sim t/t_c \sim 1$. In other words, the condition that the gas is a dilute gas must be introduced into (11.25) through the limit:

$$D \to 0 \;,\; t \to \infty \;,\; Dt = s \text{ finite} \tag{12.5}$$

which in the time scale (12.1) is equivalent to taking equation (12.3) in the limit $D \to 0$, s finite.

We shall later show that

$$\lim_{t \to \infty} K^{(j,k)}(t)\, A(\vec{p}_j . \vec{p}_k) = K_{\infty}^{(j,k)} A(\vec{p}_j . \vec{p}_k), \tag{12.6}$$

$$\lim_{t \to \infty} K^{(j,k)}(t)\, A(\vec{p}_j . \vec{p}_k)\, g_2(\vec{r}_j - \vec{r}_k\,;0) = 0, \tag{12.7}$$

where $A(\vec{p}_j . \vec{p}_k)$ is any function of the momenta of the particles j and k and $K_{\infty}^{(j,k)}$ is the Boltzmann collision operator. With these two relations, equation (12.3) may be written, in the limit $D \to 0$, as

$$\frac{\partial \tilde{f}_n(s)}{\partial s} = \sum_{j=1}^{n} K_{\infty}^{(j,n+1)} \tilde{f}_{n+1}(s) . \tag{12.8}$$

This equation is irreversible and is the final result of our derivation. To obtain the Boltzmann equation from it, it suffices to note that the solution of (12.8) with the initial condition (11.22) is

$$\tilde{f}_n(\vec{p}^{\,n};s) = \prod_{j=1}^{n} \tilde{f}_1(\vec{p}_j\,;s), \tag{12.9}$$

where $\tilde{f}_1(\vec{p}_j\,;s)$ satisfies

$$\frac{\partial \tilde{f}_1(\vec{p}_j\,;s)}{\partial s} = K_{\infty}^{(j,k)} \tilde{f}_1(\vec{p}_j\,;s)\, \tilde{f}_1(\vec{p}_k\,;s) . \tag{12.10}$$

Indeed, for a given initial condition (11.22) the equation (12.8) has a unique solution. Since (12.9) satisfies (12.8) if (12.10) is verified, the expression (12.9) is the solution, and equation (12.10) must be verified. Equation (12.10) is the Boltzmann equation.

I emphasize that (12.9) is a factorization for any time, that is, it expresses the same thing as the hypothesis of molecular chaos. The main difference between our derivation of the Boltzmann equation and the traditional one is that we do not postulate molecular chaos, but we derive it by means of the limit (12.5) which introduces the irreversibility.

Why does this limit introduce the irreversibility? Before answering this question we must prove properties (12.6) and (12.7).

Their proof will give us the answer.

13. Binary collisions

We start by writing the expression $K^{(j,k)}(t)A(\vec{p}_j;\vec{p}_k)$ in terms of the variables $(\vec{r}_b,\vec{p}_b)$ corresponding to the centre of mass of the particles j and k and of

$$\vec{r} = \vec{r}_j - \vec{r}_k \quad , \quad \vec{p} = \frac{1}{2}(\vec{p}_j - \vec{p}_k), \tag{13.1}$$

which are the relative coordinates. We obtain

$$K^{(j,k)}(t)A(\vec{p}_j.\vec{p}_k) = \int d\vec{p}_k \int d\vec{r}\, \frac{\partial\varphi(\vec{r})}{\partial\vec{r}}\cdot\frac{\partial}{\partial\vec{p}}\, e^{-itL}\hat{A}(\vec{p};\vec{p}_b) \tag{13.2}$$

where $\hat{A}(\vec{p};\vec{p}_b) = A(\vec{p}_j.\vec{p}_k)$ and

$$L = -i\,\frac{\vec{p}}{\mu}\cdot\frac{\partial}{\partial\vec{r}} + i\,\frac{\partial\varphi}{\partial\vec{r}}\cdot\frac{\partial}{\partial\vec{p}} = A + B\,, \; (\mu = \tfrac{1}{2}m)\,. \tag{13.3}$$

We use now the identity

$$e^{-it(A+B)} = e^{-itA} - i\int_0^t e^{-i\tau A}\, B\, e^{-i(t-\tau)(A+B)}\, d\tau, \tag{13.4}$$

valid for any operators A and B, to get

$$e^{-itL}\hat{A}(\vec{p};\vec{p}_b) = \hat{A}(\vec{p};\vec{p}_b) + \int_0^t e^{-i\tau(\vec{p}/\mu)\cdot(\partial/\partial\vec{r})}\frac{\partial\varphi}{\partial\vec{r}}\cdot\frac{\partial}{\partial\vec{p}}\, e^{-i(t-\tau)L}\hat{A}(\vec{p};\vec{p}_b) \tag{13.5}$$

Using the same notation as in (9.3), (9.5) and (9.6), this expression may be written as

$$\hat{A}(\vec{p}_{-t};\vec{p}_b) - \hat{A}(\vec{p};\vec{p}_b) = \int_0^t \frac{\partial\varphi(|\vec{r}-(\vec{p}/\mu)\tau|)}{\partial\vec{r}}\cdot\frac{\partial}{\partial\vec{p}}\,\hat{A}\left[\vec{P}\left(\vec{r}-\frac{\vec{p}}{\mu}\tau,\vec{p};-t+\tau\right);\vec{p}_b\right]d\tau. \tag{13.6}$$

If we choose a coordinate system such that the positive axis z is in the direction of $\vec{p}$, this equation reads

$$\hat{A}\left[\vec{P}(x,y,z;\vec{p};-t);\vec{p}_b\right] - \hat{A}(\vec{p};\vec{p}_b) =$$

$$= \frac{\mu}{|\vec{p}|}\int_{z-|\vec{p}/\mu|t}^{z} dz' \; \frac{\partial\varphi(x,y,z')}{\partial\vec{r}}\cdot\frac{\partial}{\partial\vec{p}}\;\hat{A}\left[\vec{P}(x,y,z';\vec{p};-t+\frac{z-z'}{|\vec{p}/\mu|});\vec{p}_b\right]. \qquad (13.7)$$

Due to the finite range of the forces between the particles the integrand of this expression is different from zero only in an interval of time $\tau_c \sim 2\mu\lambda/|\vec{p}|$, where τ_c is the duration of the collision. On the other hand, for the same reason

$$\vec{P}_{-t} = \vec{P}(\vec{r},\vec{p};-t) = \vec{p}^{\circ}(\vec{r},\vec{p}) \quad \text{if} \quad -\lambda \le z \le \lambda \quad \text{and} \quad t > \alpha\tau_c \qquad (13.8)$$

where α is larger than unity and $\vec{p}^{\circ}$ (independent of t) is the relative momentum before the collision. Also

$$\vec{P}_{-t} = \vec{P}(x,y,z;\vec{p};-t) = \vec{p}^{\circ}(x,y;\vec{p}) \quad \text{if} \quad \lambda < z < 2\lambda \quad \text{and} \quad t > 2\alpha\tau_c \qquad (13.9)$$

where $\vec{p}^{\circ}$ is now independent of t and z and dependent on the impact parameter $b=(x^2+y^2)^{1/2}$ and on $\vec{p}$ which is now the relative momentum after the collision. Therefore, for $\lambda < z < 2\lambda$ and $t > 2\alpha\tau_c$ according to (9.11), (13.8) and (13.9), the expression (13.7) may also be written in the form

$$\hat{A}\left[\vec{p}^{\circ}(x,y;\vec{p});\vec{p}_b\right] - \hat{A}(\vec{p};\vec{p}_b) = \frac{\mu}{|\vec{p}|}\int_{-\infty}^{\infty} dz' \; \frac{\partial\varphi(x,y,z')}{\partial\vec{r}}\cdot\frac{\partial}{\partial\vec{p}}\;\hat{A}\left[\vec{p}^{\circ}(x,y,z';\vec{p});\vec{p}_b\right]. \qquad (13.10)$$

But using (13.8), if $t > \alpha\tau_c$ we have

$$\hat{A}\left[\vec{p}^{\circ}(x,y,z;\vec{p});\vec{p}_b\right] = e^{-itL}\hat{A}(\vec{p};\vec{p}_b), \qquad (13.11)$$

and therefore, integrating both sides of (13.10) with respect to $\vec{p}_k$ x and y, the right-hand side gives $\mu/|\vec{p}|$ times the right hand side of (13.2) when $t > \alpha\tau_c$. Hence, when $t \to \infty$,

$$K_{\infty}^{(j,k)} A(\vec{p}_j,\vec{p}_k) = K^{(j,k)}(t > \alpha\tau_c)\, A(\vec{p}_j,\vec{p}_k) =$$

$$= \int d\vec{p}_k \iint dx\, dy \; \frac{|\vec{p}_j - \vec{p}_k|}{m}\left[A(\vec{p}_j^{\circ},\vec{p}_k^{\circ}) - A(\vec{p}_j,\vec{p}_k)\right]. \qquad (12.12)$$

Thus, we have proved the property (12.6).

On the other hand, by using the coordinates of the centre of mass and the relative variables (13.1), one may write

$$K^{(j,k)}(t)\, A(\vec{p}_j,\vec{p}_k)\, g_2(|\vec{r}_j-\vec{r}_k|;0) = \int d\vec{p}_j \int d\vec{r}\, \frac{\partial \varphi}{\partial \vec{r}} \cdot \frac{\partial}{\partial \vec{p}}\, \hat{A}(\vec{p}_{-t};\vec{p}_b)\, g_2(|\vec{r}_{-t}|;0), \tag{13.13}$$

where the integrand is different from zero only if $|\vec{r}| \leq \lambda$. But then, for times which are large enough, we will have

$$|\vec{r}_{-t}| > \varkappa \quad \text{if} \quad |\vec{r}| \leq \lambda \qquad \text{and} \qquad t > \beta\tau_c, \tag{13.14}$$

and, according to (11.19), $g_2(|\vec{r}_{-t}|;0) = 0$. Thus, we have proved the property (12.7), and completed the derivation of the Boltzmann equation, since together with (13.12), equation (12.10) reads

$$\frac{\partial \tilde{f}_1(\vec{p}_j;s)}{\partial s} = \int d\vec{p}_k \iint dx\,dy\, \frac{|\vec{p}_j-\vec{p}_k|}{m}\left[f_1(\vec{p}_j^{\,\circ};s)f_1(\vec{p}_k^{\,\circ};s)-f_1(\vec{p}_j;s)f_1(\vec{p}_k;s)\right]. \tag{13.15}$$

14. Molecular chaos

We have seen that the introduction of the thermodynamic limit and the limit (12.5) into the Liouville equation (or the BBGKY hierarchy), which is reversible, allows us to obtain the Boltzmann equation, which is irreversible, and moreover, the property (12.9) gives us molecular chaos at every time s. The thermodynamic limit has already been briefly discussed and we have seen how it removes the recurrences due to the Poincaré's cycles. We have also proved that the limit (12.5) removes the time reversibility of the mechanical system, but we have not discussed why this is so, and we have not yet investigated the meaning of this limit.

When we discussed the meaning of the evolution of the entropy of an isolated system we arrived at the conclusion that to obtain an increase in entropy it is necessary to introduce more and more microstates during the evolution of the system, Also, when we discussed the possibility of a generalized η-theorem, valid at all times, we saw that it is necessary to introduce new ensembles for every time. Both conclusions can be expressed by saying that <u>to obtain</u>

irreversibility it is necessary to interrupt the dynamical evolution. This interruption ignores the exact evolution and introduces in our problem a lack of information.

On the other hand let us remember, that the factor giving irreversibility in the traditional derivation of the Boltzmann equation is the hypothesis of molecular chaos, i.e. the assumption that there is no correlation between the particles. We have also said that after two particles collide, they remain correlated. Now, assuming that they are not correlated in the next collision implies that one "must forget" what happened in the former collision. In this manner, instead of having a collision constrained by the preceding ones (in the sense that the collision parameters are imposed by the earlier collisions), we have a totally independent collision. This independent collision may be realised in more ways than if it were constrained. Thus we understand how the assumption of molecular chaos leads to irreversibility; it is equivalent to introducing successively into each collision more and more arbitrariness, more and more entropy.

We may ask how in our previous mechanical derivation we could get collisions not constrained by the earlier ones. The decisive factor introducing irreversibility was the replacement of expression (13.2) by expression (13.12) in going from (12.3) to (12.8). Let us examine both expressions. Expression (13.2) allows one to follow the evolution of the particles j and k along their continuous trajectories. On the contrary, expression (13.12), which is valid only for times $t > \alpha \tau_c$, follows the evolution of the particles before and after each collision, but not during the collisions. That means that in the Boltzmann equation (13.15) the rate of change of $\tilde{f}_1(\vec{p}_j;s)$ is not given as a continuous succession of values of this function at all times, but instead as a discrete succession of these values at times when there are no collisions.

To pass from (12.3) to (12.8) we introduced the limit (12.5) into (12.3). This limit is therefore responsible for the molecular chaos at all times. It is clear that on the time scale $s = Dt \sim t/t_c$ the duration of the collisions τ_c becomes negligible when t_c is large. So that, when we are dealing with times s we can ignore intervals of time of the order of τ_c. The dynamical evolution of the system is thus interrupted. We could say that there are "more instants" of time in the Liouville equation than in the Boltzmann equation. Irreversibility is due to this loss of information.

IV. THE IRREVERSIBILITY OF OTHER EQUATIONS

15. The Prigogine-Brout equation

I will now briefly discuss in the same manner some other irreversible equations. Let us begin with the Prigogine-Brout equation.

We consider a system with a Hamiltonian function

$$H = \sum_{j=1}^{N} \frac{\vec{p}_j^{\,2}}{2m} + \lambda \frac{1}{2} \sum_{\substack{j=1 \\ j \neq k}}^{N} \sum_{k=1}^{N} \varphi(|\vec{r}_j - \vec{r}_k|) = H^0 + \lambda H'. \tag{15.1}$$

Here H^0 is the Hamiltonian function of a system of free particles and H' may be regarded as a perturbation due to the forces between the particles. We assume that the perturbation is very weak and, thus, the parameter λ measuring the strength of the perturbation is considered to be small. The Liouville operator corresponding to the Hamiltonian (15.1) may be written as

$$L = L^0 + \lambda L' \tag{15.2}$$

and the Liouville equation for this system is

$$\frac{\partial}{\partial t} \varrho(\vec{r}^N, \vec{p}^N; t) = -i(L^0 + \lambda L') \varrho(\vec{r}^N, \vec{p}^N; t). \tag{15.3}$$

We now follow Zwansig's method and introduce a projection operator

$$P\ldots = \frac{1}{V^N} \int d\vec{r}^N \ldots \tag{15.4}$$

such that the momentum distribution function $f(\vec{p}^N; t)$ is given by

$$f(\vec{p}^N; t) = \frac{1}{V^N} \int d\vec{r}^N \varrho(\vec{r}^N, \vec{p}^N; t) = P \varrho(\vec{r}^N, \vec{p}^N; t). \tag{15.5}$$

By projecting the Liouville equation with P and (1-P) and using techniques similar to those used above for deriving the hierarchy (11.14), after some calculation and using additional assumptions on the behaviour at t=0, one finds:

$$\frac{\partial f(\vec{p}^N;t)}{\partial t} = -\lambda^2 \int_0^t \mathcal{P}L' e^{-i\tau L^0} L'\mathcal{P} f(\vec{p}^N;t-\tau)d\tau + O(\lambda^3). \tag{15.6}$$

One easily sees from this equation that <u>there is a characteristic time</u> of the system related to the coupling parameter λ . Indeed, if is sufficiently small this equation shows that $f(\vec{p}^N;t)$ is a slowly varying function of t. (It is clear that in the case that $\lambda=0$ the momentum distribution function of a system with Hamiltonian function

$$H^0 = \sum_{j=1}^{N} \frac{\vec{P}_j^{\,2}}{2m} \tag{15.7}$$

does not change with time). Thus one can define for our system a time t_c such that for $t < t_c$, the momentum distribution function may be regarded as constant. This time t_c is in this case of the order of λ^{-2} and therefore it is convenient to introduce a <u>time scale</u>

$$t/t_c \sim \lambda^2 t = s \tag{15.8}$$

and deal with times s instead of t ; for small values of λ the function

$$\tilde{f}(\vec{p}^N;s) = f(\vec{p}^N;s/\lambda^2) = f(\vec{p}^N;t) \tag{15.9}$$

has a rate of change measured on the scale (15.8) larger than that measured on scale t:

$$\frac{\partial \tilde{f}(\vec{p}^N;s)}{\partial s} = -\int_0^{s/\lambda^2} \mathcal{P}L' e^{-i\tau L^0} L'\mathcal{P}\tilde{f}(\vec{p}^N;s-\lambda^2\tau)d\tau + O(\lambda). \tag{15.10}$$

Taking in equation (15.6), the <u>limit of very weak coupling</u>

$$\lambda \to 0 \ , \ t \to \infty , \quad \lambda^2 t = s \ \text{ finite}, \tag{15.11}$$

which is equivalent to taking the limit $\lambda \to 0$ in equation (15.10), one obtains

$$\frac{\partial \tilde{f}(\vec{p}^N;s)}{\partial s} = -\left(\int_0^\infty \mathcal{P}L' e^{-i\tau L^0} L' \mathcal{P}\, d\tau\right)\tilde{f}(\vec{p}^N;s) = -\Psi \tilde{f}(\vec{p}^N;s). \tag{15.12}$$

Or, written on the scale t,

$$\frac{\partial f(\vec{p}^N;t)}{\partial t} = -\lambda^2 \Psi f(\vec{p}^N;t). \tag{15.13}$$

Equations (15.6) and (15.10), which are called "pre-master equations", are reversible; equations (15.12) and (15.13), which are called "master equations" are irreversible. The irreversibility is here introduced by means of the limit (15.11) which interrupts the dynamical evolution of the system. Let us note that in the derivation of the Prigogine-Brout equation (15.12), which we have sketched, the introduction of the irreversibility was carried out in a way similar to that used in our treatment of the Boltzmann equation. Using the weak-coupling limit (15.11), we have neglected the detailed process of changing of the momenta and we consider only the rate of change of the momentum distribution function during time intervals of the order of t_c.

16. Langevin and Fokker-Planck equations

The Langevin equation for Brownian motion may be written in the form

$$\frac{d}{dt}\vec{P}_B(t) = -\gamma \vec{P}_B(t) + \vec{R}(t), \tag{16.1}$$

where $\vec{P}_B(t)$ is the momentum of the Brownian particle at time t, γ is a friction coefficient and $\vec{R}(t)$ is a rapidly fluctuating, stochastic force. This equation can be derived, by means of projection operator techniques, from Newton's equation

$$\frac{d}{dt}\vec{P}_B(t) = iL\vec{P}_B(t), \tag{16.2}$$

where L is the Liouville operator of the system consisting of the Brownian particle and the particles of the fluid. As a first step one may derive the Mori equation

$$\frac{d}{dt}\vec{p}_B(t) = -\int_0^t \overleftrightarrow{\gamma}(\tau)\cdot\vec{p}_B(t-\tau)\,d\tau + \vec{R}(t), \tag{16.3}$$

which is just equation (16.2) written in another, equivalent form. There the tensor $\overleftrightarrow{\gamma}(\tau)$ is related to the force $\vec{R}(t)$ through

$$\overleftrightarrow{\gamma}(t) = \frac{1}{m_B kT}\langle\vec{R}\,\vec{R}(t)\rangle, \tag{16.4}$$

where m_B is the mass of the Brownian particle, k is Boltzmann's constant, T the absolute temperature, and $\langle\vec{R}\vec{R}(t)\rangle$ the correlation function of R, which is given by

$$\vec{R}(t) = e^{it(1-P)L}\left(\dot{\vec{p}}_B - \frac{1}{m_B kT}\vec{p}_B\cdot\langle\vec{p}_B\,\dot{\vec{p}}_B\rangle\right). \tag{16.5}$$

Here, P is the projection operator

$$P\ldots = \frac{1}{m_B kT}\vec{p}_B\cdot\langle\vec{p}_B\ldots\rangle = \frac{1}{m_B kT}\vec{p}_B\cdot\int f^{eq}\ldots d\Gamma \tag{16.6}$$

with the integration taken over the phase space of the whole system of Brownian and fluid particles, and with f^{eq} the canonical distribution function of the whole system.

The Mori equation is reversible as is Newton's equation. The first term on the right-hand side of equation (16.3) is not a dissipative term, and the force $\vec{R}(t)$ is not stochastic.

If we use the notation $m_f/m_B = \lambda^2$, $\overleftrightarrow{\gamma}' = m_B\overleftrightarrow{\gamma}$ and $\vec{p}_B = m_B\vec{v}_B$, we may write the Mori equation as

$$\frac{m_f}{\lambda^2}\frac{d\vec{v}_B(t)}{dt} = -\int_0^t \overleftrightarrow{\gamma}'(\tau)\cdot\vec{v}_B(t-\tau)\,d\tau + \vec{R}(t). \tag{16.7}$$

Now we note that the time interval necessary for a given change in the velocity of the Brownian particle to occur (for a given mean force exerted on it by the fluid particles) is proportional to its mass. When the mass of the Brownian particle is much larger than that of the fluid particles, there is a characteristic time t_c such that for times $t < t_c$ no appreciable change of $\vec{v}_B$ is seen. Only the various collisions between the Brownian particle and the fluid particles taking place during time intervals t_c are able to change $\vec{v}_B$ appreciably. We therefore introduce a time scale $s = \lambda^2 t$ in which the typical character of Brownian motion can be exhibited. With this new time scale the Mori equation reads

$$m_f \frac{d\tilde{\vec{v}}_B(s)}{ds} = -\int_0^{s/\lambda^2} \vec{\vec{\gamma}}'(\tau)\cdot\tilde{\vec{v}}_B(s-\tau\lambda^2)\,d\tau + \tilde{\vec{R}}(s), \tag{16.8}$$

where

$$\tilde{\vec{v}}_B(s) = \vec{v}_B(s/\lambda^2) = \vec{v}_B(t) \quad \text{and} \quad \vec{R}(t) = \tilde{\vec{R}}(s). \tag{16.9}$$

We take now the Brownian limit

$$\lambda^2 \to 0 \quad , t \to \infty \quad , \lambda^2 t = s \text{ finite} \tag{16.10}$$

and obtain

$$m_f \frac{d\tilde{\vec{v}}_B(s)}{ds} = -\left(\int_0^{\infty} \vec{\vec{\gamma}}'(\tau)\,d\tau\right)\cdot\tilde{\vec{v}}_B(s) + \tilde{\vec{R}}(s) \tag{16.11}$$

or, with the former time scale

$$m_B \frac{d\vec{v}_B(t)}{dt} = -\left(\int_0^{\infty} \vec{\vec{\gamma}}'(\tau)\,d\tau\right)\cdot\vec{v}_B(t) + \vec{R}(t). \tag{16.12}$$

Note that

$$\hat{\vec{\vec{\gamma}}}' = \int_0^{\infty} \vec{\vec{\gamma}}(\tau)\,d\tau = \lim_{z\to 0} \frac{1}{2kT}\int_{-\infty}^{\infty} e^{-z\tau}\langle \vec{R}\,\vec{R}(\tau)\rangle\,d\tau \tag{16.13}$$

is a constant and therefore the first term on the right-hand side of (16.12) is a dissipative term. Equation (16.13) expresses the fluctuation-dissipation theorem for this case, and one sees that

$$\langle \vec{R}\,\vec{R}(\tau)\rangle = 2kT\,\hat{\vec{\vec{\gamma}}}'\,\delta(\tau) \tag{16.14}$$

and $\vec{R}(\tau)$ is a fastly-fluctuating force.

An analogous limit must be employed for deriving the Fokker-Planck equation

$$\frac{df_B}{dt}=\left(\frac{\partial}{\partial t}+\frac{\vec{p}_B}{m}\cdot\frac{\partial}{\partial \vec{r}_B}+\langle F_{Bf}\rangle\cdot\frac{\partial}{\partial \vec{p}_B}\right)f_B(t)=\frac{\partial}{\partial \vec{p}_B}\cdot\hat{\vec{\gamma}}'\cdot\left[\frac{\vec{p}_B}{m_B}+kT\frac{\partial}{\partial \vec{p}_B}\right]f_B(t) \qquad (16.15)$$

from the Lebowitz-Résibois equation

$$\frac{df_B(t)}{dt}=\int_0^t \frac{\partial}{\partial \vec{p}_B}\cdot\vec{\vec{\gamma}}'(\tau)\cdot\left(\frac{\vec{p}_B}{m_B}+kT\frac{\partial}{\partial \vec{p}_B}\right)f_B(t-\tau)d\tau, \qquad (16.16)$$

which may be derived from the Liouville equation by using projection operator techniques. In (16.15) and (16.16) $f_B(t)$ is the reduced distribution function for the Brownian particle.

17. The Choh-Uhlenbeck equation

Let us return now to our derivation of the Boltzmann equation. The substitution of the formal solution of (11.13) for into (11.10) allowed us to obtain (11.14). If we now substitute this solution of (11.13) into (11.14) we get

$$\begin{aligned}\frac{\partial f_n(t)}{\partial t} &= D\int_0^t \frac{\partial K_{n,n+1}(\tau)}{\partial \tau} f_{n+1}(t-\tau)d\tau + D^2\int_0^t \frac{\partial K_{n,n+2}(\tau)}{\partial \tau} f_{n+2}(t-\tau)\,d\tau \\ &+ D K_{n,n+1}(t)h_{n+1}(0) + D^2 K_{n,n+2}(t)\,h_{n+2}(0) \\ &- iD^3\int_0^t K_{n,n+2}(\tau)\lim_{V\to\infty}(1-P_{n+2})\mathcal{L}_{n+2,n+3}\,h_{n+3}(t-\tau)\,d\tau\end{aligned} \qquad (17.1)$$

where

$$K_{n,n+2}(t) = -\lim_{V\to\infty}\int_0^t P_n\mathcal{L}_{n,n+1}e^{-i\tau L_{n+1}}(1-P_{n+1})\mathcal{L}_{n+1,n+2}\,e^{-i(t-\tau)L_{n+2}}. \qquad (17.2)$$

The structure of equation (16.1) is quite similar to that of (11.14). The primary difference between them is that the terms, which for finite t are of the order of D^2, have been written here in the same way as those of order D. In addition, we have in (17.1) a term of the order of D^3. Iterating ℓ times by this procedure gives

$$\frac{\partial f_n(t)}{\partial t} = \sum_{m=1}^{\ell} D^m \left\{ \int_0^t \frac{\partial K_{n,n+m}(\tau)}{\partial \tau} f_{n+m}(t-\tau)\, d\tau + K_{n,n+m}(t)\, h_{n+m}(0) \right\}$$
$$- i D^{\ell+1} \lim_{V\to\infty} \int_0^t K_{n,n+\ell}(\tau)(1-P_{n+\ell}) \mathcal{L}_{n+\ell,\,n+\ell+1}\, h_{n+\ell+1}(t-\tau)\, d\tau, \tag{17.3}$$

where

$$K_{n,n+m}(t) = (-i)^m \lim_{V\to\infty} P_n \mathcal{L}_{n,n+1}$$
$$\times \left\{ \prod_{k=1}^{m-1} \left[\int_{\tau_{k-1}}^{t} e^{-i\tau_k L_{n+k}} (1-P_{n+k}) \mathcal{L}_{n+k,\,n+k+1}\, e^{-i\tau_k L_{n+k+1}}\, d\tau_k \right]_{ord} \right\} e^{-itL_{n+m}} \tag{17.4}$$

Here, $\tau_0 = 0$. The index "ord" denotes that the product (which is unity if $m=1$) is ordered from left to right according to increasing values of k. Equations (17.1) and (17.2) become respectively (17.3) and (17.4), for the particular case when $m=2$. Iterating indefinitely by this procedure (i.e. if $\ell=\infty$) we obtain

$$\frac{\partial f_n(t)}{\partial t} = \sum_{m=1}^{\infty} D^m \left\{ \int_0^t \frac{\partial K_{n,n+m}(\tau)}{\partial \tau} f_{n+m}(t-\tau)\, d\tau + K_{n,n+m}(t)\, h_{n+m}(0) \right\} \tag{17.5}$$

<u>All these equations are exact in the thermodynamic limit</u>.

In particular if n=1, for times $t > 2\alpha\tau_c$ and neglecting the term which for finite times is of the order of D^3, equation (17.1) takes the form

$$\frac{\partial f_1(t)}{\partial t} = \int_0^t \left[D \frac{\partial K_{1,2}(\tau)}{\partial \tau} f_2(t-\tau) + D^2 \frac{\partial K_{1,3}(\tau)}{\partial \tau} f_3(t-\tau) \right] d\tau, \tag{17.6}$$

where $K_{1,2}(t)$ is given by (11.15) and $K_{1,3}(t)$ is given by (17.2), in the case n=1. For times t such that $t > \alpha\tau_c$, $K_{1,2}(t)$ becomes the <u>Boltzmann collision operator</u> and $K_{1,3}(t)$ may be written as

$$K_{1,3}(t) = \lim_{V\to\infty} \iint d\vec{r}_2\, d\vec{p}_2 \iint d\vec{r}_3\, d\vec{p}_3 \Big\{ i\, P^{(1)} L^{(1,2)} e^{-itL^{(1,2)}} (1 - 2\, P^{(1,2)})$$

$$- i\, P^{(1)} L^{(1,2)} e^{-itL^{(1,2,3)}} + i\, P^{(1)} L^{(1,2)} e^{-itL^{(1,2)}} P^{(1,2)} \sum_{j=1}^{2} e^{-itL^{(j,3)}}$$

$$+ \int_0^{\tau_c} P^{(1)} L^{(1,2)} \left(e^{-i\tau L^{(1,2)}} - e^{-itL^{(1,2)}} \right) P^{(1,2)} \qquad (17.7)$$

$$\times \sum_{j=1}^{3} L^{(j,3)} e^{-i(t-\tau) L^{(j,3)}} d\tau \Big\} \, .$$

This operator will be called "<u>triple collision operator</u>". We see, choosing the initial conditions which we assumed in deriving the Boltzmann equation, that there is no contribution in (17.6) due to the initial conditions on the equation for the rate of change of $f_1(t)$. However <u>it cannot be proved</u> that the solutions of (17.6) satisfy exactly the hypothesis of molecular chaos at every time.

Exoanding the functions $f_2(t-\tau)$ and $f_3(t-\tau)$ involved in (17.6) in a Taylor series around $\tau = 0$, and neglecting the terms which for finite times are of the order of D^3, or smaller, equation (17.6) may be written in the form

$$\frac{\partial f_1(t)}{\partial t} = D K_{1,2}(t) f_2(t) + D^2 K_T(t) f_3(t) \qquad (17.8)$$

The operator $K_T(t)$ is the <u>Choh-Uhlenbeck operator</u>

$$K_T(t) = -i \lim_{V\to\infty} \iint d\vec{r}_2\, d\vec{p}_2 \iint d\vec{r}_3\, d\vec{p}_3\; P^{(1)} L^{(1,2)}$$

$$\times \Big\{ e^{-itL^{(1,2,3)}} + e^{-itL^{(1,2)}} - e^{-itL^{(1,2)}} \sum_{j=1}^{2} e^{-itL^{(j,3)}} \Big\} \qquad (17.9)$$

<u>Equation (17.8) is an irreversible equation</u>. The justification for neglecting the various terms in arriving at this equation is given elsewhere. The important point is that we have obtained an irreversible equation by considering only times $t > 2\alpha\tau_c$ so that <u>equation (17.8) is not valid for all times</u>. We thus interrupt the continuous evolution of the system and look into it only at discontinuous instants of time. In this case we do not take a limit such as (12.5), because

by doing so we would get the Boltzmann equation. However, in considering times $t > 2\alpha\tau_c$ we are employing a certain kind of time scale just as in the case of the irreversible equations which we studied above.

18. Final remarks

In the derivation of the irreversible equations which I have presented to you there are some common points. Let us summarize them.

The first point is that we deal with large systems. The size of the system is introduced either through the thermodynamic limit (as in the case of the Boltzmann equation) or by considering that there is an equilibrium distribution function and a temperature in the system (as in the case of the Langevin equation). By considering large systems we remove the recurrences due to Poincaré's cycles.

The second point is that there is always a characteristic time (or frequency) associated with the system. We have denoted this time by t_c. This time is related to a characteristic parameter of the system (D, λ^2 or m_f/m_B) which is assumed to take small values. The rate of change of the studied function (the momentum distribution function, or the momentum of the Brownian particle) is proportional to the characteristic parameter and thus, for small values of it, the function is a slowly varying function of time t, and it is roughly constant in any interval of time of a duration of the order of $\Delta t < t_c$. In order to see an evolution of the system it is necessary to consider times $t > t_c$.

The third point is the introduction of a new time scale $s \sim t/t_c$ Then we take the limit considering the characteristic parameter to be very small and the time t large:

$$t_c \to \infty, \quad t \to \infty, \quad t/t_c \sim s \quad \text{finite.} \tag{18.1}$$

This limit introduces the irreversibility because it interrupts the dynamical evolution of the system in a special way. Indeed in passing from one instant s to another, we do not consider in detail the exact evolution of the system. In the case of the Boltzmann equation, for instance, the passage from one time s to another is not made as a consequence of a specified collision. Perhaps this is better seen in the case of Brownian motion, in which the passage from time s to another time is not made as a consequence of one collision between the

Brownian particle and one of the fluid, but as a consequence of the many collisions between the Brownian particle and the particles of the fluid. Indeed, during the time t_c many fluid particles collide with the Brownian particle and the change of velocity of the latter is the result of these very many collisions.

Another common point in the derivation of the irreversible equations is the use of projection operator techniques. This is a matter of taste and is not required for irreversibility.

We conclude therefore that we cannot derive irreversible equations from reversible ones if we do not force the irreversibility, because the dynamical evolution cannot introduce the irreversibility by itself. This forced introduction of irreversibility has been made here by interrupting the dynamical evolution by considering a time scale such that in the passage from one instant to another some amount of information about the detailed evolution is lost. This is done by means of a hypothesis, namely that the limit (18.1) needs to be considered. (In the case of the Choh-Uhlenbeck equation this limit is not taken, but a similar hypothesis is introduced in considering only times larger than $2\alpha\tau_c$).

We have not studied here the properties which a system must have in order to behave irreversibly, i.e. the structure of the exact reversible evolution equations of the system, which when considering a limit such as (18.1), give rise to irreversible equations. Our aim here has only been to show how irreversibility must be forced and how it is a consequence of some loss of information during the dynamical evolution of the system.

I gratefully acknowledge the aid given to me by Prof. Dr. W. Smith, who read this manuscript, and by Miss P. Gandia, who prepared it.

REFERENCES

For the latest ideas of the Brussels group on the mechanical theory of irreversibility, see

I. Prigogine, Acta Phys. Austriaca, suppl X (1973) 401.

I. Prigogine, C. George, F. Henin and L. Rosenfeld, Chem. Seripta, 4 (1973) 5.

For our sections 2 to 6, see

R.C. Tolman, "The princples of Statistical Mechanics", Oxford 1962.

R. Jancel, Foundations of Classical and Quantum Statistical Mechanics Oxford, 1969.

For our sections 7 and 8, see

K. Huang, Statistical Mechanics, New York, 1963

For the thermodynamic limit, see

P. Mazur and E. Montroll, J. Math. Phys., 1 (1960) 70.

H. Wergeland, in Sitges International School of Physics, May 1974, New York, 1972.

For our sections 11 to 13, see

P. Mazur and J. Biel, Physica, 32 (1966) 1633.

For the Prigogine-Brout equation, see

R. Brout and I. Prigogine, Physica 22 (1956) 621.

R.W. Zwanzig, In Lectures on Theoretical Physics, vol. 111, New York, 1960.

For the equations of Brownian Motion, see

H. Mori, Progr. Theor. Phys. 33 (1965) 423.

J.L. Lebowitz and P. Résibois, Phys. Rev. 139A (1965) 1101.

For the Choh-Uhlenbeck equation, see

J. Biel and J. Marro, Nuovo Cimento, 20 B (1974) 25.

For a discussion of why some terms of our hierarchies may be neglected, see

J. Biel, J. Marro and L. Navarro, Phys. Lett. 44 A (1973) 41; and Nuovo Cimento 20 B (1974) 55.

For a kinetic theory of irreversibility, see

R. Balescu, in Sitges International School of Physics, May 1972, New York, 1972.

ERGODIC THEORY AND STATISTICAL MECHANICS *

Joel L. Lebowitz
Belfer Graduate School of Science
Yeshiva University, New York,
NEW YORK 10033.

I. INTRODUCTION
  1.1 Scope of lectures

II. ERGODICITY AND ENSEMBLE DENSITIES
  2.1 The ergodic problem
  2.2 Brief history of ergodic theory

III. SYSTEMS OF OSCILLATORS AND THE KAM THEOREM

IV. MIXING
  4.1 Time correlations and transport coefficients

V. K- AND BERNOULLI SYSTEMS
  5.1 The baker's transformation
  5.2 The Kolmogorov-Sinai entropy

VI. ERGODIC PROPERTIES AND SPECTRUM OF THE INDUCED UNITARY TRANSFORMATION

VII. INFINITE SYSTEMS

APPENDIX: Ergodic properties of simple model system with collisions

REFERENCES

---

* Research supported by the United States Air Force Office of Scientific Research Grant Number 73-2430A.

I. INTRODUCTION

I would like to talk about some of the progress that has been made in recent years in the mathematical theory of measure preserving transformations: ergodic theory. Since the dynamical flow in the phase space, which describes the time evolution of a Hamiltonian system, is an example of such a transformation, this work has, in my opinion, much relevance to statistical mechanics and to the question of irreversibility. While the recent progress in this field is due mostly to the work of mathematicians like von Neumann, Hopf, Kolmogorov, Sinai, Ornstein and others, the origins of the subject go back to the founding fathers of statistical mechanics; Boltzmann, Maxwell, Gibbs and Einstein. These men and their followers invented the concept of ensembles to describe equilibrium and nonequilibrium macroscopic systems. In trying to justify the use of ensembles, and to determine whether the ensembles evolved as expected from nonequilibrium to equilibrium, they introduced further concepts such as "ergodicity" and "coarse graining". The use of these concepts raised mathematical problems that they could not solve, but like the good physicists they were they assumed that everything was or could be made all right mathematically and went on with the physics.

Their mathematical worries, however, became the seeds from which grew a whole beautiful subject called "ergodic theory". Here I will describe some recent (and some not so recent)developments that partially solve some of the problems that worried the Founding Fathers. Although results are not yet well enough developed to answer all the questions in this area that are of interest to physicists, such as the derivation of kinetic equations or the general problem of irreversibility, they do make a start.

Since it has been only in the last few years that physicists have again become deeply involved in this subject, there is a big gap in the statistical mechanics literature concerning the developments in ergodic theory which have occurred in the last forty years. As a recent convert I have preached the gospel of ergodic theory many times in many places. Some of you will therefore have heard some parts of this talk before or you may have read it in some of the reviews of the subject I have written. Indeed these notes contain some (almost) verbatim transcripts of my article with Penrose in Physics Today.

I have also borrowed freely from joint works with Goldstein, Lanford and Aizenman, as well as independent works of these authors. The credit, but certainly not the blame, for what I will present here is therefore a shared one.

A partial list of references, especially suited for physicists, is given at the end.

1.1 Scope of lectures

Ergodic theory is concerned with the time evolution of Gibbs ensembles. It has revealed that there is more to the subject than the simple question of whether a dynamical system is ergodic (which means roughly, whether the system, if left to itself for long enough, will pass close to nearly all the dynamical states compatible with conservation of energy). Instead there is a hierarchy of properties that a dynamical system may have, each property implying the preceding one, and of which ergodicity is only the first (see Table 1). The next one, called "mixing" provides a formulation of the type of irreversible behaviour that people try to obtain by introducing coarse-grained ensembles. At the top of our hierarchy is a condition (the Bernoulli condition) ensuring that in a certain sense the system, though deterministic, may appear to behave as randomly as the numbers produced by a roulette wheel.

Some of the mathematical results we shall be discussing have established the positions of some model physical systems in this hierarchy. Of particular interest to physicists is the work of Ya. Sinai on the hard-sphere system, which shows that this system is both ergodic and mixing. I shall also discuss some work by A.N. Kolmogorov, V.I. Arnold and J. Moser on systems of coupled anharmonic oscillators, which shows that, contrary to a common assumption, these systems may not even reach the "ergodic" rung on the ladder. (G.H. Walker and J. Ford have described this work for physicists).

All the physical systems I shall discuss obey classical mechanics or are models of such systems. I shall consider first systems which have a finite number of degrees of freedom and are confined to a finite region of physical space. Here the concepts, if not the proofs, are basically simple. Later I shall discuss infinite systems, by which I mean the limit of a finite system as its size increases without bound. The concepts involved here are more difficult or at least may be less familiar to you. Also the basic ingredient for the study of the ergodic properties of such systems, the existence of a well-defined

Infinite systems: Ideal gas. Hard rod system, perfect harmonic crystal Baker's transformation, Geodesic flow on space of negative curvature, Particle moving among fixed convex scatterers	Bernouilli system	Equivalent to roulette wheel
Infinite system: Lorentz gas Two or more hard spheres moving in two or higher dimensions?	K-system	Essential randomness
Simple model system with collisions	Mixing system	Approach to equilibrium
One dimensional harmonic oscillator	Ergodic system	Use of microcanonical ensemble

Table 1. Hierarchy of Systems

The middle column lists the systems that will be discussed in these lectures, with the strongest at the top. Every mixing system is ergodic, every K-system is mixing and every Bernoulli system is a K-system. At the left are examples of the system and at the right physical interpretations or implications.

time evolution, has only very recently been proved by Lanford. My discussion of infinite systems will therefore be even more sketchy than for the finite case. The reason for discussing them at all is that it is only in this limit, usually referred to as the thermodynamic limit, that the distinction between microscopic and macroscopic observables, which appears essential to any complete theory of irreversibility, can be formulated precisely.

My reason for not dealing with quantum systems here is that a finite quantum system can never exhibit any of the properties higher than simple ergodicity in our hierarchy (although, of course, a large quantum system may approximate closely the behaviour characterized by these concepts). This is because the spectrum of a finite quantum system is necessarily discrete, whereas for a finite classical system the spectrum (of the Liouville operator) can be continuous. Infinite quantum systems can, and do, exhibit ergodic behaviour similar to classical systems. Some very beautiful work on such systems has just, very recently, been done by Haag, Kastler and Eva Trych-Pohlmeyer, who is here. I hope to elaborate slightly on these remarks about quantum systems during these lectures.

I should point out right here that care must be exercised in drawing analogies between the ergodic properties of finite and infinite systems, as the dependence of these properties on the interactions between the particles, and thus also their physical interpretation, may be very different in the two cases. Thus while a finite ideal gas (classical system of non-interacting point particles) is not even ergodic, the infinite ideal gas has the strongest possible ergodic properties: it is a Bernoulli system, c.f. Table 1.

The explanation of the good ergodic properties of the infinite ideal gas is simple: local disturbances 'fly off' unhindered to infinity where they are no longer observable. This means that the 'approach' or better the return to equilibrium of a large (infinite) system, which is perturbed locally away from equilibrium, may occur even in the absence of a local 'dissipative' mechanism such as is provided by collisions. It can happen simply, as it does in the ideal gas (or the perfect harmonic crystal) by the disturbance disappearing from sight by the free streaming motion of the particles (phonons).

This kind of return to equilibrium is of course not described by a kinetic or hydrodynamic equation and is therefore not the kind of irreversibility which is of interest in real physical systems. It is therefore necessary to introduce additional structure, to that provided by ergodic theory alone, to distinguish between infinite systems of the ideal gas type and more realistic physical systems, such as the Lorentz gas, where there exists a local mechanism, e.g. collisions, for the approach to equilibrium. A start in this direction has been made by S. Goldstein who considered the (generalized) ergodic properties of an infinite system under the joint group consisting of the time evolution and space translations. He showed that these two different kinds of systems can indeed be clearly distinguished with this sharper tool. The work of Haag, Kastler and Pohlmeyer, mentioned earlier, also has some bearing on this question and I hope to return to this point later.

II. ERGODICITY AND ENSEMBLE DENSITIES

Before we go on to discuss the new results, we review some mathematical definitions. If our dynamical system has n degrees of freedom, we can think of its possible dynamical states geometrically, as points in a 2n-dimensional space (phase space), with n position coordinates and n momentum coordinates. If the energy of the system is E, then its dynamical state x ($= (q_1 \ldots q_n, p_1 \ldots p_n)$) must lie on the <u>energy surface</u> $H(x) = E$, where H is the Hamiltonian function. We denote the energy surface, which is $(2n - 1)$-dimensional, by S_E or just S and assume that S is smooth and of finite extent; for example in the case of a system of harmonic oscillators, for which the Hamiltonian is a quadratic form, the energy surfaces are $(2n - 1)$-dimensional ellipsoids.

The time evolution of the system causes x to move in phase space, but since we are assuming our system to be conservative the point x always stays on the energy surface. If the system is in some state x at some time t_0 then its state at any other time $t_0 + t$ is uniquely determined by x and t (only). Let us call the new state $\phi_t(x)$. This defines a transformation ϕ_t from S onto itself. There is one such function for each value of t.

We shall want to integrate dynamical functions (that is, functions of the dynamical state) over S. When doing this it is convenient not to measure $(2n - 1)$-dimensional "areas" on the surface S in the usual way but to weight the areas in such a way that the natural motion of the system on S carries any region R (after any time t) into a region $\phi_t(R)$ of the same area. This can be accomplished by defining the weighted area of a small surface element near x, dx, to be such that $dx dE$ is the correct Euclidean 2n-dimensional volume element of a "pill box" with base dx and height dE. Formally

$$dx = d\sigma_E(x)/|\nabla H(x)|, \qquad x \in S_E$$

where $d\sigma_E(x)$ is the 'usual' surface area on S_E and $\nabla H(x)$ is the gradient of the Hamiltonian

$$|\nabla H(x)| = \left\{\sum_{i=1}^{n}\left[\left(\frac{\partial H}{\partial q_i}\right)^2 + \left(\frac{\partial H}{\partial p_i}\right)^2\right]\right\}^{\frac{1}{2}}.$$

By a <u>Gibbs ensemble</u> we mean an infinitely large hypothetical collection of systems, all having the same Hamiltonian but not necessarily the same dynamical state. We shall only consider ensembles whose systems all have the same energy, so that their dynamical states are distributed in some way over some energy surface S. It may happen that this distribution can be described by an ensemble density; by this we mean a real-valued function ρ on S such that the fraction of members of the ensemble whose dynamical states lie in any region R on the surface S is

$$\int_R \rho(x)dx$$

with dx the weighted area defined above. The simplest ensemble density on S is given by

$$\rho(x) = C \quad (\text{all } x \text{ in } S)$$

where C is a constant, which can be determined from the normalization condition $\int_S C\,dx = 1$. This is called the microcanonical ensemble on S.

The systems constituting the ensemble evolve with time, so that the ensemble density will depend on time. The rule connecting the ensemble densities ρ_t describing the same ensemble at different times t is Liouville's theorem, which can be written

$$\rho_t(x) = \rho_o\left[\phi_{-t}(x)\right] \quad (\text{all } t, \text{ all } x \text{ in } S)$$

where $\rho_o(x)$ is the ensemble density at time zero. For example, Liouville's theorem shows that the density of the microcanonical ensemble does not change with time: If

$$\rho_o(x) = C$$

for all x in S, then Liouville's theorem gives, for any t,

$$\rho_t(x) = C$$

for all x in S.

2.1 The ergodic problem

The principal success of ensemble theory has been in its application to equilibrium. To calculate the equilibrium value of any dynamical function we average it over a suitable ensemble. The same ensemble also enables us to estimate the magnitude of the fluctuations of our dynamical function. To ensure that the calculated averages are independent of time, we use an invariant ensemble; that is, one for which the fraction of members of the ensemble in every region R on the energy surface S is independent of time. We already know one invariant ensemble: the microcanonical, whose ensemble density is uniform on S. Before we can use it confidently to calculate equilibrium values, however, we would like to be sure that this is the only invariant ensemble: If other invariant ensembles exist then, in principle, they could do just as well for the calculation of equilibrium properties, and we would have to choose which to use in a particular situation.

There are two questions to settle: the first is whether there are any invariant ensembles on S that do not have an ensemble density. In general there are; for example in the case of a hard-sphere system in a box one could have an invariant ensemble where every particle moves on the same straight line being reflected at each end from a perfectly smooth parallel wall (see fig. 1).

The obviously exceptional character of this motion is reflected mathematically in the fact that this ensemble, though invariant, is confined to a region of zero "area" on S and therefore has no ensemble density. To set up such a motion would presumably be physically impossible because the slightest perturbation would rapidly destroy the perfect alignment. It is therefore natural to rule out such exceptional ensembles by adopting the principle that any ensemble corresponding to a physically realizable situation must have an ensemble density.

There remains the second part of the question: Are there any invariant ensembles on S that do have a density but differ from the microcanonical ensemble? This is equivalent to the ergodic problem in which one compares the time averages of a dynamical function f,

$$f^{*}(x) = \lim_{T \to \infty} \frac{1}{T} \int_{0}^{T} f(\phi_t(x))\,dt$$

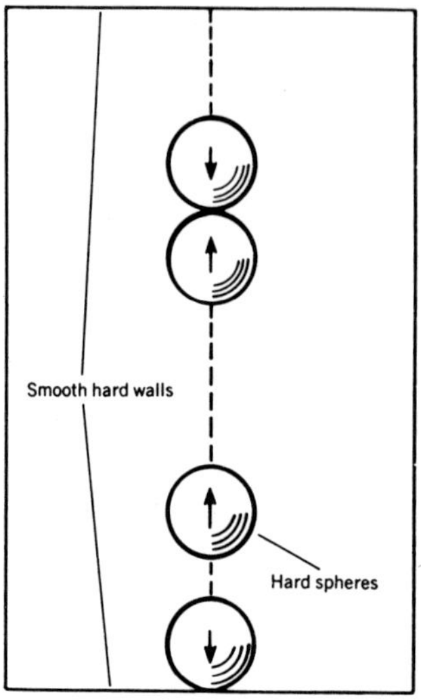

Fig. 1. Ensemble with no ensemble density.

Hard spheres move up and down the dotted line, which meets the perfectly smooth hard walls at right angles. Collisions between particles and collisions with the walls do not deflect the particles from the line if they are perfectly aligned at the start. An ensemble of such systems has no ensemble density because it is concentrated on a region on the energy surface with zero area (just as the area of a line or of a line segment in a plane is zero).

with its microcanonical ensemble average

$$\langle f\rangle = \int_S f(x)dx \Big/ \int_S dx \; .$$

A system is said to be ergodic on its energy surface S if time averages are in general equal to ensemble averages; that is, if for every integrable function f we have

$$f^*(x) = \langle f \rangle \qquad (1)$$

for almost all points x on S. "Almost all" means that if M is the set of points x for which eq. 1 is false, we have $\int_M dx = 0$. The answer to our second question is given by a theorem , which we shall not prove: the microcanonical ensemble density is the only invariant ensemble - that is, the only one satisfying $\rho[\phi_t(x)] = \rho(x)$ for all x in S - if and only if the system is ergodic on S.

The physical importance of ergodicity is that it can be used to justify the use of the microcanonical ensemble for calculating equilibrium values and fluctuations. Suppose f is some macroscopic observable and the system is started at time zero from a dynamical state x, for which $f(x)$ has a value that may be very far from its equilibrium value. As time proceeds, we expect the current value of f, which is $f[\phi_t(x)]$, to approach and mostly stay very close to an equilibrium value with only very rare large fluctuations away from this value. This equilibrium value should therefore be equal to the time average f^* because the initial period during which equilibrium is established contributes only negligibly to the formula defining $f^*(x)$. The theorem tells us that this equilibrium value is almost always equal to $\langle f \rangle$, the average of f in the microcanonical ensemble, provided the system is ergodic.

To justify the use of the microcanonical ensemble in calculating equilibrium fluctuations we proceed in a similar way. For some observable event A (such as the event that the percentage of gas molecules in one half of a container exceeds 51%) let R be the region in phase space consisting of all phase points compatible with the event A; that is the event A is observed if and only if the phase point is in R at time t. If the system is observed over a long period of time, the fraction of time during which event A will be observed is given by the time average $g^*(x_0)$, where x_0 is the initial dynamical state and g is defined by

$$g(x) = \begin{cases} 1 & \text{if } x \text{ is in } R \\ 0 & \text{if not.} \end{cases}$$

The ergodic theorem tells us that for almost all initial dynamical states this fraction of time is equal to the ensemble average of g, which is

$$\langle g \rangle = \int_R dx \Big/ \int_S dx \, .$$

This is just the "probability" of the event A as calculated in the microcanonical ensemble on S.

Another way of defining ergodicity is to say that any integrable invariant function is constant almost everywhere. That is to say, if

f is an integrable function satisfying the condition that

$$f[\phi_t(x)] = f(x)$$

for all x in S, then there is a constant c such that $f(x)$ equals c for almost all x: In other words, the set M of points x for which $f(x)$ does not equal c satisfies $\int_M dx = 0$. This has the physical interpretation that for a Hamiltonian system ergodic on S every integrable constant of the motion is constant on S. Furthermore if ergodicity holds on each S_E then there are no integrable constants of the motion other than functions of the energy E. Indeed, if there were other constants of the motion (for example angular momentum if the Hamiltonian had an axis of symmetry) we could construct invariant densities that were not microcanonical by taking $\rho(x)$ to be a function of one of these other constants of the motion, and so the system would clearly be nonergodic. (When such additional constants of the motion exist they must be taken into account in the statistical mechanics and thermodynamics of the system; the standard methods, based on the microcanonical ensemble, must then be generalized for these systems.)

These relationships between ergodicity and constants of the motion are a consequence of Birkhoff's theorem that ergodicity, as defined by the equality of the time average to the ensemble average, eq. 1, is equivalent to the energy surface being "metrically transitive". Stated precisely this means that a system is ergodic on S if and only if <u>all</u> the regions R on S left invariant by the time evolution, $\phi_t(R) = R$, either have zero area or have an area equal to the area of S.

The difficult part of Birkhoff's Theorem is to show that $f^*(x)$, which involves taking the time average over infinite times, actually exists for almost all x when $f(x)$ is an integrable function. It is then relatively easy to show that $f^*(x)$ is time invariant; that is, $f^*[\phi_t(x)] = f^*(x)$, and that ergodicity is equivalent to S being metrically intransitive.

2.2 Brief history of ergodic theory

The "ergodic hypothesis" was introduced by Boltzmann in 1871. To quote Maxwell "..(it) is that the system, if left to itself in its actual state of motion, will, sooner or later, pass through every phase which is consistent with the equation of energy." In our notation "phase" means dynamical state and the original ergodic hypothesis means that if y and x are any two points on the energy surface S_E, then $y = \phi_t(x)$ for some t. The ergodic hypothesis thus stated was proven to be false, whenever S_E has dimensionality greater than one, by A. Rosenthal and M. Plancherel in 1913. S.G. Brush gives a nice account of the early work on this problem.

The definition of an ergodic system given in eq. 1 can be shown to imply what is sometimes called the "quasi-ergodic" hypothesis, which replaces "every phase" in Maxwell's statement by "every region R on S_E of finite area", with the further qualitication that this is true for "almost all" dynamical states. Indeed as was pointed out, the fraction of time that the system will spend in R is equal, for an ergodic system, to the fraction of the area of S_E that is occupied by R.

III. SYSTEMS OF OSCILLATORS AND THE KAM THEOREM

We shall now consider some examples of ergodic and non-ergodic systems. The simplest example of an ergodic system is the simple harmonic oscillator whose Hamiltonian (in some suitable units) is

$$H(q,p) = \frac{1}{2}\omega(p^2 + q^2)$$

where ω is the angular frequency. The transformation ϕ_t for this system is a rotation through angle ωt in the (q,p) plane. The trajectories, which here coincide with the energy surfaces S_E, are circles of radius $(2E/\omega)^{1/2}$. (The surface element dx is here proportional to the ordinary length of an arc segment.) To be invariant under the transformation ϕ_t an ensemble density on S must be unaffected by rotations and is therefore a constant. It follows, then, that the only invariant density is the microcanonical density and so the simple harmonic oscillator is ergodic.

Almost as simple is the multiple harmonic oscillator (physically, say, an ideal crystal), that is, a system with two or more degrees of freedom whose potential energy is a quadratic form in the position coordinates. Unlike the simple harmonic oscillator it cannot be ergodic, because it has constants of the motion (the energies of the individual normal modes) that are not constant on the energy surfaces (the surfaces of constant total energy).

It used to be thought that this lack of ergodicity was an

accident and that any small anharmonicity (such as would inevitably be present in a real system) must make the system ergodic by permitting transfer of energy from one mode to another. In 1954, however, Kolmogorov announced results that contradicted this belief. In 1955, Enrico Fermi, J. Pasta and S.W. Ulam carried out a computer simulation of such a system. Initially, they excited one mode only, and instead of the equipartition of the energy between all modes predicted by the microcanonical ensemble they found that most of it appeared to remain concentrated in a few modes; this indicated that anharmonic oscillator systems may not be ergodic.

The lack of ergodicity was proved rigorously by Kolmogorov, Arnold and Moser. They found that if the frequencies of the unperturbed oscillators are not "rationally connected" (that is, if no rational linear combination of them is zero) then, in general, adding to the Hamiltonian an anharmonic perturbation sufficiently small compared to the total energy does not make the system ergodic. The unperturbed trajectories (possible paths of the phase point) all lie on n-dimensional surfaces in S (which has 2n-1 dimensions) called "invariant tori", and "KAM" prove that under a weak perturbation most trajectories continue to lie within smooth n-dimensional tori, so that the perturbed system is also non-ergodic. The trajectories that do not lie on the new invariant tori are, on the other hand, very erratic and may fill some $(2n - 1)$-dimensional region densely. One consequence of this very complicated behaviour is that even though the system is not ergodic the motion can no longer be decomposed into independent normal modes.

Similar results probably hold also for rationally connected frequencies (which cannot be treated rigorously, although they are of more physical interest); thus Michael Hénon and Carl Heiles carried out computer calculations for the Hamiltonian

$$H = \frac{1}{2}(p_1^2 + p_2^2 + q_1^2 + q_2^2) + (q_1^2 q_2 - \frac{1}{3}q_2^3) \qquad (3.1)$$

whose unperturbed frequencies $\omega_1 = 1$, $\omega_2 = 1$, are rationally connected since $1\cdot\omega_1 - 1\cdot\omega_2 = 0$. They found that the energy surfaces with E equal to 1/12, 1/8 and probably also 1/6 are not ergodic (see figure 2). As seen in the diagrams the fraction of the area corresponding to smooth curves (which are responsible for the non-ergodic behaviour) decreases as the energy increases.

For a general system of anharmonic oscillators, such as a real crystal, we expect similar behaviour, with the fraction of S_E corresponding to non-ergodic behaviour decreasing as E increases, and

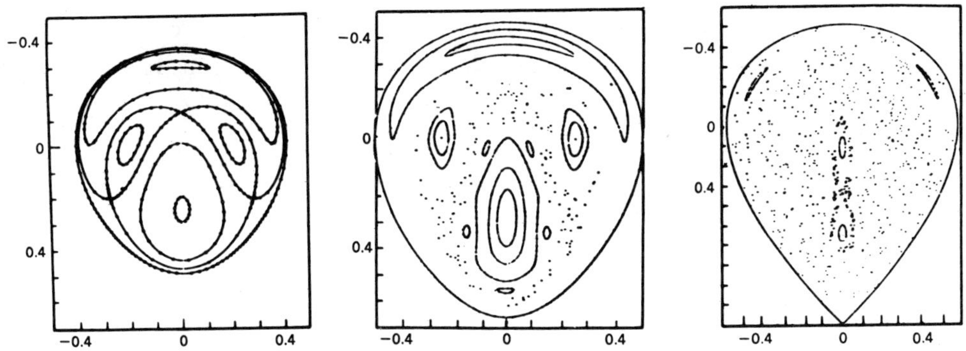

Fig. 2. Nonergodicity of an anharmonic oscillator system with rationally connected frequencies.

The Hamiltonian for this system is given in eq. (3.1). Michael Hénon and Carl Heiles did computer calculations for this system and found that energy surfaces S_E with E equal to 1/12, 1/8 and probably 1/6 are not ergodic. The planes shown here are intersections of the surfaces q_1 equal to zero with S_E for E equal to 1/12 (a), 1/8 (b) and 1/6 (c), and the points are the intersections of a trajectory with this plane. When the trajectory lies on a smooth two-dimensional invariant torus, the intersection points form a smooth curve, but the intersections of an "erratic" trajectory (one that does not lie on a smooth curve) are more or less random. Note that the fraction of area corresponding to smooth curves (which are responsible for the nonergodic behaviour) decreases with increasing energy.

probably disappearing altogether at some critical energy, above which the system would be ergodic and perhaps also show the stronger properties that we shall discuss. At present very little is known about the magnitude of this critical energy in a system with many degrees of freedom.

In the case of gases, the situation is somewhat different. If there were no interaction at all between the molecules then the energy of each molecule would be an invariant of the motion, so that the system (an ideal gas) would be nonergodic. The KAM theorem would therefore lead us to expect nonergodicity to persist in the event of a sufficiently weak interaction between the particles. The actual interactions, however, are not weak because two molecules very close together repel each other strongly; consequently the theorem does not apply. A simple model of this type is the hard-sphere gas enclosed in a cube with perfectly reflecting walls or periodic boundary conditions. Sinai has outlined a proof that this system _is_ ergodic; he has published a detailed proof, based on the same ideas, for a particle moving in a periodic box containing any number of rigid convex elastic scatterers. We shall refer again to this important result.

IV. MIXING

We have seen how to formulate a condition to ensure that the equilibrium properties of a dynamical system are determined by its energy alone and can be calculated from a microcanonical ensemble. This ergodicity condition does not, however, ensure that if we start from a non-equilibrium ensemble the expectation values of dynamical functions will approach their equilibrium values as time proceeds. An illustrative example is given by the harmonic oscillator. For the harmonic-oscillator system, Liouville's theorem shows that the ensemble density repeats itself regularly at time intervals of $2\pi/\omega$; therefore all ensemble averages also have this periodicity, and so cannot irreversibly approach their equilibrium values.

To ensure that our ensembles approach equilibrium in the way we would expect of ensembles composed of real systems, we need a stronger condition than ergodicity. To see what is required, let us start at $t = 0$ with some ensemble density $\rho_0(x)$ on S, which is supposed to represent the initial non-equilibrium state. At a later time t the ensemble density is, by Liouville's theorem, $\rho_0[\phi_{-t}(x)]$. The expectation value of any dynamical variable f at time t is therefore

$$\int_S f(x)\,\rho_0\left[\phi_{-t}(x)\right]\,dx \qquad (4.1)$$

As t becomes large we would like this integral to approach the equilibrium value of f, which is (for an ergodic system) $\int_S f(x)\,dx \Big/ \int_S dx$. A sufficient condition for this is that the system should satisfy the condition called <u>mixing</u> which is that for every pair of functions f and g whose squares are integrable on S, we require

$$\lim_{t\to\pm\infty} \int_S f(x)g(\phi_{-t}(x))dx = \frac{\int_S f(x)dx \int_S g(x)}{\int_S dx} \quad . (4.2)$$

The special case where g is ρ_0 shows that integral (4.1) will approach the equilibrium value of f for large t when the system is mixing. Another way of looking at this condition is that it requires every equilibrium time-dependent correlation function such as $\langle f(x) g[\phi_t(x)] \rangle$ to approach a limit $\langle f \rangle \langle g \rangle$ as t approaches $\pm \infty$.

The condition of mixing can be shown to be equivalent to the following requirement: if Q and R are arbitrary regions in S, and an ensemble is initially distributed uniformly over Q, then the fraction of members of the ensemble with phase points in R at time t will approach a limit as t approaches ∞; this limit equals the fraction of the area of S occupied by R.

This condition can be stated formally as follows: Let $\mu(A)$ denote the normalized area of a set A on the energy surface S, i.e. the probability in the microcanonical ensemble of finding the system in A,

$$\mu(A) = \int_A dx \Big/ \int_S dx \quad .$$

(For simplicity we shall set $\int_S dx = 1$ from now on.)

A system is mixing iff for any sets A and B

$$\mu(A_t \cap B) \xrightarrow[t \to \pm\infty]{} \mu(A)\mu(B) \,, \qquad A_t = \phi_t(A) \quad . \tag{4.3}$$

We can interpret this to mean that if we start with an ensemble at $t = 0$ such that all the systems are in region A,

$$\rho_0(x) = \begin{cases} [\mu(A)]^{-1} & \text{if } x \in A \\ 0 & \text{otherwise} \end{cases} ,$$

then the fraction of systems in this ensemble that will be in region B at time t will approach $\mu(B)$, see figure 3.

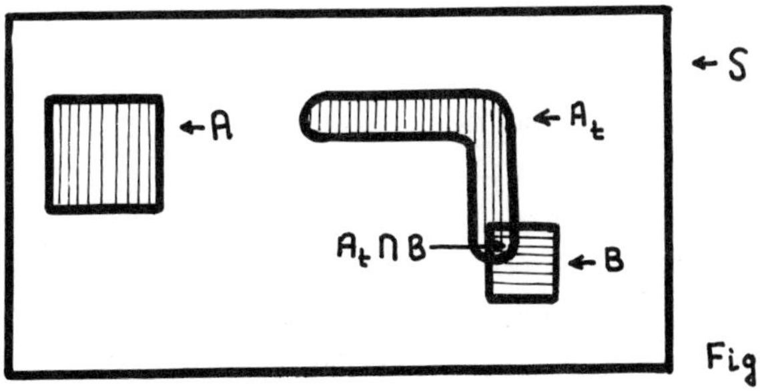

Figure 3.

Mixing is a stronger condition than ergodicity: it can easily be shown to imply ergodicity but is not implied by it, as we have seen in the case of a simple harmonic oscillator.

The proof is simple. Suppose that A is an invariant set, $A_t = A$. Then if the system is mixing we must have,

$$\mu(A_t \cap A) = \mu(A) = [\mu(A)]^2$$

so that $\mu(A) = 0$ or $\mu(A) = 1$ which, since it holds for all invariant sets, implies ergodicity as was noted earlier.

The mathematical definition of mixing was introduced by John von Neumann in 1932 and developed by E. Hopf, but goes back to J. Willard Gibbs, who discusses it by means of an analogy: "... the effect of stirring an incompressible liquid .... Let us suppose the liquid to contain a certain amount of colouring matter which does not affect its hydrodynamic properties ... (and) that the colouring matter is distributed with variable density. If we give the liquid any motion whatever ... the density of the colouring matter at any same point of the liquid will be unchanged ... Yet... stirring tends to bring a liquid to a state of uniform mixture."

Gibbs saw clearly that the ensemble density ρ_t of a mixing system does not approach its limit in the usual "fine-grained" or "pointwise" sense of $\rho_t(x)$ approaching a limit as $t \to \infty$ for each fixed x. Rather, it is a "coarse-grained" or "weak" limit, in which the average of $\rho_t(x)$ over a region R in S approaches a limit as $t \to \infty$ for each fixed R. (A similar distinction applies in defining the entropy. The fine-grained entropy $-k\int \rho_t(x)\log\rho_t(x)dx$, where k is Boltzmann's constant, retains its initial value forever, but the coarse-grained entropy $-k \int \bar{\rho}_t(x)[\log \bar{\rho}_t(x)]dx$, where $\bar{\rho}_t(x)$ is a coarse-grained ensemble density obtained by averaging $\rho_t(x)$ over cells in phase space, does change for a non-equilibrium ensemble, and approaches as its limit the equilibrium entropy value $k \log \int_S dx$).

It is sometimes argued that one cannot have a proper approach to equilibrium for any finite mechanical system because of a theorem, due to Poincaré, that every such system eventually returns arbitrarily close to its initial state. (The time involved, however, will be enormously large for a macroscopic system. Boltzmann, for example, estimated a typical Poincaré period for 100 cm^3 of gas to be enormously long compared to 10 raised to the power 10 raised to the power 10 years.) Here, however, we are considering ensembles, not individual systems, and the mixing condition guarantees that the ensemble density eventually becomes indistinguishable from the microcanonical density and remains so forever after. It is true that individual systems in the ensemble will return to their initial dynamical states, as required by Poincaré's theorem, but this will happen at different times for different systems, so that at any particular time only a very small fraction of the systems in the ensemble are close to their initial dynamical states.

The reason for the irrelevance of Poincaré recurrences in mixing systems is that the motion of the phase point is very unstable. Dynamical states that start very close to each other in phase space

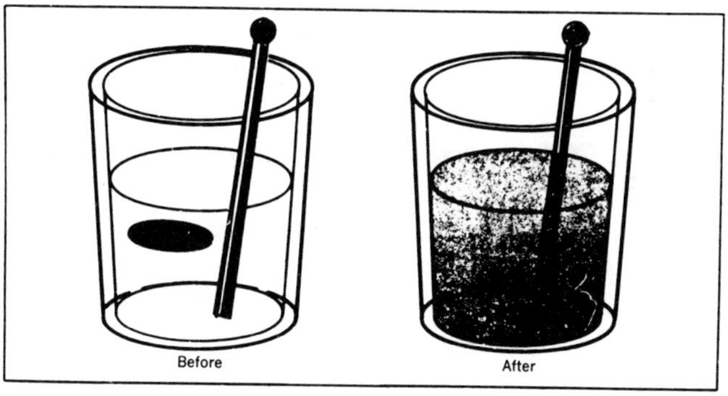

Fig. 4. A familiar example of "mixing".
According to V.I. Arnold and A. Avez, the two liquids are rum (twenty percent) and cola (eighty per cent), with the result of the mixing process known as a "Cuba libre".

become widely separated as time progresses, so that the recurrence time depends extremely sensitively on the initial conditions of the motion. (The importance of this instability in statistical mechanics was first recognized by N.S. Krylov, a Russian physicist who died in his twenties in 1947.) This type of instability appears to be characteristic of real physical systems, and leads to one sort of irreversibility: even if we could reverse the velocities of every particle in a real system that has been evolving towards equilibrium, the system would not necessarily return or even come close to its initial dynamical state with the velocities reversed because the unavoidable small external perturbations would be magnified. This instability is noticeable in molecular-dynamics calculations with hard-sphere systems:

if we numerically integrate the equations of motion from time 0 to time t and then try to recover the initial state by integrating backwards from time t to time 0, we obtain instead a completely new state. This is because the numerical integration is unstable to small rounding-off errors made during the computation, which play the same role as external perturbations in a real system.

Only a few physical systems have been proven so far to be mixing. The most important is the hard-sphere gas, mentioned above. Sinai's proof that this system is ergodic also gives the stronger result that it is mixing. Roughly, Sinai's method of proving mixing is to show that the hard-sphere system is unstable in the sense discussed above. Physically this instability comes from the fact that a slight change in direction of motion of any particle is magnified at each collision with the convex surface of another particle. The full proof for the simplest case of a "single" particle moving among fixed convex scatterers is already quite complicated and the proof for moving hard sphere has not yet appeared. That is why I put a question mark next to this system in Table 1. I have some private information however (from G. Gallavotti) that Sinai is in the process of writing up the proof and has already finished a hundred pages of manuscript. As indicated in Table 1 the proof will actually be that this system is a K-system.

A very simple model system with ′collisions′ which is mixing was considered by Goldstein, Lanford and Lebowitz. This will be described in the Appendix.

4.1 Time correlations and transport coefficients

As indicated by all the speakers here, the study of time correlation functions of the form $\langle g(t)f \rangle$ plays a central role in the statistical mechanical theory of non-equilibrium phenomena. When a system is mixing then these functions will certainly approach their uncorrelated values as $|t| \to \infty$, provided that g and f are square integrable, $\langle g^2 \rangle < \infty$, $\langle f^2 \rangle < \infty$, i.e.

$$\langle g(t)f \rangle - \langle g \rangle \langle f \rangle \longrightarrow 0 \quad \text{as} \quad |t| \to \infty . \tag{4.4}$$

A system is said to be <u>weakly mixing</u> if, under the same conditions on square integrability

$$\lim_{T\to\infty} \frac{1}{T}\int_0^T \left|\langle g(t)f\rangle - \langle g\rangle\langle f\rangle\right| dt \to 0 \quad \text{as} \quad T\to\pm\infty \tag{4.5}$$

This is clearly a weaker requirement then (4.4). It is however still stronger than ergodicity which can be shown to be equivalent to the requirement that

$$\lim_{T\to\infty} \frac{1}{T}\int_0^T \left[\langle g(t)f\rangle - \langle g\rangle\langle f\rangle\right] dt \to 0 \quad \text{as} \quad T\to\pm\infty \tag{4.6}$$

In many cases one is interested in the time integrals of these correlations. This stems from the fact, pointed out by many of the lecturers here (indeed Professor Kubo is the modern father of this subject) that linear transport coefficients, such as heat conductivity, viscosity etc., may be expressed as integrals over time (from $t = 0$ to $t = \infty$) of the time correlation of appropriate dynamical functions (Einstein-Green-Kubo). These functions represent the 'fluxes' associated with the transport processes in question. A well known example of such a 'formula' is the Einstein relation between the self-diffusion constant D and the integral of the velocity auto-correlation function of a specified particle, say particle one, $\langle v_1(t)\, v_1\rangle$, $(\langle v_1\rangle = 0)$.

It might appear that for mixing systems these transport coefficients could be defined meaningfully, without going to the thermodynamic limit of an infinite size system, as long as $[\langle g(t)f\rangle - \langle g\rangle\langle f\rangle]$ approached zero sufficiently rapidly to be integrable. Such is, however, not the case, as indeed it should not be on physical grounds. Formally this occurs because the flux functions whose time correlations are of interest for transport coefficients can generally be written as Poisson brackets with the Hamiltonian H, i.e. $f = (F,H)$, $g = (G,H)$, and for mixing systems it can be shown that when f, g, F, G are all square integrable, then

$$\lim_{T\to\infty} \int_0^T \langle f(t)g\rangle\, dt = \langle (F,H)G\rangle \quad ,$$

$$\lim_{T\to\infty}\int_0^T \langle f(t)f\rangle\, dt = 0 .$$

Thus for a finite mixing system confined by a wall,

$$\lim_{T\to\infty}\int_0^T \langle v_1(t)v_1\rangle\, dt = \lim_{T\to\infty}\langle q_1(T)v_1\rangle = 0 \quad (4.7)$$

since

$$v_1 = (q_1,H) \quad \text{and} \quad \int v_1^2\, dx < \infty , \quad \int q_1^2\, dx < \infty .$$

Note that when q_1 is an angle variable, e.g., in the case of periodic boundary conditions, then $v_1 \neq (q_1,H)$ and (4.7) need not hold. We would still have, however, $\langle v_1(t)v_1\rangle \to 0$ if the system is mixing.

When the system is not mixing, the limit $T \to \infty$ in the above integrals need not exist. It is still true however that for any finite system

$$\lim_{T\to\infty}\int\langle f(t)f\rangle\, dt = \lim_{T\to\infty}\left[-\frac{d}{dT}\langle F(T)F\rangle\right] = 0, \quad \text{if it exists.}$$

This is so since,

$$\langle F(T)F\rangle \leq \langle F(T)F(T)\rangle^{\frac{1}{2}}\langle F^2\rangle^{\frac{1}{2}} = \langle F^2\rangle ,$$

so that when F is square integrable $d/dt\langle F(T)F\rangle$ can either oscillate or approach zero.

These time correlation integrals will therefore, if they exist at all, be equal to zero in any finite system. (The interesting fact is that they do exist for mixing systems). The Einstein-Kubo type formulae for transport coefficients can therefore be mathematically meaningful only in the thermodynamic limit. It is of course possible, and even likely, that for macroscopic systems there exists values of T for which the integral in (4.7) is ´close´to the diffusion constant.

V. K- AND BERNOUILLI SYSTEMS

In order to give a physical definition of these systems we define a finite partition of the energy surface S of our dynamical system as any finite collection of n non-overlapping regions $R_0, \ldots, R_k$ which together cover the whole of S (see Fig. 5).

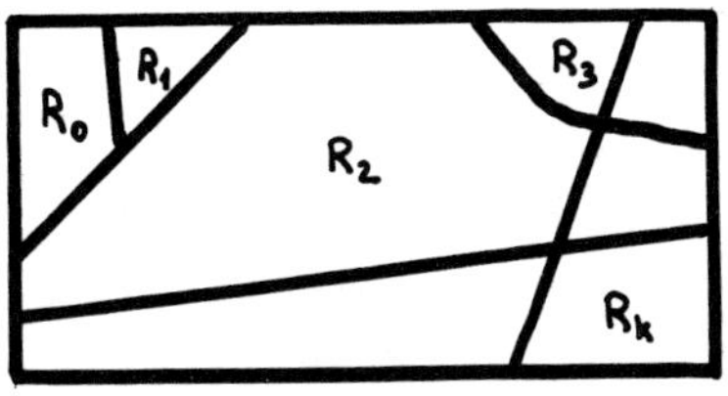

$$\sum \mu(R_i) = 1, \quad \mu(R_i \cap R_j) = 0.$$

Figure 5.

Suppose now that an experiment is made that will determine which of these regions the phase point is in at any time, but gives no information whatever about which part of the region it is in. That is, every time we use this measuring device we obtain an outcome that is a positive integer - the label of the region the phase point of the system is in at that time. As an example we may consider a system of k particles in a box and the experiment consists of measuring the number of particles in one half of the box.

Suppose we use the device repeatedly at intervals of, say, τ seconds. Its outcome will be a sequence of positive integers from the set $\{0, \ldots, k\}$, which can be extended indefinitely. In general, we would expect these integers to be correlated; that is, the microcanonical probability for each new observation depends on what has been observed before (as in a Markov process, for example). This correlation comes about because the dynamical states of the system at different times are deterministically related, through the equations of motion. Indeed, it may be possible to choose the partition and the time intervals between measurements in such a way that the outcome of later measurements is completely determined by the outcome of the earlier ones. That this is possible even in the case of ergodic systems may be seen easily by considering the simple harmonic oscillator considered earlier and making the interval between measurements equal to the period τ. The oscillator will then be found each time in the same set R_i, since $\phi_\tau R_j = R_j$ for a system with period τ.

It should be clear however from our earlier discussion that such a deterministic behaviour of successive measurements is impossible for a mixing system. For in a mixing system each set R_i will eventually become uniformly spread out over the whole energy surface, i.e.

$$\lim_{n \to \infty} \mu\left[(\phi_{n\tau} R_i) \cap R_j\right] = \mu(R_i)\mu(R_j) \quad ,$$

for every value of τ . It might still be possible however that successive measurements provide more and more information so that eventually one really knows what the result of the next measurement will be. Somewhat more formally we may let $\epsilon(n)$ be the uncertainty in the outcome of the n-th measurement (given the results of the first n-1 measurements). A K-system is then a system for which $\epsilon(n) \geq \epsilon > 0$ no matter how large n is and no matter what the partition, or τ , is. For a system which is only mixing there will be partitions for which $\epsilon(n) \longrightarrow 0$ as $n \to \infty$. It is, in this sense, that K-systems have an essential randomness in them.

This will be made more precise in the next section when we discuss the Kolmogorov-Sinai entropy of a flow. First however I shall discuss the last and highest member of our hierarchy: the Bernoulli system. This will also give me the opportunity to introduce to you the paradigm of Bernoulli systems; the baker's transformation.

A Bernoulli system is one for which it is possible to choose the regions $R_0, \ldots, R_{n-1}$ in such a way that the observations made at different times are completely uncorrelated, just like the numbers shown at different times by a roulette wheel. At the same time, the regions so chosen give enough information to discriminate between dynamical states: if two systems have different dynamical states at some time, then the observations made on them cannot yield identical results for the observations at every time. Such a partition is called a generating partition. When such regions can be chosen, we call the system a Bernoulli system, i.e. a Bernoulli system permits the construction of an independent generating partition . There is of course no requirement, and indeed no possibility, that all partitions be of this type. Since however a Bernoulli system is also a K-system, <u>every</u> partition will have the inherent randomness associated with K-systems which we discussed earlier.

Recently Gallavotti and Ornstein showed that a point particle moving (in two or higher dimensions) among fixed convex scatterers (in a box with rigid walls or on a torus; periodic boundary conditions) is a Bernoulli system. (More precisely a Bernoulli flow, which means that there exists an independent generating partition for every $\tau > 0$). The proof of Gallavotti and Ornstein utilizes the results of Sinai that this system is a K-system as well as the techniques developed by Ornstein and Weiss who showed that the geodesic flow on a space of constant negative curvature is Bernoulli.

5.1 The baker's transformation

As an illustration of a simple Bernoulli system, consider a system whose phase space is the square $0 < p \leq 1$, $0 < q \leq 1$ shown in Figure 6, and whose (non-Hamiltonian) law of motion is given by a mapping known as the baker's transformation because it recalls the kneading of a piece of dough.

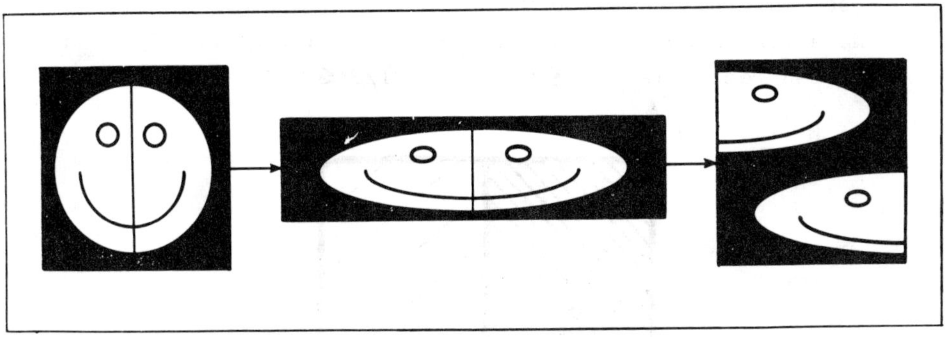

Fig. 6. The baker's transformation recalls the kneading of a piece of dough.
We first squash the square to half its original height and twice its original width, and then cut the resulting rectangle in half and move the right half of the rectancle above the left.

If the phase point is (p,q) at time t, then at time $t + 1$ it is at the point obtained by squashing the square to a $(1/2 \times 2)$ rectangle, then cutting and reassembling to form a new square as shown in the diagram. The formula for this transformation is

$$\phi(p,q) = \begin{cases} (2p, q/2) & \text{if } 0 \leq p \leq 1/2 \\ [2p - 1, \; q/2 + 1/2] & \text{if } \tfrac{1}{2} < p \leq 1 . \end{cases} \tag{5.1}$$

If p and q are written in binary notation (1/8 in binary notation is 0.00100 ..., 1/4 is 0.01000, and so on), the transformation removes the first digit after the binary point from p and attaches it to q, so that

$$\phi(0.p_1p_2 \ldots, \quad 0.q_1q_2 \ldots) \quad = \quad (0.p_2p_3 \ldots, \; 0.p_1q_1q_2 \ldots)$$

where the p_i and q_i take on the values 0 and 1. This transformation is invertible and from it we can define ϕ_{-1} as the inverse of ϕ and $\phi_{\pm t}$ as the t-th iteration of $\phi_{\pm 1}$. (Only integer values of the time are used here, rather than all real values, as in our discussion of dynamics earlier in this talk, but we do not regard this distinction as important.) Moreover, the transformation preserves geometrical area, and so the analog of the microcanonical distribution is just a uniform density.

To see how this completely deterministic system can at the same time behave like a roulette wheel, we take the regions R_0 and R_1 to be the two rectangles $0 \leq p < 1/2$, $1/2 \leq p < 1$ as shown in Fig. 7.

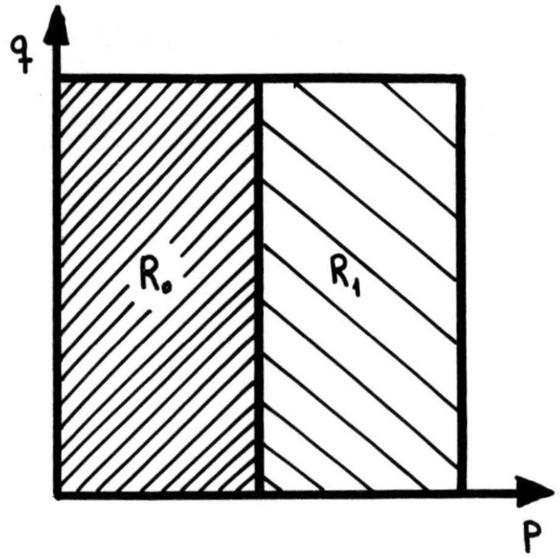

Fig. 7. Definition of the regions R_0 and R_1 used to show that the baker's transformation is a Bernoulli system.

Suppose the phase point at time 0 is

$$(p,q) \quad = \quad (0.p_1p_2 \ldots, \quad 0.q_1q_2 \ldots) \quad .$$

If p_1 is zero, the system at time 0 is in R_0; if p_1 is one, the system at time 0 is in R_1. At time 1 the phase point is

$$(0.p_2p_3 \ldots, \quad 0.p_1q_1q_2 \ldots)$$

and so we observe the phase point in region R_{p_2} . At time 2 it is in R_{p_3} and so on. Likewise, at time -1 it is in R_{q_1} , at time -2 in R_{q_2}, and so on. Each observation is determined by a different digit in the binary representation of the number pair (p,q).

Since the analog of the microcanonical ensemble for this system has a uniform density in the square it is not difficult to see that the microcanonical probability of <u>each</u> of these digits in the binary expression for (p,q) is 1/2, and is uncorrelated with all the other digits. The observations made at different times t (= integer) are therefore uncorrelated, and so the baker's transformation model is a Bernoulli system.

The baker's transformation is an example of a Bernoulli shift. Let ξ denote a point in the space of doubly infinite sequences $\xi_i \in (0, 1)$, with i an integer (positive, negative or zero). We set $\xi_i = p_{i+1}$ for $i \geq 0$, and $\xi_i = q_{-i}$ for $i < 0$. The ξ specifies a point in the unit square

$$\xi = (\ldots \xi_{-1}, \xi_0, \xi_1, \ldots)$$

and the baker's transformation is simply the shift

$$(\phi\xi)_i = \xi_{i+1} \quad .$$

Possibly one should not read too much physical meaning into this type of result, for with a more complicated dynamical system the regions $R_0, R_1, R_2, \ldots, R_n$ would probably be exceedingly complicated sets in phase space, but from a "philosophical" point of view it is very interesting to see how the same dynamical system can show perfect determinism on the microscopic level and at the same time perfect randomness on a "macroscopic" level.

It is the interplay of these two apparently incompatible levels of description that give the foundations of statistical mechanics their fascination.

5.2 The Kolmogorov-Sinai entropy

We consider as before a partition of the energy surface S into k disjoint cells A_i, $i = 1, \ldots, k$. (This is a slight change in notation). This collection of sets $\{A_i\}$ is called a <u>partition</u> $\mathcal{A}$, $\mathcal{A} = \{A_i\}$; the A_i are the 'atoms' of $\mathcal{A}$. Since $\mu(A_i)$ is the probability (in the microcanonical ensemble) of finding the system in A_i, Kolmogorov defined the 'entropy' (not to be confused with the thermodynamic entropy) of this partition $h(\mathcal{A})$, in analogy with information theory entropy, as

$$h(\mathcal{A}) = -\sum \mu_0(A_i) \ln \mu_0(A_i) \quad ,$$

Clearly, $h(\mathcal{A}) \geq 0$, with the equality holding if and only if $\mu_0(A_j) = 1$, for some j, i.e., there is <u>complete</u> certainty that $x \in A_j$. (We shall generally ignore sets of measure zero, setting $\mu_0(C) \ln \mu_0(C) = 0$ if $\mu_0(C) = 0$, and writing $A_j = S_E$ when $\mu_0(A_j) = 1$.) The maximum value which $h(\mathcal{A})$ can take is $\ln k$ corresponding to $\mu_0(A_i) = k^{-1}$ for all $i = 1, \ldots, k$.

Given two partitions $\mathcal{A} = \{A_i\}$, $i = 1, \ldots, k$ and $\mathcal{B} = \{B_j\}$, $j = 1, \ldots, m$, we denote the 'sum' of the partitions $\mathcal{A}$ and $\mathcal{B}$ by $\mathcal{A} \vee \mathcal{B}$; $\mathcal{A} \vee \mathcal{B}$ is the partition whose atoms are all (non-zero measure) sets $A_i \cap B_j$. The entropy of $\mathcal{A} \vee \mathcal{B}$ is,

$$h(\mathcal{A} \vee \mathcal{B}) = \sum_{i,j} \mu(A_i \cap B_j) \ln \mu(A_i \cap B_j) \quad .$$

The 'conditional entropy' of a partition $\mathcal{A}$, relative to a partition $\mathcal{B}$ is defined as

$$h(\mathcal{A}/\mathcal{B}) = \sum_j \mu(B_j) \left\{ \sum_i \mu(A_i/B_j) \ln \mu(A_i/B_j) \right\} ,$$

where

$$\mu(A_i/B_j) \equiv \mu(A_i \cap B_j)/\mu(B_j) \quad .$$

For a given flow operator ϕ_t, and some fixed time interval τ, we construct the sets $\phi_\tau A_i$, $\phi_{2\tau} A_i$, ... and define $\phi_\tau \mathcal{A}$ as the partition whose atoms are the sets $\{\phi_\tau A_i\}$. Kolmgorov then sets

$$h(\mathcal{A}, \phi_\tau) = \lim_{n \to \infty} \frac{1}{n} h\left(\bigvee_{j=0}^{n-1} \phi_{j\tau} \mathcal{A}\right) \quad .$$

It can be shown that $h(\mathcal{A}, \phi_{j\tau}) = j h(\mathcal{A}, \phi_\tau)$. The K-S entropy of the flow ϕ_t is defined as (S for Sinai)

$$h(\phi_\tau) \equiv \sup_{\mathcal{A}} h(\mathcal{A}, \phi_\tau) = h\tau \; ,$$

where h is now an intrinsic property of the flow. It was shown by Sinai that a system is a K-system <u>iff</u> $h(\mathcal{A},\phi_\tau) > 0$ for <u>all</u> non-trivial partitions $\mathcal{A}$, i.e., for partitions whose atoms are not all of measure zero or one.

We can now specify the precise sense in which K-systems are 'random' even when the flow is entirely deterministic. As indicated earlier the atoms of the partition $\mathcal{A}$, $\{A_i\}$, $i = 1,\dots, k$, correspond to different possible outcomes of the measurement of some dynamical function $f(x)$, i.e. if $x \in A_i$ then the result of the measurement will be α_i, etc. (Since the set of outcomes of the measurement is finite, being equal to k, $k < \infty$, the measurement is a 'gross' one. It need now however be restricted to measuring just one property of the system; we can replace $f(x)$ by a finite set of functions.) The probability (in the microcanonical ensemble) of an outcome α_i is $\mu_0(A_i) \equiv p(\alpha_i)$. Now if these dynamical functions were measured first at $t = -\tau$, and then at $t = 0$, the joint probability that the result of the first measurement is α_j and the result of the second is α_i, is equal to the probability that the dynamical state of the system x at the time of the present measurement $t = 0$ is in the set $A_i \cap \phi_\tau A_j$, i.e., $p(\alpha_i, \alpha_j) = \mu_0(A_i \cap \phi_\tau A_j)$. The conditional probability of finding the value α_i, if the result of the previous measurement was α_j, is $p(\alpha_i/\alpha_j) = \mu(A_i \cap \phi_\tau A_j)/\mu(\phi_\tau A_j) = \mu(A_i \cap \phi_\tau A_j)/\mu(A_j)$. In a similar way the probability of finding the result α_i at $t=0$, given that the results of the previous measurements at times $-\tau, -2\tau, \dots, -n\tau$ were $\alpha_{i_1}, \alpha_{i_2}, \dots, \alpha_{i_n}$,

$$p(\alpha_i/\alpha_{i_1},\dots,\alpha_{i_n}) = \Big[(A_i \cap \phi A_{i_1} \dots \cap \phi_{n\tau} A_{i_n})/ \mu_0(\phi_\tau A_{i_1} \cap \phi_\tau A_{i_2} \dots \cap \phi_{n\tau} A_{i_n})\Big].$$

It can be readily shown that

$$h(\mathcal{A},\phi_\tau) = \lim_{n\to\infty} h(\mathcal{A}/\bigvee_{k=1}^{n} \phi_{k\tau}\mathcal{A}) = \lim_{n\to\infty}\Big\{-\sum p(\alpha_{i_1},\alpha_{i_2},\dots, \alpha_{i_n})$$

$$\times \Big[\sum_{i=1}^{k} p(\alpha_i/\alpha_{i_1},\dots,\alpha_{i_n}) \ln p(\alpha_i/\alpha_{i_1},\dots,\alpha_{i_n})\Big]\Big\}.$$

Hence $h(\mathcal{A}, \phi_\tau) > 0$ for all non-trivial partitions implies that no matter how many measurements of the values of $f(x)$ we make on a system at times, $-\tau, \ldots, -n\tau$, the outcome of the next measurement is still uncertain. (N.B. the measurements are 'coarse' since $\mu(A_i) > 0$).

VI. ERGODIC PROPERTIES AND SPECTRUM OF THE INDUCED UNITARY TRANSFORMATION

It is possible, and for many purposes useful, to consider the Hilbert space L^2 of square integrable functions on the energy surface S (Koopman). The integration here is again with respect to the microcanonical ensemble density dx ; $\Psi(x) \in L^2$ is a complex valued function of $x \in S$, such that

$$\int_S |\Psi|^2 \, dx < \infty \quad .$$

The time evolution ϕ_t then induces a transformation U_t in L^2,

$$U_t \Psi(x) = \Psi[\phi_t(x)]$$

which is unitary

$$\int |U_t \Psi|^2 \, dx = \int |\Psi|^2 \, dx \quad .$$

We may therefore write $U_t = \exp[itL]$ where iL is the generator of U_t. For a Hamiltonian flow with $H = H(q_1, \ldots, q_n, p_1, \ldots, p_n)$ iL is the Liouville operator or Poisson bracket

$$iLf = (f, H) = \sum_i \left[\frac{\partial f}{\partial q_i} \frac{\partial H}{\partial p_i} - \frac{\partial f}{\partial p_i} \frac{\partial H}{\partial q_i} \right] .$$

There is an intimate connection between the ergodic properties of the flow and the spectrum of U_t which is of the form $\exp(it\lambda)$, with λ in the spectrum of L. Clearly $\lambda = 0$ is always a discrete eigenvalue of L corresponding to the eigenfunction $\Psi = \text{const.}$

The following equivalences and implications can be shown to hold: K-property $\Rightarrow$ absolute continuity of the spectrum of L with respect to Lebesgue measure (on the space orthogonal to the constants) $\Rightarrow$ mixing $\Rightarrow$ continuity of the spectrum $\Leftrightarrow$ weak mixing $\Rightarrow$

ergodicity $\Leftrightarrow \lambda = 0$ is a simple eigenvalue.

This may be a good place to note that, due to the discrete nature of the energy spectrum for finite quantum systems confined to a bounded domain V, there will be no mixing (decay of correlations) in such a system. For such quantum systems we do not therefore gain anything from the use of ensembles and we are forced to look at the infinite volume limit for signs of long time irreversibility. The remarkable thing about Sinai's result for hard spheres is that it shows that finite classical systems can and do have purely continuous spectra. (Note that when Planck's constant $h \to 0$ the number of energy levels between some fixed E and $E + \Delta E$ becomes infinite.)

VII. INFINITE SYSTEMS

Since the number of particles contained in a typical macroscopic system is very large ($\sim 10^{26}$) there is great interest from the point of view of statistical mechanics in the ergodic properties of infinite systems (corresponding to the thermodynamic limit in equilibrium statistical mechanics). As I indicated in the introduction, there are still some serious conceptual problems about the nature of the ingredients, in addition to ergodic theory, which are necessary to get the right physics. I shall therefore be extremely brief here.

The natural setting for this discussion is the more abstract form of ergodic theory. This theory deals typically with the triplet (X, μ, ϕ_t), X is a space equipped with a measure μ which is left invariant by ϕ_t. (I have left out explicit mention of Σ the collection of measurable sets). ϕ_t is a flow if t is a real variable and a discrete transformation if t is an integer (in which case $\phi_n = \phi^n$ and we can write the triplet as (X, μ, ϕ)). The triplet (X, μ, ϕ_t) is usually referred to as the dynamical system.

In our discussion of finite Hamiltonian systems we had $X = S_E$, μ = microcanonical measure (ensemble), and ϕ_t the time evolution given by the solution of the Hamiltonian equations of motion. All our analysis can be easily translated to the more general setting. Thus we say that the dynamical system (X, μ, ϕ_t) is ergodic if for any set $A \subset X$, which is left invariant by ϕ_t, $\mu(A) = 0$ or $\mu(A) = 1$. Equivalently (X, μ, ϕ_t) is ergodic if there does not exist another measure μ', which is absolutely continuous with respect to μ, and is also invariant under ϕ_t.

We say that μ' is absolutely continuous with respect to μ if $\mu(A) = 0 \Rightarrow \mu'(A) = 0$. When μ' is absolutely continuous with respect to μ we can write $d\mu' = \rho(x)d\mu$, i.e. μ' has an 'ensemble density' with respect to $d\mu$.

It should now be clear what is involved in the ergodic theory of infinite systems: X will be the space of infinite configurations (locally finite), μ will be some stationary measure under the time evolution ϕ_t assuming this can be suitably defined. It may now be much more difficult to justify a priori the use of the Gibbs measure (at a given temperature and chemical potential) and those absolutely continuous with respect to it as the only physically suitable measures (assuming there are also other stationary measures available) than it was to argue in the finite system for the use of ensembles with ensemble densities. I will leave the discussion of this to Professor Haag and only refer you now to Table 1 for some of the results known for infinite systems.

Acknowledgements

As already indicated, parts of these lecture notes come from my article with O. Penrose in Physics Today.

Appendix: Ergodic properties of simple model system with collisions*

We are interested in the ergodic properties of dilute gas systems. These may be thought of as Hamiltonian dynamical systems in which the particles move freely except during binary "collisions". In a collision the velocities of the colliding particles undergo a transformation with "good" mixing properties (cf. Sinai's study of the billiard problem). To gain an understanding of such systems we have studied the following simple discrete time model: The system consists of a single particle with coordinate $\underline{r} = (x,y)$ in a two-dimensional torus with sides of length (L_x, L_y), and "velocity" $\underline{v} = (v_x, v_y)$, in the unit square $v_x \in [0, 1]$, $v_y \in [0, 1]$. The phase space Γ is thus a direct product of the torus and the unit square. The transformation T which takes the system from a dynamical state $(\underline{r}, \underline{v})$ at "time" j to a new dynamical state $T(\underline{r},\underline{v})$ at time j+1 may be pictured as resulting from the particle moving freely during the unit time interval between j and j+1 and then

*From paper by S. Goldstein, O.E. Landford III and J.L. Lebowitz.

undergoing a "collision" in which its velocity changes according to the baker's transformation, i.e.

$$T(\underline{r},\underline{v}) = (\underline{r} + \underline{v}, B\underline{v})$$

with B the baker's transformation defined in Section 5,

$$B(v_x,v_y) = \begin{cases} (2v_x, \frac{1}{2}v_y), & 0 \leq v_x \leq \frac{1}{2} \\ (2v_x - 1, \frac{1}{2}v_y + \frac{1}{2}), & \frac{1}{2} < v_x \leq 1 . \end{cases}$$

The normalized Lebesgue measure $d\mu = dxdydv_xdv_y/L_xL_y = d\underline{r}\, d\underline{v}/L_xL_y$ in Γ is left invariant by T. We call U_T the unitary transformation induced by T on $L^2(d\mu)$, $U_T\phi = \phi \circ T$. Our interest lies then in the ergodic properties of T and in the spectrum of U_T.

We note first that the transformation B on the velocities is, when taken by itself as a transformation of the unit square with measure $d\underline{v}$, well known to be isomorphic to a Bernoulli shift. It therefore has very good mixing properties.

The ergodic properties of our system which combines B with free motion turn out to depend on whether L_x^{-1} and L_y^{-1} satisfy the independence condition (I),

$$n_xL_x^{-1} + n_yL_y^{-1} \notin \quad \text{for } n_x \text{ and } n_y \text{ integers unless } n_x = n_y = 0. \tag{I}$$

<u>Theorem 1</u>: When (I) holds, the spectrum of U_T, on the complement of the one-dimensional subspace generated by the constants, is absolutely continuous with respect to Lebesgue measure and has infinite multiplicity.

It follows from Theorem 1 that when (I) holds the dynamical system (Γ, T, μ) is at least mixing. We do not know at present whether it is also a Bernoulli shift or at least a K system.

<u>Theorem 2</u>: When (I) does not hold the system (Γ, T, μ) is not ergodic.

REFERENCES

General:

V.I. ARNOLD A.AVEZ, Egodic Problems of Statistical Mechanics Benjamin, New York, 1968.

A.S. WIGHTMAN, in Statistical Mechanics at the turn of the Decade, (E.G.D. Cohen, ed.), M. Dekker, New York, 1971

J.L. LEBOWITZ, "Hamiltonian Flows and Rigorous Results in Non-equilibrium Statistical Mechanics", in Statistical Mechanics, New Concepts, New Problems, New Applications (S.A. Rice, K.F. Freed J.C. Light eds.). U. of Chicago Press, 1972.

I. E. FARQUHAR, in Irreversibility in the Many-Body Problem, (J. Biel and J. Rae, eds.), Plenum, New York 1972.

J.L. LEBOWITZ and O. PENROSE,"Modern Engodic Theory", Phys. Today, Febr. 1973. p.23.

J. FORD, "The Transition from Analytic Dynamics to Statistical Mechanics," Advances in Chemical Physics (1973).

O.E. LANFORD III, "Ergodic theory and approach to equilibrium for finite and infinite systems", contribution in Boltzmann Equations (Theory & Applications) Proceeding, Symposium, Vienna Sept. 1972, Springer-Verlag, 1973.

Ya. G. SINAI, "Ergodic Theory", ibid.

S.G. BRUSH, Transport Theory and Stat. Phys., 1971, for history of ergodic hypothesis.

O. PENROSE, Foundations of Statistical Mechanics, Pergamon, Oxford, 1970.

Anharmonic Oscillators - KAM theorem:

E. FERMI, J. PASTA and S. ULAM, Studies of Non-Linear Problems. Los Alamos Scient. Lab. Report LA- 1940 (1955): also reprinted in Enrico Fermi: Collected Papers, Volume II. University of Chicago Press, Chicago (1965) Page 987.

M. HENON and C. HEILES, Astron.Journal, 69 73 (1964)

G. WALKER and J. FORD, Phys. Rev. 188, 416 (1969)

J. FORD and G.H. LUNSFORD, Phys. Rev. A 1, 59 (1970)

G.M. ZASLAVSKY and B.V. CHIRIKOV, Usp. Fiz. Nauk 105, 3 (1971).

Mixing, K, and Bernoulli Systems:

J.W. GIBBS, Elementary Principles in Statistical Mechanics, Yale U. Press, New Haven, 1902 (reprinted by Dover, New York, 1960)

J. von NEUMANN, Annals of Math. 33 587 (1932)

E. HOPF. J. Math and Phys, 13, 51 (1934); Ergoden Theorie, Springer, Berlin (1937)

P.R. HALMOS, Measure Theory, Van Nostrand Reinhold, New York (1950)

P.R. HALMOS Lectures on Ergodic Theory, Chelsea, New York (1956).

A.N. KOLMOGOROV, "Address to the 1954 International Congress of Mathematicians"(translated in R. Abrahams, Foundations of Mechanics, Benjamin, New York (1967) Appendix D).

D.S. ORNSTEIN, "Bernoulli shifts with the same entropy are isomorphic", Advances in Math. 4, 337 (1970).

D.S. ORNSTEIN, "The isomorphism theorem for Bernoulli flows", Advances in Math. 10, 124 (1973)

D.S. ORNSTEIN, Ergodic Theory, Randomness, and Dynamical Systems Lecture notes from Stanford University.

M. SMORODINSKY, Ergodic Theory, Entropy, Springer Lecture Notes 214 (1970).

P. SHIELDS, The Theory of Bernoulli Shifts, University of Chicago Press.

S. GOLDSTEIN, O.E. LANFORD and J.L. LEBOWITZ, "Ergodic Properties of Simple Model System with Collisions", J. Math. Phys. 14, 1228 (1973)

Hard spheres - finite Lorentz system:

Ya. G. SINAI, Sov. Math - Dokl. 4 1818 (1963); "Ergodicity of Boltzmann's Equations" in Statistical Mechanics Foundations and Applications (T.A. Bak. ed) Benjamin, New York (1967);

Ya. G. SINAI, "Dynamical systems with elastic reflections" Russ. Math. Surveys 25, 137 (1970)

G. GALLOVOTTI and D.S. ORNSTEN. Comm. Math. Physics, to appear.

Infinite Systems:

S. GOLDSTEIN, "Ergodic Theory and Infinite Systems", Thesis, Yeshiva University. N.Y. (1974)

R. HAAG, D. KASTLER and E.B. TRYCH-POHLMEYER. Comm. Math. Phys., to appear

K.L. VOLKOVYSSKII and Ya. G. SINAI, "Ergodic properties of an ideal gas with an infinite number of degrees of freedom", Funct. Anal. Appl. 5, 185 (1971)

Ya. G. SINAI, "Ergodic properties of a gas of one-dimensional hard rods with an infinite number of degrees of freedom", Funct. Anal. Appl. 6, 35 (1972).

S. GOLDSTEIN and J.L. LEBOWITZ, "Ergodic Properties of an infinite system of particles moving independently in a periodic field", Comm. Math. Phys., to appear.

O. DE PAZZIS, "Ergodic properties of a semi-infinite hard rods system", Commun. Math. Phys. <u>22</u>, 121 (1971).

D. RUELLE, <u>Statistical Mechanics - Rigorous Results</u>, Benjamin New York (1969)

S. GOLDSTEIN, "Space-time ergodic properties of a systems of infinitely many independent particles", to appear .

O.E. LANFORD and J.L. LEBOWITZ "Ergodic Properties of Harmonic Crystals", to appear.

CORRELATION FUNCTIONS IN HEISENBERG MAGNETS

M. De Leener

Université Libre de Bruxelles
Bruxelles, Belgium

I. INTRODUCTION

1.1. The model
1.2. Origin and validity of the Heisenberg Hamiltonian

II. NEUTRON SCATTERING EXPERIMENTS AND SPIN CORRELATION FUNCTIONS

III. SOME GENERAL PROPERTIES OF THE SPIN CORRELATION FUNCTIONS

3.1. Hermitian Symmetry
3.2. Detailed balance
3.3. Lattice symmetry
3.4. Spin-rotational symmetry

IV. LOW TEMPERATURE THEORY

4.1. Spin waves
4.2. Spin wave theory of the correlation functions

V. HIGH TEMPERATURE THEORY

VI. THE CRITICAL REGION

REFERENCES

I. INTRODUCTION. [1]

1.1 The model

The isotropic Heisenberg model represents a magnet as a system of N spins S, fixed at the sites of a lattice and coupled through so-called exchange forces, described by the Hamiltonian

$$
\begin{aligned}
H &= -\sum_{a\neq b} J_{ab}\, \vec{S}_a \cdot \vec{S}_b \\
&= -\sum_{a\neq b} J_{ab} \left[S_a^z S_b^z + S_a^+ S_b^- \right],
\end{aligned}
\tag{1.1}
$$

where $J_{ab} = J(\vec{r}_a - \vec{r}_b)$ is the exchange interaction between lattice points a and b ; S_a^z and $S_a^\pm = S_a^x \pm i S_a^y$ are the spin operators, which obey the usual angular momentum commutation relations:

$$
\begin{aligned}
\left[S_a^z, S_b^\pm \right] &= \pm S_a^\pm \delta_{a,b}^{Kr} \\
\left[S_a^+, S_b^- \right] &= 2 S_a^z \delta_{a,b}^{Kr}
\end{aligned}
\tag{1.2}
$$

(we set $\hbar = 1$ throughout). We recall that S_a^z is hermitian and that $S_a^+ = (S_a^-)^\dagger$.

It is sometimes convenient to express the Heisenberg Hamiltonian in terms of the Fourier transforms of the spin operators, defined as

$$
S_{\vec{q}}^\alpha = \sum_a S_a^\alpha e^{i\vec{q}\cdot\vec{r}_a}, \tag{1.3}
$$

with the inverse relation

$$
S_a^\alpha = \frac{1}{N} \sum_{\vec{q}} S_{\vec{q}}^\alpha e^{-i\vec{q}\cdot\vec{r}_a} \tag{1.4}
$$

where the wavenumber $\vec{q}$ takes N values inside the first Brillouin zone of the reciprocal lattice. (From now on, except where necessary to avoid confusion, we shall not explicitly indicate the vector character of the wavenumbers $\vec{q}$).

The operators obey the following commutation relations:

$$[S^z_q, S^\pm_{q'}] = \pm S^\pm_{q+q'}$$

$$[S^+_q, S^-_{q'}] = 2 S^z_{q+q'} \; ; \tag{1.5}$$

moreover,

$$S^z_q = (S^z_{-q})^\dagger \quad , \quad S^+_q = (S^-_{-q})^\dagger \quad . \tag{1.6}$$

The Hamiltonian (1.1) is easily transformed into

$$H = -\frac{1}{N} \sum_q J(q) \left[S^z_q S^z_{-q} + S^+_q S^-_{-q} \right], \tag{1.7}$$

where

$$J(q) = \sum_b J_{ab} \, e^{i\vec{q}\cdot(\vec{r}_a - \vec{r}_b)} \tag{1.8}$$

(with the convention: $J_{aa} = 0$). Note that the reality and inversion symmetry of J_{ab} leads to

$$J(q) = J(q)^* = J(-q) . \tag{1.9}$$

1.2 Origin and validity of the Heisenberg Hamiltonian

In order to understand, in a very naive way, the physics underlying the model Hamiltonian (1.1), let us go back to Heisenberg's original remark that exchange forces, as they appear in the elementary (Heitler-London) theory of the stability of the hydrogen molecule, can give rise to energies of the order of those observed in ferromagnetic transitions ($k_B T_c \simeq 0.1$ eV, where T_c is the critical temperature).

In this calculation, we consider the lowest-lying stationary states of a system of two electrons, 1 and 2, in the presence of two fixed protons, a and b, described by the Hamiltonian

$$H = H_{1a} + H_{2b} + H' , \tag{1.10}$$

where

$$H_{i\alpha} = \frac{p_i^2}{2m} - \frac{e^2}{r_{i\alpha}} \tag{1.11}$$

and

$$H' = \frac{e^2}{r_{ab}} + \frac{e^2}{r_{12}} - \frac{e^2}{r_{1b}} - \frac{e^2}{r_{2a}} \quad . \tag{1.12}$$

One approximates the eigenfunctions of H as products of the (1s) atomic wave functions $\varphi_\alpha(i)$, solutions of the equations

$$H_{i\alpha}\,\varphi_\alpha(i) = E\,\varphi_\alpha(i)\,. \tag{1.13}$$

To satisfy the Pauli principle, the total wave functions have to be antisymmetric with respect to permutation of the electrons; one then writes them as

$$\Psi_{\underset{(-)}{+}} = \left[\varphi_a(1)\varphi_b(2) \underset{(-)}{+} \varphi_a(2)\varphi_b(1)\right]\Phi^{\overset{A}{(S)}}(1,2), \tag{1.14}$$

where Φ^A is the antisymmetric eigenfunction, and Φ^S one of the three symmetric eigenfunctions, of the total spin $\vec{S}_T = \vec{S}_1 + \vec{S}_2$, i.e.

$$\begin{aligned} S_T^2\,\Phi^{\overset{A}{(S)}} &= l(l+1)\,\Phi^{\overset{A}{(S)}} \\ S_T^z\,\Phi^{\overset{A}{(S)}} &= m\,\Phi^{\overset{A}{(S)}}, \end{aligned} \tag{1.15}$$

with $l = 0$, $m = 0$ for Φ^A and $l = 1$, $m = +1,0,-1$ for Φ^S .

One defines the overlap integral I , Coulomb integral V and exchange integral U as

$$\begin{aligned} I &= \int d^3r_1\, \varphi_a^*(1)\,\varphi_b(1) \\ V &= \int d^3r_1\, d^3r_2\, |\varphi_a(1)|^2\, |\varphi_b(2)|^2\, H' \\ U &= \int d^3r_1 d^3r_2\, \varphi_a^*(1)\,\varphi_b^*(2)\, H'\, \varphi_b(1)\,\varphi_a(2)\,. \end{aligned} \tag{1.16}$$

Averaging the total Hamiltonian with the approximate wave functions $\Psi_{\underset{(-)}{+}}$, one finds for the singlet (+) and triplet (-) states, the energies

$$E_\pm = 2E + \frac{V \pm U}{1 \pm |I|^2} \simeq 2E \pm U \tag{1.17}$$

and the energy gap between the two states is thus

$$J = E_+ - E_- = 2\,\frac{U - V|I|^2}{1 - |I|^4} \simeq 2\,U\,; \tag{1.18}$$

J turns out to be negative and the ground state is therefore the singlet state with antiparallel spins.

Now, if we forget about the higher excited states, the energies (1.17) may be labelled according to the spin quantum numbers l and m, and obtained as eigenvalues of the following effective Hamiltonian acting only on spin variables:

$$H_{eff} = -J\,\vec{S}_1 \cdot \vec{S}_2\,. \tag{1.19}$$

Indeed,

$$H_{eff} = -\frac{J}{2}\left[(\vec{S}_1 + \vec{S}_2)^2 - S_1^2 - S_2^2\right] \tag{1.20}$$

has the eigenvalues

$$E_{l,m} = -\frac{J}{2}\left[l(l+1) - 2\,\tfrac{1}{2}\left(\tfrac{1}{2}+1\right)\right], \tag{1.21}$$

i.e.

$$E_{l=0,m=0} = \frac{3}{4}J$$

$$E_{l=1,m=+1,0,-1} = -\frac{1}{4}J\,, \tag{1.22}$$

identical to (1.17), except for an unimportant constant shift. Thus we see that, although all spin-spin interactions are neglected in this theory (see (1.10) - (1.12)), the Pauli principle enables us to specify the space part of the wave functions through their spin quantum numbers, which leads to the effective coupling (1.19).

Heisenberg generalized this model Hamiltonian to describe the lowest-lying states of a magnetic solid. In the absence of exchange, the ground state would be $(2S+1)N$ - times degenerate; in the corresponding reduced Hilbert space, the effective Hamiltonian (1.1) removes this degeneracy and leads to ferromagnetism $(J > 0)$ or antiferromagnetism $(J < 0)$. Note that one expects J_{ab} to be of short range (first or second neighbours). This generalization to real solids is of course a highly nontrivial problem: a solid is not a giant hydrogen molecule! Moreover, other mechanisms (indirect exchange, superexchange) exist, which also lead to effective Hamiltonians of the form (1.1).

The Heisenberg model is known to describe satisfactorily a large class of ionic solids which exhibit ordered magnetic behaviour, e.g., EuO , CrO_2 (ferromagnets), MnF_2, MnO, $RbMnF_3$ (antiferromagnets). However, in many cases, one has to generalize the Hamiltonian (1.1) to take account of various sorts of anisotropy; one may, for instance, describe uniaxial or planar magnets by writing

$$H = -\sum_{a \neq b} \left[J^{\parallel}_{ab} S^z_a S^z_b + J^{\perp}_{ab} S^+_a S^-_b \right]. \tag{1.23}$$

We shall not enter into these theoretical refinements and shall limit ourselves to the isotropic model. Then the Hamiltonian (1.1) depends only on scalar products of spin operators, and is therefore invariant under simultaneous rotations of all the spins. Consequently, the total spin is an invariant of the motion:

$$\left[H , \sum_a \vec{S}_a \right] = 0, \tag{1.24}$$

as may be verified from (1.1) and (1.2). Note that this does not imply that all properties of the system will be isotropic, since the spins are located at the sites of a lattice, the symmetry of which will in general be reflected at the macroscopic level. To make things as simple as possible, we shall only consider Bravais lattices with cubic symmetry. Indeed, we shall be even more specific and limit ourselves to the case of ferromagnets. These may be defined by the fact that $J(q)$ has its absolute maximum at $q = 0$, which reflects the fact that the (essentially positive) interaction J_{ab} favours the parallel orientation of the spins (this point will become evident in part VI). We shall make use later of the fact that $J(q)$ may be expanded around this maximum:

$$J(q) = J(0) \left[1 - \alpha q^2 \cdots \right]. \tag{1.25}$$

In particular, when the interaction is limited to z nearest neighbours,

$$J(q) = J \sum_{\delta} \cos(\vec{q}\cdot\vec{\delta})$$

$$\xrightarrow[q\to 0]{} zJ\left[1 - \frac{(q\delta)^2}{6}\cdots\right] , \tag{1.26}$$

where δ is the distance between nearest neighbours.

In the following section, we shall show that neutron scattering experiments allow the measurement of time- and space-dependent spin correlation functions in Heisenberg magnets. We shall then, for completeness, recall some of the general properties of these functions. Finally, their theory will be reviewed in the three temperature domains of major interest: the low temperature spin wave region $(T \ll T_c)$, the paramagnetic region $(T \gg T_c)$ and the critical region $(T \simeq T_c)$.

II. NEUTRON SCATTERING EXPERIMENTS AND SPIN CORRELATION FUNCTIONS [2]

Slow neutrons, with wavelengths λ of the order of a typical lattice spacing $(\simeq 1 \text{Å})$, have energies

$$\mathcal{E} = \frac{p^2}{2M} = \frac{\hbar^2}{2M\lambda^2} \simeq 0.1 \text{ eV} \tag{2.1}$$

of the order of $k_B T_c$. This is the reason why they are excellent probes for studying spin fluctuations in magnets. Following Van Hove [3], we shall now show that, in the Born approximation, the cross-section for the scattering of neutrons by a Heisenberg magnet is very simply related to time- and space-dependent correlation functions.

In a scattering experiment (see Fig. 1), we direct at the sample a beam of neutrons with momentum $\vec{P}_i$, described by a plane wave function

$$\Psi_i = \frac{1}{\sqrt{\Omega}} e^{i\left[\vec{P}_i\cdot\vec{R} - \frac{P_i^2}{2M} t\right]} \tag{2.2}$$

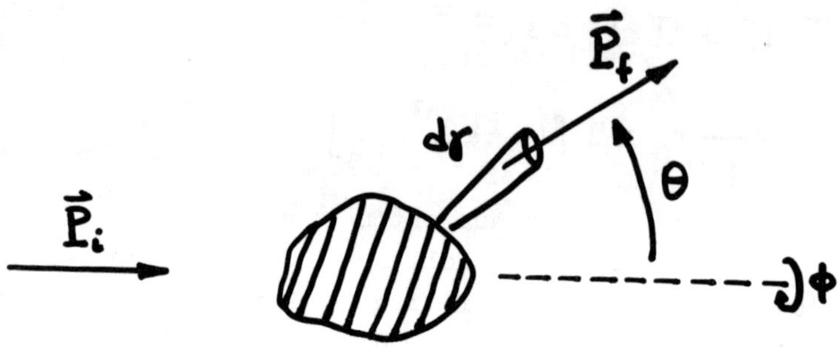

Fig. 1

(Ω is the volume of the laboratory). At large distances from the target, the outgoing wave may be analyzed as a superposition of final states

$$\Psi_f = \frac{1}{\sqrt{\Omega}} e^{i\left[\vec{P}_f \cdot \vec{R} - \frac{P_f^2}{2M} t\right]} \tag{2.3}$$

with momentum $\vec{P}_f$. Instead of $\vec{P}_f$, usual coordinates are the polar angles θ and ϕ, and the energy transfer

$$\omega = \frac{P_f^2 - P_i^2}{2M} . \tag{2.4}$$

One measures the (infinitesimal) number dN of neutrons coming out per unit time, with a momentum inside the infinitesimal element d^3P_f around $\vec{P}_f$ or, equivalently, with an energy transfer between ω and $\omega + d\omega$, in the solid angle $d\gamma = \sin\theta\, d\theta\, d\phi$. The scattering cross-section is then defined as the ratio of the number of scattered neutrons $dN/d\gamma\, d\omega$ to the incoming flux $(N/\Omega)P_i/M$:

$$\frac{d^2\sigma}{d\gamma d\omega} = \frac{M\Omega}{P_i N} \frac{dN}{d\gamma\, d\omega} \tag{2.5}$$

(N is the total number of neutrons). The number dN is N times the probability W_{if} that a neutron makes the transition from state $\vec{P}_i$ to state $\vec{P}_f$, multiplied by the number of final states in

$d^3P_f = P_f^2 dP_f d\gamma = MP_f d\omega d\gamma$. Since the density of plane wave states in momentum space is $\Omega/8\pi^3$, we arrive at

$$\frac{d^2\sigma}{d\gamma d\omega} = \frac{M^2\Omega^2}{8\pi^3}\frac{P_f}{P_i} W_{if} . \tag{2.6}$$

In the Born approximation, the transition probability is given by

$$W_{if} = \sum_{n,m} |\langle \vec{P}_i, n|\mathcal{V}|\vec{P}_f, m\rangle|^2 2\pi\, \delta(\omega_{mn} - \omega) \frac{e^{-\beta E_n}}{Z} , \tag{2.7}$$

where the sums run over the energy eigenstates of the spin system, $Z = \sum_n \exp(-\beta E_n)$ is the corresponding canonical partition function at reciprocal temperature $\beta = (k_B T)^{-1}$ and $\omega_{mn} = E_n - E_m$. This formula differs from the well-known "golden rule" of second-order perturbation theory in quantum mechanics only by the fact that one sums over all initial and final states of the spin system, weighting the initial states by the canonical probability $\exp(-\beta E_n)/Z$. The magnetic interaction between the neutron and the spins in the system may be written as

$$\mathcal{V} = \sum_a \vec{V}(\vec{R} - \vec{r}_a)\cdot \vec{S}_a \tag{2.8}$$

if the neutron beam is unpolarized and if one does not measure the eventual neutron spin polarization. Its matrix elements can be calculated as follows (see (2.2-3) and (1.3)):

$$\begin{aligned} &\langle \vec{P}_i, n|\mathcal{V}|\vec{P}_f, m\rangle \\ &= \sum_a \langle n|\vec{S}_a|m\rangle \cdot \frac{1}{\Omega}\int d^3R\, e^{i(\vec{P}_f - \vec{P}_i)\cdot\vec{R}}\vec{V}(\vec{R} - \vec{r}_a) \\ &= \sum_a \langle n|\vec{S}_a\, e^{i(\vec{P}_f - \vec{P}_i)\cdot\vec{r}_a}|m\rangle \cdot \\ &\qquad \frac{1}{\Omega}\int d^3R\, e^{i(\vec{P}_f - \vec{P}_i)\cdot(\vec{R} - \vec{r}_a)}\vec{V}(\vec{R} - \vec{r}_a) \\ &= \frac{1}{\Omega}\vec{V}_q \cdot \langle n|\vec{S}_q|m\rangle , \end{aligned} \tag{2.9}$$

where we have introduced the momentum transfer

$$\vec{q} = \vec{P}_f - \vec{P}_i \,, \tag{2.10}$$

and the Fourier transform of the interaction $\vec{V}(\vec{R})$

$$\vec{V}_q = \int d^3R \; \vec{V}(\vec{R}) \, e^{i\vec{q}\cdot\vec{R}} \,. \tag{2.11}$$

Substituting (2.7) and (2.9), we find that the scattering cross-section can be written as

$$\frac{d^2\sigma}{d\gamma\, d\omega} = \frac{N M^2}{8\pi^3} \frac{P_f}{P_i} \sum_{\substack{\alpha,\beta \\ (=x,y,z)}} V_q^\alpha \left(V_q^\beta\right)^* S^{\alpha\beta}(q,\omega) \,, \tag{2.12}$$

where all the information about the spin system is contained in the tensor

$$S^{\alpha\beta}(q,\omega) = \frac{1}{N} \sum_{n,m} \frac{e^{-\beta E_n}}{Z} \langle n| S_q^\alpha |m\rangle \langle m| S_{-q}^\beta |n\rangle \, 2\pi\, \delta(\omega_{mn} - \omega) \,. \tag{2.13}$$

To put this function in a form independent of the (unknown) energy eigenstates of the spin system, we replace the Dirac delta function by its usual representation

$$\delta(\omega) = \frac{1}{2\pi} \int_{-\infty}^{+\infty} e^{i\omega t} \, dt \tag{2.14}$$

and we get

$$S^{\alpha\beta}(q,\omega) = \int_{-\infty}^{+\infty} dt \; e^{-i\omega t} \, \Gamma_q^{\alpha\beta}(t) \,, \tag{2.15}$$

where

$$\begin{aligned} \Gamma_q^{\alpha\beta}(t) &= \frac{1}{NZ} \sum_{n,m} \langle n| e^{-\beta E_n} e^{iE_n t} S_q^\alpha e^{-iE_m t} |m\rangle \langle m| S_{-q}^\beta |n\rangle \\ &= \frac{1}{NZ} \, \mathrm{Trace} \left\{ e^{-\beta H} e^{iHt} S_q^\alpha \, e^{-iHt} S_{-q}^\beta \right\} \\ &= \frac{1}{N} \langle S_q^\alpha(t) \, S_{-q}^\beta(0) \rangle \end{aligned} \tag{2.16}$$

(we used the closure relation $\sum_m |m\rangle\langle m| = 1$) ; the brackets indicate an average over the canonical distribution and

$$S_q^\alpha(t) = e^{iHt} S_q^\alpha e^{-iHt} \tag{2.17}$$

is the Heisenberg picture of S_q^α . Going back to the localized spin operators S_a^α , we have:

$$\Gamma_q^{\alpha\beta}(t) = \frac{1}{N} \sum_{a,b} \Gamma_{ab}^{\alpha\beta}(t)\, e^{i\vec{q}\cdot(\vec{r}_a - \vec{r}_b)}, \tag{2.18}$$

with

$$\Gamma_{ab}^{\alpha\beta}(t) = \langle S_a^\alpha(t)\, S_b^\beta(0)\rangle. \tag{2.19}$$

The functions $S^{\alpha\beta}(q,\omega)$ thus turn out to be the four-dimensional Fourier transforms (over time and space) of the two-spin correlation functions $\Gamma_{ab}^{\alpha\beta}(t)$.

In the expression (2.12) for the cross-section, it is possible to separate out a part corresponding to pure forward elastic scattering. Indeed, when $t \to \infty$ or when $|\vec{r}_a - \vec{r}_b| \to \infty$, the two spins a and b in (2.19) must become statistically independent:

$$\Gamma_{ab}^{\alpha\beta}(t) \xrightarrow[|\vec{r}_a-\vec{r}_b|\to\infty]{t\to\infty} \langle S_a^\alpha(t)\rangle\langle S_b^\beta(0)\rangle \equiv \langle S^\alpha\rangle\langle S^\beta\rangle \tag{2.20}$$

($\langle S_a^\alpha(t)\rangle$ is independent of a and t). We can then separate this limiting value by writing

$$\Gamma_{ab}^{\alpha\beta}(t) = \delta\Gamma_{ab}^{\alpha\beta}(t) + \langle S^\alpha\rangle\langle S^\beta\rangle, \tag{2.21}$$

where

$$\delta\Gamma_{ab}^{\alpha\beta}(t) = \langle [S_a^\alpha(t) - \langle S^\alpha\rangle]\,[S_b^\beta(0) - \langle S^\beta\rangle]\rangle . \tag{2.22}$$

This leads to a similar separation for $S^{\alpha\beta}(q,\omega)$:

$$S^{\alpha\beta}(q,\omega) = \delta S^{\alpha\beta}(q,\omega) + 2\pi N \langle S^\alpha\rangle\langle S^\beta\rangle\, \delta_{q,0}^{Kr}\, \delta(\omega) . \tag{2.23}$$

'he second term corresponds to forward elastic scattering, which cannot be distinguished from the unscattered beam. Above the critical temperature, $\langle S^\alpha \rangle = 0$. Below T_c, if we study a single ferromagnetic domain and choose the z axis parallel to $\langle \vec{S} \rangle$, only $S^{zz}(q,\omega)$ will contain such an elastic term. In the following, we shall forget about this uninteresting part and, without indicating it explicitly, only consider the inelastic function $\delta S^{\alpha\beta}(q,\omega)$.

We have up to now considered correlation functions $\Gamma^{\alpha\beta}$ (and $S^{\alpha\beta}$) with $\alpha,\beta = x,y,z$. From now on, we shall use instead the variables (+, -, z), which have the advantage that the only non-vanishing elements of the correlation tensor $\Gamma^{\alpha\beta}$ are Γ^{zz}, Γ^{+-} and Γ^{-+}. This follows from the fact that the Heisenberg Hamiltonian conserves the total spin, which implies that, in the definition (2.16) or (2.19), the operators S^α and S^β must change the total spin in exactly opposite ways. From the definition of the raising and lowering operators $S^\pm = S^x \pm iS^y$, one readily obtains the relations between the functions $\Gamma^{\alpha\beta}$ in variables (x, y, z) and (+, -, z):

$$\begin{aligned} \Gamma^{xx} &= \Gamma^{yy} = \frac{1}{4}(\Gamma^{+-} + \Gamma^{-+}) \\ \Gamma^{xy} &= -\Gamma^{yx} = \frac{i}{4}(\Gamma^{+-} - \Gamma^{-+}) \\ \Gamma^{xz} &= \Gamma^{zx} = \Gamma^{yz} = \Gamma^{zy} = 0 \,. \end{aligned} \tag{2.24}$$

III. SOME GENERAL PROPERTIES OF THE SPIN CORRELATION FUNCTIONS [4]

3.1 Hermitian symmetry

Since the functions $S^{\alpha\beta}(q,\omega)$ are measurable quantities, they must be real:

$$S^{\alpha\beta}(q,\omega) = \left[S^{\alpha\beta}(q,\omega)\right]^* \quad (\alpha,\beta = +,-,z). \tag{3.1}$$

This can be readily proved if one starts from the definition (2.13) and uses (1.6) or

$$\begin{aligned} \langle n | S_q^z | m \rangle &= \langle m | S_{-q}^z | n \rangle^* \\ \langle n | S_q^+ | m \rangle &= \langle m | S_{-q}^- | n \rangle^* . \end{aligned} \tag{3.2}$$

Taking the proper Fourier transforms (see (2.15) and (2.18)), one obtains the corresponding relations for $\Gamma_q^{\alpha\beta}(t)$ and $\Gamma_{ab}^{\alpha\beta}(t)$:

$$\Gamma_q^{\alpha\beta}(t) = \left[\Gamma_q^{\alpha\beta}(-t)\right]^*$$

$$\Gamma_{ab}^{\alpha\beta}(t) = \left[\Gamma_{ba}^{\alpha\beta}(-t)\right]^*. \tag{3.3}$$

3.2 Detailed balance

The functions $S^{\alpha\beta}(q,\omega)$ satisfy the relation

$$S^{\alpha\beta}(q,\omega) = e^{-\frac{\omega}{k_BT}} S^{\beta\alpha}(-q,-\omega) \tag{3.4}$$

To establish this, start again from the definition (2.13), permute the summation indices n and m and use the identity

$$e^{-\beta E_m}\delta(\omega_{nm}-\omega) = e^{-\beta E_n} e^{\beta\omega_{mn}}\delta(-\omega_{mn}-\omega)$$

$$= e^{-\beta E_n} e^{-\beta\omega}\delta(\omega_{mn}+\omega). \tag{3.5}$$

Eqs. (3.1) and (3.4) show that it suffices to know the real functions $S^{\alpha\beta}(q,\omega)$ along the positive ω axis, in order to possess all of the information concerning spin correlations (and, more generally, all linear response phenomena, through the fluctuation-dissipation theorem).

3.3 Lattice symmetry

The space dependence of the correlation functions must of course have the symmetry properties of the lattice; in particular, the correlation functions are invariant under lattice translations and inversion:

$$\Gamma_{ab}^{\alpha\beta}(t) = \Gamma_{ba}^{\alpha\beta}(t) \tag{3.6}$$

only depends on the relative distance $\vec{r}_a - \vec{r}_b$; similarly,

$$\Gamma_q^{\alpha\beta}(t) = \Gamma_{-q}^{\alpha\beta}(t) \tag{3.7}$$

and

$$S^{\alpha\beta}(q,\omega) = S^{\alpha\beta}(-q,\omega) . \tag{3.8}$$

3.4 Spin-rotational symmetry

3.4.1 Sum rule

Since the total spin is an invariant of the motion (see (1.24)), the functions $\Gamma_{q=0}^{\alpha\beta}(t)$ are time-independent

$$\Gamma_{q=0}^{\alpha\beta}(t) = \frac{1}{N}\sum_{a,b}\langle S_a^{\alpha}(t)S_b^{\beta}(0)\rangle = \frac{1}{N}\sum_{a,b}\langle S_a^{\alpha}(0)\,S_b^{\beta}(0)\rangle = \Gamma_{q=0}^{\alpha\beta}(0). \tag{3.9}$$

3.4.2 Symmetry of the correlation tensor

In the paramagnetic region $(T > T_c)$, there is no spontaneous magnetization and hence no privileged direction. The full invariance of the Heisenberg Hamiltonian under simultaneous rotations of the spins (or a rotation of the spin coordinate system) then implies that, in addition to the relations (2.24), we must have:

$$\Gamma^{xx} = \Gamma^{yy} = \Gamma^{zz} , \tag{3.10}$$

the other components being zero:

$$\Gamma^{xy} = \Gamma^{xz} = \cdots = 0 . \tag{3.11}$$

Or, in the (+, -, z) variables, we have:

$$\Gamma^{zz} = \frac{\Gamma^{+-}}{2} = \frac{\Gamma^{-+}}{2} . \tag{3.12}$$

The whole $\Gamma^{\alpha\beta}$ tensor is thus reduced to a single scalar function,

$$\Gamma_{ab}(t) = \Gamma_{ab}^{zz}(t) \equiv \frac{\langle \vec{S}_a(t) \cdot \vec{S}_b(0) \rangle}{3} , \tag{3.13}$$

with corresponding definitions for $\Gamma_q(t)$ and $S(q,\omega)$, from (2.15) and (2.18).

In the ferromagnetic region $T<T_c$, the rotational symmetry is broken and one is left with the three independent functions, Γ^{zz} , Γ^{+-} and Γ^{-+} .

IV. LOW TEMPERATURE THEORY. [5]

4.1 Spin waves

4.1.1 The ground state

An essential property of the ferromagnetic Heisenberg model is that its ground state and lowest excited states are known exactly. If the choice of the spin coordinate system is such that the spontaneous magnetization is aligned along the negative z direction, the ground state eigenvector is

$$|0\rangle = |-S,-S,\dots\rangle \tag{4.1}$$

in the localized spin representation

$$|\{m_a\}\rangle = \prod_a |m_a\rangle , \tag{4.2}$$

defined by

$$S_a^z |m_a\rangle = m_a |m_a\rangle . \tag{4.3}$$

The ground state energy is easily found:

$$\begin{aligned} H|0\rangle &= -\sum_{a\neq b} J_{ab}[S_a^z S_b^z + S_a^+ S_b^-]|0\rangle \\ &= -\sum_{a\neq b} J_{ab} S^2 |0\rangle \\ &= -NS^2 J(q=0)|0\rangle ; \end{aligned} \tag{4.4}$$

in particular, if the interaction is limited to z nearest neighbours,

$$E_0 = -NS^2 z J. \tag{4.5}$$

4.1.2 Lowest excited states

To find the lowest excited states, we start from the ground state and raise one spin. Because of the translational symmetry of the lattice, we can deduce from an argument similar to Bloch's theorem in solid state physics, that the wave function must be a linear combination of the following form:

$$|1_q\rangle = C \sum_a e^{i\vec{q}\cdot\vec{r}_a} S_a^+ |0\rangle = C S_q^+ |0\rangle, \tag{4.6}$$

where C is a normalizing factor. To establish that $|1_q\rangle$ is an eigenfunction of H, it suffices to verify that

$$[H, S_q^+]|0\rangle = \omega_q S_q^+ |0\rangle, \tag{4.7}$$

because (4.7) implies that

$$\begin{aligned} H S_q^+ |0\rangle &= S_q^+ H |0\rangle + [H, S_q^+]|0\rangle \\ &= [E_0 + \omega_q] S_q^+ |0\rangle. \end{aligned} \tag{4.8}$$

From (1.1) and (1.2), we get:

$$\begin{aligned} [H, S_a^+]|0\rangle &= -\sum_{b\neq c} J_{bc} [(S_b^z S_c^z + S_b^+ S_c^-), S_a^+]|0\rangle \\ &= -2\sum_b J_{ab} (S_b^z S_a^+ - S_b^+ S_a^z)|0\rangle \\ &= 2S \sum_b J_{ab} (S_a^+ - S_b^+)|0\rangle, \end{aligned} \tag{4.9}$$

whence

$$[H, S_q^+]|0\rangle = \sum_a e^{i\vec{q}\cdot\vec{r}_a} [H, S_a^+]|0\rangle =$$

$$= 2S\Big(\sum_a e^{i\vec{q}\cdot\vec{r}_a} S_a^+ \sum_b J_{ab} - \sum_b e^{i\vec{q}\cdot\vec{r}_b} S_b^+ \sum_a e^{i\vec{q}\cdot(\vec{r}_a-\vec{r}_b)} J_{ab}\Big)|0\rangle$$

$$= 2S\,[J(0) - J(q)]\,S_q^+|0\rangle, \tag{4.10}$$

which justifies (4.7) with

$$\omega_q = 2S\,[J(0) - J(q)], \tag{4.11}$$

the Bloch spin wave dispersion relation. Note that

$$\omega_q = \omega_{-q} \geqslant 0 \tag{4.12}$$

and that ω_q only vanishes at $q = 0$, since $J(q)$ is an even function of q and has its maximum value at $q = 0$. Near this point, ω_q behaves as follows (see (1.25) and (1.26)):

$$\omega_q \simeq 2SJ(0) \propto q^2 \tag{4.13}$$

i.e., for nearest-neighbour interactions,

$$\omega_q \simeq \frac{SzJ}{3}(q\delta)^2. \tag{4.14}$$

The normalizing factor C in (4.6) is obtained from the condition

$$\langle 1_q|1_q\rangle = C^2\langle 0|S_{-q}^- S_q^+|0\rangle$$

$$= C^2\langle 0|(S_q^+ S_{-q}^- - 2S_{q=0}^z)|0\rangle$$

$$= C^2\, 2NS = 1, \tag{4.15}$$

whence

$$C = \frac{1}{\sqrt{2NS}}. \tag{4.16}$$

We thus proved that the operator

$$a_q^+ = \frac{S_q^+}{\sqrt{2NS}} , \tag{4.17}$$

acting on the ground state, "creates" a spin wave with energy ω_q.

We shall also make use of the following relation:

$$[H , S_q^z] |0\rangle = 0 , \tag{4.18}$$

which is readily established by a calculation similar to (4.9)-(4.10).

We finally remark that the same type of calculation can be made for antiferromagnets. However, in this case, the ground state and, a fortiori, the lowest excited states are not known exactly. One finds spin waves with an energy $\omega_q :: q$ for small values of q.

4.1.3 Higher excited states

It is easily verified that a two-spin-wave state

$$|1_q , 1_{q'}\rangle = a_q^+ a_{q'}^+ |0\rangle \tag{4.19}$$

is not an eigenstate of the Heisenberg Hamiltonian. For spins 1/2, for instance, this basis is overcomplete, since there exist N(N+1)/2 states of this type (N(N-1)/2 for $q \neq q'$, plus N for $q = q'$), while there can only be N(N-1)/2 states with two spins up, in the localized spin basis (4.2). Going back to the definitions (4.6) and (4.17), we see that the error comes from the fact that, in (4.19), we include unphysical situations where the same spin is raised twice. This error is of order 1/N and remains of this order in an n-spin-wave state, as long as n is finite. However, at any finite temperature, the number of spin waves is of order N and the error is not negligible in the thermodynamic limit $N \to \infty$. This difficulty has been studied by Dyson [6] ; in his very remarkable work, he showed that the independent spin wave states, although forming an overcomplete basis, remain a suitable approximation as long as the fraction of reversed spins is small (in fact, up to temperatures $T \lesssim T_c/2$). The physical reason for this rather surprising result, which will become clear when we come back to Dyson's results (see below, Eqs. (4.30), (4.45) and (4.48)), is that the corrections to the free spin wave theory may be interpreted as arising from interactions between spin waves and are therefore (essentially) proportional to the square of their density n/N.

Let us first consider the independent spin wave approximation, where one takes as eigenstates of the Heisenberg Hamiltonian the basis

$$|\{n_q\}\rangle = \prod_q \frac{(a_q^+)^{n_q}}{\sqrt{n_q!}} |0\rangle ; \tag{4.20}$$

this amounts to assuming that the commutation relations (4.7) and (4.18) are true in general:

$$[H, S_q^{\pm}] = \pm \omega_q S_q^{\pm} \tag{4.21}$$

$$[H, S_q^z] = 0. \tag{4.22}$$

In the basis (4.20), it is easily verified that the operator a_q^+ defined in (4.17), and its hermitian conjugate

$$a_q = \frac{S_{-q}^-}{\sqrt{2NS}} , \tag{4.23}$$

only have the following non-vanishing matrix elements:

$$\langle \{n_{q'}\}, n_q+1 | a_q^+ | \{n_{q'}\}, n_q \rangle = \langle \{n_{q'}\}, n_q | a_q | \{n_{q'}\}, n_q+1 \rangle$$

$$= \sqrt{n_q + 1} , \tag{4.24}$$

i.e., they behave as the creation and annihilation operators of the second quantization formalism. In particular, (4.24) implies that they obey the boson commutation relations:

$$[a_q, a_{q'}^+] = \delta_{q,q'}^{Kr} , \tag{4.25}$$

and that

$$\hat{n}_q = a_q^+ a_q \tag{4.26}$$

is the operator for the number of spin waves q.

The Heisenberg Hamiltonian is thus reduced to the free spin wave Hamiltonian

$$H_0 = E_0 + \sum_q \omega_q \hat{n}_q \,. \tag{4.27}$$

Note that, comparing (4.25) with the exact commutation relations (1.5) of S_q^+ and S_q^-, we must conclude that the operator has been approximated by its eigenvalue corresponding to the ground state $|0\rangle$:

$$S_q^z \simeq -NS\,\delta_{q,0}^{Kr} \,. \tag{4.28}$$

Dyson's results go further than this simple approximation: he has shown that the first corrections to the lowest order model (4.27) appear as an interaction between spin waves, such that the Hamiltonian becomes

$$H_D = H_0 + V_D \,, \tag{4.29}$$

where

$$V_D = \frac{1}{2N} \sum_{p,q,r} \Phi(p,q,r)\, a_{q+p}^+ a_{r-p}^+ a_r a_p \tag{4.30}$$

and

$$\Phi(p,q,r) = J(r-p) + J(q+p) - J(p) - J(r-p-q)$$

$$\equiv \frac{1}{2S}\left[\omega_p + \omega_{r-p-q} - \omega_{q+p} - \omega_{r-p}\right] \tag{4.31}$$

(see (4.11)). The low temperature expansion of the thermodynamic properties are correctly obtained from this Hamiltonian, a highly nontrivial result indeed, since it is defined in a Hilbert space which includes unphysical states.

4.2 Spin wave theory of the correlation functions

The free spin wave form of the correlation functions defined in (2.16) is easily obtained by evaluating the canonical average in the basis of the approximate eigenvectors $|\{n_q\}\rangle$ with the eigenvalues

$$E_{\{n_q\}} = E_0 + \sum_q \omega_q n_q \; . \tag{4.32}$$

We first remark that (4.22) implies that $S_q^z(t)$ (and hence $\Gamma_q^{zz}(t)$) is time-independent. From (4.28), we then get:

$$\Gamma_q^{zz}(t) = N S^2 \delta^{Kr}_{q,0} \tag{4.33}$$

and, from (2.15),

$$S^{zz}(q,\omega) = NS^2 \, 2\pi \, \delta(\omega) \, \delta^{Kr}_{q,0} \; ; \tag{4.34}$$

this is just the forward elastic scattering contribution that we decided to separate out (see (2.23)), as it cannot be observed.

To calculate Γ_q^{+-}, we again start from the definition (2.15) and go over to creation and annihilation operators, using (4.17), (4.23) and (4.21), i.e. (remember that $\partial_t S_q^+(t) = i[H, S_q^+(t)]$)

$$S_q^+(t) = e^{i\omega_q t} S_q^+ = \sqrt{2NS}\, e^{i\omega_q t} a_q^+ \, . \tag{4.35}$$

We thus find at once:

$$\Gamma_q^{+-}(t) = 2S \, e^{i\omega_q t} f_q \, , \tag{4.36}$$

where f_q is the well-known distribution function of the free Bose gas:

$$f_q = \langle a_q^+ a_q \rangle = \frac{\sum_{n=0}^{\infty} n \, e^{-\beta\omega_q n}}{\sum_{n=0}^{\infty} e^{-\beta\omega_q n}} = \frac{1}{e^{\beta\omega_q} - 1} \; . \tag{4.37}$$

A similar calculation leads to

$$\begin{aligned} \Gamma_q^{-+}(t) &= 2S \, e^{-i\omega_q t} (1 + f_q) \\ &= 2S \, e^{-i\omega_q t} e^{\beta\omega_q} f_q \, . \end{aligned} \tag{4.38}$$

Fourier transforming with respect to time, we arrive at the corresponding scattering spectral functions:

$$S^{+-}(q,\omega) = 4\pi S f_q \delta(\omega - \omega_q) \tag{4.39}$$

$$S^{-+}(q,\omega) = 4\pi S e^{\beta\omega_q} f_q \delta(\omega + \omega_q). \tag{4.40}$$

The physical origin of the singular frequency dependence of these expressions is simply the fact that, at any given wavenumber q, the free spin wave system can only absorb or emit a quantum ω_q, corresponding to the creation or annihilation of a spin wave. The expressions (1.39) and (4.40) are very useful, since they allow the dispersion relation for individual spin waves to be measured, at low temperatures, by neutron scattering techniques. Note that these expressions, in accordance with the detailed balance relation (3.4), exhibit a very strong temperature dependence. The contribution $S^{+-}(q,\omega)$ to the scattering spectrum, which corresponds to the annihilation of a spin wave, vanishes at low temperatures as the average number f_q of spin waves with energy ω_q :

$$f_q \underset{\beta\to\infty}{\simeq} e^{-\beta\omega_q} , \tag{4.41}$$

whereas the contribution from the creation of a spin wave tends to a finite limit: $1 + f_q \longrightarrow 1$.

When the temperature is raised, the interactions between spin waves, as described by Dyson's Hamiltonian ((4.29)-(4.31)), will gradually become important. This will have two effects: the spin wave frequency will be renormalized and become temperature-dependent; moreover, the spin waves will acquire a finite life-time, since they will no longer be exact eigenstates of the system. We therefore expect that, for example, the correlation function (4.36) will become

$$\Gamma_q^{+-}(t) \simeq e^{i\tilde{\omega}_q t - \Gamma_q |t|} \, \Gamma_q^{+-}(0) , \tag{4.42}$$

whence the singular delta behaviour of the spectrum (4.39) will be smoothed to a Lorentzian shape:

$$\frac{S^{+-}(q,\omega)}{\Gamma_q^{+-}(t=0)} \simeq \frac{2\gamma_q}{(\omega-\tilde{\omega}_q)^2+\gamma_q^2}\ , \tag{4.43}$$

centered around the renormalized frequency and with a width proportional to the damping. The calculation of $\tilde{\omega}_q$ and γ_q requires a careful analysis which we shall not go into; the physical meaning of the results is, however, easy to understand intuitively.

We may divide the Dyson Hamiltonian (4.29) into its diagonal and off-diagonal parts (in the free spin wave basis (4.20)); we write:

$$H_D = H^d + H^{o.d.}\ , \tag{4.44}$$

where

$$H^d = E_0 + \sum_q \omega_q \hat{n}_q + \frac{1}{N}\sum_{q,r} \Phi(p=0,q,r)\,\hat{n}_q\,\hat{n}_r\ , \tag{4.45}$$

$H^{o.d}$ being the off-diagonal part of V_D. The low temperature behaviour of $\tilde{\omega}_q$ turns out to be dominated by H^d; it may be obtained by taking the equilibrium average (over the free spin wave distribution) of the derivative of H^d with respect to $\hat{n}_q$:

$$\begin{aligned} \omega_q &= \left\langle \frac{\delta H^d}{\delta \hat{n}_q} \right\rangle \\ &= \left\langle \omega_q + \frac{2}{N}\sum_r \Phi(p=0,q,r)\,\hat{n}_r \right\rangle \\ &= \omega_q + \frac{2}{N}\sum_r \Phi(p=0,q,r)\, f_r\ , \end{aligned} \tag{4.46}$$

where f_r is the free boson distribution (4.37). Using the explicit expression (4.31) for the function Φ, one shows that the renormalized energy $\tilde{\omega}_q$ tends to the free spin wave value ω_q, when $T \longrightarrow 0$, as follows:

$$\tilde{\omega}_q \underset{T\to 0}{\simeq} \omega_q \left[1 - c\left(\frac{T}{T_c}\right)^{5/2} \right] , \tag{4.47}$$

where c is a constant.

The damping is due to the off-diagonal part of the Dyson Hamiltonian and is given by

$$\gamma_q = \frac{1}{N^2} \sum_{p,r} \left[\Phi(p,q,r) \right]^2 2\pi\, \delta(\omega_q + \omega_r - \omega_{r-p} - \omega_{q+p})$$

$$\times \left\{ f_r (1+f_{r-p})(1+f_{q+p}) - (1+f_r) f_{r-p} f_{q+p} \right\} \tag{4.48}$$

The physical meaning of this result is evident: the damping of a spin wave q is due to its (direct and inverse) scattering with another spin wave r, yielding two spin waves (r-p) and (q+p). The transition probability for this scattering is given, in the Born approximation, by the "golden rule", weighted by temperature-dependent statistical factors. The analysis of the temperature and wavenumber dependence of γ_q is a subtle problem, because of the presence of the singular delta function and because one has to consider different regimes, according to the relative values of ω_q and $k_B T$. In particular, when $\omega_q \ll k_B T \ll k_B T_c$, the damping is found to vanish as follows:

$$\gamma_q :: q^4 T^2 \ln\left(\frac{T}{T_c (q\delta)^2} \right) . \tag{4.49}$$

(:: shall be used to denote proportionality.)

V. HIGH TEMPERATURE THEORY

In the paramagnetic region ($T \geqslant T_c$), we know (see section 3.4.1) that the information contained in the correlation tensor reduces to that in the single function

$$S(q,\omega) = \frac{1}{N}\sum_{a,b} e^{i\vec{q}\cdot(\vec{r}_a-\vec{r}_b)} \int_{-\infty}^{\infty} dt\, e^{-i\omega t} \langle S_a^z(t) S_b^z(0)\rangle . \qquad (5.1)$$

Moreover, it suffices to know the part of this function which is even in the frequency variable:

$$S^+(q,\omega) = \frac{1}{2}\left[S(q,\omega) + S(q,-\omega)\right]. \qquad (5.2)$$

Indeed, the detailed balance property (3.4), together with (3.8), implies that $S^+(q,\omega)$ is related to the odd function

$$S^-(q,\omega) = \frac{1}{2}\left[S(q,\omega) - S(q,-\omega)\right], \qquad (5.3)$$

as follows:

$$S^-(q,\omega) = \frac{e^{-\beta\omega}-1}{e^{-\beta\omega}+1}\, S^+(q,\omega) = -\tanh\left(\frac{\beta\omega}{2}\right) S^+(q,\omega). \qquad (5.4)$$

In the following, we shall therefore limit ourselves to the study of $S^+(q,\omega)$ and its Fourier transforms

$$\Gamma_q^+(t) = \frac{1}{2\pi}\int_{-\infty}^{\infty} d\omega\, S^+(q,\omega)\, e^{i\omega t} \qquad (5.5)$$

$$\Gamma_{ab}^+(t) = \frac{1}{N}\sum_q \Gamma_q^+(t)\, e^{-i\vec{q}\cdot(\vec{r}_a-\vec{r}_b)}, \qquad (5.6)$$

which are real and even functions of the time (see (3.7)). We shall also find it convenient to define normalized functions $\tilde{\Gamma}_{ab}$, $\tilde{\Gamma}_q$ and $\tilde{S}$ by dividing Γ^+_{ab}, Γ^+_q and S^+ by the equilibrium correlation function $\Gamma^+_q(t=0) \equiv \Gamma_q(t=0)$, so that

$$\tilde{\Gamma}_q(t) = \frac{\Gamma_q^+(t)}{\Gamma_q(t=0)} \qquad (5.7)$$

equals 1 at t=0.

The relation (5.4) shows that the even functions (5.2), (5.5) and (5.6) contain the essential information in most interesting situations, since the factor tanh $(\beta\omega/2)$ tends to zero when $\omega \to 0$. Thus, when we are concerned with the low-frequency behaviour of $S(q,\omega)$, its odd part $S^-(q,\omega)$ becomes negligible.

We shall now consider the high-temperature limit ($\beta \to 0$), where, at all frequencies, $S^-(q,\omega) = 0$ and $\Gamma_q(t)$ and $\Gamma_{ab}(t)$ are real functions, even in the time variable.

When the temperature goes to infinity, the equilibrium correlation function is trivial to compute, since the interactions become negligible ($\exp(-\beta\mathcal{H}) \to 1$) and the spins become independent : $\langle S_a^\alpha \; S_b^\beta \rangle \to 0$ if $a \neq b$. From the definitions (2.15) and (2.19), we then find:

$$\Gamma_q(t=0) \xrightarrow[\beta\to 0]{} \frac{1}{3N} \sum_a \langle \vec{S}_a^{\,2} \rangle = \frac{S(S+1)}{3} . \tag{5.8}$$

The time dependence of the correlation functions $\tilde{\Gamma}_q(t)$ remains, however, a nontrivial N - body problem, since the whole dynamics appears in the evolution of the Heisenberg spin operators (2.17).

The classical theory of the high-temperature behaviour of the dynamic correlation function, due to de Gennes [7] , is based on the computation of the moments $\langle\omega^{2n}\rangle_q$ of $\tilde{S}(q,\omega)$, defined as follows:

$$\langle \omega^{2n} \rangle_q = \frac{\int_{-\infty}^{\infty} d\omega \, \omega^{2n} \, \tilde{S}(q,\omega)}{\int_{-\infty}^{+\infty} d\omega \, \tilde{S}(q,\omega)} . \tag{5.9}$$

From (5.5) and (5.7), one readily sees that these moments are related to derivatives of $\tilde{\Gamma}_q(t)$ at t=0:

$$\langle \omega^{2n} \rangle_q = (-i)^{2n} \frac{\partial^{2n}}{\partial t^{2n}} \tilde{\Gamma}_q(t) \Big|_{t=0} . \tag{5.10}$$

Hence, their knowledge gives us information about the short-time behaviour of $\tilde{\Gamma}_q(t)$:

$$\tilde{\Gamma}_q(t) = 1 - \langle \omega^2 \rangle_q \frac{t^2}{2!} + \langle \omega^4 \rangle_q \frac{t^4}{4!} \cdots \tag{5.11}$$

and, provided that this series expansion converges (or at least may be continued) on the whole real axis, the knowledge of $\Gamma_q(t)$ and hence $\tilde{S}(q,\omega)$.

Although the computational difficulties rapidly grow with n , the first few derivatives of

$$\tilde{\Gamma}_q(t) = \langle S_q^z(t) S_{-q}^z(0) \rangle / \langle S_q^z S_{-q}^z \rangle \tag{5.12}$$

can be computed, at t=0, by making repeated use of the equation of motion of the operator $\vec{S}_q(t)$:

$$\partial_t \vec{S}_q(t) = i\left[H, \vec{S}_q(t) \right], \tag{5.13}$$

and the moments (5.10) take the form

$$\langle \omega^{2n} \rangle_q = \frac{1}{\langle S^z_q S^z_{-q} \rangle} \langle [H,[H,\ldots,[H,S^z_q]\ldots] S^z_{-q} \rangle . \tag{5.14}$$

For example, the second moment is given by

$$\begin{aligned} \langle \omega^2 \rangle_q &= \langle [H,[H,S^z_q]] S^z_{-q} \rangle / \langle S^z_q S^z_{-q} \rangle \\ &= -\langle [H,S^z_q][H,S^z_{-q}] \rangle / \langle S^z_q S^z_{-q} \rangle \end{aligned} \tag{5.15}$$

(we used the cyclic invariance of the trace: Trace $\{ABC....D\}$ = Trace $\{BC...DA\}$ = ...). The commutators can of course be evaluated directly from (1.7) and (1.5):

$$[H,S^z_q] = \frac{1}{N} \sum_{q'} \left[J(q') - J(q-q') \right] S^+_{q-q'} S^-_{q'} . \tag{5.16}$$

The moments $\langle \omega^{2n} \rangle_q$ are thus expressed in terms of equilibrium averages of products of 2(n+1) spin operators, averages which can (in principle) be evaluated exactly, since the spins become statistically independent at infinite temperature. For later use, let us note here that, when $q \to 0$, the equation of motion (5.13) and (5.16) takes the form of a conservation equation:

$$\partial_t S^z_q \underset{q\to 0}{\simeq} i \vec{q} \cdot \vec{j}_z , \tag{5.17}$$

where $\vec{j}_z$ is the spin (or magnetization) current:

$$\vec{j}_z = \frac{1}{N} \sum_{q'} \left(\vec{\nabla}_{q'} J(q') \right) S^+_{-q'} S^-_{q'} ; \tag{5.18}$$

this clearly reflects the fact that $S^z_{q=0}$ is an invariant of the motion.

De Gennes computed the second and fourth moments and considered the two simple limiting situations of small and large wavenumbers (compared to the inverse of the lattice spacing δ).

(i) $q\,\delta \longrightarrow \infty$: *

Here, it turns out that one has approximately

$$\langle \omega^4 \rangle_{q\to\infty} \simeq 3\,\langle \omega^2 \rangle^2_{q\to\infty} \,, \qquad (5.19)$$

a relation which would be exact if $\tilde{S}(q,\omega)$ was a Gaussian function. De Gennes therefore proposed as a reasonable approximation for large values of the wavenumber q:

$$\tilde{S}(q,\omega) = \sqrt{\frac{2\pi}{\langle \omega^2 \rangle_q}}\; e^{-\frac{\omega^2}{2\langle \omega^2 \rangle_q}} \;; \qquad (5.20)$$

this prediction turns out to be in good agreement with experiment. This is not too surprising, since large wavenumbers correspond to short distances, where the dynamics should take place over short times (of the order of J^{-1}), for which the moment expansion (5.11) may be expected to converge rapidly.

(ii) $q\delta \to 0$:

In this limit, corresponding to large distances (and hence long times), de Gennes postulated, following Van Hove, that the correlation function $\tilde{\Gamma}_q(t)$ should obey the (hydrodynamic) laws of macroscopic physics, describing the decay of a local fluctuation of the magnetization (or spin) density. Since the total spin is an invariant of the motion, the macroscopic magnetization density must obey a conservation equation:

$$\partial_t M(\vec{r},t) = -\vec{\nabla}\cdot\vec{J}(\vec{r},t) \;; \qquad (5.21)$$

---

* The vector $\vec{q}$ is of course limited to the first Brillouin zone. However, if one considers polycrystalline powders, one has to average $\tilde{\Gamma}_q(t)$, and hence the moments, over the direction of $\vec{q}$, which breaks the periodicity in the reciprocal lattice.

the thermodynamics of linear irreversible phenomena then predicts that the magnetization current density $\vec{J}$ should be proportional to the gradient of the local magnetic field, itself proportional to the magnetization density:

$$\vec{J}(\vec{r},t) = -\lambda \vec{\nabla} \mathcal{H}(\vec{r},t) = -\frac{\lambda}{\chi} \vec{\nabla} M(\vec{r},t), \tag{5.22}$$

where λ is an Onsager coefficient and χ, the magnetic susceptibility. We thus arrive at the diffusion equation

$$\partial_t M(\vec{r},t) = D \nabla^2 M(\vec{r},t), \tag{5.23}$$

with $D = \lambda/\chi$. If this equation is supposed to be applicable to $\tilde{\Gamma}(\vec{r} = \vec{r}_a - \vec{r}_b, t) \equiv \tilde{\Gamma}_{ab}(t)$ * (for large distances and long times), taking the space-Fourier transform ($\vec{\nabla} \rightarrow -i\vec{q}$) leads us to conclude that $\tilde{\Gamma}_q(t)$ obeys the following equation:

$$\partial_t \tilde{\Gamma}_q(t) \underset{\substack{q \to 0 \\ t \to \infty}}{\simeq} -Dq^2 \tilde{\Gamma}_q(t), \tag{5.24}$$

whence

$$\tilde{S}(q,\omega) = 2 \operatorname{Re} \int_0^\infty dt\, e^{-i\omega t} \tilde{\Gamma}_q(t)$$

$$= 2 \operatorname{Re} \frac{1}{i\omega + Dq^2} \tag{5.25}$$

$$= \frac{2Dq^2}{\omega^2 + (Dq^2)^2} \tag{5.26}$$

(we took into account the fact that (5.24) is only valid for positive times and that $\tilde{\Gamma}_q(t)$ is an even function of t).

If one tries to fit this Lorentzian form to the known moments of

---

* This argument is more convincing if one considers the relaxation function instead of the correlation function. However, in the small wavenumber or low frequency limit, these two functions become identical.

$\tilde{S}(q,\omega)$, one runs into the (meaningful) difficulty that all moments of this function diverge (except for n = 0). De Gennes therefore decided to use (5.26) for frequencies ω smaller, in absolute value, than a cut-off ω_c and to put it equal to zero for $|\omega| > \omega_c$. The two parameters D and ω_c can then be fitted to the exact (second and fourth) moments, which yields:

$$D q^2 = \frac{\pi}{2\sqrt{3}} \left(\frac{\langle\omega^2\rangle_q^3}{\langle\omega^4\rangle_q} \right)^{1/2} :: J(q\delta)^2$$

$$\omega_c = \left(\frac{3\langle\omega^4\rangle_q}{\langle\omega^2\rangle_q} \right)^{1/2} :: J \qquad (5.27)$$

for a cubic lattice with nearest-neighbour interactions J.

It can be shown that all moments are proportional to q^2, when $q \to 0$, which supports the diffusion hypothesis (5.26). It is clear, however, that one cannot be satisfied with this moment fitting procedure, since no reliable information can be obtained from the short time expansion (5.11), concerning the (asymptotic) long time behaviour of the correlation function. More sophisticated theories have therefore been developed recently in order to justify, on a microscopic basis, the heuristic (gaussian and lorentzian) assumptions inherently necessary in the moment method, and also to furnish general methods for calculating the shape and width of the scattering function $\tilde{S}(q,\omega)$ in all interesting regions.

This problem has been approached from different points of view. In particular, Résibois and the author [8] derived a kinetic equation for $\tilde{\Gamma}_q(t)$, valid in the whole paramagnetic region ($T \gtrsim T_c$), through a perturbation expansion and resummation technique, in the limit of long range exchange forces, thus generalizing, to time-dependent correlation functions, the molecular field (or Weiss) theory of magnets. With a different and more compact method, Kawasaki [9] arrived at the same result. Although the equilibrium Weiss theory is extremely simple (as we shall see in the following section, where we shall come back to the limitations of our long range force theory), its non-equilibrium generalization is far from trivial. We shall, therefore, avoid the rather involved techniques leading to these results and rather try to show how one can arrive at them by intuitive physical reasoning.

Let us first remark that (5.25) suggests that it is interesting to write the Laplace transform of $\tilde{\Gamma}_q(t)$,

$$\bar{S}(q,\omega) = \int_0^\infty dt\, e^{-i\omega t}\, \tilde{\Gamma}_q(t), \tag{5.28}$$

in the following way:

$$\bar{S}(q,\omega) = \frac{1}{i\omega + G_q(\omega)}, \tag{5.29}$$

since the function $G_q(\omega)$ thus defined is expected to have a simpler behaviour than $\bar{S}(q,\omega)$ (or $\tilde{S}(q,\omega) = 2\,\mathrm{Re}\,\bar{S}(q,\omega)$) itself; in particular, if the diffusion assumption (5.26) is correct, $G_q(\omega)$ must in the "hydrodynamic" limit where $\omega \to 0$ and $q \to 0$, with ω/q^2 finite, tend to the (real) limit

$$G_q(\omega) \xrightarrow[\substack{q\to 0\\ \omega\to 0\\ (\omega/q^2\ \text{finite})}]{} D q^2 . \tag{5.30}$$

Eq. (5.29) is the Laplace transform equivalent of the non-markoffian kinetic equation

$$\partial_t \tilde{\Gamma}_q(t) = -\int_0^t dt'\, \tilde{G}_q(t')\, \tilde{\Gamma}_q(t-t'), \tag{5.31}$$

with the initial condition $\tilde{\Gamma}_q(t=0) = 1$ and with

$$G_q(\omega) = \int_0^\infty dt\, e^{-i\omega t}\, \tilde{G}_q(t). \tag{5.32}$$

Of course, writing (5.29) or (5.31) merely amounts to a definition of the functions $G_q(\omega)$ and $\tilde{G}_q(t)$; to go any further, one needs to establish independent rules to evaluate them.

This is the result obtained by Résibois and the author, and by Kanawasaki: they showed that the kernel $\tilde{G}_q(t)$ can be expressed in terms of the correlation function $\tilde{\Gamma}_q(t)$ itself, in such a way that (5.31) becomes a closed, but nonlinear, kinetic equation for $\tilde{\Gamma}_q(t)$. More precisely, $\tilde{G}_q$ is given by an infinite series of successive approximations:

$$\tilde{G}_q(t|\{\tilde{\Pi}_{q'}\}) = \tilde{G}_q^{(2)} + \tilde{G}_q^{(4)} + \cdots, \tag{5.33}$$

where, for example,

$$\tilde{G}_q^{(2)} = \frac{4S(S+1)}{3N} \sum_{q'} \frac{\gamma_{q-q'}\gamma_{q'}}{\gamma_q} \left[J(q') - J(q-q')\right]^2 \tilde{\Pi}_{q-q'}(t)\, \tilde{\Pi}_{q'}(t). \tag{5.34}$$

In this expression, we introduced

$$\gamma_q = \frac{3\langle S_q^z S_{-q}^z\rangle}{NS(S+1)}, \tag{5.35}$$

which goes to 1 when $T \to \infty$ (see (5.8)).

To clarify the physical meaning of these results, we first have to show that, if the diffusion assumption, (5.24) - (5.26), is correct, then the diffusion coefficient is given by a "Kubo formula" [10] :

$$D = \frac{1}{NS(S+1)\gamma_{q=0}} \int_0^\infty dt \langle \vec{j}_z(t)\cdot\vec{j}_z(0)\rangle, \tag{5.36}$$

where $\vec{j}_z$ is the magnetization current defined in (5.18). Indeed, from (5.26), we must conclude that

$$D = \lim_{\omega\to 0} \lim_{q\to 0} \frac{\omega^2}{q^2} \operatorname{Re} \int_0^\infty dt\, e^{-i\omega t}\, \tilde{\Pi}_q(t) \tag{5.37}$$

(note the order of the limits). Integrating twice by parts, we have:

$$\int_0^\infty dt\, e^{-i\omega t}\, \tilde{\Pi}_q(t) = -\frac{e^{-i\omega t}}{i\omega} \tilde{\Pi}_q(t)\Big|_0^\infty + \frac{e^{-i\omega t}}{\omega^2} \partial_t \tilde{\Pi}_q(t)\Big|_0^\infty$$
$$- \frac{1}{\omega^2} \int_0^\infty dt\, e^{-i\omega t}\, \partial_t^2\, \tilde{\Pi}_q(t). \tag{5.38}$$

The two first terms disappear, because (a) $\tilde{n}_q(t)$ and its derivative vanish when $t \to \infty$, (b) the first term is imaginary and does not contribute to (5.37) and (c) $\partial_t \tilde{n}_q|_{t=0}$ is zero since $\tilde{n}_q(t)$ is an even function of t. We are thus left with

$$D = -\lim_{\omega\to 0}\lim_{q\to 0}\frac{1}{q^2}\,\mathrm{Re}\int_0^\infty dt\, e^{-i\omega t}\,\partial_t^2\,\tilde{n}_q(t)$$

$$= -\lim_{\omega\to 0}\lim_{q\to 0}\frac{3}{NS(S+1)\Gamma_q q^2}\,\mathrm{Re}\int_0^\infty dt\, e^{-i\omega t}\langle\partial_t^2 S_q^z(t)S_{-q}^z(0)\rangle \tag{5.39}$$

(see (5.12) and (5.35)). But, through (5.13),

$$\langle\partial_t^2 S_q^z(t)S_{-q}^z(0)\rangle = -\langle[H,[H,S_q^z(t)]]\,S_{-q}^z(0)\rangle$$

$$= \langle[H,S_q^z(t)]\,[H,S_{-q}^z(0)]\rangle, \tag{5.40}$$

by the cyclic invariance of the trace, and when $q \to 0$, this average tends to (see (5.17), (5.18))

$$-\vec{q}\cdot\langle\vec{j}_z(t)\,\vec{j}_z(0)\rangle\cdot\vec{q} \equiv -\frac{q^2}{3}\langle\vec{j}_z(t)\cdot\vec{j}_z(0)\rangle, \tag{5.41}$$

if, as everywhere, we suppose the lattice symmetry to lead to an isotropic diffusion coefficient. Substituting (5.40) and (5.41) into (5.39) and taking the double limit $q \to 0$ and $\omega \to 0$ (which exists if the diffusion assumption is correct), we immediately arrive at (5.36). The transport coefficient D is thus expressed as the time integral of a "Green-Kubo integrand", which is as usual the autocorrelation function of the current $\vec{j}_z$ associated with the conserved quantity $S^z_{q\to 0}$.

Of course, the interest of this exact result is essentially formal, since the evaluation of the correlation function in (5.36) still involves an N-body problem as difficult as that of $\tilde{n}_q(t)$ itself. We shall now show that a simple decoupling (or random phase) approximation, applied to

$$\langle \vec{j}_z(t) \cdot \vec{j}_z(0) \rangle = \frac{1}{N^2} \sum_{q',q''} \left(\vec{\nabla}_{q'} J(q') \cdot \vec{\nabla}_{q''} J(q'') \right)$$

$$\langle S^+_{-q'}(t) S^-_{q'}(t) S^+_{-q''}(0) S^-_{q''}(0) \rangle, \tag{5.42}$$

leads to an expression of the diffusion coefficient, which is consistent with the first approximation (5.34) of the kernel of the kinetic equation (2.31). We write

$$\begin{aligned} \langle S^+_{-q'}(t) S^-_{q'}(t) S^+_{-q''}(0) S^-_{q''}(0) \rangle &\simeq \\ &\langle S^+_{-q'}(t) S^-_{q'}(t) \rangle \langle S^+_{-q''}(0) S^-_{q''}(0) \rangle \\ &+ \langle S^+_{-q'}(t) S^-_{q''}(0) \rangle \langle S^-_{q'}(t) S^+_{-q''}(0) \rangle . \end{aligned} \tag{5.43}$$

The first term does not contribute to (5.42), since

$$\sum_{q''} \left(\vec{\nabla}_{q''} J(q'') \right) \Pi^{+-}_{q''}(0) = 0 , \tag{5.44}$$

by symmetry: $J(q'')$ and $\Pi^{+-}_{q''}$ are even functions of the wavenumber. For the second term, we note that, by the lattice translational symmetry,

$$\begin{aligned} \langle S^+_{-q'}(t) S^-_{q''}(0) \rangle &= \delta^{Kr}_{q',q''} \Pi^{+-}_{q'}(t) \\ &\equiv \delta^{Kr}_{q',q''} \frac{2NS(S+1)}{3} \gamma_{q'} \tilde{\Pi}_{q'}(t) \end{aligned} \tag{5.45}$$

(see (3.12), (5.12) and (5.35)), the same result holding for $\langle S^-_{q'}(t) S^+_{-q''}(0) \rangle$. Collecting these results, we arrive at the following approximation for the diffusion coefficient:

$$D \simeq D^{(2)} = \frac{4S(S+1)}{9\,\gamma_{q=0}} \int_0^\infty dt \; \frac{1}{N} \sum_{q'} |\vec{\nabla}_{q'} J(q')|^2$$

$$\gamma_{q'}^2 \, \tilde{\Pi}_{q'}(t)^2 . \qquad (5.46)$$

It is now easy to verify that the limit (5.30), with the approximation (5.34) for $\tilde{G}_q(t)$, leads to exactly the same result, i.e.

$$D^{(2)} = \lim_{\substack{q \to 0 \\ \omega \to 0}} \frac{1}{q^2} \, G_q^{(2)}(\omega)$$

$$\equiv \lim_{q \to 0} \frac{1}{q^2} \int_0^\infty dt \; \tilde{G}_q^{(2)}(t) , \qquad (5.47)$$

if we expand, in $\tilde{G}_q^{(2)}$,

$$J(q') - J(q-q') \simeq \vec{q} \cdot \vec{\nabla}_{q'} J(q') \qquad (5.48)$$

and make the same isotropy assumption as in (5.41).

The mechanism by which the non-markoffian kinetic equation (5.31) tends to the markoffian diffusion equation (5.24), is then the following: When $q \to 0$, the kernel $\tilde{G}_q(t)$ tends to the form $q^2 f(t)$, where $f(t)$ decays over a q-independent time scale. The correlation function $\tilde{\Pi}_q(t)$, on the contrary, becomes a very slowly decaying function (with a time scale $\tau_q :: q^{-2}$), since $\partial_t \tilde{\Pi}_q :: q^2$. For times $t \to \infty$ (of the order of τ_q), we may then approximate (5.31) by the asymptotic form:

$$\partial_t \tilde{\Pi}_q(t) \simeq - \left(\int_0^\infty dt' \; \tilde{G}_q(t' | \{\tilde{\Pi}_{q'}\}) \right) \tilde{\Pi}_q(t) , \qquad (5.49)$$

i.e. a diffusion equation, with the definition (5.30) for D.

Note that this argument can only be valid if, in the small wave-number and low frequency limit, the behaviour of the kernel $\tilde{G}_q(\omega)$ (which depends nonlinearly on the correlation functions $\tilde{\Pi}_q$, themselves) is not dominated by the contributions of these functions for

$q' \simeq q$. As we shall see, it is the failure of this assumption near T_c which leads to the singular critical phenomena. Outside of the vicinity of the critical region, the assumption is correct, but one should not conclude that (5.49) is, in any sense, the leading term of an analytic expansion. All one can safely say is that the diffusion equation becomes exact in the limit $q \to 0$, $t \to \infty$, with $q^2 t$ finite, i.e. that $G_q(\omega)/q^2$ has a finite limit when $\omega \to 0$ and $q \to 0$. However, it has a singularity at $\omega = 0$, $q = 0$ and cannot be expanded straightforwardly around this point. This is an example of the now classical phenomenon of the "long tails" of Green-Kubo integrands 11 which, as we shall see in the following section, may be considered precursors of the critical singularities.

We shall not examine this problem in detail, but simply show, on the example of $G_q^{(2)}$ and at infinite temperature (where $\gamma_q = 1$), that the contribution to $\tilde{G}_q^{(2)}(t)/q^2$ coming from values of $q' \simeq q$, tend to zero as $t^{-5/2}$ (or q^5) in the hydrodynamic limit, this non-analytic behaviour implying of course that $G_q^{(2)}(\omega)/q^2$ is singular at the point $\omega = 0$, $q = 0$. To this aim, we separate the sum over q' in (5.34) into two parts, by writing

$$\tilde{G}^{(2)}_{q} = \tilde{G}^{(2)}_{q>} + \tilde{G}^{(2)}_{q<}, \tag{5.50}$$

where $G^{(2)}_{q>}$ contains the terms with $q' > q_0$ and $G^{(2)}_{q<}$, those with $q' < q_0$, q_0 being a cut-off wavenumber such that $q \ll q_0 \ll B$ (B represents the Brillouin zone edge); this condition can always be satisfied when $q \to 0$. For $\tilde{G}^{(2)}_{q>}$, we may proceed as before (see (5.48))and we find:

$$\frac{\tilde{G}^{(2)}_{q>}}{q^2} \xrightarrow[q \to 0]{} \frac{4S(S+1)}{9N} \sum_{q'>q_0} |\vec{\nabla}_{q'} J(q')|^2 \, \tilde{\Gamma}_{q'}(t)^2, \tag{5.51}$$

in accordance with (5.46) - (5.47). There is no reason to expect this expression to have a non-analytic behaviour for long times: for finite values of $q' > q_0$, $\tilde{\Gamma}_{q'}(t)$ may be assumed to be bounded by an exponential $A \exp(-\alpha t)$. In $\tilde{G}^{(2)}_{q<}$, however, problems might arise when $q' \simeq q$, since in the hydrodynamic limit, $\tilde{\Gamma}_{q' \simeq q}$ is expected to behave as $\exp[-Dq'^2 t]$. Expanding $J(q')$ and $J(q-q')$ for q and q' small (see (1.25))and going to the limit of an infinite system, where

$$\frac{1}{N}\sum_{q'} \longrightarrow \frac{1}{\Delta}\int d^3q' \qquad (5.52)$$

(Δ is the volume of the Brillouin zone), we arrive at the following guess for $G^{(2)}_{q<}$:

$$\tilde{G}^{(2)}_{q<} \simeq \frac{4S(S+1)}{3\Delta}\left[\alpha J(0)\right]^2 \int_{q'<q_0} d^3q' \left[|\vec{q}-\vec{q}'|^2-\vec{q}'^2\right]^2 e^{-D\left[|\vec{q}-\vec{q}'|^2+\vec{q}'^2\right]t} \quad . \qquad (5.53)$$

We then introduce dimensionless variables $\vec{y} = \vec{q}'/q$ and $\tau = Dq^2t$ and find:

$$\frac{\tilde{G}^{(2)}_{q<}}{q^2} \simeq \frac{4S(S+1)}{3\Delta}\left[\alpha J(0)\right]^2 q^5 \int_{y<\frac{q_0}{q}\to\infty} d^3y \left[|\vec{1}-\vec{y}|^2-y^2\right]^2 e^{-\left[|\vec{1}-\vec{y}|^2+y^2\right]\tau} \quad . \qquad (5.54)$$

When $q \rightarrow 0$ and $t \rightarrow \infty$, with $\tau = Dq^2t$ finite, this contribution to the Green-Kubo integrand vanishes like q^5 or $t^{-5/2}$. It is thus found to be negligible, which justifies (5.46): even if this $t^{-5/2}$ "long tail" behaviour is taken seriously, the integral (5.47) converges. But the corrections to (5.51) are non-analytic when $t \rightarrow \infty$, and the same is expected for $G^{(2)}_q(\omega)/q^2$ near $\omega = 0$ and $q = 0$.

Coming back to the general kinetic equation (5.31) and (5.33), we see that it allows, in principle, to calculate the correlation function in all domains of q and t values. Such explicit calculations are of course very difficult, since the kernel is given by an infinite series of successive approximations, each of which is a non-linear functional of the correlation function itself. Computations have been made, which justify de Gennes' qualitative predictions and lead to numerical results of the same order of magnitude.

VI. THE CRITICAL REGION

Remarkable phenomena occur when one approaches the critical temperature. Fluctuations at long wavelengths become more and more important; in particular, the equilibrium correlation function $\Gamma_q(t=0)$ diverges at q=0. Before describing the dynamical aspects of these phenomena, it is necessary that we briefly recall the classical molecular field (of Weiss) theory of these equilibrium fluctuations 12 . In this theory, every spin is treated as if it were submitted to the average field due to the others. It is easy to understand that this approximation becomes rigorous in the limit of long range forces, i.e. if one considers that every spin has a constant interaction J with z neighbours and one goes to the limit $z \to \infty$ and $J \to 0$, keeping zJ finite (we shall see that k_BT_c is proportional to zJ).

In this approximation, let us calculate the correlation function (5.35) for $T \gtrsim T_c$. To simplify the argument, we consider the case of spins S = ½, but the result turns out to be general. We first remark that the isotropy of the system implies that

$$\gamma_{ab} = \frac{3\langle S_a^z S_b^z \rangle}{S(S+1)} = 4\langle S_a^z S_b^z \rangle \tag{6.1}$$

may be calculated, for $a \neq b$, in the localized spin basis (4.2), as the average of S_b^z over the partial canonical ensemble where S_a^z is fixed, say up $(S_a^z = +\frac{1}{2})$:

$$\gamma_{ab} = 2\langle S_b^z \rangle_{a\ \mathrm{up}} . \tag{6.2}$$

Now, a spin $\vec{S}$ in a magnetic field $\vec{\mathcal{H}}$ is described by a Hamiltonian $H = -\vec{\mathcal{H}}_0 \cdot \vec{S}$, where $\vec{\mathcal{H}}_0 = g|e|\hbar\,(2mc)^{-1}\vec{\mathcal{H}}$, g being the Landé factor, equal to 2 for an electron. We can then say that, in the Heisenberg model (1.1), a spin b feels a "field" due to its neighbours, equal to

$$\vec{\mathcal{H}}_b = 2 J_{ab} \vec{S}_a + 2 \sum_{c(\neq a)} J_{cb} \vec{S}_c \tag{6.3}$$

(we have separated the term corresponding to the spin a which we fix up). The molecular field approximation then consists in evaluating (6.2) as if the spin b were in the average field

$$\vec{\mathcal{H}}_b \simeq \langle \vec{\mathcal{H}}_b \rangle_{a\ up} = \Big[J_{ab} + \sum_{c(\neq a)} J_{cb} \gamma_{ac} \Big] \vec{1}_z \ . \tag{6.4}$$

We then have:

$$\begin{aligned} \gamma_{ab} &\simeq 2 \frac{\mathrm{trace}\{S_b^z\, e^{\beta \mathcal{H}_b S_b^z}\}}{\mathrm{trace}\{e^{\beta \mathcal{H}_b S_b^z}\}} \\ &= \frac{e^{\beta\mathcal{H}_b/2} - e^{-\beta\mathcal{H}_b/2}}{e^{\beta\mathcal{H}_b/2} + e^{-\beta\mathcal{H}_b/2}} \\ &= \tanh(\beta\mathcal{H}_b/2)\ . \end{aligned} \tag{6.5}$$

But, to be consistent, we should only retain, from this result, the leading term in its expansion in powers of $(1/z)$. For $k_BT \gtrsim k_BT_c :: zJ$, one has

$$\beta\mathcal{H}_b \lesssim \frac{J}{k_BT_c} = O\left(\frac{1}{z}\right). \tag{6.6}$$

To this order, we may approximate $\tanh(\beta\mathcal{H}b/2) \simeq \beta\mathcal{H}b/2$ and forget about the exclusion $c \neq a$ in (6.4), whence

$$\gamma_{ab} = \beta J_{ab}/2 + \sum_c \gamma_{ac}\, \beta J_{cb}/2\ . \tag{6.7}$$

This equation is easily solved by a Fourier transformation and one finds

$$\bar{\gamma}_q = \sum_{b(\neq a)} \gamma_{ab}\, e^{i\vec{q}\cdot(\vec{r}_a - \vec{r}_b)} = \frac{\beta J(q)/2}{1 - \beta J(q)/2}\ . \tag{6.8}$$

The complete Fourier transform (5.35) is then obtained by adding the missing term $\gamma_{aa} = 4\langle (S_a^z)^2 \rangle = 1$; the result is

$$\gamma_q = \frac{1}{1 - \beta J(q)/2}\ . \tag{6.9}$$

For a ferromagnet, the maximum value of $J(q)$ occurs at $q = 0$. We then see that, when the temperature is lowered from $T = \infty$ (where $\gamma_q = 1$), γ_q remains finite until we reach a temperature $T_c = T(0)/(2k_B)$, where $\gamma_{q=0}$ diverges, indicating the appearance of long range order. Below

T_c, (6.9) must of course be modified. In the critical region, i.e. when $q \to 0$ and $T \to T_c$, we may expand (6.9) to the lowest order in q^2 and $\epsilon = (T-T_c)/T_c$ (see (1.25)) and obtain

$$\gamma_q \underset{\substack{q\to 0\\ T\to T_c}}{\simeq} \frac{1}{\alpha\,(q^2+\varkappa^2)}\,, \tag{6.10}$$

where $\varkappa$ is the inverse of a "correlation length"

$$\xi = \varkappa^{-1} = \sqrt{\alpha T_c}\;\epsilon^{-1/2}. \tag{6.11}$$

The Fourier transform of (6.10) gives the asymptotic form of the correlation function γ_{ab} for large distances $|\vec{r}_a - \vec{r}_b|$:

$$\gamma_{ab} \underset{\substack{|\vec{r}_a-\vec{r}_b|\to\infty\\ T\to T_c}}{\simeq} \frac{e^{-\varkappa|\vec{r}_a-\vec{r}_b|}}{|\vec{r}_a-\vec{r}_b|}\,, \tag{6.12}$$

the range of which diverges at T_c. This result is the equivalent, for the Heisenberg model, of the Ornstein - Zernicke theory of the critical opalescence in classical fluids.

It is now well known that the molecular field theories (Weiss theory for magnets, Van der Waals theory for fluids...)do not correctly describe the critical phenomena. They can only be expected to apply to a "precritical" region, where $(T - T_c)/T_c \gg 1/z$ (if T_c is taken at its molecular field value). The reason for their failure is clear: the critical phenomena appear when the macroscopic properties of the system become dominated by the fluctuations at very long wavelengths (of the order of ξ), i.e. when ξ becomes much larger than any molecular characteristic length of the problem, such as the range of the interparticle interaction. But the molecular field theories only becomes rigorous when this range is infinite, and taking this limit first evidently forbids going into the true critical region.

Modern theories of critical phenomena [13] try to describe the "critical exponents", which characterize the singular behaviour of the various thermodynamic quantities, and predict:

(i) That these exponents are largely universal, i.e. independent of the details of the microscopic interactions and only sensitive to general features of the system (dimensionality, symmetries, ...).

(ii) That they are related by simple algebraic relations, known as "scaling laws", in such a way that only two of them are independent.

These predictions are well verified experimentally and their theoretical justification has been considerably clarified recently, through the remarkable work of Wilson and others [14]. Methods for calculating the critical exponents from first principles have even been developed for particular models.

Going into these problems would lead us far outside the scope of these lectures. We shall therefore limit ourselves to stating what the scaling hypothesis says about the static correlation function γ_q, namely that in the critical region, i.e. when q and $\varkappa$ go to zero, γ_q becomes a homogeneous function of these variables

$$\gamma_q \underset{\substack{q\to 0\\ \varkappa\to 0}}{\simeq} \frac{1}{q^{2-\eta}} f\left(\frac{\varkappa}{q}\right) \tag{6.13}$$

(in three dimensions). Note that (6.10) is of this form, with the critical exponent $\eta = 0$. Experimentally, η is found to be small ($\eta \lesssim 0.1$), hence (6.10) is not far from the truth; but the dependence of $\varkappa$ on the temperature is very different from the molecular field prediction: if one expresses the divergence of $\xi = \varkappa^{-1}$ as

$$\xi :: \epsilon^{-\nu}, \tag{6.14}$$

the critical exponent ν is found to be about 0.7, instead of 1/2, as in (6.11).

We are now ready to leave equilibrium and consider the dynamic correlation function $\tilde{\Pi}_q(t)$. De Gennes has applied the moment method to this problem [7,8]. Of course, even an approximate evaluation of the moments (5.14) is extremely difficult at finite temperature, but one can show that only the denominator

$$\langle S^z_q S^z_{-q} \rangle = \frac{NS(S+1)}{3} \gamma_q \tag{6.15}$$

has a singular behaviour near the critical point. The reason for this is that the average in the numerator involves spins which are interacting (because of the commutators with H) and hence equilibrium correlations at finite distances (of the order of a few interaction ranges), which have no critical behaviour. It is then reasonable to write, as a qualitative approximation,

$$\langle \omega^{2n} \rangle_{q,T} \simeq \frac{1}{r_q} \langle \omega^{2n} \rangle_{q,T=\infty} . \tag{6.16}$$

Generalizing the diffusion assumption (5.21) - (5.24) to allow for a q-dependent diffusion constant, i.e. writing

$$\partial_t \tilde{\Gamma}_q(t) \simeq -D_q\, q^2\, \tilde{\Gamma}_q(t), \tag{6.17}$$

with

$$D_q = \frac{\lambda}{\chi_q} , \tag{6.18}$$

where χ_q is the static susceptibility at wavenumber q, de Gennes concluded that (see (5.27)) D_q should behave as follows:

$$D_q(T) \simeq \frac{D(T=\infty)}{r_q} . \tag{6.19}$$

Near the critical point, with the molecular field approximation (6.10)-(6.11) for γ_q, this theory thus predicts that $\tilde{S}(q,\omega)$, and hence the neutron scattering spectrum, should have a Lorentzian shape (see 15.26)) with a width

$$\omega_\kappa(q) \equiv q^2 D_q(\kappa) :: q^2 (q^2 + \kappa^2) , \tag{6.20}$$

i.e., at $T = T_c$,

$$\omega_{\kappa=0}(q) :: q^4 \tag{6.21}$$

and, at q = 0,

$$\omega_\kappa(q=0) :: q^2\kappa^2 :: q^2(T-T_c) . \qquad (6.22)$$

Note that we could have arrived at these qualitative predictions by purely macroscopic arguments; indeed, if one makes the generalized diffusion assumption (6.17) - (6.18) and supposes that the Onsager coefficient λ is a slowly-varying function of T near T_c, one immediately obtains (6.19), since χ_q behaves as γ_q when $q \to 0$.*

Recent experiments have clearly shown that these conjectures are wrong and that the above theory cannot be made to fit the facts by the mere replacement of the molecular field approximation for γ_q by its scaling form (6.13). What is observed is in agreement with a phenomenological description proposed by Halperin and Hohenberg [15] and known as "dynamic scaling laws" or "assumptions" (DSA). These state that, in the critical region, the correlation function $\tilde{\Gamma}_q(t)$ depends on the three variables q, t and T only through the combinations $\omega_\kappa(q)t$ and q/κ :

$$\tilde{\Gamma}_q(t) \simeq \Pi\left(\omega_\kappa(q)t, \frac{\kappa}{q}\right), \qquad (6.23)$$

where $\kappa = \xi^{-1}$ is related to the temperature trough (6.14), and that the characteristic frequency $\omega_\kappa(q)$ is a homogeneous function of q and κ:

$$\omega_\kappa(q) = q^{\zeta}\,\varphi\left(\frac{\kappa}{q}\right) \qquad (6.24)$$

(note that (6.20) is of this form, with $\zeta = 4$). Fourier-transforming (6.23) with respect to the time variable (see (2.15)),one obtains the corresponding assumption for the scattering spectrum:

---

* The fluctuation theorem relates the static susceptibility and correlation function as follows:

$$\lim_{q\to 0} \chi_q = \frac{N g^2 \mu_B^2 S(S+1)}{3 k_B T} \lim_{q\to 0} \gamma_q ,$$

where g is the Landé factor and μ_B, the Bohr magneton.

$$\tilde{S}(q,\omega) = \frac{1}{\omega_\varkappa(q)}\, \mathfrak{S}\left(\frac{\omega}{\omega_\varkappa(q)}, \frac{q}{\varkappa}\right). \tag{6.25}$$

To visualize the implications of the DSA, one may represent the critical region as in Fig. 2. In this diagram, the ordinate variable is the wavenumber q; along the positive abscissa axis, we plot $\varkappa = \varkappa_+ :: \epsilon^{\nu}$, the inverse of the correlation length for $T > T_c$; on the negative side, the variable $\varkappa$ is conventionally taken equal to minus the (positive) inverse $\varkappa_- :: |\epsilon|^{\nu'}$ of the correlation length below T_c. On the figure, three limiting regions are indicated, where a simpler behaviour may be expected to occur: region **I**, the ordered hydrodynamic region, where $q \ll \varkappa$ ($T < T_c$); region **II**, the transition region, where $q \gg \varkappa$ ($T \simeq T_c$); region **III**, the paramagnetic or disordered hydrodynamic region, where again $q \ll \varkappa$ ($T > T_c$).

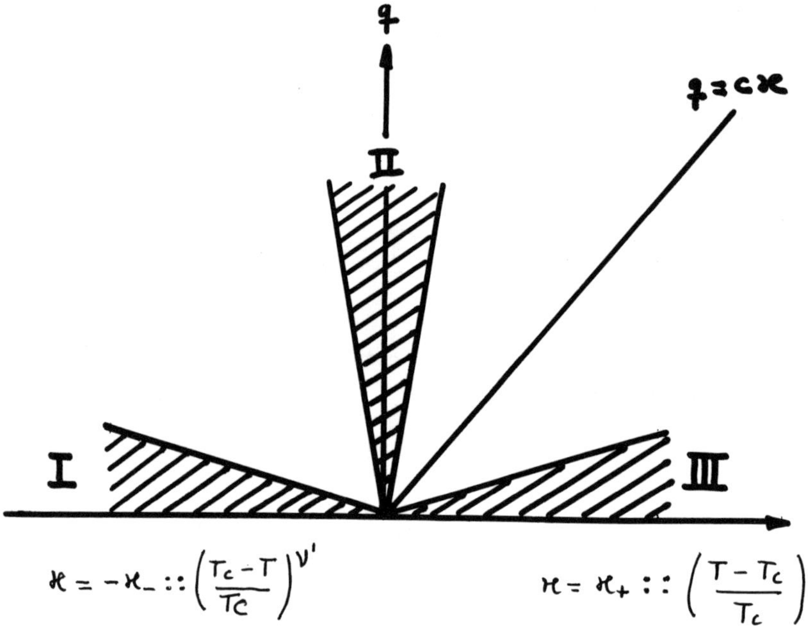

Fig. 2

Let us now consider, as indicated in the figure, a straight line $q = ck$, where c is an arbitrary constant; the assumptions (6.23) and (6.24) mean that the correlation function $\tilde{\Gamma}_q(t)$ should retain the same shape along this line, if we scale the time variable as $\tau = q^{\zeta} t$. The corresponding experimental prediction is then that, if one normalizes the scattering spectrum (6.25) to an arbitrary constant, its form should remain unchanged, except that its width should vary proportionally to q^{ζ}. When we go around the critical point $q = \varkappa = 0$, the "dynamic critical exponent" ζ is assumed not to vary and the function $\varphi(x = \varkappa/q)$, to be continuous in the whole diagram.

Of course, when we consider the left-hand-side region ($T<T_c$), we should distinguish between the transverse and longitudinal functions $\tilde{\Gamma}^{+-}_q(t)$ and $\tilde{\Gamma}^{zz}_q(t)$. The DSA say nothing about the precise shape of the function $\Gamma(x,y)$, and hence of the scattering spectrum, except that nothing spectacular happens when one crosses the $\varkappa = o$ ($T=T_c$) line, where $\tilde{\Gamma}^{+-}_q$ and $\tilde{\Gamma}^{zz}_q$ must become the unique $\tilde{\Gamma}_q$ function of the paramagnetic region.

The power of the DSA comes from the fact that (6.24) relates what happens in the whole $(q,\varkappa)$ plane and, in particular, in the three limiting regions indicated in the figure. To use this fact, Halperin and Hohenberg developed a macroscopic theory of spin waves, from which it appeared that in region I (for $q \ll \varkappa$), such waves remain well-defined excitations up to the critical point, with an energy

$$\omega^{I}_{\varkappa}(q) :: \varkappa_-^{1/2} q^2 \tag{6.26}$$

and a (negligible) damping $\gamma_q :: q^4/\varkappa^{3/2}$ (up to a possible logarithmic factor). Thus, the scattering spectrum should exhibit well-defined peaks (see (4.39),(4.40)), centered at (plus or minus) the frequency (6.26), moving towards $\omega = o$ when $T \to T_c$, as $\varkappa_-^{1/2} :: [(T_c - T)/T_c] ::$ $[(T_c - T)/T_c]^{0.35}$ (ν' is believed to be equal to ν). The scaling relation (6.24) then implies that the exponent $\zeta = 5/2$ (and that $\varphi(x = \varkappa/q)$ behaves as $(-x)^{1/2}$ when $x \to -\infty$). Here and in the following, we neglect all corrections of order η to the critical exponents, because these effects are too small to be measured with any accuracy and because the theory of dynamic critical phenomena below T_c cannot be considered today as sufficiently well founded for these subtleties to be taken into account; for instance, in isotropic ferromagnets, nothing definite can yet be said about the importance of longitudinal fluctuations.

We can now use the result $\zeta = 5/2$ in region II: when we approach the critical point by lowering q at $\kappa = 0$ ($T = T_c$), the scaling relation (6.24), with the implicit assumption that φ ($x = \kappa/q$) is continuous near x = o, immediately leads to the conclusion that the scattering spectrum, whatever its shape, must have a width proportional to $q^{5/2}$:

$$\omega_{\kappa}^{II}(q) :: q^{5/2} \quad . \tag{6.27}$$

Finally, in region III, the wavenumber q is much smaller than all other characteristic wavenumbers in the problem, including κ_+. Then, the macroscopic arguments leading to the diffusive behaviour (5.26) should be correct: the scattering spectrum should have a central lorentzian peak, with a width

$$\omega_{\kappa}^{III}(q) :: D(\kappa_+)\, q^2 \quad ; \tag{6.28}$$

(6.24) then implies that, since $\zeta = 5/2$, $\varphi(x = \kappa/q)$ must behave as $x^{1/2}$ when $x \to +\infty$, which leads to the following temperature dependence of the diffusion coefficient: $D(\kappa_+) :: \kappa_+^{1/2} :: \epsilon^{\nu/2}$.

The DSA agree remarkably well with experimental observations, not only for isotropic ferromagnets, but for essentially all critical points (provided suitable additional parameters are introduced to take eventual anisotropies into account). The lesson they teach us is that the classical theory, based on moment expansions, is intrinsically incorrect: the scaling relation (6.24) with $\zeta = 5/2$ cannot be obtained from it, even in region III, where the diffusion assumption is expected to be correct. Moreover, one does not see how it could ever describe the transition region II, where the scattering spectrum passes continuously from the two-peak spin wave shape of region I to the diffusion Lorentzian shape of region III.

It is therefore interesting that the theory based on the kinetic equation (5.31 and 33), i.e. rigorous in the limit of long range forces,permits to justify the DSA (at least in a precritical region) and, in addition, to determine the shape of the scaled correlation function (6.23) - (6.24) [16]. The situation is however only clear on the paramagnetic side of the critical region ($T \gtrsim T_c$) while, in the ordered region, it is still a subject of theoretical (and experimental) controversy. Let us stress here again that, although this theory is the non-equilibrium generalization of the very simple and "classical" molecular field theory, it is by no means trivial and not at all equivalent to the "classical" theory based on the macroscopic equations (6.17) - (6.18).

The error in this macroscopic theory lies in two places. First, the generalized diffusion equation (6.17) is only correct when q is much smaller than any other wavenumber in the problem, i.e. in region III. Second, the assumption that the Onsager coefficient in (6.18) exists and is regular near T_c is incorrect in the whole critical region [17] . This is due to the fact that, because of the divergence of the equilibrium correlation functions, the mechanism which, away from the critical point, leads to the "long tail" behaviour of the Green-Kubo integrands, has here the dramatic effect that small wavenumber (time-dependent) fluctuations dominate the critical dynamics.

To show this, let us start again from the kinetic equation (5.31), with the approximation (5.34) for the kernel $\tilde{G}_q$. To arrive at an equation of the form (6.17), we have to replace (5.31) by its markoffian approximation (5.49), which leads to the definition (see (6.18))

$$\lambda^{(2)} = \lim_{q\to o} D_q^{(2)} \chi_q = \lim_{q\to o} \frac{\chi_q}{q^2} \int_o^\infty dt\, \tilde{G}_q^{(2)}(t \mid \{\tilde{\Gamma}_{q'}\}) \quad . \tag{6.29}$$

We know that this "markoffianization" procedure is only allowed if the small wavenumber correlation functions $\tilde{\Gamma}_{q'}$ do not dominate the behaviour of $\tilde{G}_q$. We shall now show, in the particular case of the transition region (κ = o), that this assumption leads to a contradiction.

Proceeding as in (5.50), we separate, in $\tilde{G}_q^{(2)}$, the contributions to the sum over q′ in two domains separated by a cut-off q_o, with $q \ll q_o \ll B$. Since the susceptibility χ_q behaves as γ_q when $q \to o$ (see the footnote on page 279), the contribution from $q' > q_o$ gives (compare with(5.51)):

$$\lambda_>^{(2)} :: \sum_{q'>q_o} (\gamma_{q'})^2 \, |\vec{\nabla} J(q')|^2 \int_o^\infty dt\, \tilde{\Gamma}_{q'}(t)^2 \quad . \tag{6.30}$$

Since the cut-off q_o restricts the values of q′ to be outside the critical region, $\lambda_>^{(2)}$ is expected to be a finite number, even at T_c.

If this was the dominant contribution to $\lambda^{(2)}$, the classical theory would be correct and (with the molecular field approximation for γ_q), we would find $\tilde{\Gamma}_{q\to o}$ (t) to behave as $\exp[-\lambda q^2 t/\chi_q] \simeq \exp[-c\, q^4 t]$. To test this assumption, we substitute this guess for $\tilde{\Gamma}_{q'}$ in the second contribution, from $q' < q_o$. Expanding everything for q and q′ small (including the equilibrium correlation functions $\gamma_{q'}$ (see (6.10)), we thus get (compare with (5.53)):

$$\lambda_<^{(2)} :: \frac{1}{q^2} \int_{q'<q_o} d^3q' \frac{\left[|\vec{q}-\vec{q}'|^2 - q'^2\right]^2}{|\vec{q}-\vec{q}'|^2 q'^2} \int_o^\infty dt \; e^{-c\left[|\vec{q}-\vec{q}'|^4 + q'^4\right]t}$$

$$:: \frac{1}{q^2} \int_{q'<q_o} d^3q' \frac{\left[|\vec{q}-\vec{q}'|^2 - q'^2\right]^2}{|\vec{q}-\vec{q}'|^2 q'^2 \left[|\vec{q}-\vec{q}'|^4 + q'^4\right]} , \tag{6.31}$$

which, when we introduce the reduced variable $\vec{y} = \vec{q}'/q$, becomes (compare with (5.54)):

$$\lambda_<^{(2)} :: \frac{1}{q^3} \int d^3y \frac{\left[|\vec{1}-\vec{y}|^2 - y^2\right]^2}{|\vec{1}-\vec{y}|^2 y^2 \left[|\vec{1}-\vec{y}|^4 + y^4\right]} :: q^{-3} . \tag{6.32}$$

Thus, the assumption that $\lambda_>^{(2)}$ is the dominant contribution leads to the contradictory result that $\lambda_<^{(2)}$ diverges as q^{-3}. The origin of this divergence is easily traced back to the presence of the singular equilibrium correlation functions $\gamma_{q-q'}$ and $\gamma_{q'}$ in the kernel (5.34).

This proves that, on the contrary, the behaviour of $\tilde{\Gamma}_q(t)$ near the critical point is dominated by the region $q'<q_o$, i.e. that small wavenumber fluctuations determine themselves self-consistently. This implies of course that, in the kinetic equation (5.31) and (5.33) there is no clear separation between the time scales of the function $\tilde{\Gamma}_q(t)$ and of the kernel $\tilde{G}_q(t\,|\{\tilde{\Gamma}_{q'}\})$; hence, the equation will (except in region III, where $q/\varkappa \to o$) retain its non-markoffian form. We shall now show that, if this fact is properly taken into account, this kinetic equation can be shown to justify the dynamic scaling assumptions, with the value $\zeta = 5/2$ for the critical exponent.

We start once again from the approximate kernel (5.34), limit the sum over q' to the dominant region $q' \simeq q < q_o$ and expand everything for $q' \simeq q \simeq \varkappa \to o$, using (1.25) and (6.10); the kinetic equation thus becomes:

$$\partial_t \tilde{\Gamma}_q(t) = -\gamma^2 \int_{q'<q_o} d^3q' \int_o^t dt' \frac{(q^2+\varkappa^2)(|\vec{q}-\vec{q}'|^2 - q'^2)^2}{(|\vec{q}-\vec{q}'|^2+\varkappa^2)(q'^2+\varkappa^2)}$$

$$\tilde{\Gamma}_{q-q'}(t')\, \tilde{\Gamma}_{q'}(t')\, \tilde{\Gamma}_q(t-t') , \tag{6.33}$$

where

$$\gamma^2 = \frac{4\,S(S+1)\ \alpha\ J(0)^2}{3\Delta} \quad . \tag{6.34}$$

To account for the fact that $q' \simeq q \simeq \varkappa$, we go over to the reduced variables $\vec{y} = \vec{q}\,'/q$ and $K = \varkappa/q$, which leads to

$$\partial_t\, \tilde{\Gamma}_q(t) = -\gamma^2 q^5 \int_{y < \frac{q_0}{q} \to \infty} d^3y \int_0^t dt' \, \frac{(1+\varkappa^2)(|\vec{1}-\vec{y}|^2 - y^2)^2}{(|\vec{1}-\vec{y}|^2+\varkappa^2)(y^2+\varkappa^2)}$$

$$\tilde{\Gamma}_{q|\vec{1}-\vec{y}|}(t')\ \tilde{\Gamma}_{qy}(t')\ \tilde{\Gamma}_q(t-t'). \tag{6.35}$$

We then remark that the factor $\gamma^2 q^5$ in front of the integrals disappears when we introduce dimensionless time variables $\tau = \gamma q^{5/2} t$ and $\tau' = \gamma q^{5/2} t'$, in such a way that (6.35) becomes the following equation, for the function $\mathbb{\Gamma}_q(\tau) \equiv \tilde{\Gamma}_q(t)$:

$$\partial_\tau \mathbb{\Gamma}_q(\tau) = -\int d^3y \int_0^\tau d\tau' \frac{(1+\varkappa^2)(|\vec{1}-\vec{y}|^2 - y^2)^2}{(|\vec{1}-\vec{y}|^2+\varkappa^2)(y^2+\varkappa^2)}$$

$$\mathbb{\Gamma}_{q|\vec{1}-\vec{y}|}(|\vec{1}-y|^{5/2}\,\tau')\ \mathbb{\Gamma}_{qy}(y^{5/2}\,\tau')\ \mathbb{\Gamma}_q(\tau-\tau') \ . \tag{6.36}$$

The variable q now only appears as a factor in the index of every function $\mathbb{\Gamma}$. If this equation possesses a unique solution, it cannot depend on this arbitrary factor. Hence, we have reduced the kinetic equation to the following asymptotic form:

$$\partial_\tau \mathbb{\Gamma}(\tau) = -\int d^3y \int_0^\tau d\tau' \frac{(1+\varkappa^2)\ (|\vec{1}-\vec{y}|^2 - y^2)^2}{(|\vec{1}-\vec{y}|^2+\varkappa^2)(y^2+\varkappa^2)}$$

$$\mathbb{\Gamma}(|\vec{1}-\vec{y}|^{5/2}\,\tau')\ \mathbb{\Gamma}(y^{5/2}\,\tau')\ \mathbb{\Gamma}(\tau-\tau') \ , \tag{6.37}$$

for the single function

$$\Gamma_q(t) \simeq \mathbb{\Gamma}(\tau\,;\,K) = \mathbb{\Gamma}(\gamma\, q^{5/2} t\ ;\ \frac{\varkappa}{q}) \ , \tag{6.38}$$

i.e. we have recovered the DSA relations (6.23) and (6.24), with $\zeta = 5/2$, in the limit of long range forces and with the approximate kernel $\tilde{G}_q^{(2)}$.

It can be shown that this result is correct to all orders in the series (5.33) for $\tilde{G}_q$, i.e. that

(i) when all intermediate modes $q', q'', \ldots$ are limited inside the region $q' \simeq q'' \ldots \simeq q < q_0$, the introduction of reduced variables y', y'',... and τ, τ', τ'',... makes the equation dimensionless;

(ii) that the corrections to this asymptotic equation (the contributions where some wavenumbers q',... are larger than q_0) are negligible in the limit where $t \to \infty$ and $q \to 0$, with $q^{5/2}t$ finite.

As expected, one finds that the first corrections behave as $q^{3/2}$; this corresponds to the "classical" result, where $\partial_t \tilde{\Gamma}_q(t)$ is proportional to q^4 , i.e. $\partial_\tau \tilde{\Gamma}_q(t) :: \partial_{q^{5/2}t} \tilde{\Gamma}_q(t) :: q^{3/2}$.

In principle, these results go beyond the phenomenology of the dynamic scaling assumptions, since they allow to determine the correlation function Γ and the scaling function φ . Such computations have been performed, with the lowest order approximations for the kernel $\tilde{G}_q$, and the agreement with experiment is remarkably good; in particular, the scaling function $\varphi(x = \kappa/q)$ has the correct behaviour and goes to infinity as $x^{1/2}$ when $x \to \infty$ (in region III). Let us however stress again that this theory is, in principle, limited to the model of long range forces and that agreement with experimental observations is only obtained if one fits the temperature dependence of χ to its measured value. Moreover, we only discussed the paramagnetic region; below T_c, the question is still open.

REFERENCES

The literature concerning the subject of these lectures is extremely vast. We only cite here, in addition to a few fundamental papers, some general references, review articles and books, where a more complete bibliography will be found.

PART I

1) D. MATTIS: The Theory of Magnetism, (Harper and Row, 1965).

PART II

2) P.G. DE GENNES, in Magnetism, Vol. III, p. 115 (editred by G.T. Rado and H. Suhl, Academic Press, 1963).

3) L. VAN HOVE, Phys. Rev. 95 (1954), 249 and 1374.

PART III

4) W. MARSHALL and R.D. LOWDE, Rep. Progr. Phys. 31 (1968) B 705.

PART IV

5) F. KEFFER, Handb. Phys. 18 (1966)1.

6) F.J. DYSON, Phys. Rev. 102 (1956) 1217 and 1230.

PART V

7) P.G. DE GENNES, Comm. Energie Atomique (France), Rapport Nº 925 (1959).

8) P. RÉSIBOIS and M. DE LEENER, Phys. Rev. 152 (1966) 305,318; 178 (1969) 806, 819.

9) K. KAWASAKI, Ann. Phys. 61 (1970) 1.

10) R. KUBO, in Lectures in Theoretical Physics, vol. I, chap. 4 (Interscience, 1959).

11) P. MAZUR, lectures in this volume.

PART VI

12) R. BROUT: Phase Transitions, chap. 2 and 5 (Benjamin, 165).

13) H.E. STANLEY: Introduction to Phase Transitions and Critical Phenomena (Oxford University Press, 1971)

14) K.G. WILSON and J. KOGUT, Physics Reports (to be published).

15) B.I. HALPERIN and P.C. HOHENBERG, Phys. Rev. 177 (1969) 952.

16) M. DE LEENER, Physica 56 (1971) 62.
P.C. HOHENBERG, M. DE LEENER and P. RÉSIBOIS, Physica 65 (1973) 505.

17) K. KAWASAKI, J. Phys. Chem. Solids 28 (1967) 1277.

References 4, 7, 8 and 9 are also cited in Part VI.

ON THE ENSKOG HARD-SPHERE KINETIC EQUATION AND THE TRANSPORT PHENOMENA OF DENSE SIMPLE GASES

Manuel G. Velarde

Departamento de Física
Universidad Autónoma de Madrid
Canto Blanco (Madrid) Spain

I. INTRODUCTION: THE HARD-SPHERE MODEL INTERACTION

II. FROM THE BOLTZMANN APPROACH TO THE ENSKOG EQUATION

III. HYDRODYNAMIC EQUATIONS AND THE (NEW ENSKOG) COLLISIONAL (OR POTENTIAL) TRANSFER

IV. SOLUTION OF THE ENSKOG EQUATION FOR PRACTICAL PURPOSES

V. TRANSPORT COEFFICIENTS FROM THE ENSKOG EQUATION

VI. COMPARISON WITH EXPERIMENTAL DATA

VII. THE SQUARE-WELL FLUID

VIII. FINAL COMMENTS

REFERENCES

I. INTRODUCTION: THE HARD-SPHERE MODEL INTERACTION

A basic aim of statistical mechanics is to explain the macroscopic properties of matter from microscopic intermolecular forces and laws of motion. Macroscopic properties fall rather naturally into two classes, equilbrium and non-equilibrium, and each of these classes contains what might be called specific and general properties. By general properties we mean those which are more or less independent of the nature of the intermolecular potential, while by specific properties we mean those which reflect the nature of the potential sensitively.

The general equilibrium properties include, for example, the existence of the laws of thermodynamics and the existence of various phases of matter, while a prominent specific equilibrium property is the temperature dependence of the second virial coefficient. Among the non-equilibrium properties, general properties might be taken to be the existence of an approach to equilibrium as well as the existence of laws of non-equilibrium thermodynamics. On the other hand, the density, and temperature dependence of the viscosity coefficient, of the diffusion coefficient and of any other transport coefficient are specific non-equilibrium properties.

The task accomplished by Boltzmann, and Chapman and Enskog, was to indicate how the transport properties of a gas of spherically symmetric molecules may be computed from the microscopic properties of these molecules in at least a sensible first approximation, nowadays called the "dilute gas" limit. As this has been the topic of discussion of various lectures that preceded me at this School, I would rather concentrate on one pragmatic attempt to go beyond the zero-density limit. I am going to talk about what Enskog did in this direction and we will see how useful it may still be today. I'll try to avoid formalism as much as I can, as I would like to keep my feet on the ground. Hints are given, however to help you to go further on in the kinetic theory of transport phenomena.

In a paper published in 1922, David Enskog proposed a modification of Boltzmann's kinetic equation. His paper, he says "is the continuation of an earlier work on dilute gases. More specialized assumptions are made about molecular forces, so that theories valid for arbitrary densities may be given. It is necessary to introduce a new hypothesis about the number of collisions, and the theoretical treatment is new in parts. This will not lead to undue complications, since I can build on my earlier results". (See the book edited by S.G. Brush, 1972).

He then starts the second section of the paper saying "We assume that we have a gas whose molecules repel one another like hard, smooth, perfectly elastic spheres with mass m (in my lectures we shall take, for convenience, m = 1) and diameter σ."

The hard-sphere potential function is thus a model, according to which, two molecules do not interact until they just touch, when they repel each other with infinite force. The forces between molecules on collision are impulsive ($t_{coll} \equiv 0$). This angle-indepéndent potential, along with some other relevant and more realistic models, is sketched in fugure 1.1, and has the mathematical form

$$\varphi(r) = \begin{cases} \infty & \text{at} \quad r \leq \sigma \\ 0 & \text{at} \quad r > 0 \end{cases} \tag{1.1}$$

where σ is the diameter of the spheres. Besides its simplicity, in classical dynamics, the main interest of this model lies in the fact that transport properties depend essentially on the existence of collisions, and not on their nature. They thus depend primarily on the fact that some force exists between molecules, and only secondarily on the nature of the force. One could even say that it is a secondary matter for these transport properties whether the force is one of attraction or repulsion. This is not, however, the case for equilibrium properties, such as the virial coefficients of the equation of state: In this latter case, the very sign of those coefficients depends on the nature of the forces. In recent years the hard-sphere model has been extensively used for the study of many-body problems. In particular, the equation of state at high densities has been studied by numerical methods on high-speed computers and a clear suggestion of a first-order solid-fluid (melting) phase transition has been found. The computer experiments are, of course, not rigorous proof of the existence of the hard-sphere melting transition. They must be regarded only as a very suggestive indication, and judged on the basis of how this behaviour could be reproduced by infinite systems.

An interesting question, not yet answered, is whether the virial equation of state could possibly predict such a phase transition for an assembly of hard spheres. Clearly no phase transition could be predicted if, as has been suggested, all the hard-sphere virial coefficients were positive. But the equation of state problem* is not the topic of our discussions here. Thus we return to our original topic.

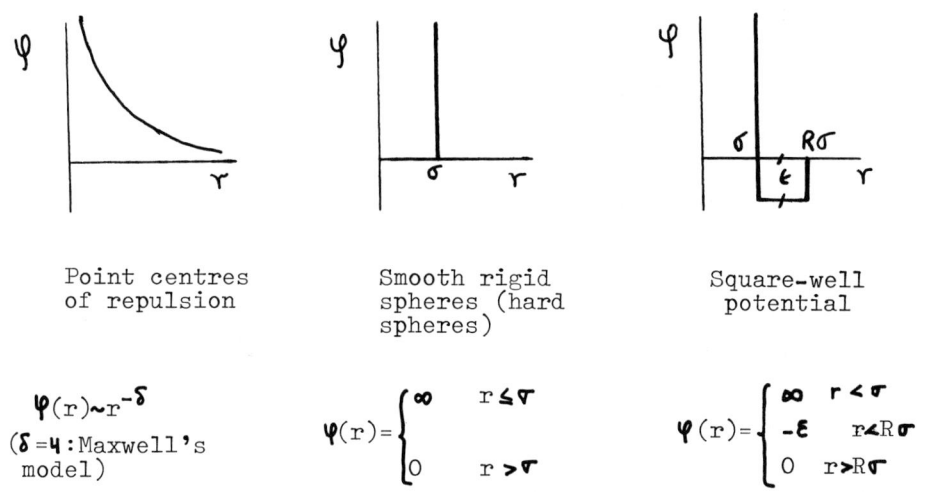

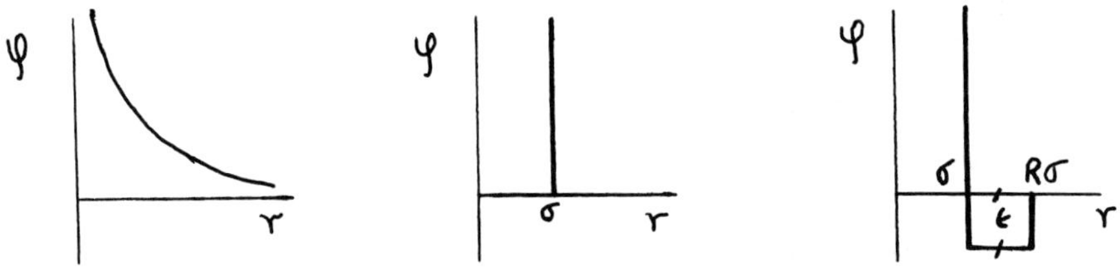

Point centres of repulsion

Smooth rigid spheres (hard spheres)

Square-well potential

$\varphi(r) \sim r^{-\delta}$

($\delta = 4$: Maxwell's model)

$$\varphi(r) = \begin{cases} \infty & r \leq \sigma \\ 0 & r > \sigma \end{cases}$$

$$\varphi(r) = \begin{cases} \infty & r < \sigma \\ -\varepsilon & r < R\sigma \\ 0 & r > R\sigma \end{cases}$$

---

*Let me point out that another approach to the equation of state of a fluid phase is that of Percus and Yevick. These authors took the opposite starting point to the virial approach. Rather than attempting a series expansion from the perfect gas end, they argued that, near the solidification curve anyway, the Debye spectrum of a fluid cannot differ greatly from that of a solid, and they looked for ways of computing the differences as a function of density. The Percus-Yevick equation of state has been obtained for a hard-sphere gas and it reproduces the equation of state of hard-sphere gas, as determined by the computer calculations, to within a few percent. We shall not, however, discuss the use of Percus-Yevick equation to obtain $g(\sigma)$ and so the Enskog transport coefficients for a <u>liquid</u>.

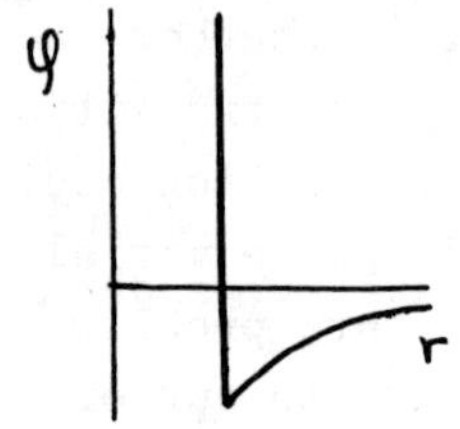

Sutherland model

$$\varphi(r)=\begin{cases}\infty & r<\sigma\\ -r^{-\delta} & r>\sigma\end{cases}$$

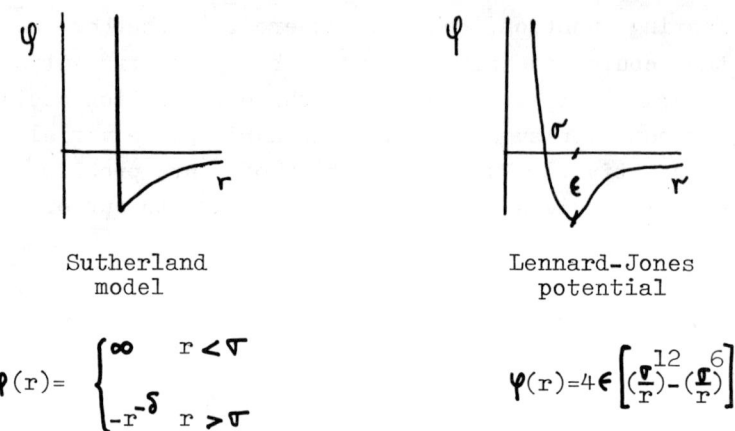

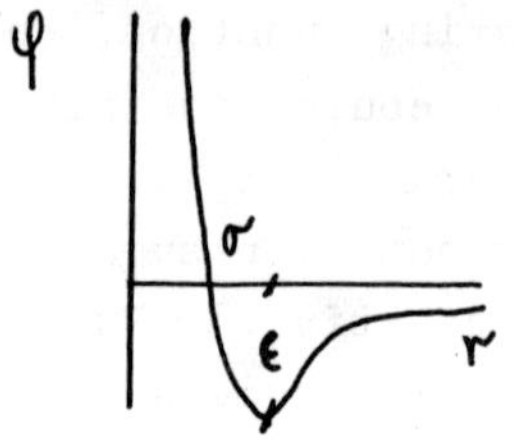

Lennard-Jones potential

$$\varphi(r)=4\epsilon\left[\left(\frac{\sigma}{r}\right)^{12}-\left(\frac{\sigma}{r}\right)^{6}\right]$$

Fig. 1.1. Pictorial representation of various spherically summetrical potential functions.

As with hard spheres, collisions are instantaneous, multiple encounters can be neglected: the probability of ternary and higher-order multiple collisions will be negligibly small; consequently, it is only necessary to consider binary encounters. This is not however, the case with real molecules; in a gas at high pressure, a molecule is in the field of force of others during a large part of its motion and multiple encounters are not rare.

The second assumption that Enskog made was a molecular chaos approximation. He claims doing this "in agreement with Jeans" (J. H. Jeans, The Dynamical Theory of Gases, Cambridge, 3rd edition, 1921, pp. 15-16 or else 4th edition, 1925, p.54). Enskog believed, on the basis of Jeans' analysts, that for hard spheres the molecular chaos assumption can remain valid at large densities. However, Jeans' argument applied strictly only to a gas in a uniform steady state. In a dense gas, even one composed of hard spheres, there may be some correlation between the velocities of neighbouring molecules, because of their recent interactions with each other, or with the same neighbours. This may be important in a gas not in a uniform steady state. "Thus what Enskog gained in mathematical simplicity is partially offset by inadequacy in the representation of physical reality" (Chapman and Cowling, Ref. 2, p. 297). We shall see, however, that Enskog's theory accounts for transport coefficients of real gases in a density range much wider than expected. "Anyone can get the right answer for the right reason; it takes a genius, or a physicist, to get the right answer for the wrong reason", says Lee A. Segel in one of his papers.

The hard sphere model is useful anyway for exploratory calculations because, at least, a nice representation of the strong, short range repulsive forces (that is that molecules have volume!) is given. In Figures 1.2 and 1.3, hard-sphere collisions are graphically displayed.

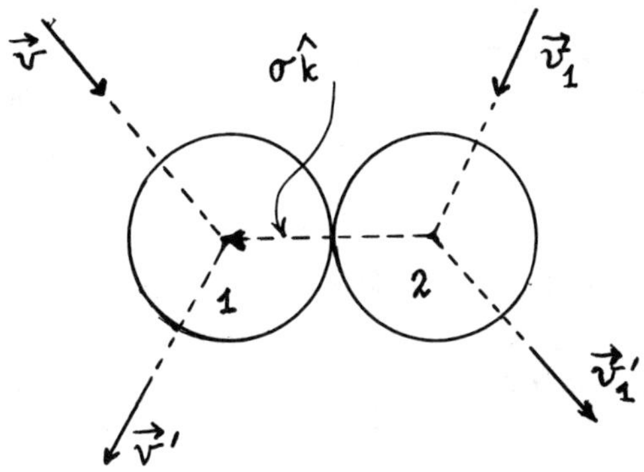

Fig. 1.2.: A collision between two hard spheres. $\vec{k}$ is a unit vector, called perihelion vector, for convenience of notation velocities of particle "2" are denoted with subscript "1".

There exists a procedure for adapting a hard-sphere potential to a more realistic potential like a Lennard-Jones function. If the "realistic" potential is φ (r), a prescription suggested on the literature (for liquid metal calculations, however) states that the hard-sphere diameters σ should be obtained through the relation

$$\varphi(\sigma) = \frac{3}{2}\, kT - \varepsilon \qquad (1.2)$$

where ε is the depth of the potential well and k is Boltzmann's universal constant. The diameters σ given by (1.2) may be thought of as "effective" diameters of the "real" particles. They correspond roughly to an average distance of approach during a collision of particles with average energy 3/2 kT - ε . We shall, in fact, come back to the problem of prescribing σ for a moderately dense gas in section VI.

In the next section we discuss in more detail Enskog's ansatz and we introduce his kinetic equation. Section III is devoted to a discussion of the collisional transfer mechanism introduced by Enskog and the hydrodynamic conservation laws. Section 4 deals with the first -order Chapman-Enskog expansion of the (asymptotic) solution of Enskog's equation. However, we ought to say that Enskog's equation is only valid for times shorter than the mean free time between collisions, as has been recently proved. We shall come back to this point in section VIII. In section V we obtain explicit expressions for the Enskog transport coefficients and we discuss, in section VI, the relevance in interpreting the experimental data. (We will be concerned with the viscosity and heat conductivity for temperatures well above critical.

Section VII deals with the square-well extension of Enskog's theory. Finally, in section 8 we briefly give comments on some general problems related to the very validity of Enskog's equation as a kinetic equation. However general derivations of the Enskog equation, as well as the controversial and very interesting quations of fluid mixtures and transport properties of polyatomic gases, are left out of these lectures.

II. FROM THE BOLTZMANN APPROACH TO THE ENSKOG EQUATION

The Boltzmann equation, already discussed at length in the course of this School, is an integrodifferential equation for the one-particle distribution function $f(\vec{r},\vec{v},t)$. In the absence of external force fields, the equation reads

$$\frac{\partial}{\partial t}f + \vec{v}.\ \frac{\partial}{\partial r} = \left(\frac{\partial f}{\partial t}\right)_{coll.} \tag{2.1}$$

where

$$\left(\frac{\partial}{\partial t}f\right)_{coll.} = \int d\vec{v}_1 \,|\vec{V}|\, p dp d\varphi \;\; f(\vec{r},\vec{v}';t)\; f\;(\vec{r},\vec{v}_1';\; t) - - f(\vec{r},\vec{v};t)\;\; f\;\;(\vec{r},\vec{v}_1;\; t) \tag{2.2}$$

Here $\vec{v}_1'$ and $\vec{v}'$ are the velocities of two molecules (2 and 1 respectively) after collision (see Figs. 1.2 and 1.3 for a graphic display in the case of hard spheres; $\vec{v}_1$ and $\vec{v}$ are the corresponding velocities before collision $\vec{V} = \vec{v}-\vec{v}_1$. The remaining factors in the integrand account for the binary collision cross-section: ρ is the impact parameter of the collision, and φ the azimuthal angle measured in the plane perpendicular to $\vec{v}-\vec{v}_1$. We shall replace here $|\vec{V}|\ \rho d\rho d\varphi$ by $(\vec{V}.\hat{k})d\hat{k}\,\sigma^2$.

Equations (2.1) and (2.2) are obtained under some very drastic assumptions that just for sake of completeness, I would like to briefly state:

i) Molecules are supposed to be point centres (see Fig. 1.1) of short range forces, say σ. The average distance travelled between collisions is large so that the finite size of the molecules can be neglected.

ii) The dilute gas limit is considered: $n \to 0, N \to \infty, \sigma \to 0$, $(N\sigma^2) < \infty$, but $(N\sigma^3) \to 0$; N is the total number of molecules and n is the average density.

iii) Molecular correlations are of dynamical origin only. Correlations of order higher than two-molecule correlations are neglected.

iv) The function $f(\vec{r},\vec{v},t)$ is a slowly varying function of position and time for intervals of order σ and t_{coll} (duration of a collision) respectively.

v) Stosszahl-Ansatz: Every time that two molecules meet, they come together uncorrelated. After the collision they are strongly correlated, however.

We have already indicated in the preceding section the reason invoked by Enskog to discuss high pressure (density) effects for hard-spheres only. In this case the collision time is vanishing, and so is the probability of multiple (more than binary) collisions.

In the dense gas case hypothesis (i) must be modified, as the ratio of core diameter σ to the mean free path between collisions becomes non-negligible. Because of the finite diameter σ the distribution function must not be taken at the same space point $\vec{r}$ at the moment of collision, but at space points which are the distance σ apart. (see Figs. 1.2, 1.3, and 2.1 for a graphic representation).

Enskog suggested that the molecular chaos assumption (v) has to allow for correlations of position at high densities. This is similar to the way in which the geometrical effects are incorporated in the van der Waals equation of state (see Fig. 2.1).

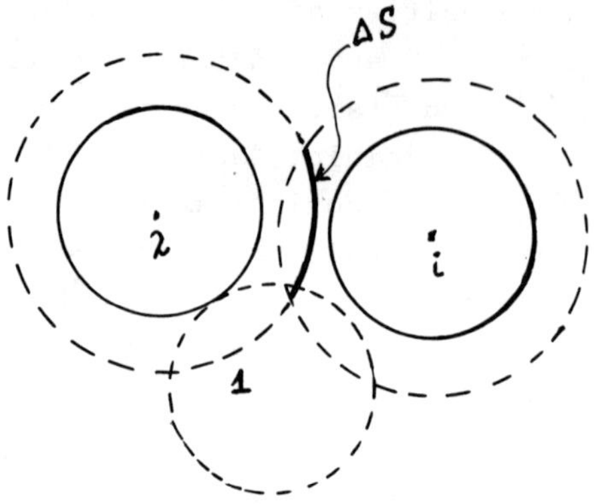

Fig. 2.1 Excluded volume and "shielding" effect of hard-sphere molecules. The center of molecule "1" is supposed to be lying on the sphere of influence of molecule "2"

For a hard sphere gas the equation of state reads

$$\frac{P}{nkT} = 1 + n\,b\,g(n) \tag{2.3}$$

(notice that the compressibility factor l.h.s. of (2.3) is always greater than unity and is a function only of the density). Here g(n) is the radial distribution function whose known virial expansion is

$$g(n) = 1 + 0.625\,nb + 0.2869\,(nb)^2 + \\ + 0.110\,(nb)^3 + 0.039\,(nb)^4 + O(nb)^5 \tag{2.4}$$

and

$$b = \frac{2\pi\sigma^3}{3} \tag{2.5}$$

Indeed one may interpret the term nbg(n) arising from excluded volume as due to the finiteness of σ (the so called co-volume in van der Waals theory). Correlations of the momenta of the colliding molecules are still neglected.*

Enskog also introduced a new physical properties transfer. (see Fig. 2.2).

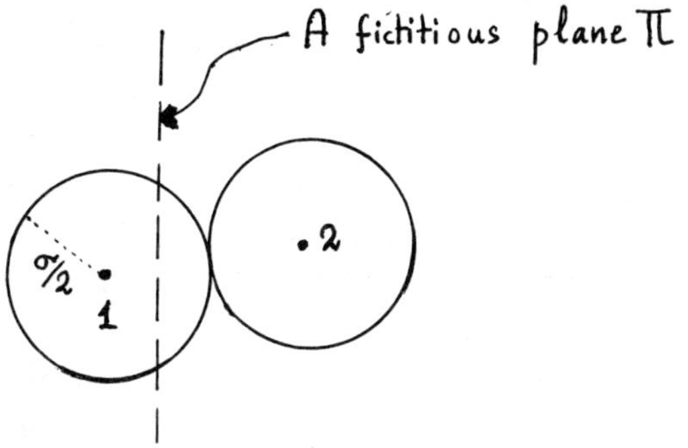

Fig. 2.2. Molecule 1 collides with molecule 2. Although the center of 1 does not cross π, momentum and energy are nevertheless transported across this plane. In the Boltzmann dilute gas limit, $\sigma \to 0$, and this effect vanishes.

In dense gases not only the translational motion between collisions will contribute to the total stress tensor $\overrightarrow{\overrightarrow{P}}$ and energy flux $\vec{J}_q$, but momentum and energy is transported also, a distance σ by an encounter. This "collisional" or "potential" transfer may be best understood if one considers a plane in space and two molecules on each side but almost in contact with each other. If they collide, momentum and energy is transported across the plane even if each molecule remains on its own side. For instance at the collision, the exchange of momentum $\vec{V}_1 - \vec{V}_1' = \vec{V}_2 - \vec{V}_2'$ is instantaneous from the centre of molecule "1" to that of molecule "2", or reciprocally (see Fig. 2.2)

In the case of hard spheres the direct collision distribution function thus becomes

$$f(\vec{r}, \vec{v}; t) f(\vec{r} - \sigma\hat{k}, \vec{v}_1; t) \tag{2.6}$$

where $\hat{k}$ is a unit vector (the perihelion vector of Fig. 1.2). This

* Enskog notices that "b is thus four times the volume of the molecule, per unit mass". And he adds, "It has been impossible to date (1922) to calculate g exactly. The theory in this paper is only complete when g has been found, but we may arrive at many quantitative results without knowing g."

vector is in fact the bisector of the $\widehat{V,V'}$ angle ($\vec{V}' = \vec{v}_1' - \vec{v}'$). The excluded volume leads to a reduced effective volume allowed to the molecules within the container. Thus the collision rate (frequency) should be higher because the distance, which two spheres have to travel in order to collide, is significantly decreased by the diameter of the spheres. At low density however, the diameter is small compared to the average distance travelled between collisions and the spheres can be considered as point particles (Boltzmann gas). In the Enskog approach this geometric factor depends on space only, as a functional of the local density. This is given by the radial distribution function) at contact,and as a functional of the density, should be determined by the equation of state (section III).

For the restituting collision (see Ref. 2) one has to replace $\vec{V}$ by $-\vec{V}$ and $+(1/2)\sigma\hat{k}$ goes to $-(1/2)\sigma\hat{k}$, for the contact point. We also note that the differential element of Eq. (2.2) is the same for both direct and restituting collisions. However, the integration in the former is for $\vec{V}.\hat{k} > 0$, whereas in the latter it is for $\vec{V}.\hat{k} < 0$.

The above described assumptions lead to the ad hoc modified Boltzmann equation

$$\frac{\partial}{\partial t} f(\vec{r},\vec{v};t) + \vec{v}\cdot\frac{\partial f(\vec{r},\vec{v};t)}{\partial t} = \sigma^2 \int (\vec{V}\cdot\hat{k})\, d\hat{k}\, d\vec{v}_1$$

$$\times \Big\{ g(\vec{r}+\tfrac{1}{2}\sigma\hat{k};t)\, f(\vec{r},\vec{v}';t)\, f_1(\vec{r}+\sigma\hat{k},\vec{v}_1';t)$$

$$- g(\vec{r}-\tfrac{1}{2}\sigma\hat{k};t)\, f(\vec{r},\vec{v};t)\, f(\vec{r}-\sigma\hat{k},\vec{v}_1;t) \Big\}. \tag{2.8}$$

This is just Enskog's kinetic equation for a dense gas of (structureless) hard spheres. It is indeed not invariant under time reversal. Also, it is evident that in this equation (2.8) we neglect the momentum correlations between successive hard-sphere collisions (in particular, a sphere is considered as always colliding with other spheres approaching in a random direction). The g-factor, however, incorporates unsystematically some three-body collision effects. Later in section 3 we shall come back to the explicit expression for $g(\sigma)$ ($g(r)$ at contact).

The centre of a molecule cannot lie within the $4\pi\sigma^3/3$ sphere of influence of another molecule (see Fig. 2.16) Thus, to a first approximation, the reduction of unit volume is to 1-2nb. Now let ΔS be the spherical surface associated with molecule "2", lying within the sphere of influence of molecule "i" (any other molecule of the gas, except "1"). Thus, the centre of molecule "1" trying to collide with "2" cannot be located on this portion ΔS. For a space separation, $R(\sigma < R < 2\sigma)$ there are $n 4\pi R^2 dR$ molecules "i".* Each "i" molecule "removes" a surface $2\pi\sigma(\sigma - R/2)$. On the average the total amount of absolute surface which is removed is (the bar means "average")

$$\overline{\Delta S} = \int_{\sigma}^{2\sigma} 2\pi\left(\sigma - \frac{R}{2}\right)\sigma\, 4\pi R^2 dR = \frac{11\pi^2}{3} n\sigma^5 \tag{2.9}$$

and the relative amount of surface removed is

$$\frac{\overline{\Delta S}}{S} = \frac{11}{12}\pi n\sigma^3 = \frac{11}{8} nb \tag{2.10}$$

Thus we end up with a collision frequency factor of

$$q = \left(1 - \frac{11}{8} nb \Big/ 1 - 2nb\right) \simeq 1 + \frac{5}{8} nb = 1 + 0.625 nb + O(nb)^2. \tag{2.11}$$

to be compared with (2.4). This result was already obtained by Clausius and Boltzmann!

---

*Notice that the probability of finding two molecules at a certain separation (say σ) where the gas is not in equilibrium is approximated by the same function at equilibrium.

III. HYDRODYNAMIC EQUATIONS AND THE (NEW ENSKOG) COLLISIONAL (OR POTENTIAL) TRANSFER

The equations of change, or formal hydrodynamic equations, are obtained in a similar way to that for the Boltzmann dilute gas case. Here, the hard sphere potential plays an essential role. It is only with this particular potential function that the energy density can be identified with the kinetic energy. There is no potential energy contribution, as the hard spheres are inpenetrable.

Let ψ be $(1, \vec{v}, v^2/2)$, a collisional invariant. Then integrating the Enskog equation, dotted with ψ, over the velocity space, one gets

$$\frac{\partial}{\partial t}\int d\vec{v}\,\psi f + \frac{\partial}{\partial \vec{r}}\cdot\int \vec{v}\,d\vec{v}\,\psi f = \int d\vec{v}\,\psi\left(\frac{\partial f}{\partial t}\right)_{coll.} \tag{3.1}$$

Notice that in contrast to the Boltzmann dilute gas problem, here the r.h.s. of equation (3.1) is non vanishing. However, if we assume that inhomogeneities are weak enough, we may approximate this r.h.s. by its Taylor expansion up to first few terms. Thus, one writes

$$\int d\vec{v}\,\psi\left(\frac{\partial f}{\partial t}\right)_{coll.} = \sum_{i=1}^{6} I_i + \mathcal{O}(\nabla^3) \tag{3.2}$$

The six terms here retained are enough to get the linear transport coefficient of the Enskog hard sphere gas, as these quantities show up at the ∇^2 order.

The quantities I_i introduced in Eq. (3.2) are:

$$I_1 \equiv g(\sigma)\int d\vec{v}\,\psi\,\sigma^2\,(\vec{V}\cdot\hat{k})\,d\vec{v}_1\,[f'f_1' - ff_1]$$

$$\equiv \int d\vec{v}\,\psi\,J_1 = 0 \tag{3.3}$$

$$I_2 \equiv \frac{1}{2}\sigma g(\sigma)\int d\vec{v}\,d\vec{v}_1\,\sigma^2(\vec{V}\cdot\hat{k})\,d\hat{k}\left[\psi(v)-\psi(v')\right]\left[\hat{k}\cdot\frac{\partial}{\partial\vec{r}}ff_1\right]$$

$$\equiv \int d\vec{v}\,\psi\,J_2 \tag{3.4}$$

$$I_3 \equiv \frac{1}{2}\sigma\int d\vec{v}\,d\vec{v}_1\,\sigma^2(\vec{V}\cdot\hat{k})\,d\hat{k}\left[\psi(v)-\psi(v')\right]\left[\hat{k}\cdot\frac{\partial}{\partial\vec{r}}g(r)\right]ff_1$$

$$\equiv \int d\vec{v}\,\psi\,J_3 \tag{3.5}$$

$$I_4 \equiv \frac{1}{2}\sigma^2 g(r)\int d\vec{v}\,d\vec{v}_1\,\sigma^2(\vec{V}\cdot\hat{k})\,d\hat{k}\left[\hat{k}\hat{k}:\frac{\partial}{\partial\vec{r}}\left(ff_1\frac{\partial}{\partial\vec{r}}\log\frac{f}{f_1}\right)\right]$$

$$\equiv \int d\vec{v}\,\psi\,J_4 \tag{3.6}$$

$$I_5 \equiv \frac{1}{4}\sigma^2\int d\vec{v}\,d\vec{v}_1\,\sigma^2(\vec{V}\cdot\hat{k})\,d\hat{k}\left[\hat{k}\cdot\frac{\partial}{\partial\vec{r}}g\right]\left[\hat{k}\cdot ff_1\frac{\partial}{\partial\vec{r}}\log\frac{f}{f_1}\right]$$

$$\equiv \int d\vec{v}\,\psi\,J_5 \tag{3.7}$$

$$I_6 \equiv \frac{1}{8}\sigma^4 \int d\vec{v}\, d\vec{v}_1\, (\vec{V}\cdot\hat{k})\, d\hat{k}\, \psi \left[\hat{k}\hat{k} : \frac{\partial}{\partial \vec{r}} \frac{\partial}{\partial \vec{r}}\, g\right]\left[f'f_1' - ff_1\right]$$

$$\equiv \int d\vec{v}\, \psi\, J_6 = 0 \; . \tag{3.8}$$

Here the J-quantities are defined, and with expressions given (3.3) through (3.8), we can write the general conservation equation (3.1) in the following formal way

$$\frac{\partial}{\partial t} \int d\vec{v}\, \psi f = -\frac{\partial}{\partial \vec{r}} \left(J_\psi^k + J_\psi^\varphi \right). \tag{3.9}$$

Here we have introduced a kinetic contribution to the ψ- current

$$J_\psi^{\,k} \equiv \int \vec{v}\, d\vec{v}\, \psi f \tag{3.10}$$

and a potential part

$$J_\psi^\varphi \equiv \frac{1}{2}\sigma g \int d\vec{v}\, d\vec{v}_1\, d\hat{k}\; \sigma^2 (\vec{V}\cdot\hat{k}) \left[\psi(v') - \psi(v)\right] ff_1 \hat{k}$$

$$+ \frac{1}{4}\sigma^2 g \int d\vec{v}\, d\vec{v}_1\, d\hat{k}\; \sigma^2 (\vec{V}\cdot\hat{k}) \left[\psi(v') - \psi(v)\right] ff_1$$

$$\hat{k}\cdot \frac{\partial}{\partial \vec{r}} \log\left(\frac{f}{f_1}\right) \hat{k} \tag{3.11}$$

Now, letting ψ be successively 1, $\vec{v}$ and $v^2/2$, we easily get the continuity equation, the Navier-Stokes equation and the energy equation in much the same manner as in the theory of dilute gases. The only difference in practice arises because of the definition of the fluxes. With the Enskog equation, and according to the general conservation law (3.9), we get two contributions for the pressure tensor

$$\overleftrightarrow{P} = \overleftrightarrow{P}^{k} + \overleftrightarrow{P}^{\varphi} \tag{3.12}$$

and for the macroscopic heat current

$$\vec{J}_q = \vec{J}_q^{\,k} + \vec{J}_q^{\,\varphi} \tag{3.13}$$

with the following identifications being made

$$\overleftrightarrow{P}^{k} \equiv \int d\vec{v}\, \vec{\xi}\vec{\xi} f, \tag{3.14}$$

$$\begin{aligned}\overleftrightarrow{P}^{\varphi} = \frac{1}{2}\sigma^3 g \int d\vec{v}\, d\vec{v}_1\, d\hat{k}\, (\vec{V}\cdot\hat{k}) \left[\vec{\xi}' - \vec{\xi}\right] f f_1 \hat{k} \\ + \frac{1}{4}\sigma^4 g \int d\vec{v}\, d\vec{v}_1\, d\hat{k}\, (\vec{V}\cdot\hat{k}) \\ \times \left[\vec{\xi}' - \vec{\xi}\right] f f_1 \hat{k}\cdot \frac{\partial}{\partial \vec{r}} \log\left(\frac{f}{f_1}\right) \hat{k},\end{aligned} \tag{3.15}$$

$$\vec{J}_q = \frac{1}{2}\int d\vec{v}\, \vec{\xi}\, \xi^2 f, \tag{3.16}$$

$$\vec{J}_q^{V} = \frac{1}{4}\sigma^3 g \int d\vec{v}\, d\vec{v}_1\, d\hat{k}\, (\vec{V}\cdot\hat{k})\, [\xi'^2 - \xi^2]\, \hat{k}\, f f_1$$
$$+ \frac{1}{8}\sigma^4 g \int d\vec{v}\, d\vec{v}_1\, d\hat{k}\, (\vec{V}\cdot\hat{k})[\xi'^2 - \xi^2] f f_1\, \hat{k}\cdot\frac{\partial}{\partial\vec{r}} \log\left(\frac{f}{f_1}\right)\hat{k}\ . \tag{3.17}$$

Here $\vec{\xi}$ is the "peculiar" velocity $\vec{\xi} = \vec{v} - \vec{u}$, and $\vec{u}$ is the macroscopic (barycentric) velocity as usually defined for dilute gases (see ref. 1 or 2).

The identification of the linear transport coefficients comes through use of the standard linear phenomenological laws. Thus, one has

$$P_{ij} \equiv p\,\delta_{ij} - 2\eta\left(\overset{\circ}{\overline{\overline{\frac{\partial}{\partial\vec{r}}\vec{u}}}}\right) - \zeta\left(\frac{\partial}{\partial\vec{r}}\cdot\vec{u}\right)\delta_{ij} \tag{3.18}$$

$$\vec{J}_q \equiv -\lambda\frac{\partial}{\partial\vec{r}}T\ . \tag{3.19}$$

In equation (3.18) we have introduced the symmetrized traceless tensor

$$\overset{\circ}{\overline{\overline{\frac{\partial}{\partial\vec{r}}\vec{u}}}} \equiv \frac{1}{2}\left[\frac{\partial}{\partial r_i}u_j + \frac{\partial}{\partial r_j}u_i - \frac{2}{3}\left(\frac{\partial}{\partial\vec{r}}\cdot\vec{u}\right)\delta_{ij}\right]. \tag{3.20}$$

Here p is the static pressure; η, ζ and λ are respectively shear, viscosity, bulk viscosity and heat conductivity. To obtain explicit expressions for these transport properties, one is faced with the integration of Enskog equation. This can be done by a Chapman-Enskog procedure in a similar manner to that for the Boltzmann equation for dilute gases. We shall come back to this in the next section.

IV. SOLUTION OF THE ENSKOG EQUATION FOR PRACTICAL PURPOSES

We are, in fact, interested here, as with the Boltzmann dilute gas problem in solving the Enskog equation in some asymptotic sense. One such way is the Chapman-Enskog procedure. As one is only interested in linear transport coefficients, and thus, in weak inhomogeneities only, one starts expanding the r.h.s. of the Enskog equation in powers of a smallness parameter δ. Later on we will set $\delta = 1$. One has

$$\frac{\partial}{\partial t} f + \vec{v} \cdot \frac{\partial}{\partial \vec{r}} f = \frac{1}{\delta} J_1(f) + \delta^0 \left[J_2(f) + J_3(f) \right] + \theta(\delta) . \qquad (4.1)$$

Notice that here none of J_i (i = 4,5,6) have appeared, as they are already linear in δ. Also, the J_1 operator is the linear Boltzmann operator up to a multiplicative constant. We shall also expand the distribution function

$$f = \sum_{n=0}^{\infty} \delta^n f^{(n)} , \quad \left[f^{(1)} = f^{(0)} \phi^{(1)} \right] \qquad (4.2)$$

At the δ^0 order, one gets the local maxwellian

$$f^{(0)} = \frac{n(\vec{r}.t)}{\left[2\pi k T(\vec{r}.t) \right]^{3/2}} \exp \frac{-\left[\vec{v} - \vec{u}(\vec{r}.t) \right]^2}{2kT(\vec{r}.t)} \qquad (4.3)$$

together with the identification of density n, macroscopic velocity $\vec{u}$, and local temperature fields T,

$$\begin{bmatrix} n \\ n\vec{u} \\ n\left(U + \frac{u^2}{2}\right) \end{bmatrix} = \begin{bmatrix} n \\ n\vec{u} \\ n\left(\frac{3}{2}kT + \frac{u^2}{2}\right) \end{bmatrix} = \int d\vec{v} \begin{bmatrix} 1 \\ \vec{v} \\ v^2/2 \end{bmatrix} f^{(0)} . \tag{4.4}$$

This leads to the standard subsidiary conditions

$$\int d\vec{v} \begin{bmatrix} 1 \\ \vec{v} \\ v^2/2 \end{bmatrix} f^{(n)} = 0 \qquad (n>0) \tag{4.5}$$

Now the gradients expansion of the l.h.s. of Enskog equation gives, <u>at $\delta^{(0)}$ -order</u>,

$$\left(\frac{\partial}{\partial t} f\right)^{(0)} + \vec{v} \cdot \frac{\partial}{\partial \vec{r}} f^{(0)} - J_2\left[f^{(0)}\right] - J_3\left[f^{(0)}\right]$$

$$= g \int d\mu \, f^{(0)} f_1^{(0)} \left[\phi'^{(1)} + \phi_1'^{(1)} - \phi^{(1)} - \phi_1^{(1)}\right] . \tag{4.6}$$

Here, for simplicity, we have introduced the differential element

$$d\mu \equiv \sigma^2 \left(\vec{V} \cdot \hat{k}\right) d\hat{k} \, d\vec{v} \, d\vec{v}_1 . \tag{4.7}$$

One ought to identify $(\frac{\partial}{\partial t} f)^{(0)}$, and this is done through the expansion

$$\left(\frac{\partial f}{\partial t}\right)^{(n)} = \sum_{m=0}^{n} \sum_{\alpha=1}^{5} \frac{\partial f^{(n-m)}}{\partial \overline{\Psi}_\alpha} \left(\frac{\partial \overline{\Psi}_\alpha}{\partial t}\right)^{(n)} \tag{4.8}$$

where $\overline{\Psi}_1 \equiv n; \overline{\Psi}_{2,3,4} \equiv \vec{u}\,;\ \overline{\Psi}_5 = T.$

Like in the Boltzmann dilute gas case, the time evolution of the macroscopic quantities is obtained through the standard hydrodynamic equations. The continuity equation gives

$$\left(\frac{\partial n}{\partial t}\right)^{(0)} = -\frac{\partial}{\partial \vec{r}}(n\vec{u}), \tag{4.9}$$

the momentum balance gives

$$\left(\frac{\partial \vec{u}}{\partial t}\right)^{(0)} = -\vec{u}\cdot\frac{\partial}{\partial \vec{r}}\vec{u} - \frac{1}{n}\frac{\partial}{\partial \vec{r}}\cdot\overleftrightarrow{P}^{(0)}, \tag{4.10}$$

and the energy balance

$$\left(\frac{\partial T}{\partial t}\right)^{(0)} = -\vec{u}\cdot\frac{\partial}{\partial \vec{r}}T - \frac{2}{3nk}\overleftrightarrow{P}^{(0)} : \frac{\partial}{\partial \vec{r}}\vec{u}\,. \tag{4.11}$$

Up to zeroth order in the gradients, we obtain from the identifications made in section 3,

$$P_{ij}^{\kappa\,(0)} = nkT\,\delta_{ij} \tag{4.12}$$

$$P_{ij}^{\varphi(0)} = \frac{1}{2}\sigma^3 g \int d\vec{v}\, d\vec{v}_1\, d\hat{k}\, (\vec{V}\cdot\hat{k})[\xi_i' - \xi_i]\xi_j$$

$$= n^2 b g\, kT\, \delta_{ij} \tag{4.13}$$

Combining now (4.12) and (4.13) we get for the total pressure tensor

$$P_{ij}^{(0)} = nkT[1 + nbg]\delta_{ij} \tag{4.14}$$

Equation (4.14 allows us to identify the factor g(r) with the expression given by the virial equation of state given in section e.q. (2.3).

We still must evaluate the J_2 and J_3 terms. After some lengthy calculations, not reproduced here, one gets

$$J_2(f^{(0)}) = -nbg f^{(0)}\Big\{\vec{\xi}\cdot\Big[\frac{2}{n}\frac{\partial n}{\partial \vec{r}} + \frac{1}{T}\frac{\partial}{\partial \vec{r}} T\Big(\frac{3\xi^2}{10kT} - \frac{1}{2}\Big)\Big] + \frac{2}{5kT}\vec{\xi}\vec{\xi} : \frac{\partial}{\partial \vec{r}}\vec{u} - \Big(1 - \frac{\xi^2}{5kT}\Big)\frac{\partial}{\partial \vec{r}}\cdot\vec{u}\Big\} \tag{4.15}$$

and

$$J_3(f^{(0)}) = -nbf^{(0)}\vec{\xi}\cdot\frac{\partial}{\partial \vec{r}} g \tag{4.16}$$

Collecting the results obtained above, one gets the following linear integral equation for $\phi(1)$

$$\int d\mu\, f^{(0)} f_1^{(0)} \left[\phi'^{(1)} + \phi_1'^{(1)} - \phi^{(1)} - \phi_1^{(1)} \right] =$$

$$= \frac{1}{g} f^{(0)} \left\{ \left[1 + \frac{2}{5} nbg\right] \frac{1}{kT} \left[\xi_i \xi_j - \frac{\xi^2}{3} \delta_{ij} \right] \cdot \frac{\partial u_i}{\partial r_j} \right.$$

$$\left. + \left[1 + \frac{3}{5} nbg\right] \left[\frac{\xi^2}{2kT} - \frac{5}{2} \right] \xi_i \frac{1}{T} \frac{\partial}{\partial r_i} T \right\} \tag{4.17}$$

Notice the striking similarity of this equation to the linearized Boltzmann equation. Indeed Eq. (4.17) reduces to the latter if one sets $g = 1$, and $b = 0$.

The remarkable feature of this equation is that if we already know the tensors $\vec{A}$ and $\vec{\vec{B}}$ which solve the Boltzmann equation, under the form

$$n\phi^{(1)}_{\text{Boltzmann}} = \vec{A} \cdot \frac{\partial}{\partial \vec{r}} \log T + \vec{\vec{B}} : \frac{\partial}{\partial \vec{r}} \vec{u} \tag{4.18}$$

we automatically know the solution of Eq. (4. 7). This solution can be written as

$$n\phi^{(1)} = \frac{1}{4} \left\{ \left[1 + \frac{3}{5} nbg\right] \vec{A} \cdot \frac{\partial}{\partial \vec{r}} \log T \right.$$

$$\left. + \left[1 + \frac{2}{5} nbg\right] \vec{\vec{B}} : \frac{\partial}{\partial \vec{r}} \vec{u} \right\} \tag{4.19}$$

Thus, the linear integral equation problem associated with the Enskog equation reduces to that of the linearized Boltzmann integral equation. One should also notice that the Enskog theory merely scales in time the solution of the Boltzmann equation. This is so because Enskog's theory amounts to assuming that a high density hard-sphere system behaves exactly like a dilute gas except that everything happens faster because of the higher rate of collision.

V. TRANSPORT COEFFICIENTS FROM ENSKOG EQUATION

In this section we shall give explicit expressions for the transport coefficients phenomenologically introduced in section 5. It suffices to explicitly evaluate the macroscopic fluxes up first order in the gradients. One may expect three different types of contribution*:

(i) a standard kinetic part (already the one and only one given by the Boltzmann equation)

(ii) a potential part due to the (new Enskog) collisional transfer, and arising from $f^{(1)}$

(iii) a potential part arising from $f^{(o)}$ and due to first order macroscopic-gradient terms in the hydrodynamic equations.

After some lengthy, but rather elementary manipulations (see, for instance, the book of Chapman and Cowling), one gets the pressure tensor:

$$P_{ij}^{(1)} = n\left(1+\frac{2}{5}nbg\right)\left[\frac{1}{n}\int d\vec{v}\,\xi_i\,\xi_j\,f^{(1)}\right] + \frac{n^2bg}{5}\left[\frac{1}{n}\int d\vec{v}\,\xi^2 f^{(1)}\right]\delta_{ij} - \frac{n^2b^2g\,k^{1/2}T^{1/2}}{\pi^{3/2}\sigma^2}\left[\frac{6}{5}\left(\overset{o}{\overline{\frac{\partial}{\partial\vec{r}}\vec{u}}}\right)_{ij} + \left(\frac{\partial}{\partial\vec{r}}\cdot\vec{u}\right)\delta_{ij}\right] \tag{5.1}$$

and the heat current

$$\vec{J}_q^{(1)} = \frac{1}{2}n\left(1+\frac{3}{5}nbg\right)\frac{1}{n}\int d\vec{v}\,\vec{\xi}\,\xi^2 f^{(1)} - \frac{3}{2}\,\frac{k\,n^2b^2g\,(kT)^{1/2}}{\pi^{3/2}\sigma^2}\,\frac{\partial}{\partial\vec{r}}T \tag{5.2}$$

*The Enskog theory for a (dense) fluid of hard discs has been developed by D.M. Gass, J. Chem. Phys. 54 (1971) 1898. We will not consider this two dimensional problem here.

Now, the dilute gas transport coefficents can be identified. Indeed they are obtained by setting b = 0. Thus, we get (a superscript B means the Boltzmann dilute gas limit):

$$P_{ij}^{(1)B} = \int d\vec{v}\, \xi_i \xi_j f^{(1)B} = -2\eta^B \left(\overset{\circ}{\overline{\overline{\frac{\partial}{\partial \vec{r}}\vec{u}}}}\right)_{ij} \tag{5.3}$$

and

$$\vec{J}_q^{(1)B} = \frac{1}{2}\int d\vec{v}\, \vec{\xi}\, \xi^2 f^{(1)B} = -\lambda^B \frac{\partial T}{\partial \vec{r}} \,. \tag{5.4}$$

They are well known definitions for dilute gases. From equations (5.3) and (5.4) we get η^B and λ^B respectively, but this is already known.

Now, incorporating η^B and λ^B, and also taking (4.14) $[P^{(0)} \equiv p_0]$

$$p_0/nkT = 1 + nbg$$

(this equation is the exact equation of state for a hard-sphere gas) and

$$\overline{\omega} = \frac{n^2 b^2 g\, (kT)^{1/2}}{\pi^{3/2}\, \sigma^2} \tag{5.6}$$

we have the complete Enskog pressure tensor again

$$P_{ij} = P_{ij}^{(0)} + P_{ij}^{(1)} = \left[p_0 - \overline{\omega}\left(\frac{\partial}{\partial \vec{r}}\cdot\vec{u}\right)\right]\delta_{ij} - \left[\frac{2\eta^B}{g}\left(1 + \frac{2}{5}nbg\right)^2 + \frac{6}{5}\overline{\omega}\right]\left(\overset{\circ}{\overline{\overline{\frac{\partial}{\partial \vec{r}}\vec{u}}}}\right)_{ij} \tag{5.7}$$

and the Enskog heat flux

$$\vec{J}_q^{(1)} = -\left[\frac{1}{g}\left(1+\frac{3}{5}nbg\right)^2\lambda^B + \frac{3}{2}k\bar{\omega}\right]\frac{\partial}{\partial\vec{r}}T \tag{5.8}$$

Notice that Eq. (5.5) is the exact equation of state for a hard-sphere gas. Approximate values of η^B and λ^B are known. They are

$$\eta^B = \frac{5}{16}\frac{(kT)^{1/2}}{\sigma^2\,\pi^{1/2}} \tag{5.9}$$

and

$$\lambda^B = 2.52\,\frac{3k}{2}\,\eta^B \tag{5.10}$$

The Enskog transport coefficients are now obtained by identification of (5.7) and (5.8) with the phenomenological laws (3.18) and (3.19) respectively. One obtains

$$\eta = \eta^B\frac{1}{g}\left(1+\frac{2}{5}nbg\right)^2 + \frac{3}{5}\bar{\omega} \tag{5.11}$$

$$\zeta = \bar{\omega} \tag{5.12}$$

and

$$\lambda = \lambda^B\frac{1}{g}\left(1+\frac{3}{5}nbg\right)^2 + \frac{3}{2}k\bar{\omega}\,. \tag{5.13}$$

We notice that the bulk viscosity ζ, given by (5.12) exists and not vanishing, in contrast to the Boltzmann dilute gas case, where it does not appear.

Enskog transport coefficients are usually discussed in terms of reduced quantities, i.e. η/η^B, ζ/η^B, and λ/λ^B. For hard spheres these ratios are temperature independent functions of nb. Also, an important feature of Enskog transport coefficients appears if a suitable new quantity is introduced. One introduces the new quantity y

$$y \equiv nbg \tag{5.14}$$

In terms of this new quantity, one has the following reduced Enskog transport coefficients*:

$$\eta/\eta^B = \left[\frac{1}{y} + 0.8 + 0.7614\,y \right] nb \tag{5.15}$$

$$\zeta/\eta^B = 1.002\, y\, nb \tag{5.16}$$

$$\lambda/\lambda^B = \left[\frac{1}{y} + 1.2 + 0.7574\,y \right] nb \tag{5.17}$$

A correction factor of the form (5.17) was already proposed on semi-empirical grounds by G. Jager in 1900! In these expressions the first term is purely kinetic. The last one is the contribution from the potential part alone. The middle term represents the cross contribution from the kinetic and potential parts of the corresponding fluxes in the correlation-function approach.

Equations (5.15) to (5.17) graphically displayed in Fig. (5.1) deserve some comments and physical interpretation. Consider, for instance, the heat conductivity (5.17).

---

*Though we do not want to discuss fluid mixtures here, just for the sake of completeness we recall that in the Enskog theory the self diffusion coefficient is given in terms of the time scaled Boltzmann dilute gas coefficient $D=D^B/g$ with $D^B = \frac{3}{8n\sigma^2}\left(\frac{kT}{\pi}\right)^{1/2}$.

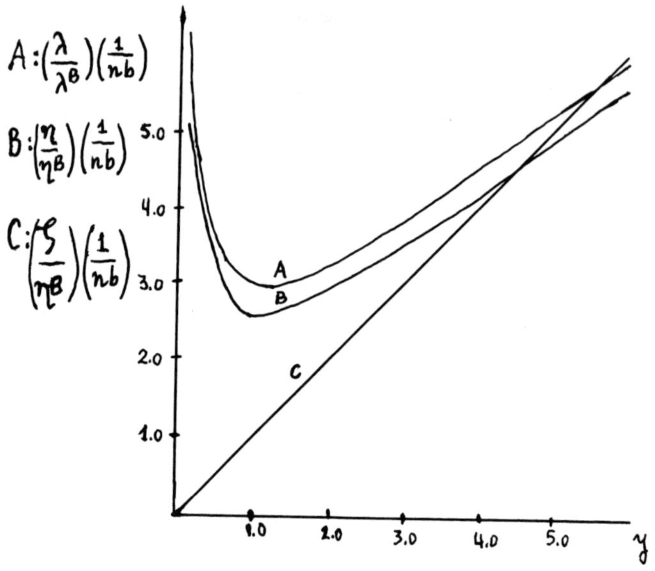

Fig. 5.1. The transport coefficients of a gas composed of hard spheres according to the theory of Enskog. A: reduced heat conductivity (eq. 5.17); B: reduced shear viscosity (eq. 5.15); C: reduced bulk viscosity (eq. 5.16) (All reduced quantities divided through nb).

The three contributions to (5.17) can be recast in a two-term decomposition,

$$\lambda = \lambda_1 + \lambda_2 \, , \tag{5.18}$$

$$\lambda_1/\lambda^B \equiv \frac{1}{y} + \frac{3}{5} \, , \tag{5.19}$$

$$\lambda_2/\lambda^B \equiv \frac{3}{5}\left(1 + \frac{3}{5}y\right) + \frac{32y}{25(1.02513)\pi} \tag{5.20}$$

Equation (5.19) shows that the (standard) translatory transport account-ted for by the λ- coefficient decreases monotonically with increasing compression (density) at constant temperature as would be expected. The collisional contribution, given by (5.20) however increases linearly with increasing compression, becoming the dominant part in the high density fluid. Both terms are plotted in Fig. 5.2. together with the sum as given by equation (5.17).

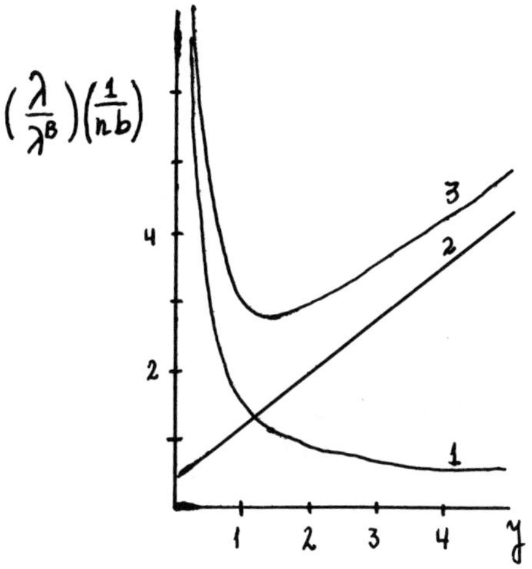

Fig. 5.2 Translatory 1, collisional 2, and total 3, heat conductivity contributions for an Enskog fluid of hard spheres.

Another remarkable feature of the reduced Enskog transport coefficients is that both the shear viscosity and the heat conductivity show a relevant minimum value as function of y, whereas the bulk viscosity does not. Obviously this latter quantity vanishes as y goes to zero. The respective minima correspond to the following values

$$[\eta/\eta^B]_{min} = 2.545\, nb \quad \text{at } y = 1.146 \tag{5.21}$$

$$[\lambda/\lambda^B]_{min} = 2.938\, nb \quad \text{at } y = 1.151 \tag{5.22}$$

This qualitative behaviour shown by Enskog transport coefficients fits nicely with the experimental data as we will see in the next section.

VI. COMPARISON WITH EXPERIMENTAL DATA

This section stems from fruitful discussions with Dr. K. Lucas, during the School, and in fact it has been written in collaboration with him.

Although the molecules of a real (simple monatomic) gas cannot be considered as hard spheres, Enskog's theory has been widely used to interpret experimental data of transport coefficients. The main reason is the lack of similarly simple expressions for the density dependence of the transport properties for more realistic models.

Let us first discuss an important qualitative feature of Enskog transport coefficients already pointed out in the preceding section. We have indicated that two of Enskog's hard-sphere transport quantities show a minimum value as functions of y, i.e., as functions of density. Is it not remarkable that experimental heat conductivity data show beautifully such a minimum (even for a polyatomic gas) ? Figure 6.1 illustrates this minimum for the heat conductivity of nitrogen. This result dates from 1953.

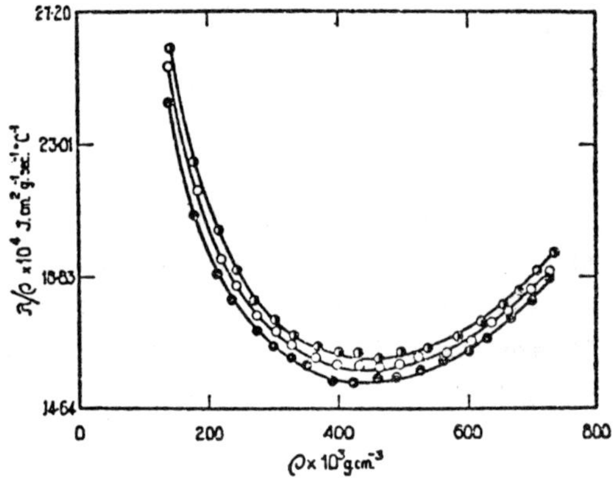

Fig. 6.1 Heat conductivity of nitrogen as a function of the density illustrating the existence of the Enskog minimum. Taken from McLaughlin (1964), where the original reference can be found.

For a quantitative comparison of the theory with experimental data, a direct procedure is to compare the density dependence of the experimental transport coefficents of a real gas with that of a hard-sphere gas with the same Boltzmann dilute fluid values for η^B and λ^B. This is achieved by calculating with equation (5.9) an effective diameter,

σ , and thus an effective value of b. The g comes from the known equation of state. Using this method the viscosity and heat conductivity of noble gases are compared with the predictions of Enskog theory. (see figures 6.2 and 6.3).

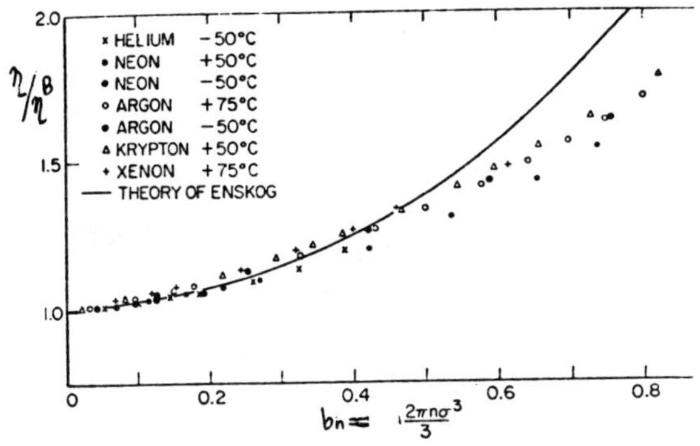

Fig. 6.2 Experimental shear viscosities of helium, neon, argon, krypton and xenon compared with the theory of Enskog. Taken from Sengers (1968) where the original references may be found. The transition for a hard-sphere fluid is bn $\simeq$ 1.8.

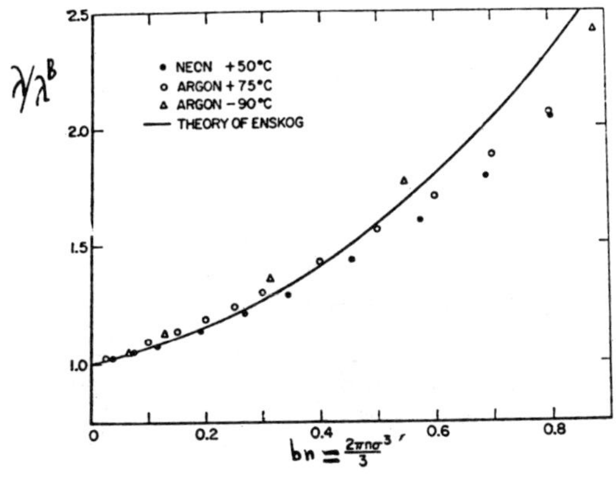

Fig. 6.3 Experimental heat conductivities of neon and argon compared with the theory of Enskog. Taken from Sengers (1968) where the original references may be found.

It is seen that the data follow the predicted behaviour up to a density of bn=0.4. This figure corresponds for helium to a density of about 650 Amagat units,for neon to about 4oo, for argon to about 160 and for xenon to 65. These are considered as moderate densities.

At higher densities, however, the theoretically expected increase is too large. Usually the experimental data are not compared with the theory of Enskog by the method just described and for this very reason we turn our attention to another procedure.*

The situation can be considerably improved in the following way. As relations (5.15) and (5.17) express the transport ratios in terms of equilibrium properties, attempts have been made to represent the transport coefficients of a real gas by the Enskog formulas using for both b and g effective values deduced from the compressibility isotherms of the real gas. One possiblity would be to derive bng from the experimental compressibility factor (p_o/nkT), using (5.5).

Since it happens that, for a hard-sphere gas, the external pressure p_o and the "thermal pressure" $T(\frac{\partial p_o}{\partial T})$ are equal**, another possibility for extending the applicability of the theory of Enskog to a real gas is to deduce explicit values of bng from the experimental thermal pressure, using the relation

$$T\left[\partial p_o/\partial T\right] = nkT\,(1+nbg) \tag{6.1}$$

with

$$\lim_{n\to 0} g(\sigma) = 1 \tag{6.2}$$

This suggestion, in fact, goes back to Enskog himself. Notice that taking (6.3) as valid for real molecules, is meant to be just an <u>ad hoc assumption</u> made only in order to see what comes out of the

*The values $(1/n)=0.54b$ ($=1.59v_o=1.12\sigma^3$; $v_o=\sigma^3/\sqrt{2}$ is the volume per sphere in a regular close-packed array of N spheres) and $(p/nkT) = 10.3$ give respectively the minimum volume and the maximum ratio of (p/T) at which the fluid phase is stable. One amagat is the density of the substance at 0^oC and 1 atm. Thus the density expressed in Amagat units is a dimensionless quantity that represents the ratio between the actual density and the density of the gas at 0^oC and 1 atm. For instance, for neon 400 Amagat units correspond to 0.35992 g/cm^3 (One Amagat is around 0.0008998 g/cm^3).

**In standard textbooks on equilibrium thermodynamics, one finds that the external pressure p_o can be written as $p_o=T(\partial P/\partial T)_V - (\partial U/\partial V)_T$ where $T(\partial p/\partial T)_T$ is called the thermal pressure and $(\partial U/\partial V)_T$ the internal pressure. The internal pressure represents the force of cohesion of the molecules. For a gas of hard spheres $(\partial U/\partial V)_T = 0$ and the external pressure is equal to the thermal pressure.

comparison of this adaptation of Enskog's theory with the experimental data. In addition to nbg, an effective value of b (the "co-volume") is needed, also. This latter parameter can be given explicit values by just looking at the limit of g as n vanishes, and the ratios (η/η^B) and (λ/λ^B) reduce to one. If the experimental (p_0,n) relation is expressed in the virial form

$$(p_0/nkT) = 1 + n\,B(T) + \theta(n^2) \tag{6.3}$$

b is related to the second virial coefficient, B, through its temperature derivative.

$$b = \frac{d(TB)}{dT} \tag{6.4}$$

This empirical <u>ad hoc</u> adaptation or modification of the Enskog thoery obtained by substituting (6.3) into (5.15) and (5.17) and thus relating the (η/η^B) and (λ/λ^B) ratios to the experimental p-V-T data, has been used by many authors. Typical results that are obtained for the transport coefficients of the noble gases are shown in figures 6.4 and 6.5 and Table I. The dotted curves represent the behaviour predicted by this procedure. In the density range shown, the difference between calculated and experimental data never exceeds 15%. We thus see that the theory of Enskog describes the main trend of the density dependence fairly well over a large density range to within about 10%,up to densities of 600-700 Amagat. The most complete comparison of this empirical modification of Enskog theory and experimental data can be found in a rent paper by Hanley, McCarty and Cohen (1972).

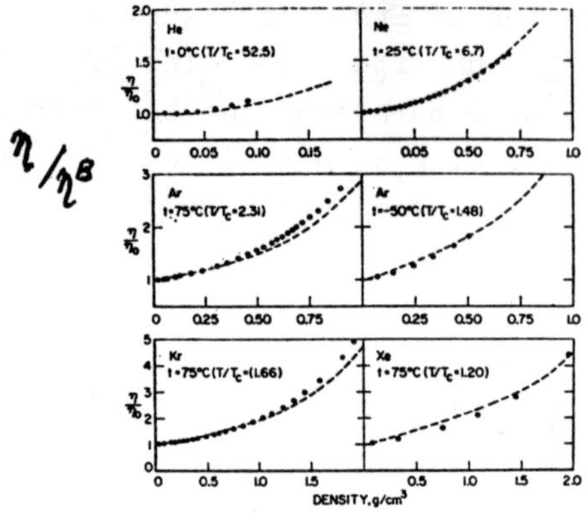

Fig. 6.4 Experimental shear viscosities of helium, neon, argon, krypton and xenon compared with the empirical ad hoc modification of the theory of Enskog. Taken from Sengers (1965, 1968), where the original references can be found.

TABLE I

	$\frac{\eta_{calc} - \eta_{exp}}{\eta_{exp}}$ (%)						
ρ (amagat)	100	200	300	400	500	600	700
Helium 0°C	0%	0%	-1%	-2%	-4%		
Neon +75°C	+1%	+1%	+1%	+1%	+1%	+1%	+1%
Neon +25°C	+1%	+1%	+1%	+1%	+1%	+2%	+2%
Argon +75°C	0%	-3%	-5%	-8%	-10%	-12%	
Argon -50°C	+5%	+5%	0%				
Xenon +75°C	+14%	+10%	0%	-7%			

	$\frac{\lambda_{calc} - \lambda_{exptl}}{\lambda_{exptl}}$ (%)						
ρ (amagat)	100	200	300	400	500	600	700
Neon +75°C	0%	0%	0%	-1%	-2%	-2%	-2%
Neon +25°C	0%	-1%	-2%	-2%	-3%	-4%	-4%
Argon +75°C	-3%	-7%	-11%	-12%	-13%	-15%	
Argon -90°C	+1%	-4%	-8%	-6%	- 4%		

Comparison of the shear viscosity and the heat conductivity of the Noble Gases with the empirical modification of Enskog's theory. Taken from Sengers (1965) where the original references can be found.

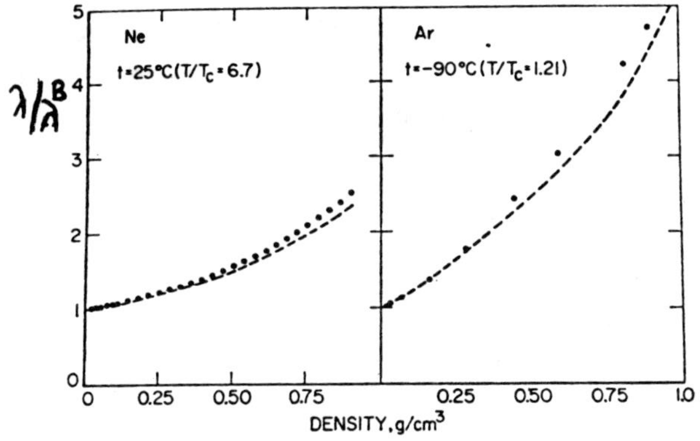

Fig. 6.5. Experimental heat conductivities of neon and argon compared with the empirical ad hoc modification of the theory of Enskog. Taken from Sengers (1965, 1968) where the original references can be found.

However, while being a practically useful recipe for predicting transport properties over a wide range of thermodynamic states, the procedures cited above should not be regarded as rigorous tests of the theory, or of Enskog's kinetic equation, or of the hard-sphere model, which has been put into it. One should not conclude on this basis of acceptable agreement with experiment, that Enskog's equation and the hard-sphere model represent the actual fluid behaviour correctly in the region studied.

A more rigorous test of the hard-sphere model has been given in a sequence of papers by Dymond, and Alder and Dymond, on the basis of the biew, that for transport properties, the hard-sphere model is equivalent to the van der Waals model of a fluid. According to this model, the molecules have a potential made up of a hard core plus a weak uniform negative part as shown in figure 6.6.

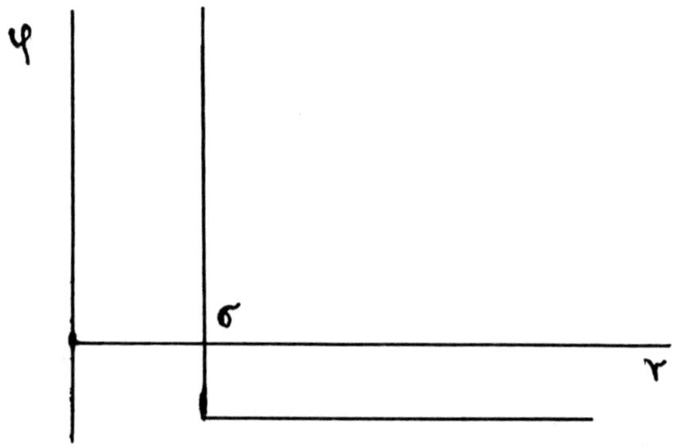

Fig. 6.6 The potential function of a van der Waals fluid

This model is highly idealised. However, as far as the attractive part goes, real system approximate this picture at densities higher than critical, where the range of the intermolecular forces can be considered large relative to the intermolecular spacing, the net resulting attractive force on a molecule between its nearest neighbours being close to zero. If furthermore, the temperature of the fluid is sufficiently high ($T > T_c$, T_c: critical temperature), so that the kinetic energy of molecules is large compared to the potential energy, the true molecular motion will approximate a succession of linear trajectories and hard core collisions. As far as the repulsive part of the potential goes, it is steep for real molecules, but not infinitely steep. Within the van der Waals model, this can be accounted for by a temperature dependent hard-sphere diameter, decreasing with increasing temperature. We can therefore expect real fluids to follow the van der Waals model at densities and temperatures above the critical values with a core site decreasing with increasing temperature.

It is known, that the p-V-T data of real fluids can be represented by a van der Waals type of equation of state at densities and temperatures above the critical values. In support of this molecular dynamics calculations for square-well molecules show that a plot of ($p^V o/NkT$) versus ($^1/T$) is straight for high densities and high temperatures, in agreement with the van der Waals equation, down to approximately the critical temperature. The equation can be written in the form

$$\frac{p}{nkT} = \left(\frac{p}{nkT}\right)_{H.S.} - \frac{a_1}{NkTV} \tag{6.5}$$

The first term of the r.h.s. accounts for the hard-sphere contribution for which an accurate equation is available, in terms of the hard-sphere diameter (see the paper by Carnahan and Starling (1969). At lower temperatures, deviations from this linear representation, and therefore deviations from the van der Waals equation becomes apparent.

When experimental data for (p/nKT) are plotted versus ($^1/T$) for density and temperatures above the critical, one gets, however, nonlinear representations up to the highest experimental temperatures. This behaviour is shown in figure 6.7.

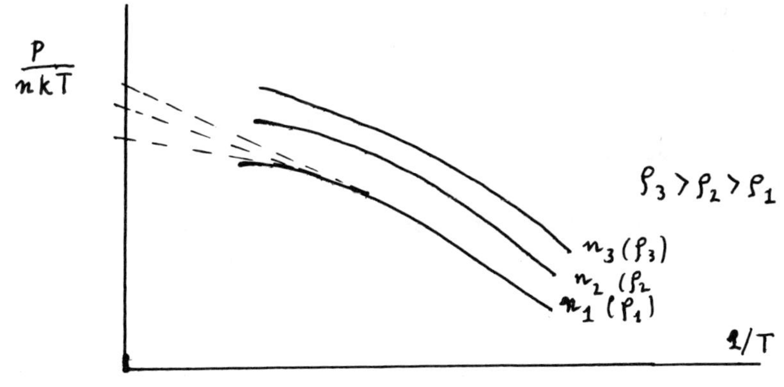

Fig. 6.7 Experimental p-v-T data as a function of $\frac{1}{T}$

This curvature is interpreted in terms of a temperature dependent hard-sphere diameter σ. At every desired temperature, the tangent is drawn. The intercepts of these (high temperatures) slopes with the (P/nKT) axis give "experimental" hard-sphere values, indeed functions of both temperature and density. These values, when interpreted in terms of the Carnahan-Sterling equation of state, lead to temperature-dependent hard-sphere diameters, σ. For densities and temperatures above critical it just happens that σ does not show any density dependence. This provides further evidence of the usefulness and validity of the van der Waals model in that region. A qualitative display of the temperature dependence of σ is given in fig. 68.

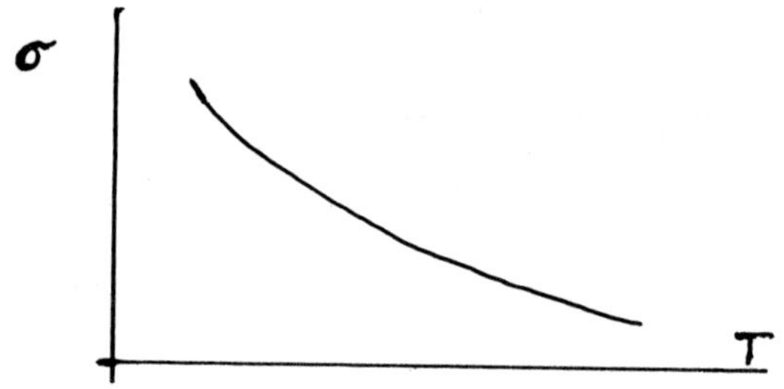

Fig. 6.8 A qualitative sketch of the dependence of σ on T

Thus, using the values of σ obtained from experimental p-V-T data, and the Carnahan-Sterling equation of state, the Enskog gas transport properties can be evaluated, and directly compared to experimental results. In doing this, one must realise however that at high densities, Enskog theory does not strictly apply. This is because of the systematic neglect of momentum correlations of the hard-spheres, in the Enskog molecular chaos assumption. Dymond and Alder (1966) have shown that the viscosity and heat conductivity of some noble gases can be predicted within 10% for densities and temperatures above the critical values, by means of Enskog's equations. The results are very satisfactory for argon, krypton and xenon, and less satisfactory for neon, where experimental values are only available at high temperatures, well above critical. The temperature dependence is also quite well reproduced. This is to be expected, as a consistent temperature dependent diameter σ has been used. Such a representation is graphically depicted in figure 6.9.

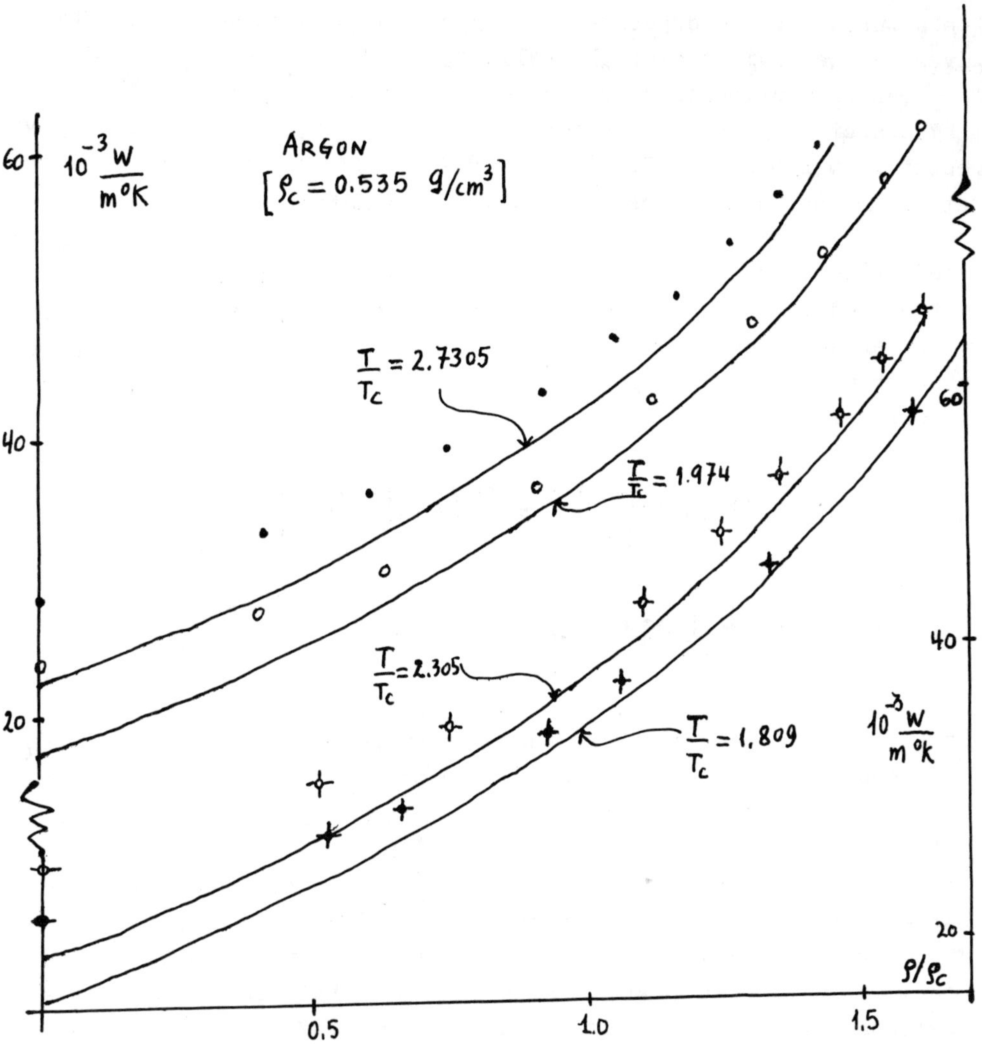

Fig. 6.9 Heat conductivity (in units of $10^{-3} \frac{W}{m°k}$) of argon at four isotherms: • (2.7305 Tc); ∘ (1.974 Tc); ✧ (2.305 Tc); ✦ (1.809 Tc); Points are calculated values using Enskog theory. The solid curves are experimental curves of Michels and Le Nendre. Notice that the two upper curves belong to the l.h. ordinate axis, whereas the two lower curves belong to the r.h. axis. Taken from unpublished results of K.Lucas

In another paper, Dymond and Alder (1968) have verified for the self diffusion coefficient, that computer calculated corrections to the Enskog predictions, by removing the molecular chaos assumption, given agreement between calculated and experimental results, within the accuracy of the experimental and machine generated data. This gives further evidence of the validity of the van der Waals model for transport properties at high densities and temperatures*.

Yet another method of testing the hard-sphere model has been given recently by Dymond (1973, 1974). P-V-T data are no longer used to determine the hard-sphere diameter. As a matter of fact, according to Dymond's suggestions, in order to test the models, there is no need of knowing the diameter, σ , in advance. Corrections for correlated molecular motions, due to molecular dynamics computations by Alder and Wainwright, are used together with Enskog's transport properties. The hard-sphere results, e, g., for viscosity, are represented in the form

$$\eta^* \equiv (\eta/\eta^B)\left(\frac{v}{v_0}\right)^{2/3} = (\eta/\eta^E)(\eta^E/\eta^B)\left(\frac{v}{v_0}\right)^{2/3}$$

$$\equiv f\left(\log[v/v_0]\right). \qquad (6.6)$$

Here η^E denotes the standard Enskog transport coefficient and (η/η^E) the computer empirical correction for correlated molecular motions. The values of η^* can also be plotted from experimental viscosity data as a function of v without any prior knowledge of the hard-sphere diameter, σ . Where both curves are superimposable, the hard-sphere theory does represent the density dependence of the data. σ or v_0, can be determined from points where the two curves coincide. Dymond then shows, that coincidence between the two curves is reached within the limits of the accuracy of the experimental data and the hard-sphere corrections for densities, from above twice the critical density down to about 1.2 times the critical density. The agreement, for heat conductivity and selfdiffusion, is similar or even better. Since σ has been determined from high-density viscosity

---

*Extension of this procedure to polyatomic fluids, even with approximately spherical molecules, do not provide results of comparable quality. Viscosity deviations are of order of 15% even at high densities, whereas such deviations in the noble gas cases are within 1 to 3%. For heat conductivity the situation is even worse because of the neglect of the influence of the internal degrees of freedom (K. Lucas and G. Ackman; unpublished results).

data no further adjustable parameter is available, and thus true _a priori_ predictions are being made. It is remarkable that the density region considered corresponds to a large band of thermodynamic states, where the transport properties change value by a factor of more than 3. Such agreement is not reached in any of the known empirical modifications of the Enskog theory on the dense region (see for instance figures 6.2 through 6.5).

Applying now the recipe to _liquid_ transport properties data (for $T<T_c$ and density above critical), one finds that the hard-sphere diameter has to be pressure-dependent as well as slightly temperature-dependent. Here again, once σ has been fixed at some given thermodynamic state of a substance from data of one property, the other transport properties are predicted without any further empirical information. One cannot, however, expect the hard-sphere model to be valid at low fluid temperatures, as there the actual non-uniformity of the attractive potential becomes important.

The procedure has also been applied to polyatomic gases by Lucas and Ackmann (unpublished results). They have verified that for a number of quasi-spherical molecules, the viscosity could be represented in terms of the corrected Enskog theory almost as well as for the monatomic simple fluids. This may be useful for practical applications, though admittedly one parameter, the hard-sphere diameter, σ , has to be adjusted to a high-density viscosity data point. One should, however, not consider this as evidence of a deep understanding of the behaviour of polyatomics in molecular collisions but merely accept it as a convenient empirical fact analogous to the situation with dilute polyatomic viscosity, according to the Chapman Enskog theory. No such good results can be found for the heat conductivity, as it is to be expected.

VII. THE SQUARE-WELL FLUID

There is one more intermolecular potential that I will mention: the square-well function. The nice feature of the square-well potential (see Fig. 1.1) is that it retains the impulsive property of the forces. The Enskog theory of square-well molecules has been developed by several authors. However, contrary to the situation with the hard-sphere gas, neither a computer calculation for square well fluids nor a time-corelation function approach seems to have been worked out. Also it should be stressed that what people have done in developing Enskog theory for a square-well gas involves some additional ad hoc approximations that are not present in the hard-sphere gas case. An equation similar to the hard-sphere Enskog equation is thus given, though the r.h.s. is somewhat arbitrarily analysed, different parts arising from the possible distinct types of binary collisions between square-well molecules. The types of collisions or "partial collisions" may be summarised as follows: There will be an impulsive force at a separation $r = R\sigma$ due to the outer edge of the well; the molecules may subsequently experience a hard-core collision at $r = \sigma$, followed by a further partial collision at $r = R\sigma$, when they separate. In writing the square-well two-body collision integral, all types of "partial collisions" give separate contributions. The molecular chaos assumption is applied to each "partial collision" independently. This is an additional assumption on top of the standard Enskog theory and should be valid, approximately, only for dense enough systems (see Davis, 1973)

Without going into any detailed description of the formal theory, we merely list here the relevant results given in the literature for a square-well model gas. The improvement proposed for the equation of state is:

$$\frac{P_0}{nkT} = 1 + nb\left[g(\sigma) + R^3 g(R\sigma)\left(1 - e^{\varepsilon/kT}\right)\right] \tag{7.1}$$

The transport coefficients are also improved in the following way: For the shear viscosity one has

$$\eta = \frac{5}{16\sigma^2}\left(\frac{kT}{\pi}\right)^{1/2}\left\{\frac{\left(1+\frac{2}{5}nb\right)\left[g(\sigma) + R^2 g(R\sigma)\alpha\right]^2}{g(\sigma) + R^2 g(R\sigma)\left[\beta + \frac{1}{6}(\epsilon/kT)^2\right]} + \frac{48}{25\pi} n^2 b^2 \left[g(\sigma) + R^4 g(R\sigma)\beta\right]\right\} . \tag{7.2}$$

The bulk viscosity is given by

$$\zeta = \frac{n^2 b^2}{\pi \sigma^2}\left(\frac{kT}{\pi}\right)^{1/2}\left[g(\sigma) + R^4 g(R\sigma)\beta\right] \tag{7.3}$$

The heat conductivity

$$\lambda = \frac{75}{64\sigma^2}\left(\frac{k^3 T}{\pi}\right)^{1/2}\left\{\frac{\left\{1 + \frac{3}{5} nb\left[g(\sigma) + R^3 g(R\sigma)\alpha\right]\right\}^2}{g(\sigma) + R^2 g(R\sigma)\left[\beta + \frac{11}{16}(\epsilon/kT)^2\right]}\right.$$

$$\left. \frac{32}{25\pi} n^2 b^2\left[g(\sigma) + R^4 g(R\sigma)\beta\right]\right\} \tag{7.4}$$

Here $g(\sigma)$ is evaluated just inside the well at $r = \sigma^+$, whereas $g(R\sigma)$ is evaluated just outside the well. We have introduced

$$\alpha = 1 - e^{\epsilon/kT} + \frac{\epsilon}{2kT}\left[1 + \frac{4}{\pi^{1/2}}\exp\left(\frac{\epsilon}{kT}\int_{\epsilon/kT}^{\sigma} x^2 \exp(-x^2)\,dx\right)\right] \tag{7.5}$$

$$\beta = e^{\epsilon/kT} - \frac{\epsilon}{2kT} - 2\int_0^\infty x^2\left[x^2 + \frac{r}{kT}\right]^{1/2}\exp(-x^2)\,dx\,. \tag{7.6}$$

The functions α and β exist tabulated in the literature.

According to David (1973) and Gubbins (1973), the Enskog square-well theory would interpret experimental data for fluids for densities above critical, rather than standard " moderately dense" gas-like data, much better than the simple hard-sphere gas theory. In particular the temperature dependence of transport coefficients seems to go better with physical reality.

VIII. FINAL COMMENTS

Even though we have found that Enskog's equation is certainly very useful in interpreting experimental data on transport properties of dense simple gases there are still fundamental questions to answer: what is, from first principles and from "exact" calculations, the domain of validity of Enskog's equation? Can we derive it from the Liouville equation, as it has been done for the Boltzman's equation? Does Enskog's equation provide some further understanding of the general irreversible behaviour of a non-equilibrium gas (and isolated system)? Is it really valid for a hydrodynamic description/dense gases?

Nobdoy has been able as yet to produce an H-theorem for the Enskog equation. The difficult lies in the definition of the non-equilibrium entropy functional for the Enskog hard-sphere gas. It is only very recently that D. Hubert (unpublished results) has been able to provide an H-theorem for an Enskog-like type of equation. He uses ideas of Prigogine which have been presented at this School (see the lectures given by Prigogine). However, it has been shown by Lebowitz et al (1969) and by some other people (see e.g. Van Beijeren and Ernst (1973), Mo and Dufty (1974) that the actual Enskog equation is a short-time kinetic equation (valid at $t = 0^+$ only). If this result is taken strictly then we are faced with the problem that Enskog's equation, valid for times shorter than the mean free time between collisions, cannot be considered seriously as a kinetic equation for the hydrodynamic regime. We are just saying that there is no convincing argument why the Enskog equation should be a good approximation for the description of a gas on the time scale on which one can describe the nonequilibrium fluid in terms of hydrodynamic equations with simple transport coefficients. There remains an open line of research which, as a matter of fact, P. Resibois and J. Lebowitz are presently exploring. J. Lebowitz has discussed this problem in one of his lectures.

One would like to define the Enskog equation in terms of some scheme that can be used to predict the (exact) high density corrections to Enskog, which are known from molecular dynamics or from general theories. With respect to computer analysis of hard-sphere systems, there is the beautiful work of Alder et al (1970). This work is a direct and unambiguous"experimental" test of the validity of the Enskog approach. Some of the results are shown in figures 8.1 and 8.2. The major assumption tested in the comparison is the molecular chaos approximation. The errors appear to be fairly small for the heat conductivity whereas for the shear viscosity, the error is within 20%, for $n\sigma^3 < 0.8$.

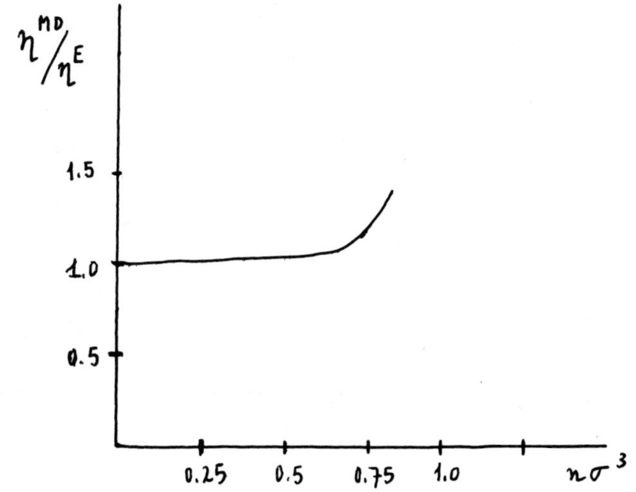

Fig. 8.1. Ratio of exact hard-sphere molecular dynamics shear viscosity to Enskog hard-sphere viscosity (a computer experiment done with 108 molecules). Taken from Alder <u>et al</u> (1970) where detailed quantitative data can be <u>found</u>. See also Gubbins (1973) paper.

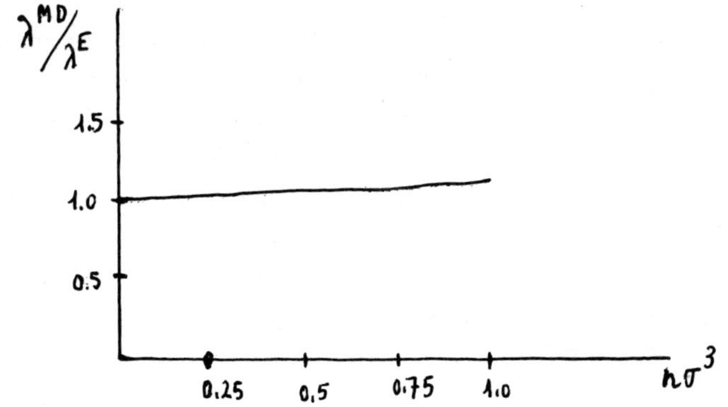

Fig. 8.2. Ratio of exact hard-sphere molecular dynamics viscosity to Enskog hard-sphre viscosity (computer experiment done with 108 molecules). Taken from Alder et al (1970), where detailed quantitative data can be found. See also Gubbins- (1973) paper.

Concerning the more basic point of view of the general theories, we know that Enskog's equation may be derived in a variety of ways from the BBGKY hierarchy, under some well defined ad hoc assumptions.

I do not want to engage here in a critical review of different formal "derivations" of Enskog's equation. However without aiming at being complete I'll merely list here a few of the papers that could be of interest: Pioneering efforts seem to be those of H.S. Green (Molecular Theory of Fluids, 1952; there exists a Dover reprint). R.F. Snider and C.F. Curtiss, Phys. Fluids 1 (1958) 122; 3 (1960) 903; J.V. Sengers and E.G.D. Cohen, Physica 27 (1961) 230. More recent papers include those of M.H. Ernst, Physica 32 (1966) 273; W.D. Henline and D.W. Condiff, J. Chem. Phys. 54 (1971) 5346; D.K. Hoffman and H.S. Green, J. Chem. Phys. 43 (1965) 4007; D.K. Hoffman and C.F. Curtiss, Phys. Fluids 8 (1965) 667. An Enskog theory for soft potentials has been recently worked out by G.B. Brinser and D.W. Condiff, J. Chem. Phys. 59 (1973) 6599.

Sengers has discussed at length the three-particle collision term in the generalized Boltzmann equation and has identified Enskog's approximation, Sengers (1973).

The discussion becomes easier if we consider a spatially homogeneous system. Retaining only the first few terms, the generalised Boltzmann equation has the form (Choh and Uhlenbeck),

$$\frac{\partial}{\partial t} f = J(ff) + K(fff). \tag{8.1}$$

Here J, K, are time-independent integral operators acting on the distribution function f. The operator J is Boltzmann's collision integral and K accounts for the effect of collisions that involve three molecules According to Enskog one can estimate, in a first approximation, the non-equilibrium correlations in configuration space by assuming that they are independent of the velocities and that they are the same as for a dense gas in equilibrium. This means that $g(\sigma)$ according to Enskog, gives the probability of finding two modules in contact ($r_{12} = \sigma$) even in the presence of gradients. This leads automatically to a replacement of the Boltzmann J operator by the approximation

$$J^E(ff) = g(\sigma)\, J(ff). \tag{8.2}$$

The radial distribution function, $g(\sigma)$, has a virial representation given by equation (2.4) or else (2.11). If we substitute this virial series into (8.2), we obtain an expansion for the collision term, which can be compared with the formal expansion (8.1) for the generalised Boltzmann equation. We conclude that the Enskog theory theory approximates the three-molecule collision term K(fff) by

$$K^{E}(fff) = \frac{5}{8} nb\, J(ff) = \frac{5}{12} n\pi\sigma^3\, J(ff). \tag{8.3}$$

One finds that for a spatially homogeneous gas the Enskog transport coefficients are given in terms of simply scaled Boltzmann values

$$\eta^{E} = \eta^{B}/g(\sigma) \quad ; \quad \lambda^{E} = \lambda^{B}/g(\sigma). \tag{8.4}$$

Use of the virial representation of $g(\sigma)$, leads to the Enskog theory estimates of the three-molecule contributions to the transport properties

$$\eta^{E} = -\frac{5}{12}\pi\sigma^3\eta^{B} = -\frac{5}{8}b\,\eta^{B}, \tag{8.5}$$

$$\lambda^{E} = -\frac{5}{12}\pi\sigma^3\lambda^{B} = -\frac{5}{8}b\,\lambda^{B}. \tag{8.6}$$

Sengers has identified these density corrections (or else (8.3)) in the full three-molecule collision problem, as the contribution of <u>double-overlapping</u> collisions i.e. two-body collisions where both colliding molecules are <u>overlapping</u> with a third molecule. The remaining contributions have also been discussed by Sengers.

Sengers has even compared the results given by the non-uniform Choh-Uhlenbeck equation and the density expansion of Enskog's properties (as given by equations (5.15) and (5.17)) and has found that the full three-molecule density corrections should be

$$\eta = \left[(-0.601 \pm 0.008) + 0.800\right] b\,\eta^{B}, \tag{8.7}$$

$$\lambda = [(-0.593 \pm 0.014) + 1.200]\, b\, \lambda^B \tag{8.8}$$

Whereas the density expansion of Enskog's theory gives (using equation (2.4))

$$\eta^E = [-0.625 + 0.800]\, b\, \eta^B, \tag{8.9}$$

$$\lambda^E = [-0.625 + 1.200]\, b\, \lambda^B. \tag{8.10}$$

Thus a fairly good agreement between Enskog predictions and more accurate results shows up. This adds one more surprising result.

A final question that I would like to comment on is a natural modification of Enskog's equation suggested recently by Van Beijeren and Ernst (1973,1974).

It is clear that the statistical factor introduced by Enskog is not exactly the local-equilibrium radial distribution function (describing the non-uniform state, the local equilibrium distribution function may be expected to involve gradients of the local density and higher powers of these gradients, as well as space derivatives of higher order than the first. In fact, the explicit form of the pair distribution function in a local-equilibrium state is well known. For hard spheres, it can be expressed as a nonlocal functional of the local density, which may be expanded around some fixed point; thus yielding explicitely all space derivatives mentioned above.

Recently Van Beijeren and Ernst have proposed to replace the function g in Eq. (2.8) by the exact local equilibrium distribution function $g(\vec{r}_1,\vec{r}_2)$, which takes into account the spatial nonuniformities in the local equilibrium state. This replacement is completely in accordance with Enskog's arguments for modifying Boltzmann's "stosszahlansatz".

The local-equilibrium pair distribution function in a non-uniform state for a gas of identical hard spheres is a functional of the local number density $n(\vec{r},t)$, and can be expressed in the form of a density expansion. We will not give here any details but will just comment on what this natural modification implies.

The usual (2.8) and the Van Beijeren and Ernst modified form of the Enskog equation both lead to the same Navier-Stokes equations and to the same expression for the corresponding transport coefficients. However, the usual Enskog equation (2.8) does not account properly for second and higher order gradients contributions, and may not be used to derive (linear and non-linear) Burnett hydro-dynamic equations, and corresponding transport coefficients, as it has been indicated in the literature. (A discussion of the inadequacy of Enskog equation for short wavelength, high frequency phenomena may be found in a paper by E.P. Gross and D. Wisnivesky, Phys. Fluids 11 (1968) 1387.) The Van Beijeren and Ernst modified Enskog equation, on the other hand, can, in principle, be used for this purpose.

The Van Beijeren and Ernst modified Enskog equation also becomes exact for times much shorter than the mean free time between collisions. That the standard Enskog, and so the van Beijeren and Ernst equation, may however be considered as astonishingly useful approximations is something that we have pointed out clearly in section VI. We thus only expect that the recent Résibois - Lebowitz approach to the Enskog equation will provide a deeper understanding of the approximation.

ACKNOWLEDGEMENTS

For the preparation of these notes I have benefited from fruitful discussions with Dr. K. Lucas, Institut für Thermo-und Fluiddymamik Ruhr-Universität, Bochum, Germany. I would also like to thank P. Résibois, J. Dufty, M.Enrst and J. Dymond, for correspondence.

REFERENCES

1. J.O. HIRSCHFELDER, C.F. CURTISS and R.B. BIRD, Molecular Theory of Gases and Liquids, Wiley, N.Y., (1954).
2. S. CHAPMAN and T.G. COWLING, The Mathematical Theory of Non Uniform Gases, Cambridge, 3rd Edition in cooperation with D Burnett, (1970).
3. D. ENSKOG, Kungliga Svenska Vetenskapsakademiens Handlingar, Ny Föld, 63, (4), (1922). In fact, I have used the English translation edited by S.G. Brush, Kinetic Theory, vol. 3. Pergamon Press, Oxford, (1972).
4. E. McLAUGHLIN, Chem. Revs. 64 (1964) 389.
5. J.V. SENGERS (a) Int. J. Heat Mass Transfer 8 (1965) 1103. (b) in Recent Advances in Engineering Science, vol. 111. A.C. Eringen, editor, Gordon and Breach, N.Y. (1968) pp. 153-196. (c) in Kinetic Equations, R.L. Liboff and N. Rostoker, editors Gordon and Breach . N.Y. (1971), pp. 137-193. (d) in The Boltzmann Equation, E.G.D. Cohen and W Thirring, editors, Springer-Verlag, Wien, 1973, pp. 177-208
6. D.E. DILLER, H.J.M. Hanley and H.M.Roder, Cryogenics 10 (1970) 286.
7. W.E. BRITTIN, editor, Kinetic Theory (Lectures in Theoretical Physics, volume 1XC, Boulder 1966) Gordon and Breach, N.Y.1967.
8. H.J.M. HANLEY, R.D. McCARTY and E.G.D. COHEN, Physica 60 (1972) 322.
9. H.T. DAVIS, in Advances in Chemical Physics, vol. XX1V, I. Prigogine and S.A. Rice, editors, Wiley, N.Y. 1973. 257-343
10. J.H. DYMOND and B.J. ALDER, J. Chem. Phys., 45 (1966) 206); ibiden 48 (1968) 343.
11. N.F. CARNAHAN and K.E. STARLING, J. Chem. Phys. 51 (1969) 635.
12. J.H. DYMOND, Proc. 6th Sym. Thermophys. Props. A.S.M.E. New York (1973) 143; Physica, 1974 (submitted for publication).
13. K.E. GUBBINS, in Statistical Mechanics (A specialist periodical Report, vol. 1.), The Chemical Society, London,(1973), pp.194-253.
14. B.J. ALDER, D.M. GASS and T.E. WAINWRIGHT, J. Chem. Phys. 53 (1970) 3813.
15. J.L. LEBOWITZ, J.K. PERCUS and J.SYKES, Phys. Rev. 188 (1969),487 There is some recent, as yet unpublished work by P. Résibois and J. Lebowitz about the "next" order correction to Enskog's equation.
16. H. van BEIJEREN and M.H. ERNST, Physica 68 (1973) 437.
17. K.C. MO and J.W. DUFTY, Physica, (1974) (to appear).

WHAT CAN ONE LEARN FROM LORENTZ MODELS?

E.H. Hauge

University of Trondheim
Trondheim, Norway

I. MODELS

1.1. Generalities
1.2. Lorentz Models

II. FROM KINETIC THEORY TO HYDRODYNAMICS

2.1. The Boltzmann Equation
2.2. Exact Solution
2.3. The Chapman-Enskog Ansatz
2.4. The Hydrodynamic Mode
2.5. The Contraction
2.6. Final Remarks

III. HIGHER DENSITY EFFECTS

3.1. The Einstein Formula for D
3.2. D from the Kinetic Equation
3.3. Cluster Expansions
3.4. Divergences
3.5. What to do about them
3.6. Tails
3.7. Abnormal Diffusion
3.8. Computer Work

IV. RIGOROUS RESULTS

4.1. The Grad Limit
4.2. Percolation Problems
4.3. Absolutely Final Remarks

REFERENCES

I. MODELS

1.1. Generalities

One of the sad facts of life for someone who is seeking to get a firm grasp on non-equilibrium statistical mechanics, is the lamentably small number of models for which exact results are available. The situation in equilibrium statistical mechanics is much brighter in this respect. In particular, there is a variety of latice models which, to a certain extent, can be exactly solved. These exact results have proved of great value as reference points in the general development of the field.

Such reference points are, perhaps, of even greater significance in non-equilibrium statistical mechanics, where there are a number of qualitative problems in need of clarification. The most famous of these is probably the basic one of reconciling microscopic reversibility with macroscopic irreversibility. But that is only the first of a long list of fundamental questions. Clearly, the usefulness of models lies less in their ability to produce numbers to be compared with experiment, than in their value as guides in qualitative matters.

Aside from their usefulness, it would be dishonest for a theoretician to suppress the fact that working with models that are exactly soluble in some sense, is quite simply a lot of fun!

Such a frivolous remark should immediately be counterbalanced by a puritanical afterthought: Models can be dangerously misleading. The fact that a model is soluble is always due to some particular simplifying features, and precisely those features can be decisive for the answers to the questions asked.

Consequently, whereas the source of worry in the case of general formalisms is the postulated existence of objects used, or the soundness of approximations introduced, the problem in the case of soluble models is shifted to the scope and relevance of the model itself. Thus, physical insight and common sense are called for, and with such subjective elements introduced, mistakes are bound to be made.

Some of these introductory generalities should be recalled in specific contexts as we proceed.

1.2. Lorentz Models

The class of models we shall consider in the following are the so-called Lorentz models. In such models classical point particles without mutual interaction move in a random array of stationary scatterers. Lorentz models are glorified pin-ball machines! The shape of the scatterers, and the dimensionality of space can be chosen freely.

It is important to realize that Lorentz models are not weird products of the imagination of idle theoreticians! In some contexts they represent quite reasonable idealisations of nature. The spherical one, in particular, is much used to study transport problems with slow neutrons in a heavy medium [1] . Also electrons scattered on impurities in a metal can be described by a (quantum mechanical)Lorentz model [2]. And finally there is the instance Lorentz himself had in mind, namely the gas mixture in which the mass ratio is very large [3] .

We shall not focus on the range of applicability in different contexts here, but the examples given show that Lorentz models are reasonably close to nature. We can therefore hope that the answers they provide to some qualitative questions are relevant in a more general context.

The crucial simplifying feature in Lorentz models is their linearity, resulting from the fact that the scatterers are not affected when hit by the moving particle. An immediate consequence is that Lorentz models can certainly <u>not</u> shed light on the important non-linear aspects of non-equilibrium theories. However, the completely linear case is still of sufficient interest in itself to warrant serious study. Also the understanding of non-linear phenomena presupposes a firm grasp of the linear ones.

The exposition that follows will not be a very systematic one. Rather, we shall see how Lorentz models in at least two general areas have proved useful. First we shall discuss the contraction from a kinetic to a hydrodynamic description. And next some central problems associated with higher density effects will be indicated. We close by advertizing Lorentz models as a challenging and fruitful field for rigorous work.

II. FROM KINETIC THEORY TO HYDRODYNAMICS

2.1. The Boltzmann Equation

Let us, for the time being, assume that the number density n_s of scatterers is sufficiently small that it makes sense to take the Boltzmann equation as the equation of motion for the one-particle distribution $f(\vec{r},\vec{v},t)$ of the moving particles. For infinite space and with no external forces, the equation reads

$$\left(\frac{\partial}{\partial t}+\vec{v}\cdot\nabla\right) f\,(\vec{r},\vec{v},t) = B\, f\,(\vec{r},\vec{v},t) \tag{2.1}$$

where B is the collision operator.

If we specialize to the three dimensional Lorentz model with spherical scatterers of radius a, the operator is particularly simple, due to the isotropy of classical scattering off a sphere (not true in two, nor in thirteen dimensions!):

$$Bf(\vec{r},\vec{v},t) = \frac{1}{4}\, n_s\, va^2 \int d\Omega_{\vec{v}'} \left[f(\vec{r},\vec{v}',t) - f(\vec{r},\vec{v},t)\right] . \tag{2.2}$$

Since the kinetic energy of a point particle is a trivial constant of the motion in Lorentz models, any velocity distribution $\phi\,(v)$, not only the Maxwellian, is an equilibrium distribution, and for simplicity we have assumed that all moving particles have the same v. The <u>direction</u> of $\vec{v}$ remains an interesting variable in $f(\vec{r},\vec{v},t)$, however, and in (2.2) the integration goes over all directions of the velocity $\vec{v}'$ after the collision with a sphere.

The nice thing about (2.2) is, aside from its linearity, that it can be written in terms of a projection operator

$$B\, f = \tau^{-1}\,(P-1)\, f \tag{2.3}$$

where

$$P\, f = \frac{1}{4\pi}\int d\Omega_{\vec{v}}\, f\,(\vec{r},\vec{v},t) \tag{2.4}$$

(clearly $P^2 = P$!), and where

$$\tau = (n_s v\, \pi\, a^2)^{-1} \tag{2.5}$$

is the mean free time, i.e., the average time between successive collisions.

Taking Fourier transforms in space and Laplace transforms in time

$$f_{\vec{k}z}(\vec{v}) = \int_0^\infty dt\, e^{-zt} \int d^3r\, e^{-i\vec{k}\cdot\vec{r}}\, f(\vec{r},\vec{v},t) \tag{2.6}$$

one obtains the following form of the Boltzmann equation

$$(z + i\vec{k}\cdot\vec{v})\, f_{\vec{k}z} = \frac{1}{\tau}(P-1)\, f_{\vec{k}z} + f_{\vec{k}}(t=0). \tag{2.7}$$

2.2. Exact solution

The beauty of (2.7) is that it can be explicitly solved. We shall use a method of solution due to J.M.J. van Leeuwen. Rewrite (2.7) as follows:

$$f_{\vec{k}z} = \frac{\tau^{-1}}{z+\tau^{-1}+i\vec{k}\cdot\vec{v}}\, P f_{\vec{k}z} + \frac{f_{\vec{k}}(t=0)}{z+\tau^{-1}+i\vec{k}\cdot\vec{v}} \tag{2.8}$$

Operate on this equation by P. The result is a closed equation for $P f_{\vec{k}z}$ since

$$P\frac{\tau^{-1}}{z+\tau^{-1}+i\vec{k}\cdot\vec{v}} = \frac{1}{2}\int_{-1}^{1} dx\, \frac{\tau^{-1}}{z+\tau^{-1}+ikvx} \tag{2.9}$$

$$= \frac{1}{kv\tau}\tan^{-1}\frac{kv\tau}{z\tau+1}\,.$$

With $Pf_{\vec{k}z}$ determined, (2.8) shows that the complete $f_{\vec{k}z}$ follows. One finds

$$f_{\vec{k}z}(\vec{v}) = \frac{\tau^{-1}}{z+\tau^{-1}+i\vec{k}\cdot\vec{v}}\left[1-\frac{1}{kv\tau}\tan^{-1}\frac{kv\tau}{z\tau+1}\right]^{-1} P\,\frac{f_{\vec{k}}(\vec{v},0)}{z+\tau^{-1}+i\vec{k}\cdot\vec{v}}$$

$$+\frac{f_{\vec{k}}(\vec{v},0)}{z+\tau^{-1}+i\vec{k}\cdot\vec{v}} \tag{2.10}$$

2.3. The Chapman-Enskog Ansatz

But having obtained the exact solution doesn't mean that there is nothing more to learn! On the contrary, we are now in the position to discuss approximation schemes with the confidence that a complete solution gives.

One old and important question in this context is: How does hydrodynamics come out of the Boltzmann equation?

The old answer to this question is provided by the Chapman-Enskog theory [3] , which is based on the following idea: After a few mean free times the distribution $f(\vec{r},\vec{v},t)$ should, as far as its time dependence is concerned, become a functional of the slowly varying hydrodynamic fields which correspond to the constants of the motion: number, momentum and energy.

For the Lorentz model, there is no momentum conservation and energy conservation adds nothing to number conservation. So the only slowly varying field is the number density $n(\vec{r},t) = 4\pi P f(\vec{r},\vec{v},t)$ for which the hydrodynamic equation is the diffusion equation

$$\frac{\partial}{\partial t} n(\vec{r},t) = D\nabla^2 n(\vec{r},t). \tag{2.11}$$

Correspondingly, the Chapman-Enskog ansatz reduces to

$$f(\vec{r},\vec{v},t) \underset{t \gg \tau}{\simeq} f(\vec{r},\vec{v}\,|\,n(\vec{r},t)) . \tag{2.12}$$

In the general non-linear case the analogous ansatz is used to construct a scheme of successive approximations where the state of local equilibrium is used as a zeroth approximation. In the first approximation Navier-Stokes hydrodynamics is obtained, together with integral equations from which the transport coefficients can be determined for a given intermolecular potential.

The fact that we are considering a linear Boltzmann equation simplifies matters. In Fourier-Laplace language, the ansatz (2.12) says that for $z\tau \ll 1$, $f_{\vec{k}z}$ must be a function of $\vec{k},\vec{v}$ and $P f_{\vec{k}z}$, and, in particular, its dependence on $P f_{\vec{k}z}$ must be linear, i.e.

$$f_{\vec{k}z} \underset{z\tau\ll 1}{\simeq} \chi_{\vec{k}}(\vec{v})\, P f_{\vec{k}z}(\vec{v}) . \tag{2.13}$$

On the other hand, we shall now show that the Chapman-Enskog ansatz in the linear case can be reformulated as an assumption on the spectrum of the eigenvalue problem associated with the Boltzmann equation (2.1), namely

$$(B - i\vec{k}\cdot\vec{v})\Psi_{\vec{k}}(\vec{v}) = \Lambda_{\vec{k}}\Psi_{\vec{k}}(\vec{v}). \qquad (2.14)$$

Conservation of the number of particles tells us that in the limit $\vec{k}\rightarrow 0$ there is an eigenfunction $\Psi_0^H(\vec{v}) = 1$ with eigenvalue $\Lambda_0^H = 0$. For small $\vec{k}$ this eigenvalue is small and negative, and the corresponding eigenfunction is called the hydrodynamic mode. The Chapman-Enskog ansatz, in this language, amounts to the assumption that all other eigenfunctions have eigenvalues with negative real parts of the order $-\tau^{-1}$. So for $t \gg \tau$ one can put

$$f_{\vec{k}}(\vec{v},t) \simeq e^{\Lambda_k t} \cdot \left\{\text{projection of } f_{\vec{k}}(\vec{v},0) \text{ on } \Psi^H_{\vec{k}}(\vec{v})\right\}$$

$$= e^{\Lambda_k t}\Psi^H_{\vec{k}}(\vec{v}) \cdot \left\{\text{function of } \vec{k} \text{ only}\right\} \qquad (2.15)$$

Laplace transformation of (2.15) then shows that the assumption made above on the spectrum of $B-i\vec{k}\cdot\vec{v}$ is equivalent to the Chapman-Enskog ansatz (2.13) with $\chi_{\vec{k}}(\vec{v}) = \Psi_{\vec{k}}(\vec{v})$

We should like to construct the projector onto the hydrodynamic mode explicitly, however. And for those of us who learned about the eigenvalue problem in quantum mechanics, it is mildly unsettling that $B - i\vec{k}\cdot\vec{v}$ is not a hermitian operator. It is symmetric but, since B is real, addition of $-i\vec{k}\cdot\vec{v}$ clearly makes it non-hermitian.

So let us make a minor digression to investigate how a projector onto an eigenstate of a non-hermitian operator is to be constructed. Suppose that A is an operator with eigenvalues a_i and "right" eigenstates $|i\rangle$:

$$A|i\rangle = a_i|i\rangle .$$

Correspondingly the adjoint operator $A^\dagger$ has the eigenvalues a_i^* and eigenstates $|i'\rangle$

$$A^\dagger|i'\rangle = a_i^*|i'\rangle$$

or, equivalently

$$\langle i'|A = a_i\langle i'| .$$

In other words, the primed states are the "left" eigenstates of A, and are in general different from the "right" ones. Their orthogonality properties are found in the usual manner

$$\langle i'|A|j\rangle = a_i \langle i'|j\rangle = a_j \langle i'|j\rangle$$

from which it follows that if $a_i \neq a_j$, $|i'\rangle$ is orthogonal to $|j\rangle$. Consequently the coefficients in the expansion of an arbitrary state $|\Psi\rangle$ in terms of the "right" eigenstates

$$|\Psi\rangle = \sum_i c_i |i\rangle$$

are, for the non-degenerate case, given by

$$c_i = \frac{\langle i'|\Psi\rangle}{\langle i'|i\rangle}$$

and the projector onto $|i\rangle$ has the form

$$\Pi_i = \frac{|i\rangle\langle i'|}{\langle i'|i\rangle}$$

Coming back to the Lorentz model, we define the scalar product as

$$\langle\phi|\chi\rangle = \frac{1}{4\pi}\int d\Omega_{\vec{v}}\, \phi^*(\vec{v})\,\chi(\vec{v})$$

$$= P(\phi^*\chi). \tag{2.16}$$

Since the operator $B-i\vec{k}\cdot\vec{v}$ is symmetric, its eigenstates and those of the adjoint operator $B+i\vec{k}\cdot\vec{v}$ are simply the complex conjugates of one another. Thus, the projector $\Pi_{\vec{k}}$, onto the hydrodynamic mode $\Psi_{\vec{k}}(\vec{v})$ (we drop superscript H from now on) is given by

$$\Pi_{\vec{k}}\, f_{\vec{k}}(\vec{v}) = \Psi_{\vec{k}}(\vec{v}) \frac{P(\Psi_{\vec{k}} f_{\vec{k}})}{P(\Psi^2_{\vec{k}})} . \tag{2.17}$$

So far the discussion has been essentially model independent. And very little would change if we turned to the general, linearized Boltzmann equation. It would be necessary to consider five hydrodynamic modes rather than one, but aside from extended book-keeping, more complicated linear cases could be treated in the same manner as above [4].

In the case of the spherical Lorentz model, however, we can do better, and replace suggestive handwaving by hard facts [5, 6]. We can go back to the exact solution (2.10) and examine the spectrum of $\left[\tau^{-1}(P-1)-i\vec{k}\cdot\vec{v}\right]$ in detail. And we can calculate the hydrodynamic mode and its eigenvalue to all orders in k. Let us start with the second task.

2.4. The Hydrodynamic Mode

Operation with P on (2.14), where now $B = \tau^{-1}(P-1)$, yields

$$-P\,(i\vec{k}\cdot\vec{v}\,\psi_{\vec{k}}) = \Lambda_k P\psi_{\vec{k}} = \Lambda_k \tag{2.18}$$

where we have chosen the normalization $P\psi_{\vec{k}} = 1$. By the same token

$$B\psi_{\vec{k}} = \tau^{-1} - \tau^{-1}\psi_{\vec{k}}$$

so that (2.14) can be written

$$\psi_{\vec{k}}\,(\vec{v}) = \frac{\tau^{-1}}{\tau^{-1}+\Lambda_k + i\vec{k}\cdot\vec{v}} \tag{2.19}$$

With $\Lambda_{\vec{k}}$ given by (2.18) this is really a non-linear integral equation for ψ_k, but the trick is to take the spherical average of (2.19) and appeal to (2.9) with the result

$$1 = \frac{1}{kv\tau}\tan^{-1}\frac{kv\tau}{\tau\Lambda_k+1}\quad . \tag{2.20}$$

This equation has a solution for the eigenvalue Λ_k, <u>provided</u> that $kv\tau \leqslant \pi/2$, and it reads

$$\Lambda_k = \tau^{-1}\left[-1 + kv\tau \quad \cot\,(kv\tau)\right] \tag{2.21}$$

The hydrodynamic mode then follows from (2.19) as

$$\psi_{\vec{k}}\,(\vec{v}) = \frac{1}{kv\tau\ \cot\,(kv\tau) + i\vec{k}\cdot\vec{v}\tau}\quad . \tag{2.22}$$

What is the meaning of the eigenvalue Λ_k ? Take the spherical average of (2.15) to get

$$n_{\vec{k}}\,(t) \simeq e^{\Lambda_k t}\cdot\left\{\text{function of } \vec{k}\right\}$$

Consequently, for $t \gg \tau$, $n_{\vec{k}}(t)$ satisfies the differential equation

$$\begin{aligned}\frac{\partial n_{\vec{k}}(t)}{\partial t} &= \Lambda_k\, n_{\vec{k}}(t)\\ &= -\,k^2 D_k\, n_{\vec{k}}(t)\end{aligned} \tag{2.23}$$

where D_k is a perfectly nice, regular diffusion kernel from which the ordinary diffusion constant $D = \frac{1}{3} v^2\tau$ follows, in the limit $k \to 0$. Note that whereas the power series expansion of D_k converges up to $kv\tau = \pi$, the hydrodynamic mode does not exist beyond $kv\tau = \pi/2$. The sweeping conclusion is: Convergence radii shouldn't be taken all that seriously in physics!

Let us finally write down the Hydrodynamic Equation in all its glory by inverting (2.23) with (2.21) to get

$$\left[\frac{\partial}{\partial t} + \frac{1}{\tau} \sum_{p=1}^{\infty} \frac{B_p}{(2p)!} (-4\lambda^2 \nabla^2)^p\right] n(\vec{r},t) = 0 \qquad (2.24)$$

Here $\lambda = v\tau$ is the mean free path, and B_p are the Bernoulli numbers ($B_1 = 1/6$, $B_2 = 1/30$, $B_3 = 1/42$, etc.) in terms of which the infinite sequence of higher order diffusion coefficients is expressed.

2.5. The Contraction

It remains to discuss the contraction from the description in terms of the Boltzmann equation to the simpler hydrodynamic description. To do this we must consider the non-hydrodynamic part of the spectrum of $B - i\vec{k}\cdot\vec{v}$. Or, equivalently, we must study the singularities in the z-plane of the exact solution $f_{\vec{k}z}$ as given by (2.10).

These singularities are seen to be the following

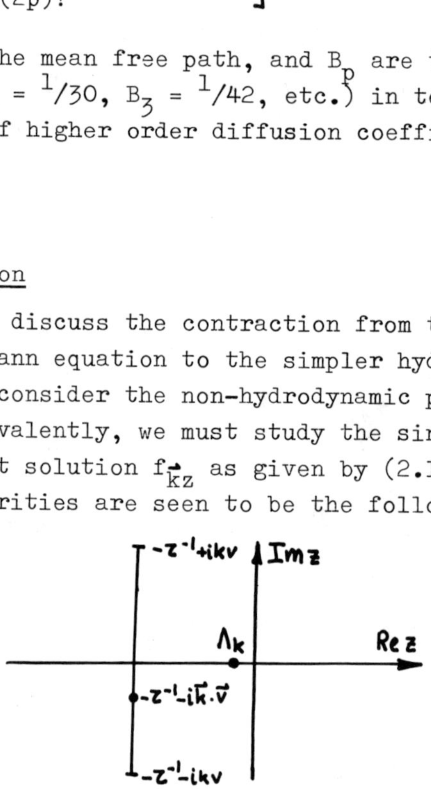

(a) The hydrodynamic pole at $z = \Lambda_k = -k^2 D_k$. It stays on the negative real axis (corresponding to pure exponential decay). For small k it behaves like $-k^2 D$, and it moves to $z = -\tau^{-1}$ as $k \to \pi/2\lambda$. For $k\lambda > \pi/2$ it does not exist.

(b) A pole at $z = -\tau^{-1} - i\vec{k}\cdot\vec{v}$.

(c) A cut from $-\tau^{-1} - ikv$ to $-\tau^{-1} + ikv$ due to the integral

$$P\left(\frac{f_{\vec{k}}(\vec{v},0)}{z + \tau^{-1} + i\vec{k}\cdot\vec{v}}\right) .$$

This cut is the degenerate version of the continuous part of the spectrum generally encountered in connection with the operator $B-i\vec{k}\cdot\vec{v}$.

The contributions from (b) and (c) decay roughly like $\exp(-t/\tau)$ so that when $kv\tau \ll \pi/2$, two time scales are widely separted: The mean free time τ , and the hydrodynamic time (of the order $\tau(\lambda k)^{-2}$) which grows indefinitely as $k \to 0$.

So, on this model, the Chapman-Enskog ansatz is explicitly substantiated. After a few mean free times one is clearly justified in (i) neglecting all modes with $k\lambda > \pi/2$ (spatial coarse graining) and (ii) for $k\lambda < \pi/2$, keeping the hydrodynamic mode only.

A little care is needed in handling the initial value problem, though. It was in fact an old puzzle, that in the Boltzmann equation the entire distribution $f(\vec{r},\vec{v},0)$ is needed as initial data, whereas if one relies on the Chapman-Enskog assumption, only the initial value for the hydrodynamic moment $n(\vec{r},0)$ seems called for. This has sometimes been called the Hilbert paradox .

In the language of the eigenvalue problem, however, this "paradox" is neatly resolved. The information needed to determine the asymptotic decay is clearly <u>not</u> the spherical part $n_{\vec{k}}(0)=4\pi P f_{\vec{k}}(\vec{v},0)$ of the initial data, but the projection $\Pi_{\vec{k}} f_{\vec{k}}(\vec{v},0)$ of $f_{\vec{k}}(\vec{v},0)$ on the slow hydrodynamic mode. A formal definition of $\Pi_{\vec{k}}$ was already given in (2.17) and with the hydrodynamic mode (2.22) it can be written down explicitly

$$\Pi_{\vec{k}} f_{\vec{k}} = \begin{cases} \Psi_{\vec{k}}(\vec{v}) \left(\frac{k\lambda}{\sin k\lambda} \right)^2 P(\Psi_{\vec{k}} f_{\vec{k}}) , & k\lambda \leq \pi/2 \\ 0 , & k\lambda > \pi/2 \end{cases} \qquad (2.25)$$

It follows that the correct initial [7,6] data to be used in the context of the Hydrodynamic Equation (2.23) is not

$$n_{\vec{k}}(0) = 4\pi \, P f_{\vec{k}}(\vec{v},0)$$

but

$$n'_{\vec{k}}(0) = 4\pi \, P(\Pi_{\vec{k}} f_{\vec{k}}(\vec{v},0)) . \qquad (2.26)$$

How big is the difference? For small k, (2.22) and (2.25) show that in general $\Pi_{\vec{k}} = 1 + \mathcal{O}(\vec{k})$. Consequently

$$n_{\vec{k}}(t) = n'_{\vec{k}}(0) \cdot \left[1 + \mathcal{O}(\vec{k})\right] \exp\left[- Dk^2 t - \mathcal{O}(k^4 t)\right] . \qquad (2.27)$$

This means that <u>if</u> one is interested in the time scale where

$Dk^2t = \mathcal{O}(1)$ (surely the most interesting one) and if one uses the ordinary diffusion equation, i.e., if D_k is replaced by D, then $\Pi_{\vec{k}}$ can be legitimately replaced by 1. If one goes beyond the ordinary diffusion equation, however, $\Pi_{\vec{k}} - 1$ can not be neglected, in general.

This conclusion is not peculiar to the Lorentz model, it holds for ordinary fluids also: If one does not go beyond the Navier-Stokes level of hydrodynamics, the actual initial values of the hydrodynamic fields can be used as initial data, otherwise not. This remark has proved relevant in the derivation of the long time tails of the Green-Kubo integrands [8] .

2.6. Final Remarks

As we have seen, the Chapman-Enskog ansatz in the linear case amounts to an assumption on the spectrum of the operator $B-i\vec{k}\cdot\vec{v}$. The beauty of the spherical Lorentz model is that all the general statements in this connection can be backed up by explicit calculation.

In the case of the full linearized Boltzmann equation the amount of information available depends, of course, on the type of interaction assumed. But the literature on the spectral properties of $B-i\vec{k}\cdot\vec{v}$ is growing, and many important results have been obtained [9].

The precise status of the Chapman-Enskog ansatz in the non-linear case is less clear, and remains a challenge!

Finally, a grain of salt. The simple picture we have presented here is a little too good to be true. When higher density effects are included, the separation of time scales will no longer be quite as clean as the Boltzmann equation predicts. The reason is the long time tail [21, 22, 8] in the velocity autocorrelation function. As we shall see, the tail is weaker in the Lorentz model [23] than in a fluid, decaying as $t^{-5/2}$ (in 3 dimensions) rather than $t^{-3/2}$. But it is there.

Nevertheless, although important modifications are necessary (like cuts along the negative real z-axis), and although the status of higher order hydrodynamics becomes questionable, some of the basic wisdom of the Chapman-Enskog theory as presented here may be expected to survive. The borderline problems beyond Navier-Stokes hydrodynamics (or beyond the classical diffusion equation) represent a very active field at the moment!

III. HIGHER DENSITY EFFECTS

3.1. The Einstein Formula for D

We shall now use the Lorentz model to illustrate some of the difficulties encountered when one wants to go beyond the low density regime where the Boltzmann equation is adequate. To simplify things we shall concentrate on the diffusion constant D, but its relation to the generalized kinetic equation will be stressed.

As a starting point, then, we need a general formula for D, valid for all densities n_s of the scatterers. For the other transport coefficients, the shear and bulk viscosities and the heat conductivity, such formulas are known as Green-Kubo formulas. They were derived about 20 years ago [10], and although few people doubt the results, the derivations contain subtleties that should not be passed over lightly [11].

The "Green-Kubo" formula for the diffusion constant, however, goes back to Einstein [12] , and its derivation is as transparent as one could wish. Define the mean square displacement as $\Delta(t) = \langle [\vec{r}(t) - \vec{r}(0)]^2 \rangle$. According to the diffusion equation one has asymptotically

$$\Delta(t) \simeq 2d\, D\, t \tag{3.1}$$

where d is the space dimensionality. On the other hand, however,

$$\frac{d\Delta}{dt} = 2\langle \vec{v}(t) \cdot [\vec{r}(t) - \vec{r}(0)] \rangle$$
$$= 2\int_0^t dt' \langle \vec{v}(t) \cdot \vec{v}(t') \rangle .$$

Since the equilibrium average $\langle\ \rangle$ depends only on the difference of the time arguments, one can write with $\tau = t - t'$

$$\frac{d\Delta}{dt} = 2\int_0^t d\tau \langle \vec{v}(0) \cdot \vec{v}(\tau) \rangle \tag{3.2}$$

and comparing this with the result (3.1) from the diffusion equation one finds that D, if it exists, must be given by

$$D = \frac{1}{d} \lim_{t\to\infty} \int_0^t d\tau \langle \vec{v}(0) \cdot \vec{v}(\tau) \rangle \quad .$$

From the derivation it is immediately clear that if the particle is enclosed in a <u>finite</u> box, $\Delta(t)$ cannot grow beyond finite bounds and D, as defined above, vanishes. It is therefore tacitly assumed that one has passed to the limit of infinite systems <u>before</u> taking the limit $t\to\infty$. The necessity of passing to the limits in this order is a general feature of <u>all</u> Green-Kubo formulas. With this in mind we simply write D as the time integral of the velocity autocorrelation function

$$D = \frac{1}{d} \int_0^\infty dt \langle \vec{v}(0) \cdot \vec{v}(t) \rangle \quad . \tag{3.3}$$

3.2. <u>D from the Kinetic Equation</u>

Next we shall show the connection between D and the generalized kinetic equation. Since the Lorentz model is linear, such an equation, if it exists at all, must be of the form

$$\left(\frac{\partial}{\partial t} + \vec{v} \cdot \nabla\right) f(\vec{r},\vec{v},t) = \int_{-\infty}^{t} dt' \int d\vec{r}'\, K(\vec{r}-\vec{r}', t-t')\, f(\vec{r}',\vec{v},t') \tag{3.4}$$

where $K(\vec{r},t)$ is an operator acting on (the direction of) the velocity $\vec{v}$, and depending on the parameters $\vec{r}$ and t. In general, then, the collision operator is expected to be non-Markovian and non-local, in contrast to the Boltzmann operator which is Markovian and local, i.e., in the notation of (3.4), contains the factor $\delta(\vec{r}-\vec{r}')\,\delta(t-t')$.

(3.3) shows that, to compute D, all we need is the average of $\vec{v}(t)$, conditional on $\vec{v}(0) = v_0 \cdot \mathbf{1}$, i.e. we can restrict ourselves to the spatially homogeneous case and study

$$\frac{\partial f(\vec{v},t)}{\partial t} = \int_0^t dt'\; K(t-t')\, f(\vec{v},t') \tag{3.5}$$

with $f(\vec{v},0) = \delta(\vec{v}-\vec{v}_0)$.

Introducing Laplace transforms one can write the formal solution of (3.5) as

$$f_z\,(\vec{v}) = (z - K_z)^{-1}\;\delta(\vec{v}-\vec{v}_o) \tag{3.6}$$

and the diffusion constant follows as

$$D = \lim_{z\to o}\; d^{-1}\int d\vec{v}_o\vec{v}_o \cdot \int d\vec{v}\;\vec{v}\;(z-K_z)^{-1}\;\delta(\vec{v}-\vec{v}_o)\Big/\int d\vec{v}_o \tag{3.7}$$

Here it is again understood that $|\vec{v}| = v$ is a fixed parameter and both the integration over $\vec{v}$ and the δ-function refer to the space of possible <u>directions</u> of $\vec{v}$. (This space can be continuous or discrete, as we shall see.)

We now perform the integration over $\vec{v}_o$ in (3.7) and note that the resulting integrand is independent of the (allowed) direction of $\vec{v}$. As a consequence (3.7) can be written as

$$D = \lim_{z\to o}\; d^{-1}\;\vec{v}\cdot\,(z-K_z)^{-1}\;\vec{v}\,.$$

Now, K_z is a scalar operator, so $K_z\vec{v}$ must be a vector (possibly negative) in the direction of $\vec{v}$. Furthermore, the only eigenfunction of K_z with vanishing eigenvalue is 1, with is orthogonal to $\vec{v}$. (Again a consequence of number conservation). So there is no problem with the inverse and we get

$$D^{-1} = \lim_{z\to o}\; dv^{-4}\;\vec{v}\cdot\,(z-K_z)\vec{v}$$

$$= -\,dv^{-4}\;\vec{v}\cdot K_o\vec{v} \tag{3.8}$$

Let us check this formula for the three dimensional spherical Lorentz model in the Boltzmann limit where $K_o = \tau^{-1}(P-1)$. We find

$$D^{-1} = \frac{-3}{v^4}\;\vec{v}\cdot\left(-\frac{1}{\tau}\,\vec{v}\right) = \frac{3}{v^2\tau}$$

as before.

3.3. Cluster Expansions

Very little has been achieved so far. All we have done is to establish the connection between the operator K(t), in a generalized kinetic equation, and the diffusion constant. What has to be done next is to think of some approximation scheme in which we can calculate K_o in successive orders. (And if we are able to calculate K_o, the chances are that we can cope with K_z and its generalization to the spatially inhomogeneous case also).

Since the Boltzmann equation has the status of a low density law in non-equilibrium statistical mechanics, similar to that of the ideal gas in equilibrium, it is tempting to try some sort of virial expansion of K_o, i.e., of D^{-1}. The Boltzmann term, of $\mathcal{O}(n)$ (from now on we drop the subscript s on the density of scatterers), is determined by the collision of the moving particle with one isolated scatterer in infinite space. In analogy with equilibrium cluster expansions one would expect the term of $\mathcal{O}(n^{\ell})$ to be determined by the collisions of the moving particle with ℓ scatterers in infinite space [13] .

There is one immediate source of complications, however. In equilibrium, correlations are weighted by combinations of Boltzmann factors with the interaction in the exponent. In addition to these "statistical" correlations, non-equilibrium theories must cope with "dynamical" correlations due to the "memory" of the moving particles. And this memory usually extends over length scales much larger than the equilibrium correlation length. As a consequence, outside the Boltzmann regime, the dynamical correlations are the ones that cause most of the difficulties.

We shall not go into the formalism which develops the collision kernel K(t) from first principles, with due care taken in dealing with both sources of correlations [14 - 16] . It would simply mean too much hard work under the circumstances. In fact, since the most important results have a certain intuitive appeal, we shall be content with basing the rest of the discussion on those results, stated without proof.

Neglecting for the moment the "statistical" correlations, we can formulate the outcome of a cluster expansion of the operator K_o in terms of the following diagrammatic rules:

(a) Construct an irreducible event where the moving particle collides with ℓ scatterers in infinite space located at $\vec{Q}_1, \ldots \vec{Q}_\ell$. An event is irreducible if its trajectory has no point between the first and the last collision such that if cut at that point, the ℓ-cluster falls apart.

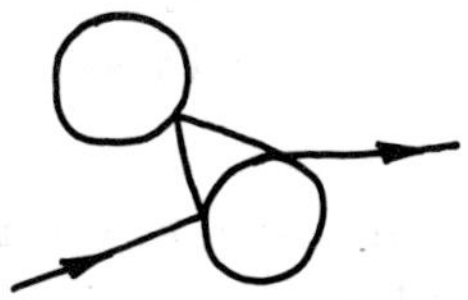

irreducible event

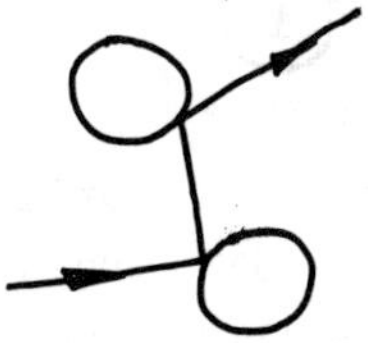

reducible event

(b) Collisions can be of two types: Real collisions governed by the laws of classical mechanics; and virtual collisions, where the particle moves through the scatterer as if it were not there.

(c) The contribution of an irreducible event, with ℓ scatterers and a given collision sequence containing m virtual collisions, to $v^{-3}\ \vec{v} \cdot K_o\ \vec{v}$ is the following:

$$n^{\ell}\ (-1)^m\ \hat{v}_i \cdot \hat{v}_f\ \text{(phase integral)} \tag{3.9}$$

where $\hat{v}_i$, $\hat{v}_f$ are unit vectors in the initial and final directions of the velocity, and the phase integral is obtained by fixing scatterer nº 1 at the origin, integrating over the collision "cylinder" of the initial collision, and integrating over the positions $\vec{Q}_2 \ldots \vec{Q}_\ell$ of the remaining scatterers, with the constraints imposed by the given sequence of real and virtual collisions.

(d) To find $v^{-3}\vec{v} \cdot K_o \vec{v}$ and thus D^{-1}, sum over all possible irreducible events and add them to the Boltzmann term.

3.4. Divergences

It will be instructive to include, in the following discussion, three different Lorentz models: the 2- and 3-dimensional ones with (circular) spherical scatterers and the 2-dimensional wind-tree model, introduced by Ehrenfest, in which the scatterers are identically oriented squares and the particles are allowed to move in the 4 directions parallel to the diagonals only. ("Tree"=scatterer, and "wind"=moving particle, in Ehrenfest's terminology).

Let us first look at a characteristic term with two scatterers, i.e. the recollision event

The most important factor in the contributions from these events to D^{-1}, is the phase integral. In the circular and spherical models (d=2,3) the contributions from large separation ℓ of the two scatterers are estimated to be

$$\lim_{L\to\infty} a^{d-1}\int^{L} d\ell \;(a\theta)^{d-1} \sim \lim_{L\to\infty} a^{3d-3}\int^{L}\frac{d\ell}{\ell^{d-1}}$$

$$\sim \begin{cases} a^3 \lim\limits_{L\to\infty} \ln L & ; d = 2 \\ a^5 \cdot \text{const} & ; d = 3 \end{cases} \tag{3.10}$$

This looks pretty bad. Although the integral remains finite in 3 dimensions, it diverges logarithmically in 2. Of course, (3.10) is just an estimate of one term out of a sum of terms all of $\mathcal{O}(n^2)$. So at this point one could still hope that the diverging terms would cooperate to destroy each other. But they won't. The simplicity of the Lorentz models makes a direct calculation feasible and the conclusion is that the divergence remains [14] .

How about Ehrenfest's wind-tree model? No divergence occurs here, since the geometry of the scatterers ensures that scatterer nº 2 stays close to nº 1. But that is not the end of the story. Look at a typical term of $\mathcal{O}(n^3)$, i.e., with 3 scatterers

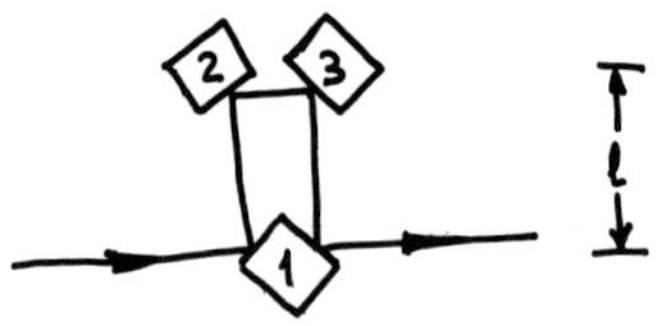

An estimate gives

$$\lim_{L\to\infty} a^4 \int^{L} d\ell \sim a^4 \lim_{L\to\infty} L \quad . \tag{3.11}$$

Its divergence is even stronger! Altogether, such estimates yield

d	shape	$\mathcal{O}(n^1)$	$\mathcal{O}(n^2)$	$\mathcal{O}(n^3)$	$\mathcal{O}(n^s)$
2	◇	finite	finite	L	L^{s-2}
2	○	-"-	ℓnL	L	L^{s-2}
3	○	-"-	finite	ℓnL	L^{s-3}

This divergence difficulty is <u>not</u> a pathological feature of Lorentz models, but is in general a stumbling block when standard cluster expansion methods are used in non-equilibrium problems [17]. In fact, it serves as an excellent example of the usefulness of Lorentz models. As you have just seen, the divergence difficulty immediately presents itself! In the general case, however, when all particles move, the formalisms were so impenetrable that almost 20 years had elapsed after Bogoliubov proposed his formal expansion scheme [18] before the difficulties were definitely uncovered [17]. The moral, particularly pertinent in non-equilibrium statistical mechanics, is clear: Don't believe in a general scheme until it has been successfully tested on reasonable models!

3.5. What to do about them

In this case the Lorentz model is not only useful in diagnosing the ills, it also points to the cure. What is wrong with the cluster expansions is clearly that events with long straight trajectories are being overemphasized. A straight segment of the path should be weighted with the probability $\exp(-\ell/\lambda)$ that it stays unbroken over a length ℓ. To lowest order in the density the mean free path is $\lambda = (n\sigma)^{-1}$ where σ is the total cross section of the scatterers.

But such a damping on the straight trajectories changes the picture completely! Take first the logarithmic divergence (3.10) in the circular, 2-dimensional model (with $\sigma = 2a$):

$$\lim_{L \to \infty} n^2 a^3 \int^L d\ell \, \frac{e^{-2an\ell}}{\ell} \underset{(na^2 \ll 1)}{\sim} n^2 a^3 \ell n (na^2) + \dots \quad (3.12)$$

Or the linear divergence in the terms with 3 scatterers

$$\lim_{L\to\infty} n^3 a^4 \int^{L} d\ell \; e^{-2an\ell} \sim \frac{1}{2} n^2 a^3 + \dots \tag{3.13}$$

The damping on the straight segments cures the divergence, but it also decreases the order in n! It is easy to see that all of the most diverging pieces of every formal order ($\sim n^s L^{s-2}$) are renormalized to finite contributions of $\mathcal{O}(n^2)$ in 2 dimensions, i.e. they contribute to the first correction to the Boltzmann result. In 3 dimensions the difficulties occur one formal order higher, and thus all the most divergent pieces contribute to $\mathcal{O}(n^3)$.

It remains to classify (and calculate!) the "most divergent diagrams". The most important class is that of the ring diagrams [19] with collision sequences of the type 1 2 3 ... ℓ 1. This is the only class of most divergent diagrams in a fluid, but it turns out that in the Lorentz models there can be additional ones. In the wind-tree model these additional diagrams can even be responsible for a qualitative change in the diffusion process! We shall come back to that later.

At this stage a comment should be made on the logarithmic term. It is clearly not of fundamental importance here. The divergence difficulty is a direct consequence of the cluster expansion method, which is fine with short range correlations, but which fails in our case due to the long range dynamical correlations. The existence of the logarithm, however, is due to a small subclass of diverging terms [14] and depends on the differential cross section being a smoothly varying function of the scattering angle, in particular at back scattering. Thus, in the wind-tree model it is absent.

Another illustration of the usefulness of Lorentz models is related to the previous point. After the discovery of the divergences, it was repeatedly claimed that they would disappear if the problem was treated by quantum, rather than classical, mechanics. The point was settled by Résibois and Velarde [20] , and the result is intuitively appealing: They showed on the 2-dimensional Lorentz model that the logarithmic divergence remains if the differential cross section stays finite for back scattering. Thus, if a quantum treatment based on cluster expansions is free of divergences, it must be due, either to additional approximations used (such as the Born approximation), or to some very special feature of the model treated.

Before we proceed, let me stress again that the handwaving discussion I have given here can be, and has been, backed up by formal arguments, and to a large extent, by explicit calculations [14 - 16] . Very little has been proved rigorously, however. Moreover, it seems quite difficult to do so along the lines followed here, since the expansions used produce immediately series that are not uniformly convergent. And since infinite resummations are indispensible, tact is required. Rigor would therefore be most welcome. There is little doubt, however, that the results obtained are correct as far as they go.

3.6. Tails

Recently the existence of long time tails in the Green-Kubo integrands has received much attention. During their computer studies on the hard sphere fluid, Alder and Wainwright [21] first discovered that the velocity autocorrelation function of a tagged particle decays like $t^{-d/2}$ in d dimensions. Many derivations of this (and related results) have appeared since [21, 8, 22]. The one of relevance here is that of Dorfman and Cohen [22] who obtained the tails from kinetic theory, in particular from the sum of ring diagrams.

But in the Lorentz model, as we have seen, the ring diagrams play the same role of (the most important class of) "most divergent diagrams" as in a fluid, so the immediate question is: Does the velocity autocorrelation function in the Lorentz model have a long time tail? The answer, which is "yes, but", was given by Ernst and Weijland [23] . We shall reproduce their derivation here since it contains, in a simplified form, most of the arguments used in the case of a fluid.

Let us take the spherical Lorentz model in 3 dimensions and study the small z behavior of the ring contribution to the Laplace transform C_z of the velocity autocorrelation function $\langle \vec{v}(0)\cdot\vec{v}(t)\rangle$.

Comparing with (3.3) and mildly generalizing (3.7) and (3.8) one easily finds, to lowest order in the density, that the ring contribution to C_z is

$$C_z^R \simeq \frac{9D^2}{v^2} \; \hat{v} \cdot K_z^R \, \hat{v} \tag{3.14}$$

where $D = (1/3)v^2\tau$ is the Boltzmann diffusion constant. The sum over all rings can be put in the form (to leading order in the density)

$$\hat{v} \circ K_z^R \, \hat{v} = \frac{1}{n}\int \frac{d^3k}{(2\pi)^3} \; \hat{v} \cdot B \, G_{\vec{k}z} \, B \, \hat{v} \tag{3.15}$$

Here $B = \tau^{-1}(P-1)$ is the Boltzmann operator and $G_{\vec{k}z}$ is the corresponding propagator

$$G_{\vec{k}z} = (z + i\,\vec{k}\cdot\vec{v} - B)^{-1} \quad . \tag{3.16}$$

The two Boltzmann operators in (3.15) clearly describe the two collisions with scatterer nº 1, but how did the propagator $G_{\vec{k}z}$ get into it? The point is that all intermediate collisions in a ring event are with scatterers that are hit only once. i.e., they are uncorrelated, and the intermediate excursion is consequently described by the Boltzmann equation. Since, furthermore, the particle must return to (roughly)where it started, we need the $\vec{k}$-integral over the propagator $G_{\vec{k}z}$.

Since $B\hat{v} = -\tau^{-1}\hat{v}$ and since, similarly, the $\vec{k}$-integral over $G_{\vec{k}z}\hat{v}$ must be a vector along $\hat{v}$, (3.15) immediately reduces to

$$\hat{v}\cdot K_z^R \hat{v} = \frac{1}{n\tau^2}\int \frac{d^3k}{(2\pi)^3}\quad \hat{v}\cdot G_{\vec{k}z}\,\hat{v} \quad . \tag{3.17}$$

We are interested in the small z, long time, behavior and we know that asymptotically everything is dominated by the hydrodynamic mode. So just like in (2.15) we write [24] with (2.17)

$$G_{\vec{k}z}\,\hat{v} \simeq \frac{1}{z-\Lambda_k}\,\Pi_{\vec{k}}\,\hat{v}$$

$$= \frac{1}{z-\Lambda_k}\quad \Psi_{\vec{k}}(\hat{v})\;\frac{P(\Psi_{\vec{k}}\hat{v})}{P(\Psi^2_{\vec{k}})} \tag{3.18}$$

If we now operate on (3.17) with P (which makes no difference since it is a number) and introduce the normalized mode

$$\overline{\Psi}_{\vec{k}} = \Psi_{\vec{k}}\left[P(\Psi^2_{\vec{k}})\right]^{-1/2} \tag{3.19}$$

(3.17) can, with (3.18), be written in the suggestive form

$$\hat{v}\cdot K_z^R\,\hat{v} \simeq \frac{1}{n\tau^2}\int \frac{d^3\,k}{(2\pi)^3}\;\frac{\langle \hat{v}|\overline{\Psi}_{\vec{k}}\rangle^2}{z-\Lambda_k} \tag{3.20}$$

where the scalar product $\langle \hat{v}|\Psi\rangle = P(\hat{v}\Psi)$ as in (2.16).

This is as close as we can get to a mode-mode formula [25] in the Lorentz model. The important difference, of course, is that in the Lorentz model we have only a single hydrodynamic mode to play with, whereas in a fluid there are five. Associated with this is the fact that

$$\langle \hat{v} | \overline{\Psi}_{\vec{k}} \rangle^2 = \left[P\,(\hat{v}\,\overline{\Psi}_{\vec{k}}) \right]^2 = -\left(\frac{\tau \Lambda_k}{\sin k\lambda} \right)^2 \qquad (3.21)$$

vanishes as k^2 in the limit $k \to o$. In the fluid, however, the "currents" in the Green-Kubo formulas are not orthogonal to all bilinear combinations of modes, even in the $k \to o$ limit.

Using (3.14) and (3.20), and inverting the Laplace transform, we find

$$\langle \vec{v}(0) \cdot \vec{v}(t) \rangle \simeq \frac{v^2}{n} \int \frac{d^3k}{(2\pi)^3} \langle \hat{v} | \overline{\Psi}_{\vec{k}} \rangle^2 \, e^{\Lambda_k t} \; . \qquad (3.22)$$

The asymptotics is determined by the small k behavior of the integrand, and keeping only the leading terms we get, with (3.21) and (2.21)

$$\langle \vec{v}(o) \cdot \vec{v}(t) \rangle \simeq -\frac{D^2}{n} \int \frac{d^3k}{(2\pi)^3} \, k^2 \, e^{-k^2 Dt}$$

$$= -\frac{6\pi D^2}{n(4\pi Dt)^{5/2}} \; . \qquad (3.23)$$

So there is a tail, even in the Lorentz model. But the extra factor k^2 from (3.21) reduces it from $t^{-3/2}$ to $t^{-5/2}$. Generalization to d dimensions immediately yields [23] $t^{-(d/2+1)}$. So even for $d = 2$, the velocity autocorrelation function is integrable! There is an additional qualitative difference from the asymptotics in a fluid: The Lorentz tail is negative. But the similarities outweigh the differences. In both cases the ring diagrams are at the heart of the mode (-mode) formula, and in both cases the "Green-Kubo" integrands decay like powers rather than exponentially.

3.7. Abnormal Diffusion

We now turn to some special features of the wind-tree model [16]. In addition to the ring events this model has another important class of "most divergent diagrams" that has to do with the following:

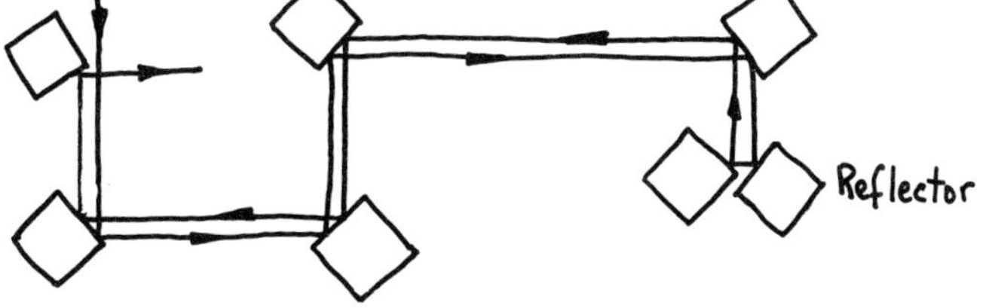

Two trees very close together can form a reflector such that the moving particle starts retracing its earlier steps. The narrower the double path created by the reflector, the longer it is likely to become, since the only way of splitting it again is by squeezing in a corner of a tree.

Clearly events of this sort will tend to slow down the diffusion process. The question is: Will they change it qualitatively? The answer turns out to depend on what is meant by a "random" distribution of scatterers. Or put differently: It depends on the "statistical correlations" which we neglected earlier.

The crucial point is how easily reflectors are generated. The two trees in a reflector are bound to stay close together and thus their mutual interaction becomes important. If they are distributed as hard squares, i.e., if they are non-overlapping, it becomes increasingly difficult to have them act as reflectors, the narrower (and thus, longer) the double paths one considers. The result is that for non-overlapping trees, these events don't cause qualitative changes, they just reduce the diffusion constant somewhat.

However, if the trees are allowed to overlap freely, reflectors for arbitrary width of the double path are easily generated, and it turns out that the mean square displacement no longer grows linearly with t, but (for small densities, $na^2 \ll 1$) behaves like [26] .

$$\Delta(t) \simeq 4\, Dt \cdot (t/\tau)^{-(4/3)\, na^2} . \tag{3.24}$$

Here D and τ are the diffusion constant and mean free time obtained from the Boltzmann equation. Thus, the diffusion process is qualitatively slowed down by the retracing events in the overlapping case.

In the formalism this is reflected by the "most divergent diagrams", associated with the retracing events, summing up to infinity [16] , even after each event has been renormalized by the mean free path cutoff as in (3.12-13). This divergence doesn't reflect a weakness in the formalism. On the contrary, from (3.2) and (3.24) it is evident that the diffusion constant does not exist in this case.

3.8. Computer Work

Finally, let me briefly mention that the wind-tree model has been extensively studied by Wood and Lado [27] . They made molecular dynamics calculations with a "forest" of 8192 trees, followed the trajectories up to 25.000 collision times, and averaged over a huge number of configurations and trajectories. The qualitative difference between the non-overlapping and the overlapping case is strikingly demonstrated by their results. They also verified the asymptotic law (3.24) to high accuracy.

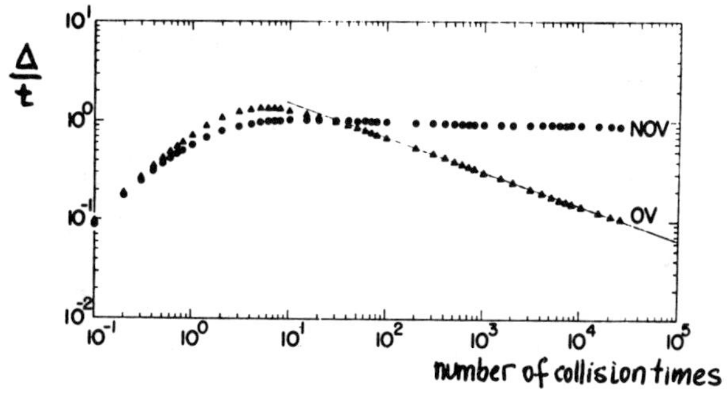

Recently, Bruin [28] published results of a similar study of the 2-dimensional Lorentz model with circular scatterers. The existence of a logarithmic term in the density dependence of the diffusion constant is here verified" experimentally" for the first time. The t^{-2} tail of the velocity autocorrelation function cannot be verified by his results,however. Whether this is due to the fact that the computer results become shaky beyond 8 collision times is hard to say. We should not forget that the $t^{-(d/2+1)}$-tail was calculated by concentrating on the ring diagrams. We did not prove that those are the only important events for the asymptotics, although the evidence from the fluid supports this conjecture!

IV. RIGOROUS RESULTS

4.1. The Grad Limit

The heading looks impressive. But the truth is that I shall present very few rigorous results, 29 and shall prove none. This is really a salesman's talk: My thesis is that Lorentz models are so simple that it should be possible to do much more rigorous work on them than I, at least, am aware has been done. On the other hand, Lorentz models are sufficiently complex. There are truly non-trivial statements to shoot at!

A valuable start along these lines has been made by Gallavotti [30] and one of the problems he considered was that of proving the Boltzmann equation in the Grad limit. Grad first raised the question of whether there is a limit in which the Boltzmann equation is exact. Letting $n \to o$ with the interactions fixed is not a very attractive possibility, since everything goes to zero or to infinity in that limit. So Grad instead suggested the following limit [31] : Keep the mean free path constant, i.e. fix the combination na^{d-1} where a is the range of the interaction. With na^{d-1}= const., let $n \to \infty$ and $a \to o$. Having the number density grow indefinitely looks ominous, but remember that it is the dimensionless density, $\rho = na^d$, that counts. And $\rho \to o$ in the Grad limit!

Since the model is linear, it is sufficient to study the Green's function $G(\vec{r},\vec{v},t|\vec{v}_o)$ defined as the probability that a moving particle has the phase $(\vec{r},\vec{v})$ at time t, given that the phase was $(o,\vec{v})$ at $t=o$. One has to say precisely how the scatterers are distributed (for example: Are they allowed to overlap or not?), but details of this sort are expected to become irrelevant in the limit.

As Gallavotti [30] has suggested, the way to proceed is to sum over all paths that lead from $(0,\vec{v}_o)$ to $(\vec{r},\vec{v})$ in a time t. What has to be proved is that the only paths that survive in the Grad limit, are those that don't intersect themselves, and that don't contain more than one collision with any given scatterer. Thereby the Stosszahlansatz is proved, and the Boltzmann equation for $G(\vec{r},\vec{v},t \mid \vec{v}_o)$ is an immediate consequence for $t > 0$.

It is easily seen that any given "non-Boltzmann" event does

indeed get a vanishing weight in the Grad limit, but I think it remains to be proved that also the sum of such events becomes negligible. There is no doubt what the result is, but as a warming up exercise I suggest that you fill in the gaps in the proof!

In so doing you should look for the necessary and sufficient condition on the interactions for the theorem to hold. It is not true for all interactions: Consider the following type [32] of oriented, hard scatterers:

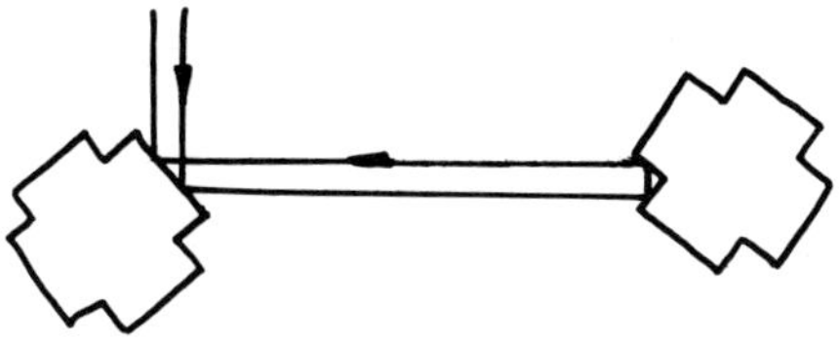

Such a weird shape would not give the Boltzmann equation, even in the Grad limit!

At this point it is natural to raise the old objection: How can it be possible to derive the irreversible Boltzmann equation from reversible mechanics, with or without a limiting procedure? Well, $t=0$ is a very special time here, namely the time when we average over all possible configurations of the scatterers. So you shouldn't expect time reversal symmetry except at $t=0$. And at $t=0$ there is symmetry. For $t<0$ the Green's function obeys the Boltzmann equation with $t \to -t$!

Away from the Grad limit life becomes more complicated, but also more interesting. I shall not make a long list of unproven statements about Lorentz models with higher density effects. Some have already been made, and as an example, try the following: Prove that the diffusion constant exists under suitable restrictions. Or more ambitiously: Prove an existence criterion that distinguishes between the non-overlapping and overlapping wind-tree models.

4.2. Percolation Problems

Let me finally mention that for very high density, the Lorentz models give rise to interesting dynamical percolation problems [16, 33]. As an example, take the wind-tree model where the trees can freely overlap. Then there is a finite probability that the moving particle is trapped in a finite volume. The simplest way of trapping it is by using 4 trees

So for small densities the probability, P_T, of being trapped is of $\mathcal{O}(n^4)$.

What happens as the density is increased? In analogy with the percolation problem on lattices, it is reasonable to guess [16] at something like this

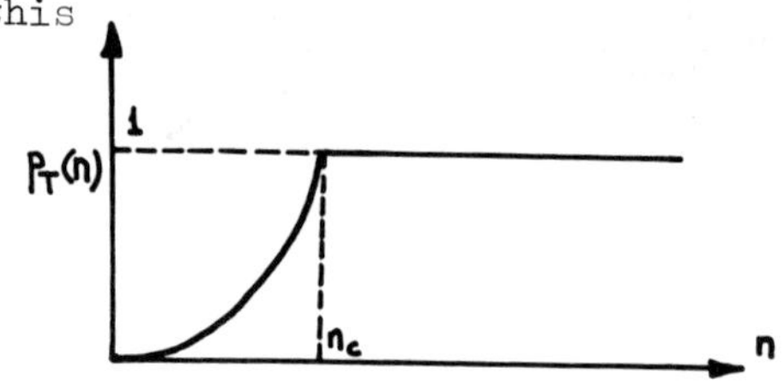

i.e., beyond the critical density n_c, the particle is surely trapped in a finite volume. The fact that $p_T(n)=1$ beyond a certain finite n has actually been proved. It is not even difficult, if one appeals to well known theorems on the site and the bond problems in percolation theory on a square lattice. In this way the following upper bound on the critical density has been established [34]

$$n_c\, a^2 < 2 \ln 2 \quad . \tag{4.1}$$

It is much more difficult to prove that $n_c > \text{const} > 0$, i.e., to prove that for small enough densities, the particle does <u>not</u> have to be trapped. The difficult part of the proof is related to this: Even if there is a hole in the "box" where the particle moves around, how do we know that the particle <u>finds</u> it, and gets out?

Clearly this problem belongs in the category of <u>ergodic</u> problems [29] , and it can be formulated like this. Construct an arbitrary box with straight lines and right angles.

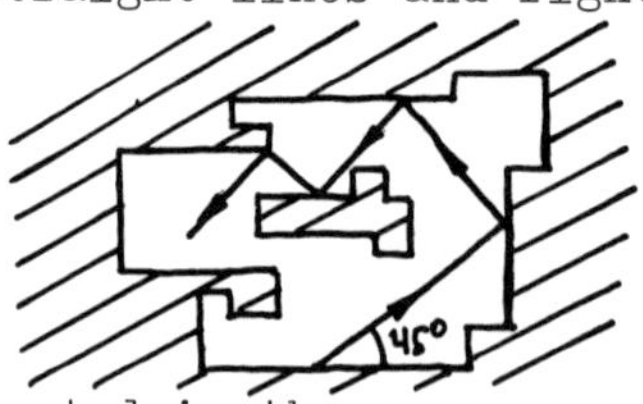

Allow islands constructed in the same way. Let a point particle start anywhere, and in a direction 45° to the sides. Prove that the trajectory is <u>dense</u>, i.e., prove that the particle sooner or later comes arbitrarily close to any point in the box.

If the lengths of the sides of the box are rationally connected, i.e., if there are relations of the type

$$\sum_{i=1}^{N} m_i \ell_i = 0 \quad , \qquad (4.2)$$

where ℓ_i is the length of the i'th side, and m_i is an integer, it is easy to see that the statement need not be true.

On the other hand, under the assumption that there are no relations of the type (4.2), the theorem has recently been proved by J.F. Aarnes [35] . On the basis of this theorem, the construction of a positive lower bound on n_c is comparatively straight forward [34] .

Needless to say, all this tells us nothing about the nature of the singularity at n_c!

4.3. Absolutely Final Remarks

In these lectures we have used Lorentz models as a testing ground for certain problems in non-equilibrium statistical mechanics. Some of the problems treated, like the relation between kinetic theory and hydrodynamics, are of obvious relevance in a more general context. Others, like the special properties of the wind-tree model with overlapping scatterers, are hardly of immediate concern to experimentalists. In fact, one of the really hard problems facing the model fan is: When to stop?

Before I come to a halt, let me nevertheless stress the obvious once more: All one can learn from Lorentz models has to do with linear effects. To learn about non-linear aspects of non-equilibrium problems, one needs a simple, non-linear and in some sense solvable model. This may sound self-contradictory. But the fact is that progress has been made in this direction lately [36] . The title of that story would be: "What can one learn from Pomeau models?"

REFERENCES

1) K.M. CASE and P.F. ZWEIFEL, Linear Transport Theory, (Addison-Wesley), Reading, Mass. (1967), Chap. 7

2) See R. PEIERLS, this volume; R. KUBO, this volume; W. KOHN and J.M. LUTTINGER, Phys. Rev. 108, 590 (1957).

3) S. CHAPMAN and J.G. COWLING, Mathematical Theory of Non-Uniform Gases, (Cambridge University Press), London, 3rd ed. (1970).

4) M. BIXON, J.R. DORFMAN and K.C. MO, Phys. Fluids 14, 1049 (1971) P. RESIBOIS, J. Stat. Phys. 2, 21 (1970), and in Irreversibility in the Many-Body Problem, J. Biel and J. Rae, eds. (Plenum Press) New York, (1972).

5) H.P. McKEAN Jr., J. Math. Phys. 8, 547 (1967)

6) E.H. HAUGE, Phys. Fluids, 13, 1201 (1970)

7) H. GRAD, Phys. Fluids, 8, 147 (1963).

8) M.H. ERNST, E.H. HAUGE and J.M.J. VAN LEEUWEN, Phys. Rev. Letters 25, 1254 (1970); Phys. Rev. A4, 2055 (1971).

9) See, for example, J.A. McLENNAN, Phys. Fluids 8, 1580 (1965). I. KUSCER and M.M.R. WILLIAMS, Phys. Fluids 10, 1922 (1967). Ø.O. JENSSEN, Phys. Norvegica 6, 179 (1972).

10) M.S. GREEN, J.Chem.Phys. 20, 1281 (1952); 22, 398 (1954). R. KUBO, J. Phys. Soc. Japan, 12, 570 (1957); R. KUBO, this volume.

11) N.G. VAN KAMPEN, Phys. Norvegica 5, 279 (1971).

12) A. EINSTEIN, Ann. Phys. 17, 549 (1905).

13) E.G.D. COHEN, Physica 28, 1025; 1045; 1061 (1962).

14) J.M.J. VAN LEEUWEN and A. WEIJLAND, Physica 36, 457 (1967); 38, 35 (1968).

15) W. HOEGY, thesis, University of Michigan, Ann Arbor, Mich. (1967).

16) E.H. HAUGE and E.G.D. COHEN, "Det Fysiske Seminar i Trondheim" № 7, (1968); J. Math. Phys. 10, 397 (1969)

17) J. WEINSTOCK, Phys. Rev. 132, 454 (1963); 140A, 460 (1965). J.R. DORFMAN and E.G.D. COHEN, Phys. Letters 16, 124 (1965), J. Math. Phys. 8, 282 (1967). R. GOLDMAN and E.A. FRIEMAN, Bull. Amer. Phys. Soc., 10, 531 (1965); J. Math.Phys. 1, 2153 (1966); 8, 1410 (1967).

18) N.N. BOGOLIUBOV in Studies in Statistical Mechanics, vol. I, J. de Boer and G.E. Uhlenbeck, eds., (North-Holland), Amsterdam (1962).

19) K. KAWASAKI and I. OPPENHEIM, Phys. Rev. 139, A 1763 (1965).

20) P. RESIBOIS and M.G. VELARDE, Physica 51, 541 (1971).

21) B.J. ALDER and T.E. WAINWRIGHT, Phys. Rev. A1, 18 (1970)

22) J.R. DORFMAN and E.G.D. COHEN, Phys. Rev. Letters 25, 1257(1970); Phys. Rev. A6, 776 (1972)

23) M.H. ERNST and A. WEIJLAND, Phys. Letters 34A, 39 (1971).

24) Y. POMEAU, Phys. Rev. A3, 1174 (1971); J. Math. Phys. 12, 2286 (1971)

25) L.P. KADANOFF and J. SWIFT, Phys. Rev. 166, 89 (1968). K. KAWASAKI, Ann. Phys. (New York), 61, 1 (1970). R.A. FERRELL, Phys. Rev. Lett. 24, 1169 (1970).

26) H. VAN BEYEREN and E.H. HAUGE, Phys. Letters 39A, 397 (1972).

27) W.W. WOOD and F. LADO, J. Comp. Phys. 7, 528 (1971).

28) C. BRUIN, Phys. Rev. Letters 29, 1670 (1972).

29) Some important results on the ergodic properties of Lorentz models with convex scatterers (Example: Spheres. Counterexample: Square trees) are presented by J.L. Lebowitz in this volume.

30) G. GALLAVOTTI, Phys. Rev. 185, 308 (1969).

31) H. GRAD, in Handbuch der Physik, S. Flügge, ed. (Springer-Verlag), Berlin (1958), Vol. XII, p. 214.

32) H. VAN BEYEREN, private communication.

33) D.J. GATES, J.Math. Phys. 13, 1005 (1972); 13,1315 (1972).

34) E.H. LIEB and E.H. HAUGE, to be published.

35) J.F. AARNES, to be published

36) J. HARDY and Y. POMEAU, J. Math. Phys. 13, 7 (1972), 13, 1042 (1972). J. HARDY, Y. POMEAU and O. de PAZZIS, Phys. Rev. Letters 31, 276 (1973), J. Math. Phys. 14, 1746 (1973).

CONDUCTIVITY IN A MAGNETIC FIELD

R. B. Stinchcombe

Department of Theoretical Physics
12 Parks Road, Oxford, England

I. INTRODUCTION
- 1.1 Natural representation

II. DERIVATION OF THE BOLTZMANN EQUATION IN A MAGNETIC FIELD
- 2.1 Density matrix equation of motion
- 2.2 Ordering of equations in weak scattering limit
- 2.3 Boltzmann equation for arbitrary $\omega_c\tau$
- 2.4 Corrections

III. SOLUTION OF THE BOLTZMANN EQUATION
- 3.1 Isotropic systems
- 3.2 Anisotropic systems

IV. QUANTUM EFFECTS
- 4.1 Landau representation
- 4.2 Longitudinal case
- 4.3 Transverse case
- 4.4 Discussion

V. COLLISIONS BETWEEN CARRIERS

VI. COLLISIONS WITH PHONONS
- 6.1 Transport equations with phonon scattering
- 6.2 Magnetophonon resonance

VII. CONCLUDING REMARKS

I. INTRODUCTION

These lectures will be concerned with classical and quantum electrical transport in the presence of magnetic fields, typically in a metal or a semiconductor.

As is well known (see for example Peierls, 1955) in the absence of magnetic fields the electrical resistance of a material results from deviations of the crystal lattice from perfect periodicity. Such deviations are caused by:

(a) Lattice vibrations.

(b) Impurities, boundaries and lattice imperfections. It is usual to separate these from the perfect lattice effects, which are accommodated in the usual Bloch states.

For a non zero field H, it is again appropriate to separate off the perfect lattice. The conduction transverse to the field is finite even in the absence of (a) or (b), but (as in the zero field case) to limit the conduction along the field H it is essential to include some collision processes. Elastic collisions, e.g., with impurities, provide the simplest model to examine. The idealisation usually made is to neglect interactions between the charge carriers. We shall at first do this, only later returning to a discussion of the features so omitted.

In classical situations transport may be discussed in terms of a distribution function measuring the number of particles likely to be in a given place with a given velocity at a given time. The Boltzmann equation determines the steady state distribution function by requiring that its total rate of change, due to external fields and collisions, should vanish. The magnetic field enters through the effects of the Lorentz force.

In the absence of a magnetic field the distribution function can be redefined in a quantum mechanical form as the occupation probability $n(\underline{k})$ of the Bloch state $\underline{k}$. The usual transport equation then takes the form

$$\left(\frac{\partial n}{\partial t}\right)_{field} + \left(\frac{\partial n}{\partial t}\right)_{collisions} = 0. \tag{1.1}$$

Here

$$\left(\frac{\partial n}{\partial t}\right)_{field} = \frac{\underline{F}}{\hbar}\frac{\partial}{\partial \underline{k}}\, n(\underline{k}) \tag{1.2}$$

$\underline{F}$ being the Lorentz force (for zero $\underline{H}$). $\left(\partial n/\partial t\right)_{collision}$, the rate of change of n due to collisions, is usually obtained by combining transition probabilities.

For non-zero H, the transport equation is usually generalised by writing

$$\underline{F} = -e\left(\underline{\mathcal{E}} + \frac{1}{c}\underline{v} \times \underline{H}\right) \tag{1.3}$$

where $\underline{v}$ is the electron velocity in state $\underline{k}$.

Such a description needs justification for several reasons. First, as is well known, even in zero field the description in terms of n(k) is incomplete, since n(k) corresponds to just the diagonal components of the density matrix ρ_T; n(k) is therefore sufficient only for obtaining average values of operators diagonal in k-representation unless the density matrix is diagonal at all times. A "repeated random phase assumption" to make it so was a serious weakness of early derivations of (1.1). It was later suggested (Peierls 1955) that in the case of impurity scattering the assumption could be replaced by assuming the scatterers to be randomly distributed, since in k-representation the off-diagonal elements of ρ_T would be expected to be small in a system which is on average homogeneous. Such properties of the scattering were exploited by Kohn and Luttinger (1957) and by Greenwood (1958) in a derivation of the zero field Boltzmann equation from the equation of motion of the density matrix. More powerful techniques developed by Van Hove (1955) were used by Chester and Thellung (1959) to give a more general treatment for the field free case.

When a magnetic field is present it is not clear in what "natural representation" the density matrix is approximately diagonal at all times. In particular the density matrix of a system which is in absence of the field on average homogeneous will no longer be diagonal in momentum representation. There would at first sight seem to be no reason why in the presence of a magnetic field the system should be described by a stationary distribution function n(k) satisfying an equation of the Boltzmann type, except in some weak field limit.

The original derivation (Jones and Zener 1934) of the Boltzmann equation (1.1)-(1.3) in the presence of a magnetic field used the repeated random phase assumption within a wave-packet analysis using plane waves as a basis. Such a derivation requires

$$\omega_c \tau \lesssim 1 \tag{1.4}$$

where ω_c is the cyclotron frequency eH/mc and τ is a collision time. (1.4) is essentially the condition that the electron should not deviate appreciably from a straight line between scatterings. The Boltzmann equation has however been applied successfully to situations in which $\omega_c\tau \gg 1$ indicating that a method of derivation not depending on (1.4) should be possible, and we shall discuss one such method shortly.

Nevertheless the field should not be too strong if the Boltzmann equation is to be applicable: (1.3) neglects the fact that since the electron can (for $\omega_c\tau \gtrsim 1$) describe closed orbits in the plane at right angles to the field direction, the energy levels will be in part discrete. Effects, e.g., the de Haas-Schubnikov oscillations, associated with such a quantization would be expected to show up for $\omega_c\tau \gtrsim 1$ when the separation of the levels exceeds K_BT:

$$\hbar\omega_c \gtrsim K_BT \qquad (1.5)$$

Even more extreme quantum effects can be expected if the level separation approaches or exceeds the fermi energy

$$\hbar\omega_c \gtrsim \zeta \qquad (1.6)$$

since then only the lowest levels will be occupied ("extreme quantum limit" - Argyres and Adams 1956).

Such effects are altogether absent from the Boltzmann equation and require a quantum approach. We shall return to this point later.

We first discuss the derivation (Stinchcombe 1961) of the Boltzmann equation along the lines employed by Kohn and Luttinger and by Greenwood for the field free case. The derivation will require certain limitations (not (1.4)) on the strength of the magnetic field, and for simplicity we consider weak scattering by an irregular static potential (e.g., impurity but not phonon scattering). We shall want later to consider other scattering mechanisms and to discuss galvanomagnetic phenomena special to them.

The derivations by Kohn and Luttinger, Greenwood, Van Hove and Chester and Thellung for the field free case rely on the fact that for an initially homogeneous system evolving under the influence of a homogeneous perturbation the density matrix remains diagonal at all times. The perturbation to be considered is the electric field and the internal scattering mechanism. The latter is only homogeneous in some average sense and so the argument requires also an averaging procedure. This gives rise to the dissipative nature of the final equation and the density matrix turns out to be not exactly diagonal, but the off-diagonal part is small if the number of degrees of freedom is large.

1.1 Natural representation

To generalise the derivation to the case where a magnetic field is also present we need first to find a representation which diagonalises the density operator of a homogeneous system in a magnetic field, and in which the average properties of the scattering mechanism can be exploited. The Landau state representation (or its generalisation for the case of electrons in a periodic potential) is a possible candidate since it does diagonalise the density matrix of a uniform system, but it would have to be checked whether it leads to the required properties of the averaged scattering matrix elements. This will be considered later. An obvious disadvantage of the Landau representation is that it is not easy to extract from it low field limits and in particular it provides no hope of a simple relationship to the k label in n(k) in the Boltzmann equation.

We proceed instead to find a possible representation by generalizing to the case of $H \neq 0$ the usual arguments which for $H = 0$ lead to a Bloch theorem. We first consider the configuration space matrix element $\langle \underline{x} | D(\mathcal{H}_0) | \underline{x}' \rangle$ of any function D of the Hamiltonian $\mathcal{H}_0$ describing non-interacting electrons in the perfect lattice in the presence of a uniform magnetic field in the z direction. Taking vector potential $\underline{A}:(0,Hx,0)$ the Schrodinger equation for the wave function $\langle \underline{x} | \alpha \rangle$ of a single electron has the same dependence on y and z as in the absence of H. The wave function therefore has the same dependence on y and z as the usual Bloch wave. It may therefore be written

$$\langle \underline{x} | \alpha \rangle = \langle x y z | k_y k_z \nu \alpha' \rangle = \exp[ik_y y + ik_z z]\, \Phi_{k_y k_z \nu \alpha'}(x,y,z) \tag{1.7}$$

where Φ has the periodicity of the lattice with respect to y and z. ν is a band index and α' is the quantum number necessary to complete the specification of the state. By using the fact that D is diagonal in this representation it may then be shown that

$$\langle x, y+\tau_y, z+\tau_z | D | x' y' z' \rangle = \langle x y z | D | x', y'-\tau_y, z'-\tau_z \rangle \tag{1.8}$$

where $\underline{\tau}$ is any lattice vector.

Information about the x-dependence of the wave function can be obtained by first solving the Schrodinger equation in the gauge $\underline{A}$:

(-Hy,0,0) and then transforming back to the original gauge. The gauge transformation corresponds to the addition of a rotation-free part (the gradient of the scalar f= H x y) to the vector potential and so the wave function changes by a phase factor $\exp[-i(e/\hbar c)f]$. In the gauge **A**: (-Hy,0,0) the x-dependence of the wave function is that of a Bloch wave. Hence in the original gauge the wave function is

$$\langle \underline{x}|\alpha\rangle = [\exp ik_x X - i\epsilon xy]\,\Psi_{k_x\alpha''}(\underline{x}), \tag{1.9}$$

where $\Psi_{k_x\alpha''}$ has the periodicity of the lattice with respect to its dependence on x, and $\epsilon = (eH/\hbar c)$. By again writing $\langle \underline{x}|D|\underline{x}'\rangle$ in terms of the matrix elements $\langle\alpha|D|\alpha\rangle$ it is a straightforward matter to obtain the behaviour of $\langle \underline{x}|D|\underline{x}'\rangle$ when x and x' each change by τ_x. In this way one finally arrives at

$$e^{i\epsilon\tau_x(y+\tau_y)}\langle \underline{x}+\underline{\tau}|D|\underline{x}'\rangle = e^{i\epsilon\tau_x y'}\langle \underline{x}|D|\underline{x}'-\underline{\tau}\rangle. \tag{1.10}$$

The general solution of this equation is

$$\langle \underline{x}|D|\underline{x}'\rangle = \exp[-i\epsilon X\eta]\,\mathcal{D}(\underline{\xi},\underline{X}) \tag{1.11}$$

where $\mathcal{D}$ has the periodicity of the lattice with respect to its dependence on $\underline{X}$, and

$$\underline{X} = \tfrac{1}{2}(\underline{x}+\underline{x}') = (X,Y,Z),$$
$$\underline{\xi} = \underline{x}-\underline{x}' = (\xi,\eta,\zeta). \tag{1.12}$$

The property (1.11) of the matrix element is unchanged when a uniform electric field $\mathcal{E}^t$ is added to the system. This may be easily seen by introducing the field by the addition of a time-dependent part $-c\int^t \mathcal{E}^t dt$ to the vector potential.

To show that the property is preserved when a scattering mechanism is introduced is more difficult since it is only expected to be maintained in some average sense. We consider scattering by a random distribution of N static impurities with total scattering potential

$$W = \sum_n^N V(\underline{r}-\underline{r}_n). \tag{1.13}$$

Consider a product of matrix elements of the scattering potential in some representation $|\alpha\rangle$:

$$\langle\alpha_0|W|\alpha_1\rangle\langle\alpha_1|W|\alpha_2\rangle\ldots\langle\alpha_{i-1}|W|\alpha_i\rangle$$

$$= \int d^3q_1\ldots d^3q_i\, V(q_1)\ldots V(q_i)\langle\alpha_0|e^{i\underline{q}_1\cdot\underline{r}}|\alpha_1\rangle\ldots\langle\alpha_{i-1}|e^{i\underline{q}_i\cdot\underline{r}}|\alpha_i\rangle$$
$$\sum_{n_1\ldots n_i}^{N}\exp\left[i\underline{q}_1\cdot\underline{r}_{n_1}+i\underline{q}_2\cdot\underline{r}_{n_2}+\cdots+i\underline{q}_i\cdot\underline{r}_{n_i}\right]. \qquad (1.14)$$

The randomness of the distribution can be exploited by averaging over all distributions. Then the multiple sum vanishes unless the exponent in the summand can be split into groups $\sum_s i\underline{q}_s\cdot\underline{r}_{n_s}$ in any one of which all r_{n_s}'s are equal and for which $\sum_s \underline{q}_s = 0$. Thus the average of (1.14) vanishes unless the sum of all the $\underline{q}$'s is zero. Suppose that one of the quantum numbers in the complete set α refers to a component of momentum or quasi-momentum. The difference of the values of this component in the states $|\alpha_0\rangle$ and $|\alpha_i\rangle$ will be a component of $\Sigma\,\underline{q}$ which vanishes because of the averaging.

Now if D' is any function of the Hamiltonian $\mathcal{H}_0$+W of the system with scattering, in the presence of the magnetic field, D' can be expanded in powers of W about its zero-scattering value D (using the usual S-matrix methods) and the resulting product of operators can be evaluated in the representation $|\alpha\rangle$ given by (1.7) and (1.9). Making use of the property deduced above of the averaged products of W's it is evident that the large part of D' is diagonal with respect to the wave vector quantum numbers appearing in the labelling of (1.7) and (1.9). (The replacement of the product of matrix elements by its average introduces errors of order $^1/N$.) Hence in the same way as before equations of the type (1.8), (1.10), (1.11) can be derived for the large part of D'.

We remark that D' could be the density matrix ρ_T of the system at some arbitrary time after the scattering and electric field have been applied. The averaging procedure makes it plausible that the functional form (1.11) is possessed by a large part of D' at all times. The large part of D', or more particularly ρ_T, then simplifies in the representation defined by

$$\rho_T(l_1;l_2) = \int e^{i\epsilon\eta X}\,\psi^*_{l_1}(\underline{x})\langle\underline{x}|\rho_T|\underline{x}'\rangle\psi_{l_2}(\underline{x}')\,d^3\underline{x}\,d^3\underline{x}' \qquad (1.15)$$

where $\psi_l(\underline{x})$ is the Bloch function for electrons in the perfect lattice in zero H, and ℓ denotes the complete label ($\underline{k},\nu$) in the reduced zone scheme:

$$\left\{\rho_T(\ell_1;\ell_2)\right\}_{\text{large part}} = \rho_T(\underline{k}_1\nu_1;\underline{k}_1\nu_2)\,\delta_{\underline{k}_1,\underline{k}_2} \tag{1.16}$$

The method we have used to discuss the diagonality is similar to the arguments of Van Hove (1955) for the zero field case but is not claimed to be as rigorous in its treatment of the scattering. The method tells us nothing about the band suffices ν_1, ν_2. By analogy with the zero field case we suppose that a rigorous treatment will confirm that $\rho_T(\ell_1;\ell_2)$ is not exactly diagonal in the label $\underline{k}$ but that the off-diagonal part is small. This will be seen to be the case in deriving the Boltzmann equation by the Kohn and Luttinger-Greenwood method.

Before embarking on such a derivation it is important to note how average values of currents can be obtained from $\rho_T(\ell;\ell')$. The average value $\overline{j_\mu}$ of the μth component of the current is $\mathrm{Tr}[\rho_T j_\mu]$. Taking the trace in $\underline{x}$-representation and obtaining $\langle\underline{x}|\rho_T|\underline{x}'\rangle$ in terms of $\rho_T(\ell;\ell')$ by using the inverse of (1.15) it can be shown (without the use of (1.16) that $\overline{j_\mu}$ reduces exactly to

$$\overline{j_\mu} = \int d^3\underline{k} \sum_{\nu\nu'} \rho_T(\underline{k}\nu;\underline{k}\nu')\langle\underline{k}\nu'|\,j_\mu(H=0)\,|\underline{k}\nu\rangle \tag{1.17}$$

where the matrix element is of the zero field current operator between Bloch wave states (i.e. of the field-free system). A similar result holds for the average energy.

Apart from the off-diagonal character of the band labels $\rho_T(\underline{k}\nu;\underline{k}\nu)$ behaves in (1.17) as though it were the occupation probability of a (zero field) Bloch state, but it is obvious from (1.15) that $\rho_T(\ell;\ell')$ is a much more complicated function. In the free electron model (no crystal lattice) (1.15) may be written as follows:

$$\left\{\rho_T(k_1k_2)\right\}_{\text{large part}} = \delta_{k_1k_2}\int d^3\underline{\xi}\,\exp\left[-i\left(\underline{k}_1-\frac{e}{\hbar c}\underline{A}(\underline{X})\right)\cdot\underline{\xi}\right]\langle\underline{X}+\tfrac{1}{2}\underline{\xi}\,|\rho_T|\,\underline{X}-\tfrac{1}{2}\underline{\xi}\rangle \tag{1.18}$$

which corresponds to the Wigner representation (Wigner 1932) for a function of the two non-commuting operators $\underline{r}$ and

$$\underline{\Pi} = \underline{p} - \frac{e}{c}\underline{A}\,. \tag{1.19}$$

$\underline{\Pi}$ corresponds to the classical velocity $\underline{v}$ and it is significant that $\underline{v}$ is the variable that appears in the classical Boltzmann equation which is valid at arbitrarily high fields.

Finally we note that our discussion, starting from (1.14), of the effect of impurity scattering on diagonality properties could be applied to other representations than that considered here. We shall later investigate the Landau representation in a related way.

II. DERIVATION OF THE BOLTZMANN EQUATION IN A MAGNETIC FIELD

2.1 Density matrix equation of motion

The homogeneity of the system (in an average sense) can be expected to produce a sharp distinction between the diagonal and the off-diagonal elements of the density matrix in the representation introduced above. Therefore, following Kohn and Luttinger and Greenwood's method for the H = 0 case we shall split the equation of motion of the density matrix into diagonal and off-diagonal parts and eliminate the off-diagonal parts to obtain a Boltzmann equation for the diagonal parts (Stinchcombe 1961).

The system considered is one of non-interacting electrons moving in a perfect periodic lattice under the influence of electric and magnetic fields, and scattered by a random distribution of static impurity centres. The total Hamiltonian for each electron is

$$\mathcal{H}_T = \mathcal{H}_0 + W + H_F \tag{2.1}$$

where

$$\mathcal{H}_0 = \frac{1}{2m}\left(p - \frac{e}{c}A\right)^2 + V_0 , \tag{2.2}$$

$$H_F = -e\,\underline{\mathcal{E}}^t \cdot \underline{x} , \tag{2.3}$$

and W is the interaction (1.13) with the impurities. V_0 is the lattice potential.

It is assumed that in the infinite past the system is in equilibrium in the absence of the electric field, and that the field is switched on in the following manner:

$$\underline{\mathcal{E}}^t = \underline{\mathcal{E}}\, e^{st}, \qquad s > 0 \tag{2.4}$$

Then the density matrix of the complete system may be written

$$\rho_T(t) = \rho + \rho_F(t) \tag{2.5}$$

where ρ_F is the change in the density matrix caused by the electric field ($\rho_F(-\infty)=0$) and ρ is the fermi function of $(\mathcal{H}_0+W)$. For the Ohmic conductivity ρ_F is only required to first order in $\underline{\mathcal{E}}$. To this order ρ_F is determined by the terms linear in $\underline{\mathcal{E}}$ in the equation of motion of the density matrix. By taking out a factor e^{st} from ρ_F and H_F, the linearised equation of motion becomes

$$i\hbar s f = [\mathcal{H}_0 + W, f] - C \tag{2.6}$$

where

$$\rho_F = f\, e^{st} \tag{2.7}$$

and

$$C = [\rho, H_1] \tag{2.8}$$

with

$$H_1 = -e\,\underline{\mathcal{E}}\cdot\underline{x}\,. \tag{2.9}$$

f is time-independent. It is the correction to the density matrix at time t=0, when the field has reached the value $\underline{\mathcal{E}}$.

We now write f and C in the representation defined by (1.15), denoting the resulting elements by $f(\ell_1;\ell_2)$ and $C(\ell_1;\ell_2)$. By writing (2.6) in the $\underline{x}$-representation and then using the transformation (1.15) and its inverse, the equation of motion for the density matrix is converted into the following equation for $f(\ell_1;\ell_2)$:

$$i\hbar s f(\ell;\ell') + C(\ell;\ell') = \sum_{\ell_1\ell_1'} \left[\mathcal{H}(\ell;\ell'|\ell_1;\ell_1') + W(\ell;\ell'|\ell_1;\ell_1')\right] f(\ell_1;\ell_1') \tag{2.10}$$

where

$$\mathcal{H}(\ell;\ell'|\ell_1;\ell'_1) = \int d^3\underline{x}\, d^3\underline{x}'\, d^3\underline{x}''\, e^{i\epsilon\eta X}\, \psi^*_\ell(\underline{x})\psi_{\ell'}(\underline{x}')$$
$$\left\{ \begin{array}{l} \langle \underline{x}|\mathcal{H}_0|\underline{x}''\rangle\, e^{-i\epsilon\eta''' X'''}\, \psi_{\ell_1}(\underline{x}'')\psi^*_{\ell'_1}(\underline{x}') \\ -\langle \underline{x}''|\mathcal{H}_0|\underline{x}'\rangle e^{-i\epsilon\eta^{o''}X^{o''}}\, \psi_{\ell_1}(\underline{x})\psi^*_{\ell'_1}(\underline{x}'') \end{array} \right\} \tag{2.11}$$

and $\mathcal{W}$ is given similarly in terms of matrix elements of W. In (2.11),

$$\xi^{o''} = x - x'', \qquad X''' = \tfrac{1}{2}(x'' + x'), \text{ etc}$$

All the terms on the right hand side of (2.10) come from the commutator of f with $(\mathcal{H}_0+W)$ and (c.f. (2.11)) each consists of two terms which can be shown to be complex conjugates of each other. Note also that from (1.11) it is possible to write

$$\langle \underline{x}|\mathcal{H}_0|\underline{x}'\rangle = e^{-i\epsilon\eta X}\, h(\underline{\xi}, \underline{X}) \tag{2.12}$$

where h has the periodicity of the lattice with respect to its $\underline{X}$-dependence.

Our earlier considerations suggest that because of the randomness of the scattering centres $f(k\nu;k\nu')$ should be much larger than $f(k\nu;k'\nu')$. We shall also find that, except when bands may overlap, $f(k\nu;k'\nu)$ is much larger than $f(k\nu;k'\nu')$. Proceeding with this in mind we split (2.10) into diagonal ($\ell' = \ell$) and off-diagonal ($\ell' \neq \ell$) parts. The equations obtained in this way can be written symbolically as

$$i\hbar s f_d + C_d = (\mathcal{H} + \mathcal{W})^d{}_d f_d + (\mathcal{H} + \mathcal{W})^d{}_{nd} f_{nd} \tag{2.13}$$

$$i\hbar s f_{nd} + C_{nd} = (\mathcal{H} + \mathcal{W})^{nd}{}_d f_d + (\mathcal{H} + \mathcal{W})^{nd}{}_{nd} f_{nd} \tag{2.14}$$

where the subscripts d, nd denote diagonal and non-diagonal parts of f's or C's. The subscripts and superscripts on the integral operators $\mathcal{H}$ and $\mathcal{W}$ refer to the character of the f's they connect. For example $\mathcal{W}^d{}_d$ means $\sum_{\ell_1}\mathcal{W}(\ell;\ell|\ell_1;\ell_1)$ ... to act on $f(\ell_1;\ell_1)$.

2.2. Ordering of equations in weak-scattering limit

We attempt to solve these complicated coupled equations in the weak scattering limit. In the treatment by Kohn and Luttinger of the zero field case it is shown that in this limit the equations have a solution which is independent of the rate s at which the electric field is switched on provided $s \ll {}^1/\tau$ where τ is of the order of the relaxation time of the system. We shall take the weak scattering limit in such a way that s is about equal to ${}^1/\tau$ (τ depends inversely on the scattering) and only later make use of the fact that in practice the rate of application of the electric field is always such that $s\tau \ll 1$.

It will turn out that the restriction $s \sim {}^1/\tau$ is not the only one that has to be imposed. In a similar way the size of the magnetic field has to be introduced into the limiting process. To show this we demonstrate how an attempt to solve the equations in the weak scattering limit for arbitrary fields breaks down.

Replacing W by λ_W W, where λ_W is a dimensionless parameter measuring the strength of the scattering, compared to the fermi energy, since $\tau \propto \lambda_W^{-2}$ our restriction on s requires that $s = \lambda_s^2 s_0$ where $\lambda_s \sim \lambda_W$ and s_0 is some reference frequency independent of λ_W. The integral operators and inhomogeneous terms can then be ordered with respect to λ_W:

$$\mathcal{H}^{\cdots}_{\cdots} \sim \lambda^0_W$$
$$\mathcal{W}^{\cdots}_{\cdots} \sim \lambda^1_W$$
$$C_{\cdots} \sim \lambda^0_W \tag{2.15}$$

(were it not for the band labels, C_{nd} would vanish to zeroth order in λ_W using the property (1.11) of functions of the Hamiltonian $\mathcal{H}_0$).

Solving (2.14) formally for f_{nd} in terms of f_d, and substituting into (2.13) shows that, to lowest order in λ_W, f_d is of order λ_W^0 and is determined by

$$\mathcal{H}^d{}_d f_d = C_d - \mathcal{H}^d{}_{nd} (\mathcal{H}^{nd}{}_{nd})^{-1} [C_{nd} - \mathcal{H}^{nd}{}_d f_d]. \tag{2.16}$$

Now in the limit of zero field the integral operators $\mathcal{H}^d{}_d$, $\mathcal{H}^{nd}{}_d$, $\mathcal{H}^d{}_{nd}$, and $\mathcal{W}^d{}_d$ vanish. For in this limit the general kernel becomes

$$[\mathcal{H}(l;l'|l_1;l_1') + \mathcal{W}(l;l'|l_1;l_1')]_{H=0}$$
$$= (E_l - E_{l'})\delta_{ll_1}\delta_{l'l_1'} + \langle l|W|l_1\rangle\delta_{l'l_1'} - \langle l_1'|W|l'\rangle\delta_{ll_1}, \tag{2.17}$$

where the matrix elements are between Bloch states and $E_l = \langle l|\mathcal{H}_{0\,H=0}|l\rangle$ is the zero field Bloch energy.

Hence in zero field the coefficient of f_d in (2.16) is zero. The equation only gives a solution valid in the weak scattering limit provided the effect of the field H is not too small. Then

$f_d \sim \lambda_W^{\,o}$.

In the absence of the magnetic field, the weak scattering treatment (Kohn and Luttinger 1957) gives $f_d \sim \lambda_W^{-2}$.

These different dependences reflect the different mechanisms which limit the momentum acquired by the electrons under the influence of the electric field: when H = 0 the scattering is entirely responsible; if there is a magnetic field perpendicular to the electric field the electrons execute closed orbits under the influence of the Lorentz force and this mechanism dominates in the weak-scattering limit. This is the physical justification of the "high field" equation (2.16). It indicates that this equation, and similar ones representing successive orders in a perturbation development in powers of λ_W, provides a basis for evaluating the transverse magneto conductance in high fields.

We concern ourselves here with the development of a method which can correctly include the zero field limit. Such a method must take into account the vanishing of $\mathcal{H}^d{}_d$, $\mathcal{H}^{nd}{}_d$, $\mathcal{H}^d{}_{nd}$ as the field goes to zero. This can be accomplished by ordering (2.13) and (2.14) not just with respect to λ_W, but also with respect to a parameter λ_H measuring the size of the field. We assume that each of λ_W and λ_H are small compared to 1 but not in relationship to each other. The limiting process $\lambda_H \ll 1$ is the extra condition referred to earlier. We define λ_H by $\hbar\omega_c = \lambda_H^2 \bar{E}$ when $\bar{E}$ is the smallest λ_W- independent energy typical of the system. The possible energies are the Fermi energy ζ, the thermal energy k_BT and e.g., an energy $\epsilon_S \sim \hbar^2/(2ma^2)$ associated with the range a of the potential of a single scatterer. The energy $\hbar/\tau$ is λ_W- dependent.

(2.13), (2.14) are to be evaluated in the limit

$$\lambda_S \sim \lambda_W \ll 1 \; ; \qquad \lambda_H \ll 1 . \tag{2.18}$$

The integral operators $\mathcal{H}^d{}_d$, $\mathcal{H}^d{}_{nd}$, $\mathcal{H}^{nd}{}_d$, $\mathcal{W}^d{}_d$ can be shown to go linearly to zero with the field; that is, they are of order $\lambda_H^{\,2}$. The other integral integral operators and the inhomogeneous terms are all of order $\lambda_H^{\,o}$:

$$\begin{aligned}
&\mathcal{H}^{nd}{}_{nd}\,,\ C_d\,,\ C_{nd} \sim \lambda_W^0 \lambda_H^0 \\
&\mathcal{W}^d{}_{nd}\,,\ \mathcal{W}^{nd}{}_d\,,\ \mathcal{W}^{nd}{}_{nd} \sim \lambda_W \lambda_H^0 \\
&\mathcal{H}^d{}_d\,,\ \mathcal{H}^{nd}{}_d\,,\ \mathcal{H}^d{}_{nd} \sim \lambda_W^0 \lambda_H^2 \\
&\mathcal{W}^d{}_d \sim \lambda_W \lambda_H^2 .
\end{aligned} \tag{2.19}$$

Using (2.19), it is now possible to deduce from (2.13), (2.14) the leading order dependences of f_d, f_{nd} on λ_W, λ_S, λ_H in the limit (2.18). Eliminating f_{nd} and discarding all but the leading order terms in the resulting coefficient of f_d and in the inhomogeneous term gives:

$$\left[it_H s - \mathcal{H}^d{}_d + \mathcal{W}^d{}_{ud} (\mathcal{H}^{ud}{}_{ud})^{-1} \mathcal{W}^{ud}{}_d \right] f_d = -C_d . \tag{2.20}$$

In (2.20), the order of the operator applied to f_d is $\ldots \lambda_S^2 + \ldots \lambda_H^2 + \ldots \lambda_W^2$. C_d is of order 1.

We can now consider three possibilities for the relative order of λ_W, λ_H :

(i) $\lambda_W^2 \ll \lambda_H^2$ $(\ll 1)$. Then $f_d \sim 1/\lambda_H^2$. In this scheme the lowest order equation contains no scattering terms. The scattering can be introduced by a perturbation development in (λ_W/λ_H). This is the high field approach already mentioned, but here it contains the additional restriction $\lambda_H \ll 1$.

(ii) $\lambda_H^2 \ll \lambda_W^2 \ll 1$. Then $f_d \sim 1/[\lambda_W^2 + \lambda_S^2]$ and the lowest order equation does not contain the field. This is introduced as a perturbation (expansion in powers of λ_H/λ_W). This method is limited to very weak fields ($\omega_c \tau \ll 1$).

(iii) The third possibility is the introduction of a limiting process with

$$\lambda_H^2 \sim \lambda_W^2 \sim \lambda_S^2 \ll 1 . \tag{2.21}$$

Then the terms of leading order in both the field and the scattering are retained. This method should be valid for arbitrary $\omega_c \tau$.

2.3. Boltzmann equation for arbitrary $\omega_c \tau$

We adopt the third possibility above, imposing the limiting process (2.21). Then

$$\begin{aligned} f_d &= \lambda_W^{-2} \text{ function of } (\lambda_H/\lambda_W, \lambda_S/\lambda_W) \\ f_{nd} &= \lambda_W^{-1} \text{ function of } (\lambda_H/\lambda_W, \lambda_S/\lambda_W) \end{aligned} \tag{2.22}$$

(consistent with the results of Kohn and Luttinger for the zero field case). Under the limiting process (2.21) equations (2.13) and (2.14) are to be replaced by

$$i\hbar s f_d + \hat{C}_d = \hat{\mathcal{H}}^d{}_d f_d + \hat{\mathcal{W}}^d{}_{nd} f_{nd} \tag{2.23}$$

$$i\hbar s f_{nd} = \hat{\mathcal{W}}^{nd}{}_d f_d + \hat{\mathcal{H}}^{nd}{}_{nd} f_{nd} \tag{2.24}$$

where the circumflex indicates that the integral operators and inhomogeneous terms are to be evaluated to leading order in λ_W and λ_H. The term $i\hbar s\, f_{nd}$ is of higher order than the other terms in (2.24) but it cannot be discarded as it later appears in a limiting form for a delta function. It is a simple matter to solve (2.24) for f_{nd} in terms of f_d, since $\hat{\mathcal{W}}^{nd}{}_d$ and $\hat{\mathcal{H}}^{nd}{}_{nd}$ are the zero field values of $\mathcal{W}^{nd}$, $\mathcal{H}^{nd}{}_{nd}$. The result is

$$f(\ell;\ell') = \frac{\langle \ell|W|\ell'\rangle\,[f(\ell';\ell') - f(\ell;\ell)]}{i\hbar s - (E_\ell - E_{\ell'})} . \tag{2.25}$$

(2.25) shows that $f_{nd} \sim \lambda_W f_d$ as long as there is no band overlap.

The last term in (2.23) can be simply reduced by inserting (2.25) for f_{nd} and recognizing the limiting form for a delta-function:

$$\hat{\mathcal{W}}^d{}_{nd} f_{nd} = 2\pi i \sum_{\ell_1} |\langle \ell|W|\ell_1\rangle|^2 \,[f(\ell_1;\ell_1) - f(\ell;\ell)]\,\delta(E_{\ell_1} - E_\ell) . \tag{2.26}$$

Using (2.26), equation (2.23) becomes an equation for f_d alone. The inhomogeneous term $\hat{C}_d$ is the zero field limit of C_d and reduces to

$$\hat{C}_d(\ell;\ell) = i e\, \underline{\mathcal{E}} \cdot \frac{\partial}{\partial \underline{k}} f^\circ(\ell;\ell) \tag{2.27}$$

where

$$f^\circ(\ell;\ell) = \langle \ell|\rho\{\mathcal{H}_{o\,H=0}\}|\ell\rangle . \tag{2.28}$$

The integral operator $\hat{\mathcal{H}}^d{}_d$ is the part of $\mathcal{H}^d{}_d$ linear in the magnetic field. It turns out that $\mathcal{H}^d{}_d$ consists of the linear term only. This is most easily seen by using the diagonal part, $\mathcal{H}^d{}_d$, of (2.11) and performing partial integrations to take derivatives from $\delta(\underline{x}-\underline{x}'')$. After the partial integrations the phase factors exp($i\,\epsilon\,\eta$ X - ...)

become unity and the terms of order H^o, H^1, H^2 can be separated. The terms of order H^o, H^2 vanish, leaving

$$\mathcal{H}^d_d(l_i l | l_i ; l_i) = \mathrm{Im} \frac{i\hbar c}{mc} \int d^3\underline{x}\, d^3\underline{x}'\, \Psi^*_{l_i}(\underline{x}') \Psi_l(\underline{x}') \Psi_{l_i}(\underline{x}) \underline{H}\cdot\left(\underline{\xi}\times\frac{\partial}{\partial \underline{x}}\right) \Psi^*_l(\underline{x}). \tag{2.29}$$

Inserting this for $\hat{\mathcal{H}}^d$d and using properties of Bloch state matrix elements yields

$$\sum_{l_i} \hat{\mathcal{H}}^d_d(l_i l | l_i ; l_i) f_d(l_i ; l_i) = -\frac{ie}{c} (\underline{v}_l \times \underline{H}) \cdot \frac{\partial}{\partial \underline{k}} f(l ; l) \tag{2.30}$$

where

$$\underline{v}_l = \hbar^{-1} \frac{\partial E_l}{\partial \underline{k}} . \tag{2.31}$$

The equation obtained by inserting the detailed forms of the kernels and inhomogeneous term into (2.23) contains, in the collision term (2.26), the square modulus of the matrix element $\langle l | W | l_1 \rangle$ of the interaction of an electron with all impurities. A similar form occurs in treatments of the zero field equation. The usual method of reducing this is as follows. We have

$$\langle l | W | l_1 \rangle = \frac{V_{ll_1}}{\Omega} \sum_i^N \exp\left[i(\underline{k} - \underline{k}_1)\cdot \underline{r}_i\right] \tag{2.32}$$

where the r_i are the locations of the impurity centres, assumed uncorrelated and always to be situated at a lattice site, and

$$V_{ll_1} = \int_\Omega \exp\left[-i(\underline{k} - \underline{k}_1)\cdot \underline{r}\right] u^*_l u_{l_1} V(r)\, d^3\underline{r} . \tag{2.33}$$

u_l is the periodic part of the Bloch function $\langle l | x \rangle$. The sum $\sum_{k_1}$ in the collision term may be broken up into regions R of momentum space so small that the only term that varies appreciably in this region is $|\langle l | W | l_1 \rangle|^2$. It may then be proved (Kohn and Luttinger 1957)) that $\sum_{k_1 \text{ in } R} |\langle l | W | l_1 \rangle|^2$ may be replaced by its average over all the different distributions of the impurities, provided that N is large. (The same is not true for $|\langle \ell | W | \ell_1 \rangle|^2$.) This average is shown to be $|V_{\ell\ell_1}|^2 N/\Omega^2$. Since the transition probability of an electron going from state ℓ to ℓ_1 is, to second order in the scattering, just

$$W_{\ell\ell_1} = \frac{2\pi N}{\hbar} \frac{|V_{\ell\ell_1}|^2}{\Omega^2} \delta(E_\ell - E_{\ell_1}) \quad , \tag{2.34}$$

the collision term can be written in its usual form

$$\sum_{\ell_1} \left\{ W_{\ell\ell_1} f(\ell_1;\ell_1) - W_{\ell_1\ell} f(\ell;\ell) \right\} . \tag{2.35}$$

Recognizing that for the physically interesting rates at which the electric field could be applied s is very much less than the inverse of the relaxation time, the term $i\hbar s\, f(\ell;\ell)$ in (2.23) may be discarded, and the final equation is obtained:

$$\frac{e}{\hbar} \underline{\mathcal{E}} \cdot \frac{\partial}{\partial \underline{k}} f^c(\ell;\ell) + \frac{e}{\hbar c} (\underline{v}_\ell \times \underline{H}) \cdot \frac{\partial}{\partial \underline{k}} f(\ell;\ell)$$
$$+ \sum_{\ell} \left\{ W_{\ell_1\ell} f(\ell;\ell) - W_{\ell\ell_1} f(\ell_1;\ell_1) \right\} = 0 \tag{2.36}$$

This is the usual Boltzmann equation Jones and Zener 1934, Peierls 1931, 1932.

The current is determined by inserting into (1.17) ρ_T in terms of f. Assuming non-overlapping bands (c.f. (2.25)), the result takes the familiar form

$$\overline{J_\mu} = \sum_{\nu} \int d^3\underline{k}\, f(\ell;\ell) \langle \ell | j_\mu (H=0) | \ell \rangle . \tag{2.37}$$

In both (2.36) and (2.37) the meaning of $f(\ell;\ell)$ is much less obvious than is usually suggested.

2.4. Corrections

A detailed investigation of corrections to (2.36) confirms that the result does not depend on $\omega_c\tau$ being small.

The first corrections of higher order in λ_W are similar to those met in the problem of conduction in zero field and can be dealt with by the arguments developed for that case. Such corrections arise from the term $W^{nd}{}_{nd}\, f_{nd}$ in (2.14) which gives an additional term in (2.24). Iterating the resulting equation gives

$$f_{nd}(l;l') = \frac{\langle l|W|l'\rangle\,[f(l';l')-f(l;l)]}{i\hbar s-(E_l-E_{l'})}$$

$$+\sum_{l_1}\langle l|W|l_1\rangle\langle l_1|W|l'\rangle\left[\frac{f(l';l')-f(l_1;l_1)}{i\hbar s-(E_{l_1}-E_{l'})}-\frac{f(l;l)-f(l_1;l_1)}{-i\hbar s-(E_{l_1}-E_l)}\right]+\cdots \tag{2.38}$$

where $l \neq l'$. The first term (c.f. (2.25)) was included in the derivation of the Boltzmann equation. The leading correction is one order of λ_W smaller but since it is a sum over many terms it is not obvious that it really can be neglected. If, in the leading correction, we break up the sum $\sum_{k_1}$ into small regions R of momentum space and perform the averaging procedure presented in I over all distributions of impurities, we may then replace $\sum_{k_1 \text{ in } R}\langle l|W|l_1\rangle\langle l_1|W|l'\rangle$ by its ensemble average. This average has a form reflecting the overall homogeneity of the system:

$$\overline{\sum_{k_1 \text{ in } R}\langle l|W|l_1\rangle\langle l_1|W|l'\rangle} \sim \delta_{kk'}\sum_{k_1 \text{ in } R}\langle l|W|l_1\rangle\langle l_1|W|l'\rangle \tag{2.39}$$

This is the "diagonal singularity" property exploited by Van Hove (1955) in the derivation of a zero field transport equation. In our derivation it must be used to take care of the troublesome corrections of higher order in λ_W. In the limit of an infinite system (2.39) becomes exact and then in (2.38) the first correction contributes only for $k = k'$. In the case of a single band, $l \neq l'$ implies $k \neq k'$ and the first correction term disappears. More generally $l \neq l'$ and $k = k'$ are possible together only if $\nu \neq \nu'$ and the first correction terms can again be shown to be small. The properties of the scattering potential ensure that the same is true for all higher iterations and for correction terms of higher order in λ_W in general. Such corrections are small provided $\hbar/\tau \ll \zeta$ (c.f. Chester and Thellung 1959, for the zero field case).

Corrections from terms of higher order in λ_H could arise from the field-dependence of the integral operators $\mathcal{W}$. But because the perturbing potential is diagonal in coordinate representation it may be shown that

$$\mathcal{W} = \widehat{\mathcal{W}}, \tag{2.40}$$

i.e. there are no magnetic field-dependent corrections to $\hat{W}$. Similarly (c.f. (2.29)) there are no corrections to $\hat{\mathcal{K}}^d d$. The principal field-dependent corrections arise from the other integral operators $\mathcal{K}$, from the inhomogeneous term C_d, and from higher order terms in the iteration of (2.13), (2.14). Such corrections may be shown to be small provided

$$\hbar\omega_c \ll \min\{\zeta, k_BT, \varepsilon_s = \hbar^2/(2ma^2), \dots\} \tag{2.41}$$

where the right hand side is the smallest λ_W-independent energy typical of the system.

Among the effects of higher order corrections are

(i) Quantum effects occurring when $\hbar\omega_c$ approaches k_BT. These show up in particular in modifications of the inhomogeneous term (2.27). For the case of free electrons obeying Boltzmann statistics the modification due to the field is obtained by replacing in (2.27) $f^o(k;k)$ by

$$\delta^{-1}\gamma^{3/2}\exp[-\beta E_k\gamma] \tag{2.42}$$

where $\beta = 1/(k_BT)$, $\beta\hbar\omega_c\gamma = \tanh\beta\hbar\omega_c$, $\beta\hbar\omega_c\delta = \sinh\beta\hbar\omega_c$ (Sondheimer and Wilson, 1951). For the degenerate case the corresponding (more complicated) replacement gives rise to the Schubnikov-de Haas oscillations in the conductivity. These oscillations have period $e\hbar/(mc\zeta r)$, where r is an integer, and appear when $\hbar\omega_c \gtrsim k_BT$, $\hbar/\tau$.

(ii) Extreme quantum limit effects occuring in the same way as (i), when $\hbar\omega_c$ approaches ζ.

(iii) Relaxation of the strict energy conservation in (2.26) due to decay of the intermediate states.

(iv) Replacement of $V_{\ell\ell}$ in (2.34) by a t-matrix element giving the complete effect of a single scatterer (in zero field). This is a desirable resummation to make since scattering potentials are not in general weak, and the weak scattering assumption would be better replaced by the assumption of low concentration of scatterers.

(v) Effects associated with the curvature of the electron trajectory during scattering, when $\hbar\omega_c$ approaches ϵ_s.

Although we have arrived at a static (also linearised) equation (2.36), the effect of an electric field oscillating with frequency ω can be treated just as easily by replacing s by s-iω throughout. Writing $\omega = \lambda^2 s_o$ the ordering $\lambda_H^2 \sim \lambda_W^2 \sim \lambda_s^2 \sim \lambda^2 \ll 1$ then gives the usual equation in which (2.36) is generalised by the addition of a term $\hbar\omega f(\ell;\ell)$ on the left hand side. Such an equation can handle cyclotron resonance and other non-static effects.

III. SOLUTION OF THE BOLTZMANN EQUATION

3.1. Isotropic systems

We have several times used the term relaxation time. Just as in the zero field case (Peierls 1955) a precise meaning to the term can be given for isotropic systems with elastic scattering in which the transition probability $W_{\underline{kk}_1}$ depends only on the angle Θ between $\underline{k}$ and $\underline{k}_1$. (We omit band labels for simplicity).

For this case the Boltzmann equation (2.36) can be solved by expanding $f(\underline{k},\underline{k})$ in spherical harmonics:

$$f(\underline{k},\underline{k}) = \sum_{\ell m} n_{\ell m}(E_k)Y_{\ell m}(\Omega). \tag{3.1}$$

Ω here denotes the spherical polar angles (θ,φ) specifying the direction of $\underline{k}$. The polar axis has been taken along the direction of the magnetic field. If (θ_1,φ_1) similarly defines the direction of $\underline{k}_1$, and $d\Omega_1$ is a corresponding element of solid angle, the collision term (2.35) can be reduced as follows:

$$\sum_{k_1} W_{kk_1}\left(f(k_1,k_1)-f(k,k)\right) \equiv \int d\Omega_1\ w(\Theta)\left[f(k_1,k_1)-f(k,k)\right]$$

$$= -\sum_{\ell m} n_{\ell m}Y_{\ell m}(\Omega)\,\frac{1}{\tau_\ell} \tag{3.2}$$

where

$$\frac{1}{\tau_\ell} = \int d\Omega_1\left[1-P_\ell(\cos\Theta)\right]w(\Theta). \tag{3.3}$$

$w(\Theta)$ is the differential scattering probability and τ_ℓ is the (zero field) relaxation time associated with the rate of change, due to collisions, of the part of f proportional to $Y_{\ell m}$.

Similarly, because $\underline{H}.\underline{k}\wedge\partial/\partial\underline{k} = H\,\partial/\partial\varphi$, the term (2.30) giving the rate of change of f due to the magnetic field can be written

$$-\frac{e}{\hbar c}(\underline{v}_k\wedge\underline{H})\cdot\frac{\partial}{\partial\underline{k}}f(\underline{k}\,\underline{k}) = \sum_{\ell m} n_{\ell m}Y_{\ell m}\,im\,\tilde{\omega}_c \tag{3.4}$$

where

$$\tilde{\omega}_c(E_k) = \frac{eH}{\hbar c}\frac{|v_k|}{k}. \tag{3.5}$$

If $\Omega_\mathcal{E}$ denotes the polar angles ($\theta_\mathcal{E}, \varphi_\mathcal{E}$) determining the direction of the electric field (with $\theta_\mathcal{E}$ again measured from the magnetic field direction) the inhomogeneous term in (2.36) is

$$-\frac{e}{\hbar}\,\underline{\mathcal{E}}\cdot\frac{\partial}{\partial \underline{k}} f^{\circ}(kk) = N(E_k)\sum_{m'=-1}^{1} Y_{1m'}(\Omega)\,Y^{*}_{1m'}(\Omega_\mathcal{E}) \tag{3.6}$$

where $N(E_k)$ is a coefficient independent of the magnetic field:

$$N(E_k) = -e\,\mathcal{E}\,|v_k|\,\frac{\partial f^{\circ}}{\partial E_k}\,. \tag{3.7}$$

The Boltzmann equation therefore gives

$$n_{lm}\left[\frac{1}{\tau_l} - im\tilde{\omega}_c\right] = N\,\delta_{l,1}\sum_{m'=-1}^{1} Y_{1m'}(\Omega_\mathcal{E})\,\delta_{mm'}\,. \tag{3.8}$$

This determines n_{lm} for all (l,m) except (0,0). n_{oo} can be shown to be zero if some inelastic collisions are allowed for. The only non-zero n_{lm} are those with l=1 and m=0 or $\pm$1. The usual zero field lifetime τ_1 appears but it is now combined for m = $\pm$ 1, with the term $im\tilde{\omega}_c$ due to the field.

For the particular ("longitudinal") case where $\underline{\mathcal{E}}//\underline{H}$, only the term with m' =0 occurs on the right hand side of (3.8) and the effect of the magnetic field completely disappears. This is the well known result that isotropic systems have no longitudinal magneto resistance.

For general orientations the component of current in the direction specified by the polar angles Ω_j: (θ_j, φ_j) is obtained by inserting f_{lm} into (2.37):

$$\overline{J}(\Omega_j) = \int dE\; B(E)\sum_{m=-1}^{1}\frac{Y^{*}_{1m}(\Omega_j)\,Y_{1m}(\Omega_\mathcal{E})}{1 - im\,\tilde{\omega}_c(E)\,\tau_1(E)} \tag{3.9}$$

where

$$B(E) = -2\,\frac{\partial Z}{\partial E}\,e^2\,\mathcal{E}\,|v_k|^2\,\frac{\partial f^{\circ}}{\partial E}\,\tau_1(E) \tag{3.10}$$

and $\partial Z/\partial E$ is the density of states in energy.

If both $\underline{\mathcal{E}}$ and $\underline{j}$ are perpendicular to H ($\theta_{j,\mathcal{E}} = \pi/2$) the sum over m reduces to

$$\left[1 + (\tilde{\omega}_c\tau_1)^2\right]^{-1/2}\cos(\varphi_j - \varphi_\mathcal{E} + \varphi_H) \tag{3.11}$$

where φ_H is the "Hall angle" given by

$$\tan \varphi_H(E) = \tilde{\omega}_c(E)\,\tau_1(E). \tag{3.12}$$

In metals the factor $\partial f^o/\partial E$ in (3.10) fixes all the energies in (3.9) at the fermi energy ζ. The average current is therefore a maximum in the direction making an angle $\varphi_H(\zeta)$ with the direction of $\underline{\mathcal{E}}$. The ratio of this maximum current to the component $\mathcal{E}\cos\varphi_H$ of $\underline{\mathcal{E}}$ in the same direction is independent of H, which is the statement of no transverse magneto-resistance for an isotropic system with one type of carrier. The ratio of the maximum current to the component of $\underline{\mathcal{E}}$ perpendicular to it (and to $\underline{H}$) is $(RH)^{-1}$ where R is the Hall coefficient. R gives a measure of the density N of conduction electrons. For the case under consideration it is given by

$$\frac{1}{R} = H\sigma_o \cot\varphi_H(\zeta) = Nec \tag{3.13}$$

where σ_o is the zero field conductivity of the system.

The slightly more complicated case where two different isotropic bands each contribute a term of the form (3.9) to the conduction current gives rise to a non-vanishing trans-verse magneto-resistance, but the longitudinal magneto-resistance remains zero. The resulting trans - verse magneto-resistance (the relative increase in the resistivity for conduction perpendicular to $\underline{H}$) is initially proportional to H^2 but except in special cases saturates when the field is such that $\tilde{\omega}_c\tau \gtrsim 1$ for each band. For particular relation-ships between the parameters of the bands, the transverse magneto-resistance can remain zero (equal mobilities) or fail to saturate (equal and opposite Hall coefficients).

3.2 Anisotropic Systems

The fact that such simple (single band) systems as the alkali metals show magneto-resistance makes it clear that the effects of anisotropy cannot be neglected even for such cases.

In the case of a single band, f is completely specified by the three components of $\underline{k}$; or equivalently (choosing the z-axis in the direction of H) by k_z, E_k and a third parameter α labelling the position on the curve (the classical phase space trajectory) defined by E = const., k_z = const.

If $dk_{//}$ is an element of arc length along the curve and $v_{\perp}$ is the component of velocity (i.e. of grad E_k) perpendicular to the curve, a convenient choice of α is the phase variable

$$\alpha = -\frac{2\pi c}{eHT}\int \frac{dk_{//}}{v_{\perp}} \tag{3.14}$$

where

$$T = -\frac{c}{eH}\oint \frac{dk_{//}}{v_{\perp}} = -\frac{c}{eH}\frac{\partial S}{\partial E} \tag{3.15}$$

T is the time taken to go around the curve, here considered closed and of area S, under the influence of the magnetic field alone; α is a dimensionless measure of the time since a fixed point on the classical orbit was passed, and increases by 2π for each complete orbit. Then

$$\frac{e}{c}(\underline{v}\times\underline{H})\cdot\frac{\partial}{\partial \underline{k}} f(E, k_z, \alpha) = \frac{2\pi}{T}\frac{\partial f}{\partial \alpha} \; . \tag{3.16}$$

We write the collision term (2.35) as $(^1/\tau)W\{f\}$, where τ is some average time between collisions introduced to make $W\{f\}$ dimensionless and $|W| \sim 1$. Putting $f = -e\tau\,\underline{\Psi}\cdot\underline{\mathcal{E}}$, the Boltzmann equation (2.36) becomes:

$$\frac{\partial \underline{\Psi}}{\partial \alpha} + \gamma W\{\underline{\Psi}\} = \gamma\,\underline{v}\,\frac{\partial f^{\circ}}{\partial E} \, , \tag{3.17}$$

where $\gamma = T/(2\pi\tau)$. Except in a special case to be considered later, $^1/\gamma$ is essentially $\omega_c\tau$. For metals, the energy E has to be the fermi energy because of the factor $\partial f^{o}/\partial E$. Using (2.37) the conductivity tensor is

$$\sigma_{ij} = -\frac{e^2 H}{2\pi c}\int dk_z \int dE \oint d\alpha\, T\tau\, v_i\, \Psi_j \tag{3.18}$$

where i and j denote particular components of $\underline{v}$ and $\underline{\Psi}$.

Though the parametrization introduced above provides a convenient framework in which to analyse anisotropic systems, the equation (3.17) is still extremely difficult to solve except in high and low field limit limits. For low fields an expansion in powers of $(^1/\gamma)$ leads, in general, to a magneto resistance proportional to H^2.

For high fields ($\omega_c\tau \gg 1$) an expansion in inverse powers of $\omega_c\tau$ is required. Since this reveals some interesting features we shall investigate this limit in somewhat greater detail, following Lifshitz and co-workers (Lifshitz, Azbel, and Kaganov 1957; Lifshitz and Peshchanskii 1959, 1960).

Since $\underline{\Psi}$ and $\underline{v}$ are each periodic in α, they may be expanded in Fourier series:

$$\underline{\Psi} = \sum_{n=-\infty}^{\infty} \underline{\Psi}_n e^{in\alpha}$$

$$\underline{v} = \sum_{n=-\infty}^{\infty} \underline{v}_n e^{in\alpha}. \tag{3.20}$$

$\underline{v}_0$ has only a z-component.

The Boltzmann equation then becomes

$$in\underline{\Psi}_n + \gamma W_{nn'}\{\underline{\Psi}_{n'}\} = \gamma \frac{\partial f^0}{\partial E} \underline{v}_n. \tag{3.21}$$

The general solution of this equation depends on a knowledge of the Fourier components $W_{nn'}$ of the integral operator W. The features which we now wish to emphasise are not critically dependent on the form of W and so for simplicity we make the collision time assumption W = 1. It should however be remarked that, for anisotropic systems, no justification can be given for such an assumption. With W = 1

$$\underline{\Psi}_n = \underline{v}_n \frac{\partial f^0}{\partial E} \frac{\gamma}{in+\gamma}. \tag{3.22}$$

This yields the conductivity:

$$\sigma_{ij} = e^2 z \iint dE\, dk_z \frac{\partial f^0}{\partial E} \sum_{n=-\infty}^{\infty} \left\{ \frac{v_{ni} v_{-nj} + \frac{in}{\gamma} v_{ni} v_{-nj}}{(n/\gamma)^2 + 1} \right\}. \tag{3.23}$$

In the limit $\gamma \ll 1$, this gives

$$\sigma_{ij} \propto \begin{pmatrix} (\frac{1}{\omega_c\tau})^2 & \frac{1}{\omega_c\tau} & \frac{1}{\omega_c\tau} \\ \frac{1}{\omega_c\tau} & (\frac{1}{\omega_c\tau})^2 & \frac{1}{\omega_c\tau} \\ \frac{1}{\omega_c\tau} & \frac{1}{\omega_c\tau} & 1 \end{pmatrix}. \tag{3.24}$$

The entries here denote only the asymptotic dependence on the field and are not intended to suggest that the coefficients multiplying these powers are all the same. The corresponding resistivity is

$$\rho = \sigma^{-1} \propto \begin{pmatrix} 1 & \frac{1}{\omega_c\tau} & 1 \\ \frac{1}{\omega_c\tau} & 1 & 1 \\ 1 & 1 & 1 \end{pmatrix}. \tag{3.25}$$

The constant diagonal matrix elements of ρ (in general different from the zero field resistivities and from each other) indicate saturation of the transverse and longitudinal magneto resistance in high fields.

This behaviour has been verified almost without exception for the longitudinal magnetoresistance of metals, but it is not always observed for the transverse case. Poly-crystalline samples of many comparatively simple metals (Au, Cu, Ag, Sn, ...) show a transverse magneto resistance linear at high fields. This behaviour was for many years unexplained. A clue to its origin is provided by the complicated angular dependences shown in the magneto resistance of single crystals of the same metals: for some directions of the crystal axes with respect to the magnetic field a saturating high field behaviour is observed, and for others a quadratic dependence is seen up to the highest fields used (Alekseevskii and Gaidukov 1959). This results from the character of the k-space constant energy surface of such a metal. Its Fermi surface extends throughout k-space, typically more like the surfaces of intersecting undulating cylinders than of separate distorted spheres. If the axes of such a cylinder makes an angle $\frac{1}{2}\pi - \theta$ with the field direction z, the intersection of the surface with the plane k_z = const. becomes a very extended ellipse for k small. The period of rotation T is then typically of order $2\pi/(\omega_c\theta)$ and

$$\gamma \sim 1/(\theta\omega_c\tau). \tag{3.26}$$

The expansion in powers of γ is not permissible when

$$\theta \sim \frac{1}{\omega_c\tau} << 1. \tag{3.27}$$

Instead a partial expansion of the type

$$\sigma = \sum \left(\frac{1}{\omega_c\tau}\right)^n \sigma^{(n)}\left(\frac{1}{\omega_c\tau\theta}\right) \tag{3.28}$$

may be used. In the limit (3.27) the resulting contribution of the extended orbits to the resistivity is

$$\rho_{ii} \propto \frac{\cos^2\beta}{\theta^2 + \left(\frac{1}{\omega_c\tau}\right)^2} \tag{3.29}$$

where β is the angle made by the current direction with the major axis of the "elliptical" orbit. Thus, as the direction θ=0 is approached there appears a quadratic rise of the transverse resistivity with H. The extended orbits do not affect the longitudinal resistivity ($\beta = \pi/2$) in this way.

Averaging (3.29) over angles in a range $\delta\theta$ such that $1 >> \delta\theta >> 1/(\omega_c\tau)$ gives an average resistivity

$$\bar{\rho} \propto H\cos^2\beta \tag{3.30}$$

(Lifshitz and Peschanskii 1959, 1960, Ziman 1958). This is thought to be the basis of the linear dependence on H of the transverse magneto resistance of polycrystalline specimens, where the averaging over angles is associated with the different orientations of the crystallites.

It is not obvious that averaging the resistivity in such a simple way is adequate since in practice the current in a polycrystal will follow complicated leakage paths. An improved though still approximate treatment can be given, based on an "effective medium" method and similar in spirit to the coherent potential approximation used in the theory of disordered electronic, lattice, and spin systems. The method is to consider a single crystallite, most simply taken to be a sphere, in an effective homogeneous medium approximating in some sense the properties of the remaining crystallites. With this effective medium is associated a conductivity tensor $\sigma_m(H)$, later to be determined by a self -consistency condition. The solution is then obtained for the current flow in the system consisting of the single crystallite (whose conductivity tensor is known, c.f. (3.23)) inside the homogeneous medium. The relationship of the current inside the crystallite to the external field can be represented by a conductivity tensor $\sigma(\theta,\varphi;\ H;\ \sigma_m(H))$ depending on the directions θ,φ of the crystallite axes as well as on the field H and the conductivity σ_m. The self-consistency condition determining σ_m is then

$$\frac{1}{4\pi}\int \sigma \sin\theta \, d\theta \, d\varphi = \sigma_m . \tag{3.31}$$

The polycrystalline specimen is then assumed to have conductivity tensor σ_m.

Such a procedure gives an effective high field transverse resistivity

$$\rho_m \propto H^{2/3} \tag{3.32}$$

(Stachowiak, to be published). This approach includes some of the effects omitted in the averaging used to obtain (3.30). But the problem of conduction in polycrystalline materials awaits a more satisfactory treatment.

IV. QUANTUM EFFECTS

Up to now we have completely neglected effects due to quantisation of the electron states. It is difficult to include them within a formalism which goes over conveniently to the low field limit. If we are prepared to lose sight of that limit and also to specialise to the case of free electrons, the use of Landau state representation suggests itself (Argyres 1958a.b).

4.1. Landau representation

The Landau states are eigenfunctions $\phi_{n\underline{k}}$ for a single electron in a uniform magnetic field:

$$\phi_{n\underline{k}} = \Phi_n(x-x_0)\exp[ik_y y + ik_z z](L_y L_z)^{-1/2} \tag{4.1}$$

The gauge A: (0,Hx,0) has here been used, and the system has been taken to be infinite in the x-direction and of lengths L_y, L_z in the remaining directions. Φ is the Hermite function of order n with argument $\epsilon^{\frac{1}{2}}(x-x_0)$ where $x_0 = k_y/\epsilon$. The associated energy eigenvalue is

$$E_{n\underline{k}} = \frac{\hbar^2 k_z^2}{2m} + \hbar\omega_c(n+\tfrac{1}{2}) \tag{4.2}$$

We now consider whether a generalised Boltzmann equation can be derived by rephrasing the development of section II in Landau representation.

The linearised equation (2.6) for the density matrix then takes the form

$$i\hbar s f_{\ell\ell'} + C_{\ell\ell'} = (E_\ell - E_{\ell'})f_{\ell\ell'} + \sum_{\ell''}(W_{\ell\ell''}f_{\ell''\ell'} - f_{\ell\ell''}W_{\ell''\ell'}) \tag{4.3}$$

where ℓ denotes $n\underline{k}$ and, for example, $f_{\ell\ell'}$ denotes $\int \phi^*_\ell f \phi_{\ell'}\, d^3\underline{x}$. The form of the matrix element $W_{\ell\ell'}$ of the scattering potential and of the inhomogeneous term are crucial in what follows.

(4.3) may be separated into diagonal and non-diagonal components as in (2.13), (2.14) but now in Landau representation. With subscripts D, ND representing respectively diagonality and non-diagonality in this representation the separated equations are formally

$$i\hbar s\, f_D + C_D = \mathcal{W}^{D}{}_{ND}\, f_{ND} \tag{4.4}$$

$$i\hbar s\, f_{ND} + C_{ND} = \mathcal{W}^{ND}{}_{D}\, f_D + (\mathcal{H} + \mathcal{W})^{ND}{}_{ND}\, f_{ND}\,. \tag{4.5}$$

The behaviour of the inhomogeneous terms in the weak coupling limit depends on the relative orientation of the fields $\underline{\mathcal{E}}$ and $\underline{H}$:

for $\underline{\mathcal{E}} /\!/ \underline{H}$ (longitudinal case)

$$C_D \sim \lambda_w^0$$
$$C_{ND} \sim \lambda_w^1 \tag{4.6}$$

while for $\underline{\mathcal{E}} \perp \underline{H}$ (transverse case)

$$C_D \sim \lambda_w^1$$
$$C_{ND} \sim \lambda_w^0\,. \tag{4.7}$$

This is because C involves $\underline{\mathcal{E}}\cdot\underline{x}$ (2.9); the longitudinal component of $\underline{x}$ is diagonal in the Landau representation, while the transverse components only link Landau states whose principal quantum numbers n differ by ± 1. The diagonal components of f can only be expected to play a dominant role in the weak coupling limit for the longitudinal case. For the transverse case the matrix elements of f between states differing by ± 1 will be most important.

4.2. LONGITUDINAL CASE

We consider first the longitudinal case $\underline{\mathcal{E}} /\!/ \underline{H}$. The λ_W ordering is used in the manner employed earlier (without the need for a simultaneous λ_H ordering, since terms like $\mathcal{H}^D{}_D$ do not appear). The term $\mathcal{W}^{ND}{}_{ND}\, f_{ND}$ in (4.5) is discarded, being of higher order in λ_W. Solving formally for f_{ND} in terms of f_D and inserting into (4.4) then gives an equation for f_D. The equation contains a complicated inhomogeneous term of which the dominant part in the weak coupling limit is just C_D:

$$\left[i\hbar s - \mathcal{W}^D{}_{ND}\,(i\hbar s - \mathcal{H}^{ND}{}_{ND})^{-1}\,\mathcal{W}^{ND}{}_D \right] f_D = - C_D . \qquad (4.8)$$

Writing the explicit forms for the integral operators and for C_D this becomes

$$\frac{e}{\hbar}\, \mathcal{E}_z \frac{\partial f^\circ_\ell}{\partial k_z} + \frac{2\pi}{\hbar} \sum_{\ell''} |W_{\ell\ell''}|^2 \left[f_{\ell\ell} - f_{\ell''\ell''} \right] \delta(E_{\ell''} - E_\ell) = 0 \qquad (4.9)$$

where

$$f^\circ_\ell = \frac{1}{\exp[\beta(E_\ell - \varsigma)] + 1} \qquad (4.10)$$

The diagonal singularity assumption, corresponding to (2.39) but now in Landau representation, is now needed for two purposes. One is to reduce the collision term in (4.9) to a form involving the transition probability for scattering by separate impurities (c.f. (2.32)-(2.35)). The second is to make the corrections to the Boltzmann equation (4.9) small. These corrections come solely from the terms of higher order in λ_W, the field having been completely included. By an argument completely analagous to that used in discussing (2.38) these terms are small provided the diagonal singularity property holds in Landau representation.

By averaging over all distributions of impurities this can be shown to be the case for scatterers whose range is very much less than the Larmor radius $\epsilon^{-\frac{1}{2}}$. For this case, the quantum transport equation (4.9) can be used to discuss the longitudinal effects.

4.3 Transverse case

For the transverse case $\mathcal{E} \perp \underline{H}$ it becomes appropriate to distinguish the matrix elements of C and of f between states $n\underline{k}$ and $n'\underline{k}'$ with $\underline{k}=\underline{k}'$ and $n=n'\pm 1$. Such a matrix element we denote by subscript Δ, any other matrix elements being denoted by $N\Delta$. Separating equation (4.3) in this way for the case $\underline{\epsilon}\perp\underline{H}$ leads to

$$i t_s f_\Delta + C_\Delta = (\mathcal{H}+\mathcal{W})^\Delta{}_\Delta f_\Delta + \mathcal{W}^\Delta{}_{N\Delta} f_{N\Delta} \tag{4.11}$$

$$i t_s f_{N\Delta} + C_{N\Delta} = \mathcal{W}^{N\Delta}{}_\Delta f_\Delta + (\mathcal{H}+\mathcal{W})^{N\Delta}{}_{N\Delta} f_{N\Delta}. \tag{4.12}$$

Now $C_\Delta \sim \lambda_W^0$, while $C_{N\Delta} \sim \lambda_W^1$. However C_Δ, $C_{N\Delta}$ and $\mathcal{H}^\Delta{}_\Delta$ each vanish with H and it becomes necessary to take note of this in order to include all terms which may dominate when H is small as well as large. For this reason we introduce a nominal λ_H parametrization, but will not limit ourselves to any small H regime though we would like to be able to include it. Then

$$\mathcal{H}^\Delta{}_\Delta \sim \lambda^2_H \;;\; \mathcal{H}^{N\Delta}{}_{N\Delta} \sim 1 \;;\; \mathcal{W} \sim \lambda_W$$

$$C_\Delta \sim \lambda_H \;;\; C_{N\Delta} \sim \lambda_W \lambda_H . \tag{4.13}$$

With the weak scattering assumption, the term in (4.12) involving $\mathcal{W}^{N\Delta}{}_{N\Delta}$ can be discarded; $f_{N\Delta}$ is then found in terms of f_Δ. Inserting the result into (4.11) and extracting the dominant (weak coupling) part of the inhomogeneous term gives the equation for f_Δ:

$$\left[i t_s - \mathcal{H}^\Delta{}_\Delta - \mathcal{W}^\Delta{}_{N\Delta}(i t_s - \mathcal{H}^{N\Delta}{}_{N\Delta})^{-1}\mathcal{W}^{N\Delta}{}_\Delta \right] f_\Delta = -C_\Delta . \tag{4.14}$$

The detailed form of (4.14) is

$$(E_l - E_{l'}) f_{l'l} + \frac{i\pi}{\hbar} \sum_{l''} f_{l'l} \left(|W_{ll''}|^2 \delta(E_{l''} - E_l) + |W_{l'l''}|^2 \delta(E_{l''} - E_{l'}) \right)$$

$$= e\hbar (2m\hbar\omega_c)^{-1/2} [f^0_{l'} - f^0_l] \left[(\mathcal{E}_x + i\mathcal{E}_y) n^{1/2} \delta_{n',n-1} + (\mathcal{E}_x - i\mathcal{E}_y)(n+1)^{1/2} \delta_{n',n+1} \right]$$

(4.15)

where $l = n\underline{k}$, $l' = n'\underline{k}'$ with $\underline{k}' = \underline{k}$ and $n' = n \pm 1$.

The validity of the development leading to (4.15) requires a diagonal singularity in the generalised form

$$\overline{\langle l | WDW | l' \rangle} \sim \delta_{ll'} \overline{\langle l | WDW | l \rangle}.$$

(4.16)

Here D is a function of $\boldsymbol{x}_0$, or a component of $\underline{x}$ multiplied by functions of $\boldsymbol{x}_0$. As in the longitudinal case, (4.16) can be verified for a random distribution of scatterers with range small compared to the Larmor radius. (4.16) makes the correction term to (4.15) small, and also allows the collision term to be rewritten in terms of transition probabilities for scattering by separate impurities.

(4.9) and (4.15) are the quantum transport equations for the cases $\underline{\mathcal{E}} \parallel \underline{H}$ and $\underline{\mathcal{E}} \perp \underline{H}$ respectively. They apply only to the free electron system with sufficiently short range scattering.

4.4. Discussion

(4.9) and (4.15) can be used to determine the currents and conductivity components using

$$\overline{j_\mu} = \sigma_{\mu\nu} \mathcal{E}_\nu = \mathrm{Tr}\, f\, j_\mu \qquad (4.17)$$

j_z is diagonal in the Landau representation, while j_x, j_y link states whose quantum numbers n differ by 1. From (4.9), (4.15) the diagonal matrix elements of f are generated by the longitudinal field $\mathcal{E}_z$ and the matrix elements with n differing by 1 are generated by the transverse field $\mathcal{E}_x$, $\mathcal{E}_y$. Hence

$$\sigma_{xz} = \sigma_{yz} = \sigma_{zx} = \sigma_{zy} = 0$$

and σ_{zz} can be discussed using (4.9), while $\sigma_{xy}(=-\sigma_{yx})$ and σ_{xx} $(=\sigma_{yy})$ require the use of (4.15) only.

For the particular case of zero range (delta-function) scatterers, the equations are easy to solve since the linked terms in the collision operators (e.g. the term in (4.9) involving $f_{\ell''\ell''}$ as a factor) then vanish. For the longitudinal case the result for the ratio of the resistivity to its zero field value is

$$\frac{\sigma_{zz}(0)}{\sigma_{zz}(H)} = 1 + \pi^2\sqrt{2}\,\frac{k_BT}{\hbar\omega_c}\left(\frac{\hbar\omega_c}{\zeta_0}\right)^{1/2}\sum_{r=1}^{\infty}\frac{(-1)^r r^{1/2}\cos\left(2\pi r\frac{\zeta_0}{\hbar\omega_c}-\frac{\pi}{4}\right)}{\sinh(2\pi^2 r k_BT/\hbar\omega_c)} \qquad (4.18)$$

where ζ_0 is the fermi energy in zero field and we have taken $\hbar\omega_c \ll \zeta_0$. A similar result is obtained for the transverse case (Argyres 1958a,b).

The oscillating terms have a form similar to the Landau Peierls terms in the free electron diamagnetism, and have the same origin - the Landau levels passing through the Fermi level as their separation varies with magnetic field. This is a quantum effect omitted from the treatment of Section II. For observation it is necessary that k_BT and the collision broadening of the states should be small compared to their spacing.

(4.18) and the corresponding result for the transverse case show no steady magneto-resistance, in agreement with the conclusions of 3.1 for isotropic systems. For $\hbar\omega_c \gtrsim \zeta$ (the quantum limit) further quantum effects occur (e.g. the suppression of the transverse current) associated with the condensation of all electrons into the n=0 oscillator state. This situation is only of academic interest for metals because of the high fermi energy but is experimentally attainable for semiconductors with a low concentration of carriers.

In Section II the possibility of treating high field transverse effects by expansion in powers of λ_W was mentioned. (4.15) is amenable to such a treatment because in the terms on the left hand side the factors multiplying f are respectively of order $\hbar\omega_c$ and $\hbar/\tau$. The expansion in powers of λ_W (actually in powers of $^1/(\omega_c\tau)$) is obtained by iteration, treating the second term on the left hand side of (4.15) as small. A more complete development of the $^1/\omega_c\tau$) expansions for the transverse case is obtained directly from (4.11), (4.12) Adams and Holstein (1959) or from a corresponding Kubo formula (Kubo, Hasegawa, and Hashitsume 1959) without the need for the diagonal singularity required for (4.15). It is clear that no such expansion method can be applied to the longitudinal equation (4.9) nor to the transverse equation generalised to deal with an oscillating electric field of frequency $\omega = \omega_c$ (cyclotron resonance). In each case the response to the electric field is limited solely by the scattering.

The leading order term in the iteration of the (static) equation (4.15) gives the following contribution to the transverse current:

$$\overline{j_x + i j_y} = \frac{e}{\hbar\omega_c} \mathrm{Tr}\left\{ (j_x + i j_y)[\rho, \underline{\mathcal{E}} \cdot \underline{x}] \right\} = \frac{Nec}{H} (-i\mathcal{E}_x + \mathcal{E}_y). \qquad (4.19)$$

($j_x + i j_y$ is related to the ladder operator which takes a Landau state into the corresponding state with n increased by 1).

(4.19) shows that to leading order in the $1/\omega_c\tau$ expansion the Hall coefficient for the quantum case has the classical value $1/(Ne\,c)$, and that the transverse conduction vanishes. The next order in the expansion gives non-vanishing contributions of order $1/(\omega_c^2\tau)$ to all the transverse components of σ, all of which contain quantum effects. In this high field situation the migration along $\underline{\mathcal{E}}$ ($\perp \underline{H}$) is only made possible by the collisions.

The quantum transport equations (4.9), (4.15) and the $1/(\omega_c\tau)$ expansions of the type discussed are in principle applicable to the more realistic case where the lattice is also present. The Landau states are then replaced by the wave functions of the electrons in the periodic lattice in the presence of a magnetic field. The approach is therefore in practice limited to systems in which the periodic potential can be included into an equivalent Hamiltonian (Blount 1962), e.g. by the introduction of an effective mass.

V. COLLISIONS BETWEEN CARRIERS

In the zero field case collisions between free electrons have no effect on the current because in each such collision the total momentum is conserved. Electron collisions can however modify the effects of impurity or phonon scattering, or can by themselves give rise to a resistance for electrons moving in a perfect lattice.

It is of some importance to see how such statements have to be modified when a magnetic field is also included.

We first consider the case of electrons interacting only with each other in the presence of a magnetic field. The same argument as in the field-free case clearly shows that the longitudinal resistance is zero. For the transverse case it is not so obvious how the result without electron collisions, given by (4.19), will be changed. We shall show that the electron-electron interation does not alter those results.

To discuss this we find it most convenient to use the Kubo formula (Kubo 1957) for the Ohmic conductivity:

$$\sigma_{\mu\nu} = \int_0^\infty dt \int_0^\beta d\lambda \, \langle J_\nu(0) \, J_\mu(t + i\hbar\lambda) \rangle , \tag{5.1}$$

where

$$\langle \ldots \rangle = \mathrm{Tr}\left[e^{-\beta \mathcal{H}} \ldots \right] / \mathrm{Tr}\left[e^{-\beta \mathcal{H}} \right] , \tag{5.2}$$

and $\underline{J}$ is the total current operator. The Hamiltonian $\mathcal{H}$ is

$$\mathcal{H} = \sum_{i=1}^{N} \frac{m}{2e^2} \, j_i^2 + U , \tag{5.3}$$

where

$$\underline{j}_i = \frac{e}{m} \left(\underline{p}_i - \frac{e}{c} \underline{A}(\underline{r}_i) \right) \tag{5.4}$$

is the current operator for the ith electron in the absence of interactions. The electron-electron interaction is

$$U = \sum_{ij} u(\underline{r}_i - \underline{r}_j) . \tag{5.5}$$

Because U is a function only of coordinates, the total current is

$$\underline{J} = \sum_i \dot{\underline{j}}_i \,. \tag{5.6}$$

Also, since the scatterings represented by U conserve total momentum, the commutator of U with J vanishes and the equation of motion of J is the same as when U = 0. Hence the combinations

$$J_{\pm} = J_x \pm i J_y \tag{5.7}$$

evolve harmonically with frequency $\pm\omega_c = \pm\, eH/(mc)$:

$$[J_{\pm}, \mathcal{H}] = \pm\,\omega_c J_{\pm} \,. \tag{5.8}$$

It follows that

$$\begin{aligned} 4i\,\sigma_{xy} &= \int_0^{\infty} dt \int_0^{\beta} d\lambda \left[\langle J_+(0)\,J_-(t+i\hbar\lambda)\rangle - \langle J_-(0)\,J_+(t+i\hbar\lambda)\rangle\right] \\ &= \langle J_+(0)J_-(0)\rangle\, \mathcal{J}(\omega_c) - \langle J_-(0)J_+(0)\rangle\, \mathcal{J}(-\omega_c)\,, \end{aligned} \tag{5.9}$$

where

$$\mathcal{J}(\omega) = [\pi\delta(\omega) - i/\omega][e^{\beta\hbar\omega} - 1]/\hbar\omega \,. \tag{5.10}$$

However

$$\langle J_+(0)J_-(0)\rangle = \langle J_-(0)\,J_+(i\hbar\beta)\rangle = e^{-\hbar\omega_c\beta}\langle J_-(0)J_+(0)\rangle \tag{5.11}$$

and the commutator of J_+ with J_- is a c-number:

$$[J_+, J_-] = -2Ne^2\hbar\omega_c/m . \tag{5.12}$$

These relations are together sufficient to determine the average on the right hand side of (5.9) with the result that

$$\sigma_{xy} = \frac{Nec}{H} \tag{5.13}$$

In a precisely similar way we find $\sigma_{xx} = \sigma_{yy} = 0$ and the Hall coefficient becomes

$$R = [H\sigma_{xy}]^{-1} = 1/(Nec) . \tag{5.14}$$

These results (c.f. (4.19) are independent of many body effects. They are also independent of statistics ; there are no oscillations in the Hall coefficient. All the above results can alternatively be obtained from the equation of motion for the density matrix of the many particle system.

The time evolution (5.8) of $J_{\pm}$ has been used by Kohn to show that the electron-electron collisions do not alter the cyclotron resonance frequency of the system. We recover this result by inserting a factor exp(iωt) into the integrand of (5.1) to describe the oscillations of the electric field. The first factor in (5.10) then becomes $[\pi\delta(\omega \pm \omega_c) - i/(\omega \pm \omega_c)]$ which yields a sharp resonance at $\omega = \pm\omega_c$.

If, in addition to the electron-electron collisions, scattering due to impurities or phonons is considered, the above development still applies for the high field behaviour ($\omega_c\tau \gg 1$): the electron-electron interaction does not alter the high field Hall coefficient.

It is not possible to dispose of many-body effects in the same way when a lattice is present, or when $\omega_c\tau$ is not large. To the extent to which the low field equation (2.36) can be applied ($\hbar\omega_c \ll k_BT, \zeta$) the effectiveness of electron-electron collisions will be measured by a collision frequency which is the same as in the zero field case. Then the usual arguments (Peierls 1955) based on the exclusion principle suggest that this collision frequency is of order

$$\left(\frac{e^2}{a}\right)^2 \frac{(k_B T)^2}{\zeta^3} , \tag{5.15}$$

and unimportant in most situations (Langer 1960, 1961, 1962a,b). At higher fields further effects, such as the modification by the field of screening by the electrons can occur (Horing 1969).

VI. COLLISIONS WITH PHONONS

In relatively pure materials the electron scattering is predominantly by phonons, except at very low temperature. The collisions of the electrons with the phonons will drive the phonon system out of equilibrium unless phonon-phonon scattering is sufficiently frequent to maintain equilibrium. We shall later restrict our discussion to that case which applies for instance, at high temperatures.

6.1. Transport equations with phonon scattering

With only phonon scattering the interaction W of Section II (2.1) or Section IV is to be replaced by the electron-phonon interaction

$$V_{elph} = \sum_q \left(\xi_q a_q e^{i\underline{q}\cdot\underline{r}} + \xi_q^* a_q^+ e^{-i\underline{q}\cdot\underline{r}} \right), \tag{6.1}$$

where a_q^+ is the creation operator for a phonon of wave vector q, and $\underline{r}$ is the electron coordinate. $\mathcal{H}_0$ (2.2) has also now to include a phonon Hamiltonian.

The formalism of Sections II and IV then applies except that in addition to the label ℓ describing the "state" of an electron, we have to add a label N to describe the state of the phonon system. N is the set of quantum numbers N_q describing the excitation of the separate modes.

For illustration we consider the quantum case (Argyres 1958 a,b). The system is then labelled by N,ℓ where ℓ is the Landau state label $n\underline{k}$. The linearised equation of motion for the density matrix is (c.f. (4.3))

$$i\hbar s\langle N\ell|f|N'\ell'\rangle + \langle N\ell|C|N'\ell'\rangle$$

$$= (E_{N\ell}-E_{N'\ell'})\langle N\ell|f|N'\ell'\rangle + \sum_{N''\ell''}\left\{\begin{array}{l}\langle N\ell|V_{elph}|N''\ell''\rangle\langle N''\ell''|f|N'\ell'\rangle \\ -\langle N\ell|f|N''\ell''\rangle\langle N''\ell''|V_{elph}|N'\ell'\rangle\end{array}\right. \tag{6.2}$$

$E_{N\ell}$ denotes the eigenvalue of the unperturbed Hamiltonian $\mathcal{H}_0$ for the eigenstate with N phonons, and the electron in state ℓ :

$$E_{N\ell} = \sum_q (N_q+\tfrac{1}{2})\hbar\omega_q + E_\ell = E_N + E_\ell \,. \tag{6.3}$$

In the limit of zero scattering the inhomogeneous term C is diagonal in the phonon labels and links electron states $n\underline{k}$, $n'\underline{k}$ with $\underline{k}=\underline{k}'$ but with $n-n'=0$ or ±1 depending on whether $\underline{\mathcal{E}} /\!/ \underline{H}$ or $\underline{\mathcal{E}}\perp\underline{H}$. It becomes appropriate, as in Section IV to discuss two different types of diagonality. In Section IV these were denoted by subscripts D, Δ ($n-n' = 0$ or ±1 respectively) applying to the longitudinal and transverse cases respectively. In the present extension of that work to electron phonon collisions diagonality also means no change in the phonon labels:

$$D : \langle N\ell| \quad |N'\ell'\rangle, \text{ with } N'=N,\ \underline{k}'=\underline{k},\ n'=n\ (\underline{\mathcal{E}} /\!/ \underline{H})$$

$$\Delta : \langle N\ell| \quad |N'\ell'\rangle, \text{ with } N'=N,\ \underline{k}'=\underline{k},\ n'=n\pm1\ (\underline{\mathcal{E}} \perp \underline{H}). \tag{6.4}$$

Consider for simplicity the longitudinal case ($\underline{\mathcal{E}} /\!/ \underline{H}$). We separate the equation (6.2) into diagonal and non-diagonal parts, and apply the ordering corresponding to weak electron phonon interaction; we then eliminate the nondiagonal components of f (neglecting for the moment the term $\mathcal{W}^{ND}{}_{ND}\, f_{ND}$). The resulting equation (c.f. (4.8)) is, in the weak coupling limit:

$$i\hbar s\langle Nl|f|Nl\rangle - \sum_{l''N''} |\langle Nl|V_{elph}|N''l''\rangle|^2 \delta(E_{Nl} - E_{N''l''}) \{\langle N''l''|f|N''l''\rangle - \langle Nl|f|Nl\rangle\}$$

$$= - \langle Nl|C|Nl\rangle. \tag{6.5}$$

The sum over N'' can be reduced immediately since the matrix elements of V_{elph} are each sums over q of matrix elements involving the creation and annihilation operators a_q^+, a_q. These operators only link states in which N_q differs by ± 1. In the collision term in (6.5) no interference between the phonons of different q's nor between emission and absorption processes then occurs. The contribution of the emission process to the collision term is for example:

$$\sum_{l''q} |\xi_q|^2 |\langle l|e^{i\underline{q}\cdot\underline{r}}|l''\rangle|^2 |\langle \ldots N_q \ldots |a_q| \ldots N_q+1 \ldots\rangle|^2 \delta(E_{l''} - E_l + \omega_q) \left\{ \begin{array}{l} \langle \ldots N_q+1 \ldots l''|f|\ldots N_q+1 \ldots l''\rangle \\ -\langle \ldots N_q \ldots l|f|\ldots N_q \ldots l\rangle \end{array} \right\}. \tag{6.6}$$

In order to average operators such as the current, depending only on the electron variables, we require only

$$\langle l|f|l'\rangle = \sum_N \langle lN|f|l'N\rangle. \tag{6.7}$$

We now assume that at all times the lattice vibrations are in thermal equilibrium at absolute temperature T. Then there are no phase relationships between probability amplitudes of different states $|N\rangle$ and the probability of finding the lattice in any one of these states is

$$P(N) = e^{-E_N/k_BT} \Big/ \sum_{N'} e^{-E_{N'}/k_BT}.$$

(6.8)

It follows that

$$\langle l N | f | l' N' \rangle = \langle l | f | l' \rangle \, P(N) \, \delta_{NN'} .$$

(6.9)

Thus summing (6.5) over all N and denoting by $\bar{N}_q$ the average value of $a_q^+ \, a_q$ we finally obtain

$$\sum_{qn''} \left[\langle n'' k+q | f | n'' k+q \rangle - \langle nk | f | nk \rangle \right] W_{nn''}(k,q) = \frac{e}{\hbar} \mathcal{E} \frac{\partial}{\partial k_z} f_l^{\circ}$$

(6.10)

where

$$W_{nn''}(k,q) = \frac{2\pi}{\hbar} |\xi_q|^2 \, |\langle n\underline{k} | e^{i\underline{q}\cdot\underline{r}} | n'' \, k+q \rangle|^2$$

$$\left[\bar{N}_q \, \delta(E_{n''k+q} - E_{nk} + \omega_q) + (\bar{N}_q + 1) \, \delta(E_{n''k+q} - E_{nk} - \omega_q) \right).$$

(6.11)

This is the Boltzmann equation for the longitudinal case (Argyres 1958b). A similar generalisation of the procedure of 4.3 for the transverse case results in the weak-coupling equation for the elements $\langle n\underline{k} | f | n' \, \underline{k} \rangle$ with $n' = n \pm 1$:

$$(E_{nk} - E_{n'k}) \langle nk | f | n'k \rangle + \frac{i}{2} \sum_{qn''} \left[W_{nn''}(k,q) + W_{n'n''}(k,q) \right] \langle nk | f | n'k \rangle$$

$$= e\hbar \, (2m\hbar\omega_c)^{-1/2} \left[f^{\circ}_{n'k} - f^{\circ}_{nk} \right] \left[(\mathcal{E}_x + i\mathcal{E}_y) n^{1/2} \, \delta_{n',n-1} + (\mathcal{E}_x - i\mathcal{E}_y)(n+1)^{1/2} \, \delta_{n',n+1} \right].$$

(6.12)

The validity of such a procedure again requires a diagonal singularity property of the type (4.16) but now to be possessed by $V_{e\ell ph}$. Wave vector conservation automatically provides a delta function relating the wave vector labels of the Landau states but the required relationship between the quantum numbers n, n' only occurs in the situation in which the effective potential due to exchange of the phonon is short range compared to the Larmor radius. This occurs with, for example either

(i) elastic acoustic phonon scattering with

$$\hbar\omega_q \ll E_\ell, k_B T \quad \text{and} \quad \omega_q \propto q, \ \xi_q \propto q^{1/2} \qquad \text{(Argyres 1958b)}$$

or (ii) optical phonon scattering with ξ_q and ω_q independent of q. (These are related to the well known conditions for effectively s-wave scattering in the zero field case.)

If quantum effects are unimportant, the formalism of Section II can be applied in an analogous manner to the phonon scattering case. The required singularity property (c.f. (2.39)) is automatically provided by the wave vector conservation, and the result is a Boltzmann equation analogous to (2.36) but with $W_{\ell\ell_1}$ replaced by the usual zero field phonon collision term.

6.2. Magneto phonon resonance

Magneto phonon resonance (Gurevich and Firsov 1961, Firsov and Gurevich 1962, Firsov et al. 1964) is one quantum effect which can occur when scattering is by optical phonons. The effect arises from an enhancement of the scattering when the phonon frequency is an integral multiple of the cyclotron frequency:

$$\omega_q \sim \omega_0 = N\omega_c . \tag{6.13}$$

This is because the density of Landau states is $^1/k_z$; when (6.13) is satisfied it is possible to satisfy the energy conservation delta-functions in (6.11) with $|n''-n|=N$, and k_z and k_z+q_z both zero thus obtaining a large transition rate because of the large density of states factors. This results in a conductivity which oscillates as a function of H. (6.13) is the condition for maxima in the transverse conductivity at high fields and minima for the longitudinal conductivity. The above argument is incomplete for the longitudinal case because it relates to values of k_z close to zero, which do not contribute appreciably to the longitudinal current. Other scattering, particularly by acoustic phonons can change the longitudinal minima into maxima (Gurevich and Firsov 1964, Kharus and Tsidilovskii 1971).

The amplitude of the transverse conductivity calculated from (6.12) is divergent on resonance. A modification of the weak-coupling scheme to take account of the finite lifetime of intermediate states is required to calculate the on-resonance amplitude. This is not needed for the longitudinal case which is anyway finite because of the small k_z factors from the current matrix elements.

The broadening of the intermediate states is given by the imaginary part of an appropriate self-energy. If the effect arises only from the optical phonons the self-energy associated with state nk is to leading order in the interaction

$$\Sigma_{nk}(\omega) = \sum_{n''q} |\xi_q|^2 \, |\langle nk| e^{iq\cdot r} | n'' k+q\rangle|^2 \left[\frac{\bar{N}_q + 1 - f^0_{n''k+q}}{E_{n''k+q} - \omega + \omega_q} + \frac{\bar{N}_q + f^0_{n''k+q}}{E_{n''k+q} - \omega - \omega_q} \right] \tag{6.14}$$

(Palmer 1970). Since this is required on resonance, the broadening has also to be included in the intermediate states in (6.14), and the self-energy determined self-consistently (Palmer 1970), More usually the finite lifetime of the intermediate states is largely accounted for by other scattering mechanisms, including collision broadening (Barker 1970).

The field at which the resonance should occur can be shifted from that given by (6.13) by self-energy effects (Palmer 1970, Nayakama 1969). The real part of the self-energy (6.14) gives rise to a relative displacement of the two Landau levels involved, displacing the resonance to higher fields which are then given, instead of (6.13), by

$$\omega_0 = \left[E_{n+N,0} + \Sigma_{n+N,0}(E_{n+N,0})\right] - \left[E_{n,0} + \Sigma_{n,0}(E_{n,0})\right]. \tag{6.15}$$

If the two lowest levels (n=0, n+N=1) are involved, equation (6.15) for the field at which the shifted resonance is seen reduces to

$$\omega_0 \sim \omega_c (1 - 0.42\,\alpha) \tag{6.16}$$

where α is the Frohlich coupling constant

$$\alpha = e^2 \left(\frac{m}{\omega_0}\right)^{1/2} \left(\frac{1}{\varepsilon_\infty} - \frac{1}{\varepsilon_0}\right) \tag{6.17}$$

giving a dimensionless measure of $|\xi_q|^2$ for polar modes. The shift given by (6.16) has been measured (Mears et al. 1968) and is satisfactorily accounted for by the theory. The shift is a polaron effect related to the polaron effective mass correction (Larsen 1970): the weak-coupling correction to the effective mass (in a magnetic field) is given by the coefficient of k_z^2 in the expansion of (6.14).

VII. CONCLUDING REMARKS

No discussion has been given here of a number of topics in magnetoconductivity, of which we list the following

(i) strong-coupling regimes
(ii) non Ohmic effects
(iii) derivation of coupled electron-phonon Boltzmann equations
(iv) effect of electron interactions in the presence of a lattice, or other interactions
(v) high field effects in the presence of a lattice.

A great deal of work has been done on some of these (notably (i) in the low concentration regime, (iv) using e.g. the random phase approximation and (v) using effective Hamiltonian methods). But important aspects of them all are not completely understood (many of the difficulties are not special to the magnetic field case). Also some of the topics discussed (e.g. the derivation of quantum transport equations) require less restrictive treatments. Meanwhile a great deal of experimental work is currently being done in the high field regime, particularly on high purity semiconductors, making the need for further theoretical advances all the more apparent.

REFERENCES

Adams E.N. and Holstein T.D., 1959, J. Phys. Chem. Solids 10, 254.

Alekseevskii N.E. and Gaidukov Yu.P., 1959, JETP 8, 383.

Argyres P.N., 1958a, J.Phys. Chem. Solids 4. 19.

Argyres P.N., 1958b, Phys. Rev. 109, 1115.

Argyres P.N., and Adams E.N., 1956, Phys. Rev. 104, 900

Barker J.R., 1970, Phys. Lett. 33A, 516.

Blount E.I., 1962, Phys. Rev. 126, 1636.

Chester G.V., and Thellung A., 1959, Proc. Phys. Soc. 73, 745.

Firsov Yu. A. and Gurevich V.L., 1962, JETP 14, 367.

Firsov Yu. A., Gurevich V.L., Parfenev P.V. and Shalyt S.S., 1964, Phys. Rev. Lett. 12, 660.

Greenwood D.A., 1958, Proc. Phys. Soc. 71, 585.

Gurevich V.L. and Firsov Yu.A., 1961, JETP 13, 137.

Gurevich V.L. and Firsov Yu. A., 1964, JETP 20, 489.

Horing N.J., 1969, Annals of Physics 54, 405.

Jones H. and Zener C., 1934, Proc. Roy. Soc. A144, 101.

Kharus G.I. and Tsidilovskii I.M., 1971, Soviet Phys. Semi conductors 5, 603.

Kohn W. and Luttinger J.M., 1957, Phys. Rev. 108, 590.

Kubo R., 1957, J. Phys. Soc. Japan 12, 570.

Kubo R., Hasegawa H., and Hashitsume N., 1959, J. Phys. Soc. Japan 14, 56.

Langer, J.M., 1960, Phys. Rev. 120, 714.

Langer J.M., 1961, Phys. Rev. 124, 1003.

Langer J.M., 1962a, Phys. Rev. 127, 5.

Langer J.M., 1962b, Phys. Rev. 128, 110.

Larsen D.M., 1970, Proc. Int. Conf. Physics of Semiconductors, Boston, p. 145.

Lifshitz I.M., Azbel M.Ia., Kaganov M.I., 1957, JETP 4, 41.

Lifshitz I.M. and Peschanskii V.G., 1959, JETP 8, 875.

Lifshitz I.M. and Peschanskii V.G., 1960, JETP 11, 137.

Mears A.L., Stradling R.A., and Inall E.K., 1968, J. Phys. C 1, 821.

Nayakama M., 1969, J. Phys. Soc. Japan 27, 636.

Palmer R.J., 1970, D.Phil. Thesis (Oxford Univ.)

Peierls R.E., 1931, Ann. Physik 10, 97.

Peierls R.E., 1932, Ergebn. D. Exakt. Naturw. 11, 264.

Peierls R.E., 1955, Quantum Theory of Solids (University Press-Oxford)

Sondheimer E.H. and Wilson A. H., 1951, Proc. Roy. Soc. A210, 173.

Stinchcombe R.B., 1961, Proc. Phys. Soc. 78, 275.

Van Hove L., 1955, Physica 21, 517.

Wigner E., 1932, Phys. Rev. 40, 749.

Ziman J.M., 1958, Phil. Mag. [8] 3, 1117.

TRANSPORT PROPERTIES IN GASES IN THE PRESENCE OF EXTERNAL FIELDS

J. J. M. Beenakker

Kamerlingh Onnes Laboratorium
University of Leiden, Netherland

I. INTRODUCTION

II. THE NON-EQUILIBRIUM POLARIZATIONS

The thermal conductivity
The shear viscosity

III. THE LIMITATION OF THE ONE MOMENT DESCRIPTION

IV. THE EFFECTIVE CROSS SECTIONS AND THEIR BEHAVIOUR

V. FIELD EFFECTS IN THE RAREFIED GAS REGIME

The transverse viscomagnetic heat flux

REFERENCES

I. INTRODUCTION

We will start a discussion of the influence of an external field on the transport properties of gases with a short survey of its history. For this we will have to go back to 1930, when Senftleben [1] discovered that the thermal conductivity of gaseous O_2 is influenced by a magnetic field. Further studies (Engelhard-Sack) showed that the viscosity was also affected. The fact that under the influence of the field NO behaved in the same way as O_2 suggested that it was a property of paramagnetic gases. This phenomenon became known as the Senftleben effect. It was extensively studied for nearly a decade. The results of this work may be summarized in the following way:

(i) In the presence of a magnetic rield, H, the transport coefficients decrease slightly (0.5 - 1%).

(ii) The effect is even in H.

(iii) At constant temperature it depends only on the ratio $\frac{H}{p}$, with p the gas pressure.

(iv) In mixtures with nonparamagnetic gases the effect is proportional to the concentration.

These results led Gorter [2] to a qualitative interpretation based on the change in the mean-free-path of an O_2 molecule caused by the magnetic field. This idea was elaborated more quantitatively by Zernike and Van Lier (1939) [3]. They assumed, following Gorter, that a quickly rotating paramagnetic diatomic molecule can be imagined as a disc with a magnetic moment, μ, perpendicular to this disc, i.e. in the direction of the axis of rotation. The cross-section, σ_0, depends on the orientation of the axis with respect to the direction of motion. Normally the direction of the axis of rotation is conserved between two collisions and the mean-free-path expression for, say, the viscosity will become

$$\eta \sim nm\bar{c}\bar{l} \quad \text{with} \quad \bar{l} \sim \left(\frac{1}{\sigma(\theta)}\right) ,$$

where the averaging is performed over all possible orientations of the molecules. If σ is of the type σ_0 $(1 + \beta P_2 \cos \theta)$, one sees immediately that the nonsphericity contribution to η is of the order β^2. In the presence of a magnetic field, however, the situation is different. The magnetic moment and, therefore, the axis of rotation

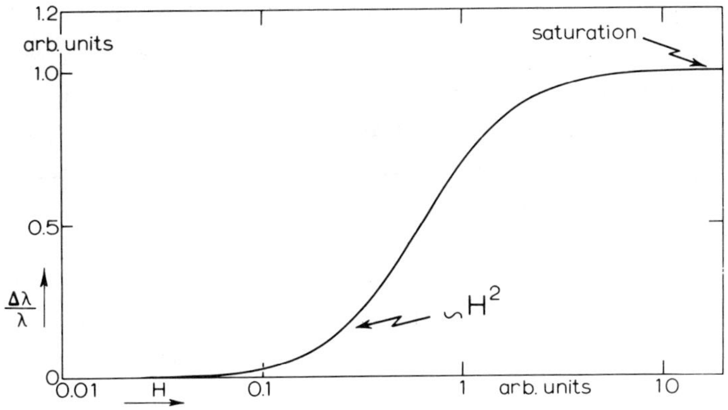

Fig. 1. Schematic diagram of the behaviour of the thermal conductivity of a paramagnetic gas in a magnetic field.

coupled to it will, in this classical picture, precess around the field direction with the Larmor angular-frequency ω_p. The precession gives rise to a periodic change of the collision cross-section of the molecule during its free flight. So one is no longer permitted to treat the axis as fixed in direction during this time. As a result, one has, in the mean-free-path picture, first to average the cross-section of every molecule over the precession and subsequently one must average over all molecules, i.e. $\bar{l} \sim \overline{(1/\bar{\sigma}_{prec})}^{all\ mol}$. This extra averaging introduces changes in the term in β^2, resulting in a decrease in the transport property. It is directly related to the fact that $(\overline{1/\sigma}) \neq (1/\bar{\sigma})$ e.g. consider two groups of molecules with $(\sigma + x)$ and $(\sigma - x)$. Then $\frac{1}{\bar{\sigma}} < \frac{1}{2}\left(\frac{1}{\sigma + x} + \frac{1}{\sigma - x}\right)$. The number of precessions between two successive collisions determines the extent of this averaging and it is clear that the effect will show saturation when the averaging is complete. Since the Larmor frequency is proportional to μH and the time between two collisions τ_f is $\sim \frac{1}{p}$ the degree of averaging will be described by the quantity $\omega_p \tau_f$ and is a function of the ratio $\frac{H}{p}$ only. Saturation takes place when $\omega_p \tau_f \gg 1$. The linear concentration dependence follows also immediately from this picture. The mean free path picture suffers, however, from all the shortcomings of the type of approach and, although it remains attractive for its simplicity, it is by now mainly of historical interest. It was not until 1961 that Kagan and Maksimov [4] pointed out how one could encompass the Senftleben effect in a Chapman-

Enskog type of theory. From Gorter's picture it is clear that to do so one has, in solving the Boltzmann equation, to retain explicitly the orientation of the molecules. That for non-spherical molecules such a solution is possible was already realized, as early as in 1922, by Pidduck [5] . Chapman and Cowling mention this possibility explicitly in their book when they treat the transport properties of rough sphere molecules. The nonequilibrium distribution function can depend vectorially on both the molecular velocity $\underline{c}$ and the molecular internal angular momentum $\underline{J}$. They neglected, however, the possible importance of the resulting anisotropy of f in $\underline{J}$. In this they were followed by many authors. The reason is that one saw no physical reason for polarization of nonspherical molecules as long as the gas was not so dense that the head and the tail of a molecule felt a different force. This last phenomenon is well known, for example, in systems containing macromolecules and gives rise there to such properties as flow-birefringence. Anisotropy in $\underline{J}$ can be important because the collision cross-section is angle-dependent and so the free lifetime of a molecule depends on the orientation of $\underline{J}$ with respect to $\underline{c}$, resulting in the possibility of polarization by a preferential absorption-like mechanism. Although its effect on the transport properties is so small that it usually gets lost in the background of the other contributions, it manifests itself directly in the case of the field effect. As Gorter pointed out, in a magnetic field the molecular axis of rotation remains no longer fixed in direction between collisions but precesses around the field direction, consequantly the angular momentum polarization produced by the collisions is partially destroyed in the time between collisions. The destruction of the angular momentum polarization will couple back on the anisotropy of the distribution function in velocity space and hence change the transport coefficients. Loosely speaking one measures directly the effect of the angular momentum polarization on the transport phenomena, or, more exactly, one measures the effect of going over from a distribution function of the form: $f(\underline{c}, \underline{J})$ to a situation where the form is $f(\underline{c}, J_H)$.

Note: the polarization is only partially destroyed as J_H remains a constant under the precession of $\underline{J}$ over a cone. Total destruction is possible in an NMR configuration, where the combination of the static field and a crossed oscillating field will induce m_j transitions (see Borman and Gorelik [6]).

For quite a long time (this includes the paper by Kagan and Maksimov) the field effect was considered to be an exclusive property of paramagnetic gases. Starting from Gorter's simple mean free path picture, Beenakker, Scoles, Knaap and Jonkman [4] showed in 1962 that this limitation is not real and that the magnetic effects are a general property of rotating molecules. They confirmed their ideas by presenting measurements of the effect of a magnetic field on the viscosity of N_2 and CO.

To explain this one has to realize that the field effect is essentially just the result of the destruction of the angular momentum polarization, and that its magnitude at saturation is not determined by the magnetic field but by the amount of polarization as caused by the nonspherical part of the molecular interaction. Consequently if one is able to destroy the polarization in N_2, one expects a change in the transport coefficients roughly equal to the one observed by Senftleben for O_2. So all one needs is a handle on the polarization. Now it is well known that every rotating molecule has a small but nonzero magnetic moment caused by its rotation. The exact theory of this magnetism is rather complicated, the magnetic moment arises from the fact that the electrons and the nuclei in a rotating molecule have not the same radius of gyration. The resulting magnetic field acts as a perturbation on the movement of the electrons. The final result of the movement of nuclei and perturbed electrons gives the rotational magnetic moment. Fermi was the first to perform such calculations. Measurements of such magnetic moments are performed either by studying Zeeman splitting in microwave spectroscopy or by molecular beam methods. The magnetic moment is given by $\underline{\mu} = g_J \underline{J} \mu_N$ where J is the rotational quantum number, μ_N is the nuclear magneton and g_J the rotational Lande factor. g_J can have either a positive or negative sign: see Table 1. The larger the molecule the smaller in general is g_J. Nor N_2 at room temperature with J of the order 8 one has a value of μ of about 3 nuclear magnetons. So this magnetic moment is around 100 x smaller than in a paramagnetic gas. Although the magnetic moment is much smaller than in the Senftleben case, the effect of the field will be the same, since it functions only as a handle to make the molecule precess in a field. It has, as such, nothing to do with how large the magnetic effect can become at saturation, this depends only on the nonsphericity. It is true that the very small magnetic moment makes it more difficult to

Table I

Molecule	g_J
H_2	+ .88
HD	+ .66
D_2	+ .44
N_2	- .28

have the molecule precessing rapidly enough to compete with the collision process that tends to maintain the polarization. This difficulty can, however, be overcome by increasing the field and so ω_p, or by decreasing the pressure and so increasing τ_f till $\omega_p \tau_f$ is again of order 1. For many molecules, fields of 50.000 ∅ and pressures of the order of a few mm are sufficient. The combination of a better understanding of the physics behind the field effects, and the realization of its general occurrence, has opened a new field of research in the study of transport properties.

Gorelik and Sinitsyn extended the Leiden work on the viscosity to the heat conductivity [8] . Since then, a large amount of experimental data has become available dealing with molecules like: N_2, CO, CO_2, and with the hydrogen isotopes, CH_4, CF_4 and SF_6; while for more complicated molecules some data is also available [9, 10] .

From the foregoing, it is clear that an electric dipole moment has the same effect in an electric field, provided that it has a non-zero component along the axis of rotation of the molecule. So all molecules having the electric analogue of the Zeeman effect (the linear Stark effect) will also give an electric effect. Molecules, such as CO, which have no net component along the rotational axis, although the dipole moment is not zero, will show a different behaviour: the perturbation by the field has to cause a dipole component along the axis of rotation before the effect of precession can occur. As a result the effect is of higher order in the field $\frac{E^2}{p}$ instead of $\frac{E}{p}$. In spectroscopy this is known as the second order Stark effect. After some unsuccessful attempts in the thirties, Senftleben [11] obtained the first successful measurements of the effect of an electric field on the thermal conductivity of polar gases (1965). Following an earlier but not completely convincing attempt by Cioara, the first reliable results for the viscosity in an electric field were obtained by Gallinaro et al., in Genoa [12]. One experimental difficulty is the fact that while a low pressure gas can be subjected to any amount of magnetic field, this is not the case for an electric field because sparkling occurs at relatively low fields. This makes the experiment rather difficult.

For the sake of completeness, one has to realize that practically every molecular anisotropy can give rise to a precession under suitable conditions. One can think of electric quadrupole moments in an inhomogeneous electric field, and of anisotropic electric or magnetic polarizability or susceptibility in, resp., electric and magnetic fields. As far as one can see, however, the fields necessary to cause a reasonable amount of precession are too high to make practical applications possible at least in the dilute gas regime.

Before continuing this introduction it is important to take an inventory of the different nonequilibrium phenomena that can occur in a gas in the presence of an external field. As an illustration the thermal conductivity will be treated in some detail.

From nonequilibrium thermodynamics we know that for small deviations from equilibrium, the general expression for the heat flux, $\underline{q}$, is given by $\underline{q} = -\underline{\underline{\lambda}} \cdot \underline{\nabla T}$, here $\underline{\underline{\lambda}}$ is a second rank tensor. Physically, this means that a heat flow in, say, the x-direction depends not only on the temperature gradient in that direction but also on the gradients in both the y and z-directions. The coefficients are not all independent. The space symmetry of the system will reflect itself in the properties of $\underline{\underline{\lambda}}$. This is known as Curie's principle. So one has, e.g., in an isotropic medium $\underline{\underline{\lambda}} = \lambda \underline{\underline{U}}$. The time reversal invariance gives rise to the Onsager relations which give that in general in the absence of a magnetic field the heat conductivity tensor is symmetric. Consider a gas in a magnetic or electric field along the z-axis. By the presence of the field the symmetry of the system is lowered. It is clear that the situation still has rotational symmetry around the field. Physically, this means that we may rotate our measuring set-up around the z-axis without changing the measured values.

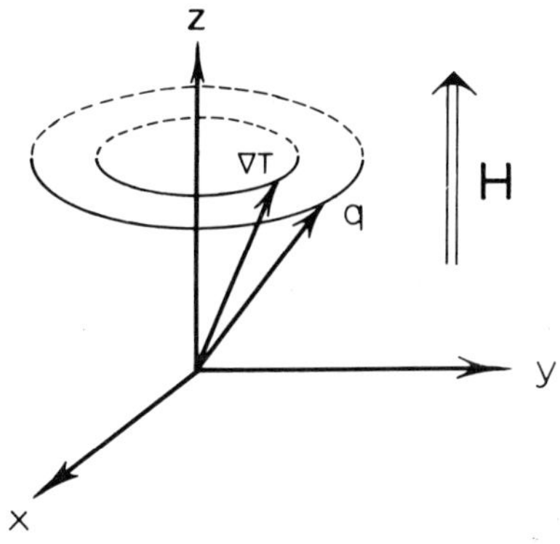

Fig. 2. Schematic diagram of the effect of axial symmetry on a measurement of the thermal conductivity.

The physical consequence of this invariance is that for equal applied gradients we will measure equal fluxes as in the non-rotated situation. So we see that the rotational invariance gives the condition $R_z \underline{\underline{\lambda}} = \underline{\underline{\lambda}}$, where R_z is the operator for a rotation around the z-axis. As a consequence the thermal conductivity has to be of the form

	$(\nabla T)_x$	$(\nabla T)_y$	$(\nabla T)_z$
q_x	$-\lambda_{xx}$	$-\lambda_{xy}$	0
q_y	λ_{xy}	$-\lambda_{xx}$	0
q_z	0	0	$-\lambda_{zz}$

Further information can be obtained from inversion symmetry. From parity invariance we get as $\Pi \lambda(\underline{E}) = \lambda(-\underline{E})$, the condition: $\underline{\underline{\lambda}}(\underline{E}) = \underline{\underline{\lambda}}(-\underline{E})$. All coefficients have to be even in the applied electric field. We know further that if we rotate the set up around the x- or y-axis over 180^0 the physical situation is again unchanged provided that we also invert the direction of the field. In this way one obtains that off diagonal elements in $\underline{\underline{\lambda}}$ are odd in the field. This condition is in conflict with the one obtained from inversion symmetry, so that the off diagonal elements have to be zero. As a result, the thermal conductivity tensor in an electric field has the form

$$\underline{\underline{\lambda}}(\underline{E}) = \begin{pmatrix} \lambda_{xx} & 0 & 0 \\ 0 & \lambda_{xx} & 0 \\ 0 & 0 & \lambda_{zz} \end{pmatrix}$$

For a magnetic field the situation is different as $\underline{H}$ is an axial vector: $\Pi \lambda(\underline{H}) = \lambda(\underline{H})$. This condition does not require the elements of $\underline{\underline{\lambda}}$ to be even in the field. From the 180^0 rotation we obtain as in the electric case:

$$\lambda_{ii}(+\underline{H}) = \lambda_{ii}(-H) \qquad \text{and} \qquad \lambda_{xy}(\underline{H}) = -\lambda_{xy}(-H) .$$

Hence the diagonal elements are, as in the electric case, even in the applied field. The off diagonal ones are odd in the field. In the presence of a magnetic field the scheme becomes [13] :

$$\underline{\underline{\lambda}}(H) = \begin{pmatrix} \lambda_{xx} & \lambda_{xy} & 0 \\ -\lambda_{xy} & \lambda_{xx} & 0 \\ 0 & 0 & \lambda_{zz} \end{pmatrix} = \begin{pmatrix} \lambda^{\perp} & -\lambda^{tr} & 0 \\ \lambda^{tr} & \lambda^{\perp} & 0 \\ 0 & 0 & \lambda^{\parallel} \end{pmatrix}$$

Application of the Onsager relations gives no further information in the case of either $\underline{E}$ or $\underline{H}$.

Note, however, that these general considerations tell only what might but not necessarily what will happen. The coefficients that are allowed to be non-zero can still be zero in a specific case (e.g. a noble gas in a magnetic field). So it is of some importance to see whether we can predict, on the basis of the molecular interaction, that the transverse coefficients can be expected to be non-zero. This is not a trivial task as is shown by the fact that although already in the thirties these symmetry considerations were known, the transverse effects were erroneously ruled out by considering a mean free path picture (von Laue). Let us consider a disc-like molecule moving with

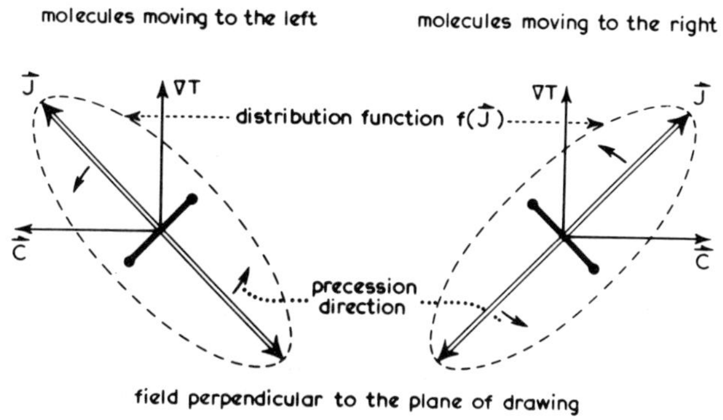

Fig.3. The origin of the transverse heat flow.

a velocity $\underline{c}$ in the direction perpendicular to a temperature gradient. As we will see later the polarization of angular momentum will be of the form: $(\underline{c}.\underline{J})\ \underline{J}.\ \underline{\nabla T}$ i.e. there is a deviation from the random orientation of $\underline{J}$ with respect to $\underline{c}$, with a preferred direction at 45^{0} with respect to $\underline{c}$ and $\underline{\nabla T}$. This is represented in the figure. Let us now see what happens if we switch on a magnetic field in the third direction. The molecular axis will start to precess - in our case in the plane of the paper. This precession will be the same for the molecules moving to the left and to the right. But as a consequence of the precession the molecule moving to the right precesses so that its disc lies along the velocity direction and so has a decreased cross-section. A molecule moving to the left has, by the precession, its disc facing the direction of motion, so that such a molecule has an increased cross-section. Consequently, the resistances for right and left heat flow are no longer equal and a net transverse flow can originate. This corresponds to the presence of non-zero off-diagonal elements in the heat conductivity tensor: i.e. heat will be transported perpendicularly to both gradient and field. (In an electron gas these coefficients are named after Righi and Leduc. These authors studied this type of phenomenon in bismuth at the end of the last century, 1885). Experimentally, this phenomenon will show up in the following way. The lateral boundaries will cause the transverse

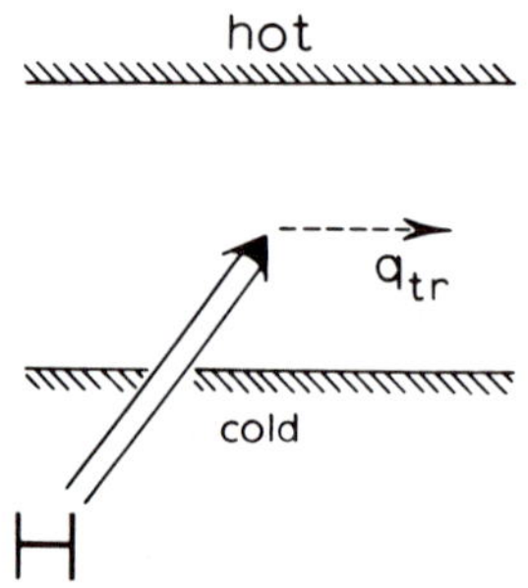

Fig. 4. Schematic diagram of a measurement of the transverse heat flow.

heat flow to heat up the wall at the right, and cool it at the left; this situation continuing until the normal heat conduction balances this effect. Compare this with the Hall-effect in metals where the resulting EMF, rather than the Hall current, is measured. For an electric field the transverse effect is obviously absent as this vector does not introduce a distinction between left and right in the problem. So that the transverse transport can not tell which direction to choose (the left-right symmetry is not broken).

Similar transverse transport phenomena occur also for viscous flow in a magnetic field. The transverse transport shows up, for example, in the following way: The boundary conditions convert the momentum flux into a pressure difference Δp across the slit arrangement. These symmetry considerations, prompted by discussions with

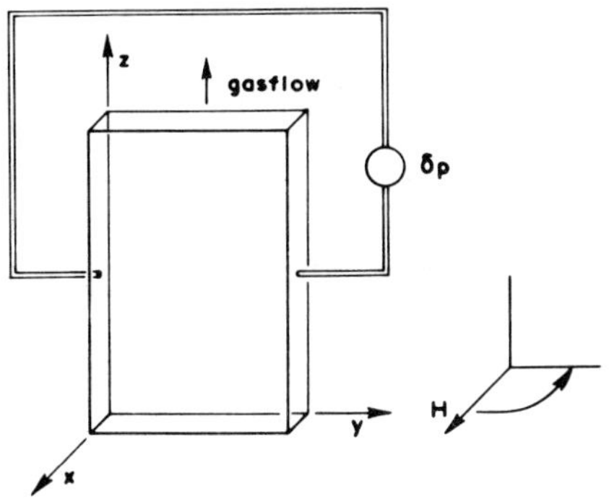

Fig. 5. Schematic diagram of a measurement of the transverse momentum flow.

Mazur, stood at the origin of the first observations of such transverse effects. They were first observed in Leiden in the viscous flow by Korving et al. [14] . This was soon followed by similar results for the thermal conductivity (Hermans et al., Leiden [15]; Gorelik et al., Moscow [16]). In so far as the sign in the problem is determined both by $\underline{H}$ and $\underline{\mu}$, it is clear that changing the sign of $\underline{\mu}$ will have the same effect as changing the direction of $\underline{H}$. So the sign of ΔT or Δp is directly related to the sign of $\underline{\mu}$. This situation is analogous to the Hall effect. There the sign of the EMF gives information about the sign of the charge carrier, here it gives the sign of the molecular magnetic moment.

In the case of viscous flow one starts from the general expression for momentum transport: $\underline{\underline{\Pi}}^s = -2\underline{\underline{\eta}} : \underline{\underline{\nabla v}}^s$, where $\underline{\underline{\Pi}}^s$ is the symmetric pressure tensor; $\underline{\underline{\eta}}$ is the general, 4th rank, viscosity tensor. Similar considerations as for the thermal conductivity reduce here the 81 elements of $\underline{\underline{\eta}}$ to 7 independent ones. Five of these are even and two odd in the applied magnetic field. The odd in field coefficients are zero in the case of an electric field. The expressions for the even coefficients were partially given already in 1939 by Von Laue, who pointed to the analogy with the problem of elastic deformation in crystals. The first general treatment was given by Hooyman, De Groot and Mazur. They introduced the following

	$(\overline{\nabla v})_{zz}$	$(\overline{\nabla v})_{xx}$	$(\overline{\nabla v})_{yy}$	$(\overline{\nabla v})_{xy}$	$(\overline{\nabla v})_{yz}$	$(\overline{\nabla v})_{zx}$	$\underline{\nabla}\cdot\underline{v}$
$\overline{\Pi}_{zz}$	$-2\eta_1$	0	0	0	0	0	-2ζ
$\overline{\Pi}_{xx}$	0	$-2\eta_2$	$-2(\eta_1-\eta_2)$	$-2\eta_4$	0	0	ζ
$\overline{\Pi}_{yy}$	0	$-2(\eta_1-\eta_2)$	$-2\eta_2$	$+2\eta_4$	0	0	ζ
$\overline{\Pi}_{xy}$	0	η_4	$-\eta_4$	$2\eta_1-4\eta_2$	0	0	0
$\overline{\Pi}_{yz}$	0	0	0	0	$-2\eta_3$	$-2\eta_5$	0
$\overline{\Pi}_{zx}$	0	0	0	0	$2\eta_5$	$-2\eta_3$	0
Π	-2ζ	ζ	ζ	0	0	0	$-\eta_v$

scheme: $\underline{H} = (0, 0, H)$. The coefficients $\eta_1, \eta_2, \ldots, \eta_5$ connect the components of the symmetric traceless tensor $\overline{\underline{\underline{\Pi}}}$ to the components of $\overline{\underline{\nabla v}}$. They can therefore be called coefficients of shear viscosity. The coefficient η_v connects the traces Π and div $\underline{v}$ and is therefore the volume viscosity. The seventh coefficient, ζ, describes a cross-effect between shear and volume viscosity. This notation was extensively used in the presentation of the earlier experimental data. The way in which the different coefficients are defined has disadvantages in that the notation is not self-evident.

A more elegant presentation can be obtained starting from a spherical tensor description. For the shear viscosity one has by C_∞ symmetry:

$$\Pi^{(2)\mu} = -2\eta_\mu [\nabla v]^{(2)\mu} \qquad \mu = 0, \pm 1, \pm 2.$$

Here use is made of Curie's principle that fluxes are only coupled to forces that belong to the same irreducible representation of the symmetry group of the equilibrium state. As $\eta_{-\mu} = \eta_\mu^*$, η_0 is real, while the other coefficients are in general complex.

Following Coope and Snider [17] we make the link to the conventional viscosity coefficients by introducing the real coefficients $\eta_\mu^\pm$ by:

$$\eta_\mu = \eta_\mu^+ + i\,\eta_\mu^- \qquad \mu \geq 0$$

η^+ and η^- are resp. even and odd in the field. In terms of these coefficients one has

$$\Pi^{(2)\mu\pm} = -2\left\{\eta_\mu^+ [\nabla v]^{(2)\mu\pm} \mp \eta_\mu^- [\nabla v]^{(2)\mu\mp}\right\}$$

In cartesian tensor notation one gets:

$$\begin{pmatrix} (3/2)^{1/2}\,\overline{\Pi}_{zz} \\ \overline{\Pi}_{xz} \\ \overline{\Pi}_{yz} \\ 1/2(\overline{\Pi}_{xx}-\overline{\Pi}_{yy}) \\ \overline{\Pi}_{xy} \end{pmatrix} = -2 \begin{pmatrix} \eta_0^+ & \cdot & \cdot & \cdot & \cdot \\ \cdot & \eta_1^+ & -\eta_1^- & \cdot & \cdot \\ \cdot & \eta_1^- & \eta_1^+ & \cdot & \cdot \\ \cdot & \cdot & \cdot & \eta_2 & -\eta_2^- \\ \cdot & \cdot & \cdot & \eta_2^- & \eta_2^+ \end{pmatrix} \begin{pmatrix} (3/2)^{1/2} S_{zz} \\ S_{xz} \\ S_{yz} \\ 1/2(S_{xx}-S_{yy}) \\ S_{xy} \end{pmatrix}$$

where S_{rs} has been written for

$$S_{rs} \equiv (\overline{\underline{\nabla v}})_{rs} = \frac{1}{2}\frac{\partial v_r}{\partial x_s} + \frac{1}{2}\frac{\partial v_s}{\partial x_r} - \frac{1}{3}\underline{\nabla}\cdot\underline{v}\ \delta_{rs}$$

and where the dots indicate those tensor elements which vanish because of the axial symmetry. The relation between $\eta^{\pm}_{\mu}$ and the coefficients of De Groot and Mazur is given in Table II, η^{+}_{μ} corresponds to even in field effects, and η^{-}_{μ} to transverse effects.

η_0^+	η_1		
η_1^+	η_3	η_1^-	η_5
η_2^+	$2\eta_2-\eta_1$	η_2^-	$-\eta_4$

Table II.

Apart from changing the field free transport coefficients the change in symmetry caused by the external field can also introduce cross effects: i.e. a gradient in one property may in the presence of a field cause a flux in a different quantity. The difference in what $\underline{H}$ and $\underline{E}$ can do stems from their different behaviour under inversion. Loosely speaking the even parity of $\underline{H}$ implies that it will not couple effects that were not coupled in the absence of a field; the odd parity of $\underline{E}$ works the other way around. Consequently an electric field can introduce a cross effect between a momentum flux and an energy flux or particle flux, and vice versa.

II. THE NON-EQUILIBRIUM POLARIZATIONS

The question to be answered in this section is: What do we know about the structure of the polarizations induced by the presence of a flux of momentum or energy? To this end, we will first treat as an illustrative example the shear viscosity in a dilute gas with rotating molecules. As a starting point one has the linearized Boltzmann equation

$$2\,\overline{\underline{W}\,\underline{W}} : \overline{\underline{\nabla v}} + \frac{i}{\hbar}\left[\mathcal{H}, \phi\right] = -n\,R_0\,\phi$$

where $\underline{W}$ is the reduced molecular velocity $W^2 = \frac{\frac{1}{2}mc^2}{kT}$. $\mathcal{H}$ is the Zeeman Hamiltonian and R_0 the linearized Waldmann-Snider collision operator. Furthermore one has $\phi = -\underline{\underline{B}} : \overline{\nabla \underline{v}}$, where $\underline{\underline{B}}$ is a second rank tensor made up from $\underline{J}$ and $\underline{W}$. Following Kagan [1,2] one writes:

$$\underline{\underline{B}} = \sum \underline{\underline{B}}^{pqrs} \odot [\underline{W}]^p \, [\underline{J}]^q \, S_r \, R_s$$

where S_r is a polynomial in W^2 corresponding to the Sonine polynomials in the treatment of spherical molecules, while R_s is a polynomial in J^2, $S_0 = R_0 = 1$*. $\underline{\underline{B}}^{pqrs}$ is a coupling tensor of rank $p + q + 2$, $\odot$ is a $p + q$ fold contraction. Note that parity considerations restrict p to even values. In our illustrative example we will consider only:

$$\underline{\underline{B}}^{20}, \quad \underline{\underline{B}}^{02} \quad \text{and} \quad \underline{\underline{B}}^{22} \quad \text{with} \quad r = s = 0 .$$

By taking moments of the linearized Boltzmann equation, one obtains three equations:

$$\frac{1}{n}\underline{\underline{\Delta}} = \langle \overline{\underline{W}\,\underline{W}} \; R_0 \; \overline{\underline{W}\,\underline{W}} \rangle_0 : \underline{\underline{B}}^{20} + \langle \overline{\underline{W}\,\underline{W}} \; R_0 \; \overline{\underline{J}\,\underline{J}} \rangle_0 : \underline{\underline{B}}^{02}$$
$$+ \langle \overline{\underline{W}\,\underline{W}} \; R_0 \; \overline{\underline{W}\,\underline{W}} \; \overline{\underline{J}\,\underline{J}} \rangle_0 \overset{4}{\odot} \underline{\underline{B}}^{22}$$

$$0 + \text{field term} = \langle \overline{\underline{J}\,\underline{J}} \; R_0 \; \overline{\underline{W}\,\underline{W}} \rangle_0 : \underline{\underline{B}}^{20} + \langle \overline{\underline{J}\,\underline{J}} \; R_0 \; \overline{\underline{J}\,\underline{J}} \rangle_0 : \underline{\underline{B}}^{02}$$
$$+ \langle \overline{\underline{J}\,\underline{J}} \; R_0 \; \overline{\underline{W}\,\underline{W}} \; \overline{\underline{J}\,\underline{J}} \rangle_0 \overset{4}{\odot} \underline{\underline{B}}^{22}$$

$$0 + \text{field term} = \langle \overline{\underline{J}\,\underline{J}} \; \overline{\underline{W}\,\underline{W}} \; R_0 \; \overline{\underline{W}\,\underline{W}} \rangle_0 : \underline{\underline{B}}^{20}$$
$$+ \langle \overline{\underline{J}\,\underline{J}} \; \overline{\underline{W}\,\underline{W}} \; R_0 \; \overline{\underline{J}\,\underline{J}} \rangle_0 : \underline{\underline{B}}^{02}$$
$$+ \langle \overline{\underline{J}\,\underline{J}} \; \overline{\underline{W}\,\underline{W}} \; R_0 \; \overline{\underline{W}\,\underline{W}} \; \overline{\underline{J}\,\underline{J}} \rangle_0 \overset{4}{\odot} \underline{\underline{B}}^{22}$$

---

* As we will see later there are indications that an expansion based on the unit vector in the $\underline{J}$ direction is more appropriate. One can take this into account by absorbing the corresponding factor in J^2 in R_s. For $Y^{(2)}$ instead of $\overline{\underline{J}\,\underline{J}}$ one has $R_0 = [J^2(J^2 - 1)]^{-\frac{1}{2}}$.

with $\underline{\underline{\Delta}}$ the isotropic fourth rank tensor $\Delta_{ijkl} = \frac{1}{2}\delta_{ik}\delta_{jl} + \frac{1}{2}\delta_{il}\delta_{jk} - \frac{1}{3}\delta_{ij}\delta_{kl}$. In the situation we consider the external field is always so small that its effect on R_o can be neglected. Hence the quantities $\langle -- R_o -- \rangle_o$ are isotropic and are thus proportional to $\underline{\underline{\Delta}}$. In this way one has, for example,

$$\langle \underline{W}\,\underline{W} \quad R_o \quad \underline{W}\,\underline{W} \rangle_o \equiv [20]\,\underline{\underline{\Delta}}$$

and
$$\langle \underline{J}\,\underline{J} \quad R_o \quad \underline{W}\,\underline{W} \rangle_o \equiv \begin{bmatrix} 02 \\ 20 \end{bmatrix} \underline{\underline{\Delta}}$$

Here $\begin{bmatrix} p & q \\ p' & q' \end{bmatrix}$ are collision integrals corresponding to the square bracket integrals in the treatment of spherical molecules.

Similarly $\langle \overline{\underline{W}\,\underline{W}} \quad R_o \quad \overline{\underline{W}\,\underline{W}} \quad \overline{\underline{J}\,\underline{J}} \rangle_o$ is proportional to the sixth rank isotropic unit tensor $\underline{\underline{D}}$: $D_{ijklmn} = \sum_{\alpha\beta\gamma} \Delta_{ij\alpha\beta}\,\Delta_{kl\alpha\gamma}\,\Delta_{mn\beta\gamma}$

For $\langle \overline{\underline{J}\,\underline{J}} \quad \overline{\underline{W}\,\underline{W}} \quad R_o \quad \overline{\underline{W}\,\underline{W}} \quad \overline{\underline{J}\,\underline{J}} \rangle_o$ the situation is more complicated as one can have more than one isotropic unit tensor of rank 8 with the correct symmetry in the indices. At this point one makes an approximation known as the <u>spherical approximation</u> and uses only <u>one</u> of these tensors:

$$\langle \overline{\underline{J}\,\underline{J}} \quad \overline{\underline{W}\,\underline{W}} \quad R_o \quad \overline{\underline{W}\,\underline{W}} \quad \overline{\underline{J}\,\underline{J}} \rangle \simeq [22]\,\underline{\underline{\Delta}}\,\underline{\underline{\Delta}} \; .$$

In this way one obtains:

$$\frac{1}{n}\,\underline{\underline{\Delta}} = [20]\,\underline{\underline{B}}^{20} + \begin{bmatrix} 20 \\ 02 \end{bmatrix} \underline{\underline{B}}^{02} + \begin{bmatrix} 20 \\ 22 \end{bmatrix} \underline{\underline{D}} \overset{4}{\odot} \underline{\underline{B}}^{22}$$

$$\text{Field term} = \begin{bmatrix} 02 \\ 20 \end{bmatrix} \underline{\underline{B}}^{20} + [02]\,\underline{\underline{B}}^{02} + \begin{bmatrix} 02 \\ 22 \end{bmatrix} \underline{\underline{D}} \overset{4}{\odot} \underline{\underline{B}}^{22}$$

$$\text{Field term} = \begin{bmatrix} 22 \\ 20 \end{bmatrix} \underline{\underline{D}} : \underline{\underline{B}}^{20} + \begin{bmatrix} 22 \\ 02 \end{bmatrix} \underline{\underline{D}} : \underline{\underline{B}}^{02} + [22]\,\underline{B}^{22}$$

In the field free case $\underline{\underline{B}}^{pq}$ itself is isotropic, hence

$$\underline{\underline{B}}^{20} = B^{20}\,\underline{\underline{\Delta}}$$

$$\underline{\underline{B}}^{02} = B^{02}\,\underline{\underline{\Delta}}$$

$$\underline{\underline{B}}^{22} = B^{22}\,\underline{\underline{D}} \; .$$

So one has:

$$\frac{1}{n} = [20]\,B^{20} + \begin{bmatrix} 20 \\ 02 \end{bmatrix} B^{02} + \begin{bmatrix} 20 \\ 22 \end{bmatrix} B^{22}\,\frac{7}{12}$$

$$0 = \begin{bmatrix}02\\20\end{bmatrix} B^{20} + [02]\, B^{02} + \begin{bmatrix}02\\22\end{bmatrix} B^{22}\, \frac{7}{12}$$

$$0 = \begin{bmatrix}22\\20\end{bmatrix} B^{20} + \begin{bmatrix}22\\02\end{bmatrix} B^{02} + [22]\, B^{22} .$$

The most drastic approximation one can make $\begin{bmatrix}ij\\k\ell\end{bmatrix} = 0$ results in $B^{20} = \frac{1}{n\,[20]}$. One has in general $\overline{\Pi} = -2\underline{\underline{\eta}} : \overline{\nabla \underline{v}_0}$ with $\underline{\underline{\eta}}$ a fourth rank tensor. For $\underline{\underline{\eta}}$ one finds: $\underline{\underline{\eta}} = \frac{1}{2} n\, kT\, \underline{\underline{B}}^{20}$. In the field free case this reduces to $\underline{\underline{\eta}} = \eta_0 \underline{\underline{\Delta}}$ or: $\overline{\Pi} = -2\eta_0 \overline{\nabla \underline{v}}$. If all polarizations involving $\underline{J}$ are neglected one obtains $\eta_{sph} = \frac{1}{2} \frac{kT}{[20]}$ as in the case of spherical molecules. The usual approach in our case is to assume $\begin{bmatrix}20\\k\ell\end{bmatrix} \neq 0$, while the other off-diagonal square brackets are set equal to zero. This is known as the <u>diagonal approximation</u>. In this way one obtains in the field free case:

$$B^{20} = \frac{1}{n\,[20]} \left\{ 1 - \frac{\begin{bmatrix}20\\02\end{bmatrix}\begin{bmatrix}02\\20\end{bmatrix}}{[20]\,[02]} - \frac{7}{12} \frac{\begin{bmatrix}20\\22\end{bmatrix}\begin{bmatrix}22\\20\end{bmatrix}}{[20]\,[22]} \right\}^{-1}$$

and

$$\eta_0 \approx \eta_{sph} \left\{ 1 + \frac{\begin{bmatrix}20\\02\end{bmatrix}^2}{[20]\,[02]} + \frac{7}{12} \frac{\begin{bmatrix}20\\22\end{bmatrix}^2}{[20]\,[22]} \right\} .$$

In this approximation every polarization gives its separate contribution to the viscosity. These corrections are quadratic in the square bracket describing the coupling. Use has been made of the fact that $\begin{bmatrix}p & q\\p' & q'\end{bmatrix} = \begin{bmatrix}p' & q'\\p & q\end{bmatrix}$. This is true if no polarizations which odd in $\underline{J}$ are involved ($q + q'$ even). For the case when $q + q'$ is odd, one has $\begin{bmatrix}p & q\\p' & q'\end{bmatrix} = -\begin{bmatrix}p' & q'\\p & q\end{bmatrix}$. This is related to the time reversal properties of the collision operator. Since the diagonal square brackets are positive definite, corresponding to the fact that a system tends to equilibrium, one has the important conclusion: Polarizations even in $\underline{J}$ increase and those odd in $\underline{J}$ decrease the viscosity.

We will here omit the derivation of the behaviour of $\underline{\underline{\eta}}$ in the presence of an external magnetic field as this is treated elsewhere in this course. Three important conclusions can be drawn from the field free situation:

(i) Every polarization in $\underline{J}$ gives in the diagonal approximation its own contribution to the field effect.

(ii) The sign of this contribution depends on whether the

polarization destroyed is odd or even in the field. The viscosity decreases in a magnetic field for even and increases for odd in $\underline{J}$ polarizations.

(iii) The magnitude of the field effects is quadratic in the coupling integral.

As shown by Snider in his lectures one has further:

(iv) The viscosity tensor in a magnetic field contains two types of elements:

(a) Elements that are even in the field. These show a field dependence that is a superposition of curves of the type $f^+(\mu\xi)$, with $\mu = 0, 1 \ldots q$ where q is the rank of the polarization in $\underline{J}$.

$$f^+(x) = \frac{x^2}{1 + x^2} .$$

(b) Elements that are odd in the field. These show a field dependence that is a superposition of curves of the type $f^-(\mu\xi)$, with $\mu = 1 \ldots\ldots q$.

$$f^-(x) = \frac{x}{1 + x^2} .$$

For the polarization considered here one has the following scheme.

	$\overline{J\,J}$	$\overline{J\,J}\quad\overline{W\,W}$
$\frac{\Delta\eta_0^+}{\eta}$	0	$-\Psi_{22}\left[2f^+(\xi_{22}) + 8f^+(2\xi_{22})\right]$
$\frac{\Delta\eta_1^+}{\eta}$	$-\Psi_{02}\ f^+(\xi_{02})$	$-\Psi_{22}\left[7f^+(\xi_{22}) + 6f^+(2\xi_{22})\right]$
$\frac{\Delta\eta_2^+}{\eta}$	$-\Psi_{02}\ f^+(2\xi_{02})$	$-\Psi_{22}\left[6f^+(\xi_{22}) + 4f^+(2\xi_{22})\right]$
$\frac{\eta_1^-}{\eta}$	$-\Psi_{02}\ f^-(\xi_{02})$	$-\Psi_{22}\left[+5f^-(\xi_{22}) - 6f^-(2\xi_{22})\right]$
$\frac{\eta_2^-}{\eta}$	$-\Psi_{02}\ f^-(2\xi_{02})$	$-\Psi_{22}\left[6f^-(\xi_{22}) + 4f^-(2\xi_{22})\right]$

Table II

Here ξ_{pq} and Ψ_{pq} are related to the collision integrals, ξ is further proportional to the quantity $g\,\mu_N\,\frac{H}{p}$. Similar conclusions hold for the other transport coefficients like thermal conductivity etc.

Now that we know how the different non-equilibrium polarizations show up in the field effect, we can set ourselves to the task of deciding on the type of polarization present. The decrease or increase of the viscosity in a field will tell immediately whether the dominant polarization is even or odd in $\underline{J}$. To know whether more than one polarization is present asks for a more detailed study. It appears to be difficult to decide on the presence of more than one polarization from the shape of the $\frac{H}{p}$ curve alone. It is necessary to study in detail the different elements of the transport coefficient tensor. This means that one has to perform measurements as a function of the orientation of the magnetic field. In the following we will discuss the situation for the heat conductivity and the shear viscosity. The discussion will be centred around the behaviour of diatomic, at most weakly polar molecules; at the end we will make some remarks about more complicated molecules.

The thermal conductivity

The structure of the $\underline{\underline{\lambda}}$ tensor is rather simple and allows, without great difficulties, a direct determination of $\lambda^{\parallel}$, $\lambda^{\perp}$ and λ^{tr}. The experimental arrangements are given schematically in Fig. 6.

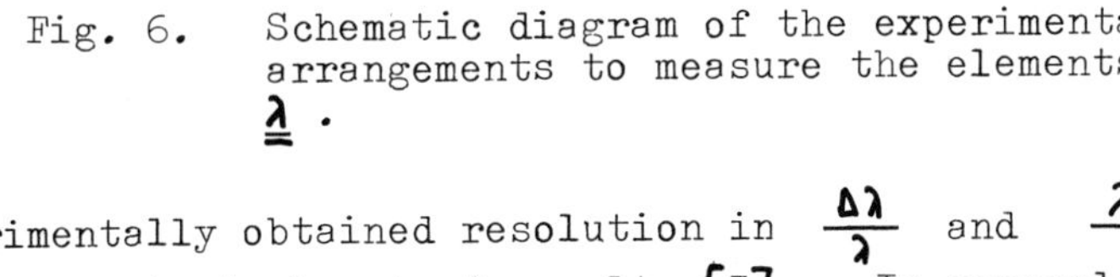

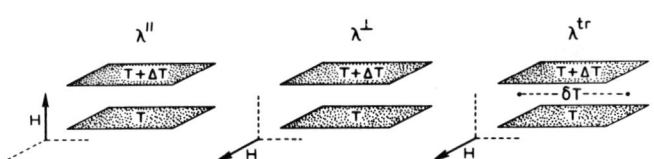

Fig. 6. Schematic diagram of the experimental arrangements to measure the elements of $\underline{\underline{\lambda}}$.

The experimentally obtained resolution in $\frac{\Delta\lambda}{\lambda}$ and $\frac{\lambda^{tr}}{\lambda}$ is 10^{-5}. Fig. 7 gives a typical set of results [3]. In general the thermal conductivity decreases in a field pointing to a dominant polarization even in $\underline{J}$. The curves drawn in Fig. 7 correspond to a $\underline{W}\,\overline{\underline{J}\,\underline{J}}$ contribution fitted to the $\frac{\Delta\lambda^{\perp}}{\lambda_0}$ results. The shaded areas indicate the experimental uncertainties. A more sensitive test for the presence of other polarizations is found by studying the ratio $(\frac{\Delta\lambda^{\perp}}{\Delta\lambda^{\parallel}})$ at saturation. This is experimentally the best determined quantity

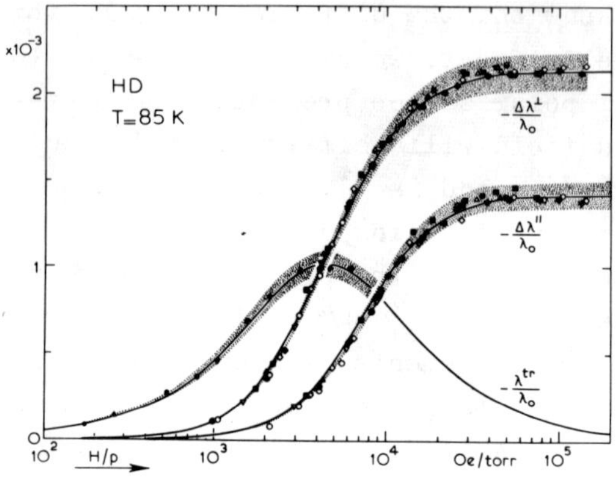

Fig. 7. $\frac{\Delta\lambda^{\perp}}{\lambda}$, $\frac{\Delta\lambda^{\parallel}}{\lambda}$ and $\frac{\lambda^{tr}}{\lambda_o}$ versus $\frac{H}{p}$ for HD at 85 K. Drawn lines represent W J J contribution. Shaded areas are the estimated experimental uncertainties.

and it is theoretically more sensitive than the ratio $(\frac{\Delta\lambda^{\perp}}{\lambda^{tr}})_{max}$ as is shown in Fig. 8.

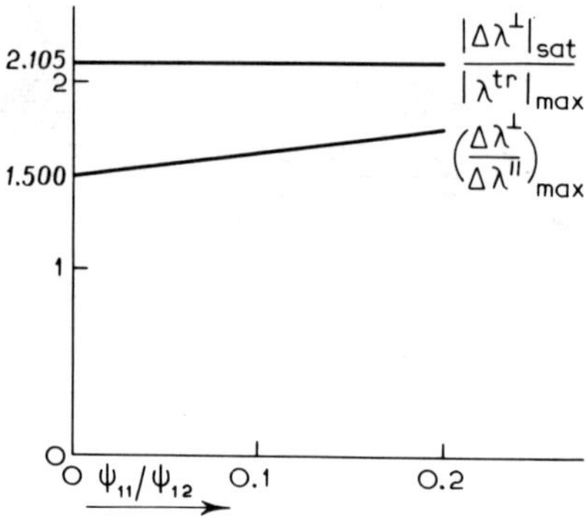

Fig. 8. Theoretical dependence of $|\Delta\lambda^{\perp}|_{sat}/|\lambda^{tr}|_{max}$ and $(\Delta\lambda^{\perp}/\Delta\lambda^{\parallel})_{max}$ on the relative strength of the W J contribution.

The result of such an analysis is given in the table III:

Gas	300 K	85 K
N_2	1.57 ± 0.01	1.50 ± 0.03
CO	1.52 ± 0.01	1.49 ± 0.03
HD	1.51 ± 0.01	1.49 ± 0.03
pH_2	1.50 ± 0.10	
oD_2	1.60 ± 0.10	

It is seen that the ratio $\left(\frac{\Delta\lambda^{\perp}}{\Delta\lambda^{\parallel}}\right)_{sat}$ differs very little from the value 1.5, corresponding to a $\underline{W}\,\overline{\underline{J}\,\underline{J}}$ polarization. One can safely conclude that for diatomic at most weakly polar molecules, $\underline{W}\,\overline{\underline{J}\,\underline{J}}$ is by far the dominant polarization present.

The shear viscosity

The analysis of the situation for the shear viscosity is not so simple. In the presence of a field the field free Navier Stokes equation $\rho \frac{d\mathbf{v}}{dt} + \underline{\nabla} p = \eta_0 \nabla^2 \underline{v} + (\frac{1}{3}\eta_0 + \eta_v)\underline{\nabla}\,(\underline{\nabla}\cdot\underline{v})$ becomes far more complicated, containing seven instead of two transport coefficients! Consequently one has to seek simple experimental situations i.e. situations where in the absence of the field only a limited number of second derivatives $\frac{\partial^2 v_\alpha}{\partial x_\beta \partial x_\gamma}$ are of importance. This is found in the case of a straight capillary with rectangular cross-section (length ℓ, width w and thickness t, see Fig. 9). When $\ell >> w >> t$ only one velocity component (along ℓ) and one velocity-gradient component (along t) are of importance. The influence of the magnetic field on the flow through this slit can be studied as a function of the orientation of the field. In the absence of a field all second derivatives of the velocity in the slit can be neglected except $\partial^2 v_\kappa / \partial i^2$. As all field effects are small, at most 1% of η_0 this is still true in the presence of a field. In polar coordinates one gets for the pressure gradients along and across the slit as a function of the orientation of the field [4] :

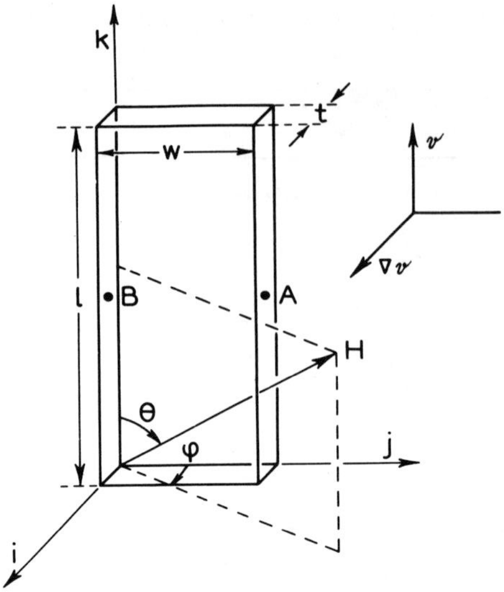

Fig. 9. Schematic diagram of capillary with rectangular cross-section.

$$\frac{\partial p}{\partial k} = \left\{ \sin^2\phi \left[\left(\frac{3\eta_0^+ + \eta_2^+}{4} \right) \sin^2 2\theta + \eta_1^+ \cos^2 2\theta \right] + \cos^2\phi \left[\eta_2^+ \sin^2\theta + \eta_1^+ \cos^2\theta \right] \right\} \frac{\partial^2 v_k}{\partial i^2}$$

$$\frac{\partial p}{\partial j} = \left\{ \sin\phi \sin\theta \left[\eta_1^- - \eta_2^- - (2\eta_1^- - \eta_2^-) \sin^2\phi \sin^2\theta \right] + \cos\phi \cos\theta \sin\theta \left[\eta_1^+ - \eta_2^+ - 4 \left(\eta_1^+ - \frac{3\eta_0^+ + \eta_2^+}{4} \right) \sin^2\phi \sin^2\theta \right] \right\} \frac{\partial^2 v_k}{\partial i^2} ,$$

with ϕ and θ as given in Fig. 9. It is seen from these equations that the pressure gradient $\frac{\partial p}{\partial k}$ ("longitudinal effect") is coupled to a linear combination of $\frac{3\eta_0^+ + \eta_2^+}{4}$, η_1^+ and η_2^+. The pressure gradient $\frac{\partial p}{\partial j}$ in the "transverse effect" equation can be split into two parts: a combination of the odd-in-field coefficients η_1^- and η_2^- and a combination of differences of the even-in-field coefficients. In this experimental set up the volume viscosity does not appear.

In the following scheme we give the situation for which these expressions take their simplest form:

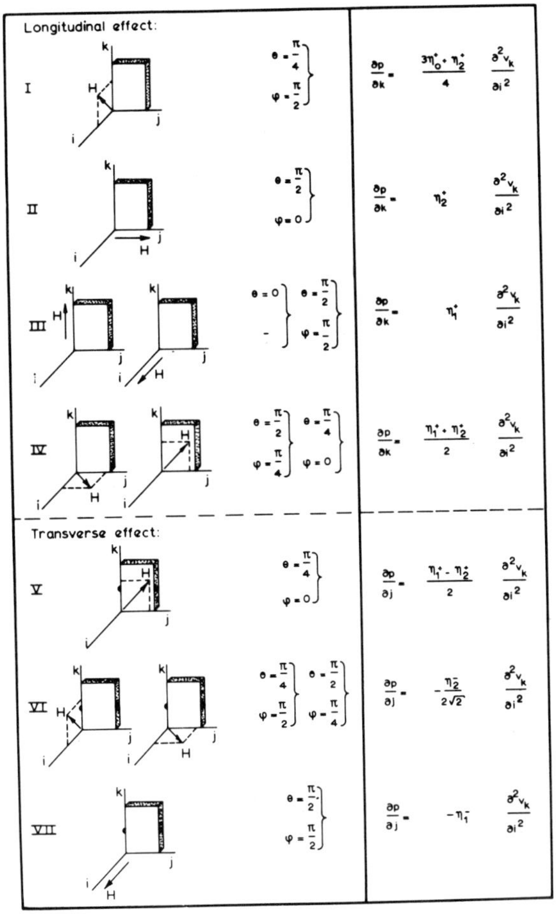

Fig. 10. The shear viscosity coefficients obtained for different orientations of the field.

Experimental conditions for accuracy are optimal if the measurements are performed in one and the same slit arrangement. Furthermore the orientation of the magnetic field can only be changed conveniently by turning a magnet around the apparatus i.e. by varying the orientation of the field in a fixed (horizontal) plane. It is convenient to choose this plane in such a way that it contains two of the simplest situations illustrated in the foregoing scheme.

For the longitudinal effects one arrives in this way at the plane containing I and II, i.e. the slit is inclined under 45^0 to the horizontal plane in which $\underline{H}$ rotates. This is illustrated in Fig. 11.

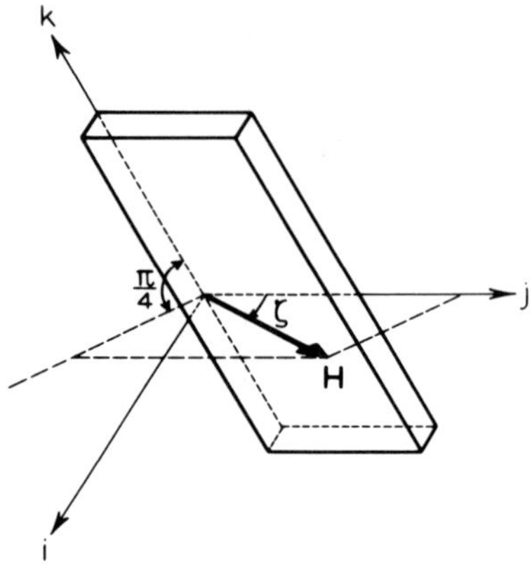

Fig. 11. Schematic diagram of capillary arrangement to measure the longitudinal viscous effects.

In this situation one has

$$\frac{\partial p}{\partial k} = \left[\frac{3\eta_0^+ + \eta_2^+}{4} \sin^4\varsigma + \eta_2^+ \cos^2\varsigma + \eta_1^+ \cos^2\varsigma \sin^2\varsigma\right] \frac{\partial^2 v_k}{\partial \iota^2}$$

For the transverse effects one combines situation VI and VII. This is illustrated in Fig. 12.

For this case one has:

$$\frac{p_A - p_B}{W(\underline{\nabla} p)_{AB}} = \frac{\eta_1^- \cos 2\chi \sin\chi - \frac{1}{2}\eta_2^- \cos\chi \sin 2\chi}{\eta_0} \quad ;$$

here $(\underline{\nabla} p)_{AB}$ is the pressure gradient along the length of the capillary at the plane AB. Figs. 13 and 14 give the results for N_2 and CO at room temperature.

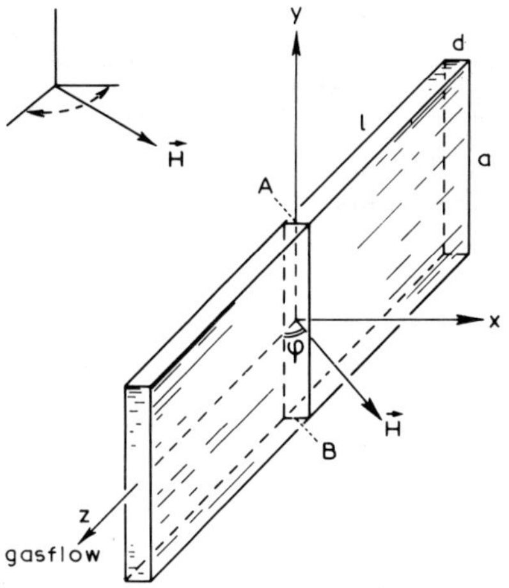

Fig. 12. Schematic diagram of capillary arrangement to measure the transverse viscous effects.

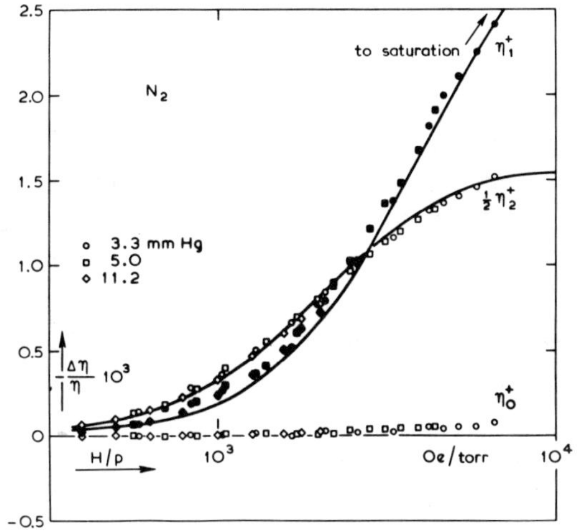

Fig. 13. $\Delta\eta_0^+$, $\Delta\eta_1^+$ and $\Delta\eta_2^+$ as functions of $\frac{H}{p}$ for N_2 at room temperature. ——: $\overline{\underline{J}\ \underline{J}}$ contribution.

The decrease of the viscosity in a field points again to a polarization even in $\underline{J}$. Further analysis shows that only minor deviations from a behaviour given by a $\overline{\underline{J}\,\underline{J}}$ polarization occur.

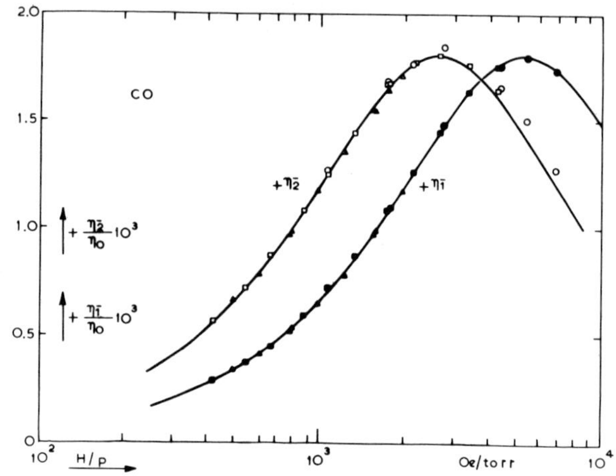

Fig. 14. The transverse viscosity coefficients η_1^- η_2^- for CO at room temperature.
—— : $\overline{\underline{J}\,\underline{J}}$ contribution.

In this respect especially, the information from $\Delta\eta_0^+$ is significant, because of the fact that $\Delta\eta_0^+$ is zero is directly related to the tensorial structure of the polarization and is insensitive to the inclusion of further scalar dependence on W^2 and/or J^2. We will come back to this point later in discussing the conclusions that can be drawn from the small deviations from a simple $\overline{\underline{J}\,\underline{J}}$ behaviour. An interesting consequence of the structure of the $\overline{\underline{J}\,\underline{J}}$ polarization is that it can be directly observed by studying flow-birefringence. (See Hess [5]). As the refractive index n of dilute gases is only slightly different from one and as the amount of $\overline{\underline{J}\,\underline{J}}$ polarization remains always small, the birefringence is small. It is, however, possible to observe it; a typical result [6] is shown in Fig. 15.

So far we have limited our considerations to simple diatomic molecules, that are at most weakly polar. We will now consider briefly the situation for more complicated molecules. For the viscosity the situation is not changed too much. In symmetric and spherical tops like CH_3F and CH_4 [7] the dominant polarization has the tensorial structure of $\overline{\underline{J}\,\underline{J}}$. Only in the case of NH_3

does the field free effect have the opposite sign, pointing to the dominance of an odd in $\underline{J}$ polarization [8] . The presence of the inversion combined with a large rotational level splitting makes this molecule a special case. For the heat conductivity the situation is different. For symmetric top molecules a change of sign of the field effect with increasing dipole moment is observed [9] .

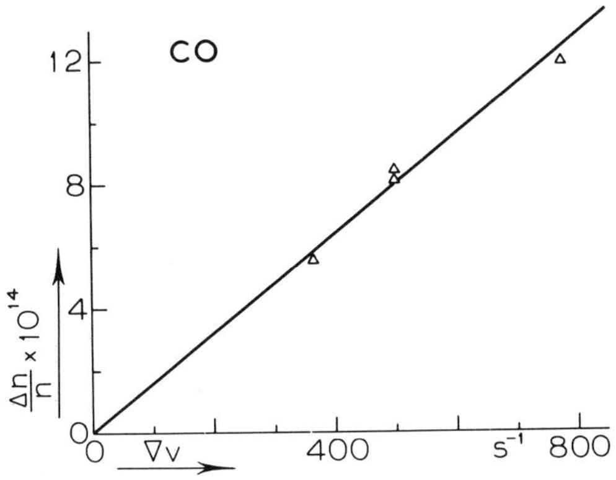

Fig. 15. The flow birefringence $\Delta n/n$ for CO as a function of the applied velocity gradient.

This points to the presence of an odd in $\underline{J}$ polarization in the heat flow in strongly polar (symmetric top) molecules. Measurements on the angular dependence are needed to decide on the structure of this polarization. Furthermore it is necessary to perform measurements on strongly polar diatomic molecules to investigate how far the symmetric top structure is the cause of this effect. This work is now in progress. Note that in the case of symmetric top molecules a third molecular vector apart from $\underline{W}$ and $\underline{J}$ can play a role, i.e. the orientation of the figure axis with respect to $\underline{J}$. Classically, this can be described by J_ζ the projection of $\underline{J}$ on the figure axis. Polarization can now also involve J_ζ . J_ζ has the character of a pseudoscalar, odd under inversion. The interesting point is that in a heat flow, also the polarization $J_\zeta \underline{J}$ is now allowed.

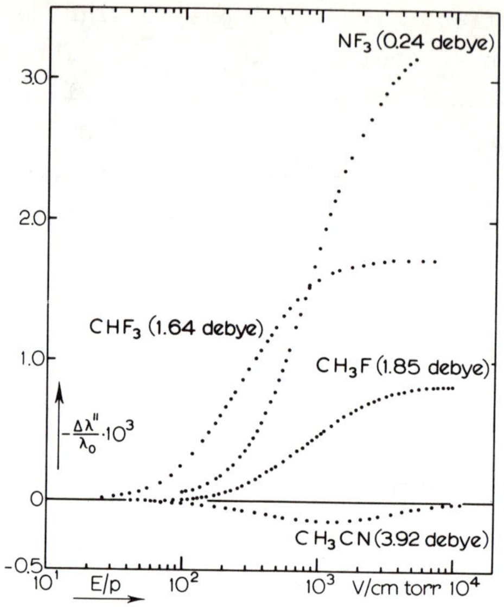

Fig. 16. $\Delta\lambda''/\lambda_0$ as a function of $\frac{E}{p}$ for some polar gases.

As was pointed out by Waldmann and Hess [10] , such a polarization corresponds to an orientation of the electric dipole moment. Measurements to detect the corresponding thermo-electric effect have so far not been successful.

III. THE LIMITATION OF THE ONE MOMENT DESCRIPTION

In the foregoing section two points have been clarified. We know the tensorial structure of the dominant polarization, while furthermore the question was answered as to how far the effect of this polarization can be described by one time constant i.e. by a curve of the form $f^{\pm}(\mu\omega\tau)$. Note, however, that on the basis of the analysis given no distinction can be made between say $\overline{\underline{J}\,\underline{J}}$ and $\overline{\underline{J}\,\underline{J}}\,R\,(J^2)$. Both the angular dependence of η and its behaviour as a function of $\frac{H}{p}$ will be the same in both cases. Furthermore deviations from the simple one time scale behaviour, if found, can be attributed to two main causes. 1) Within the same tensorial structure of the polarization more than one polynomial in W^2 or J^2 has to be taken into account. The physics behind such a situation is that the dependence of say the reorientation cross section on J^2 is such that one average is not able to represent this distribution adequately. 2) The presence of other types of polarization say $\underline{J}\,\overline{\underline{W}\,\underline{W}}$ and $\overline{\underline{J}\,\underline{J}}\,\overline{\underline{W}\,\underline{W}}$. As terms odd and even in $\underline{J}$ have a contribution of different sign, it is nearly always possible to make such a combination of two additional polarizations that the deviation from the simple $\frac{H}{p}$ behaviour can be explained. To be able to come to a more definite conclusion it is necessary to use the information obtained from other sources. A study of the Depolarized Rayleigh Line (DPR) is in this respect most promising for two reasons [1,2] :

(i) From the physics of DPR one knows the type of orientational polarization involved.

(ii) A study of the lineshape allows one to determine in how far more than one time scale is present in the decay of this polarization.

The relevant quantity for the description of light scattering is the polarizability tensor $\underline{\underline{\alpha}}$. For linear molecules it can be written in the form

$$\underline{\underline{\alpha}} = (\tfrac{2}{3}\alpha_{\parallel} + \tfrac{1}{3}\alpha_{\perp})\,\underline{\underline{U}} + (\alpha_{\parallel} - \alpha_{\perp})\,\overline{\underline{u}\,\underline{u}}\,,$$

$\underline{u}$ being a unit vector along the internuclear axis. The depolarized Rayleigh line is associated with that part of the tensor operator $\overline{\underline{u}\,\underline{u}}$ that is diagonal in the rotational quantum number j. This part, $\overline{\underline{u}\,\underline{u}}\,(\underline{J})$, is related to the tensor $\overline{\underline{J}\,\underline{J}}$ by

$$\overline{\underline{u}\,\underline{u}}\,(\underline{J}) = -\tfrac{1}{2}\,\frac{\overline{\underline{J}\,\underline{J}}}{J^2 - \frac{3}{4}}\quad .$$

Due to the orientational fluctuations the local instantaneous mean value of this tensor is in general non-zero and this gives rise to the occurrence of depolarized Rayleigh light scattering. The depolarized Rayleigh line profile is determined by the correlation function C (t) describing the decay of these fluctuations [3].

$$c\,(t) = \frac{\left\langle \dfrac{\overline{\underline{J}\,\underline{J}}}{J^2 - \frac{3}{4}}(o) : \dfrac{\overline{\underline{J}\,\underline{J}}}{J^2 - \frac{3}{4}}(t)\right\rangle_0}{\left\langle \dfrac{\overline{\underline{J}\,\underline{J}}}{J^2 - \frac{3}{4}}(o) : \dfrac{\overline{\underline{J}\,\underline{J}}}{J^2 - \frac{3}{4}}(o)\right\rangle_0}\quad .$$

The spectral function $R(\nu)$ describing the depolarized line profile is related to C (t) by

$$c\,(t) = \frac{1}{I_o}\int_{-\infty}^{+\infty} R(\nu)\, e^{-2\pi\nu t}\, dt\quad .$$

If one assumed that the decay of $\overline{\underline{J}\,\underline{J}}/(J^2 - \frac{3}{4})$ can be described with one rate coefficient (one-moment approximation) one has

$$\frac{\partial\left\langle \dfrac{\overline{\underline{J}\,\underline{J}}}{J^2 - \frac{3}{4}}\right\rangle}{\partial t} = -n\,\langle v\rangle_o\,\sigma_{DPR}\left\langle \frac{\overline{\underline{J}\,\underline{J}}}{(J^2 - \frac{3}{4})}\right\rangle$$

and the correlation function becomes:

$$C(t) = e^{-n\langle v\rangle_0 \sigma_{DPR} t} .$$

The effective reorientation cross section σ_{DPR} is given by the collision integral for orientational polarization:

$$\sigma_{DPR} = \frac{1}{\langle v\rangle_0} \frac{\left\langle \frac{\overline{\underline{J}\,\underline{J}}}{J^2 - \frac{3}{4}} : R_0 \frac{\overline{\underline{J}\,\underline{J}}}{J^2 - \frac{3}{4}} \right\rangle_0}{\left\langle \frac{\overline{\underline{J}\,\underline{J}}}{J^2 - \frac{3}{4}} \quad \frac{\overline{\underline{J}\,\underline{J}}}{J^2 - \frac{3}{4}} \right\rangle_0} .$$

The experimental results show pronounced deviation from this behaviour as is shown in fig. 17.

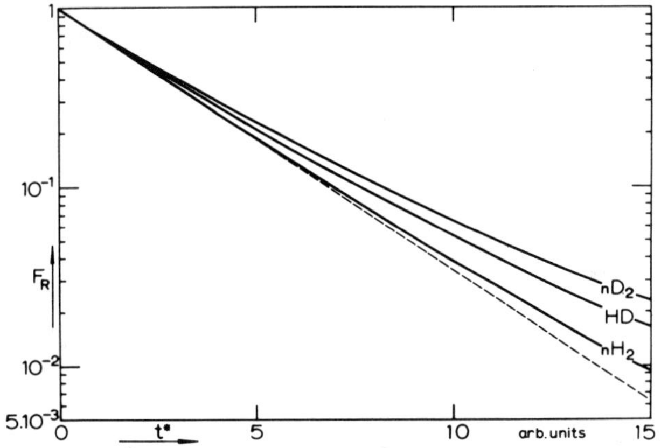

Fig.17. The deviation from a Lorentzian shape of the depolarized Rayleigh line for the gases nH_2, HD and nD_2.

It is clear that more than one time scale is present. From a study of the behaviour of the hydrogen isotopes one comes to the conclusion that the origin ff this deviation is to be found in the fact that the microscopic cross sections involving changes in $\underline{J}$ are strongly dependent on the quantum number j. Under such conditions it becomes plausible that one correlation time or one moment is not able to fully take into account the consequences of this situation. Hence one is lead to a description involving also higher moments $\left\langle R_s \frac{\overline{JJ}}{J^2 - \frac{3}{4}} \right\rangle$, where R_s is a polynomial in J^2. The moment equations for this situation are:

$$\frac{\partial}{\partial t}\left\langle R_i \frac{\overline{JJ}}{J^2 - \frac{3}{4}} \right\rangle =$$

$$-n\langle v\rangle_o \sigma_{ii}\left\langle R_i \frac{\overline{JJ}}{J^2 - \frac{3}{4}} \right\rangle - \sum_{i\neq j} n\langle v\rangle_o \sigma_{ij}\left\langle R_j \frac{\overline{JJ}}{J^2 - \frac{3}{4}} \right\rangle$$

with $R_1 = 1$. $\underline{\underline{\sigma}}$ is an effective cross section matrix with elements given by the corresponding matrix elements of the collision operator. σ_{ij} describes the coupling between the different moments. For the correlation function c_{ik} one obtains (cf. Hess)[4] the expression:

$$c_{ik}^{(t)} = (e^{-n\langle v\rangle_o \underline{\underline{\sigma}} t})_{ik} \quad .$$

For the description of DPR one needs C_{11} (t), and one has:

$$c_{11}(t) = (e^{-n\langle v\rangle_o \underline{\underline{\sigma}} t})_{11} \quad .$$

For short times this can be written as:

$$c_{11}(t) = 1 - n\langle v\rangle_o \sigma_{11} t + \frac{n^2\langle v\rangle_o^2}{2!} \sum_j \sigma_{ij} \sigma_{ji} t^2 \quad .$$

Hence in the limit for $t \to o$ the several moment description reduce to the one moment situation, and the initial slope of C_{11} (t) can be identified as $\sigma_{11} = \sigma_{DPR}$. An interesting measure of the importance of the distribution of correlation times is given by the area under the correlation function: $\int_0^\infty C(t)\,dt \equiv \tilde{\sigma}^{-1}$. Note that in the presence of only one correlation time $\tilde{\sigma} = \sigma_{DPR}$. A survey of the situation is given in the table:

	$\frac{\sigma-\tilde{\sigma}}{\sigma}$		$\frac{\sigma-\tilde{\sigma}}{\sigma}$
nH_2	0.00	N_2	0.08
pH_2	0.015	CO	0.065
HD	0.08	CO_2	0.06
nD_2	0.11	OCS	0.00

At this point it is interesting to compare the values of σ_{DPR} with what one gets by analysing the field effect on the viscosity with a one moment description. As shown in the following table large differences are found.

Å^2 / gas	σ_{DPR}	σ_{FE}
N_2	34.4 ± 0.6	23.7 ± 0.9
CO	43 ± 1	32.5 ± 0.8
CO_2	88 ± 2	69 ± 1.5

One is tempted to conclude that σ_{FE} is clearly not associated with $\frac{\overline{\underline{J}\,\underline{J}}}{J^2-\frac{3}{4}}$ but with say $\overline{\underline{J}\,\underline{J}}$. A more plausible solution to this problem is found when one takes the spread in correlation times also into account, when analysing the viscosity data. It is instructive to compare in this respect the situation encountered in DPR with that in FBR. In both cases the observable is related to $\left\langle \frac{\overline{\underline{J}\,\underline{J}}}{J^2-\frac{3}{4}} \right\rangle$. In DPR one looks at its time dependence and as we have seen one can find a region where the effects of higher moments can be neglected. The situation is, however completely different for FBR. As this is a steady state effect one measures also the effects of all other moments. In fact one has $\frac{\Delta n}{n} \sim \int_0^\infty C_{\eta_T}(t)\,dt$ [4] where $C_{\eta_T}(t)$ is the correlation function: $\langle \overline{\underline{W}\,\underline{W}}(0) : \overline{\underline{J}\,\underline{J}}(t)/(J^2-3/4)\rangle_0$

This means that the distribution of correlation times manifested by the overall behaviour of DPR will show up in the actual value of $\frac{\Delta n}{n}$. A similar situation holds for $(\frac{\Delta \eta}{\eta})_{sat}$. Hence one expects that both the FBR and the saturation of the FE will be related more to $\tilde{\sigma}_{DPR}$ than to σ_{DPR}.

The moment equations one has to solve to obtain $C_{\ell T}(t)$ simplify considerably if one assumed that the collsions couple the deformation in velocity space $\langle \overline{\underline{W}\,\underline{W}} \rangle$ only to one moment in $\underline{J}$ space. Such a situation is suggested by work by Snider. In fact he showed that in lowest order distorted wave Born approximation $\langle \overline{\underline{W}\,\underline{W}} \rangle$ is coupled to $\langle Y^{(2)} \rangle = \left\langle \frac{\overline{\underline{J}\,\underline{J}}}{J^2 (J^2 - \frac{3}{4})^{\frac{1}{2}}} \right\rangle$ and not to higher moments. In this approximation the moment equations read:

$$\frac{\partial \langle \sqrt{2}\,\overline{\underline{W}\underline{W}} \rangle}{\partial t} = -n\langle v \rangle_0\, \sigma_{00} \langle \sqrt{2}\,\overline{\underline{W}\underline{W}} \rangle - n\langle v \rangle_0\, \sigma_{01} \langle Y^{(2)} \rangle - \sqrt{2}\,\overline{\underline{\nabla}\underline{v}}$$

$$\frac{\partial \langle Y^{(2)} \rangle}{\partial t} = -n\langle v \rangle_0\, \sigma_{10} \langle \sqrt{2}\,\overline{\underline{W}\underline{W}} \rangle - n\langle v \rangle_0\, \sigma_{11} \langle Y^{(2)} \rangle - n\langle v \rangle_0 \sum_{j \neq 1} \sigma_{1j} \langle R_j Y^{(2)} \rangle$$

$$\frac{\partial \langle R_i Y^{(2)} \rangle}{\partial t} = -n\langle v \rangle_0\, \sigma_{ii} \langle R_i Y^{(2)} \rangle - n\langle v \rangle_0 \sum_{j \neq i} \sigma_{ij} \langle R_j Y^{(2)} \rangle \qquad i > 1$$

The σ matrix reads:

$$\sigma = \begin{array}{c|cccc} \sigma_{00} & \sigma_{01} & 0 & 0 & \\ \hline \sigma_{10} & \sigma_{11} & \sigma_{12} & - & - \\ & \sigma_{21} & \sigma_{22} & - & - \\ 0 & - & - & - & - \\ 0 & - & - & - & - \end{array}$$

Note that for large values of J one has:

$$Y^{(2)} = \frac{\overline{\underline{J}\,\underline{J}}}{[J^2 (J^2 - \frac{3}{4})]^{\frac{1}{2}}} \simeq \frac{\overline{\underline{J}\,\underline{J}}}{(J^2 - \frac{3}{4})} .$$

Under these conditions the quantities σ_{ij} ($i > 1$) occuring in the description of the FBR are identical to the ones encountered in DPR. So the matrix bounded by the solid lines is identical to the one in DPR.

For the correlation function $c_{\eta T}(t)$ one has:

$$c_{\eta T}(t) = c_{01}(t) = (e^{-n\langle v\rangle_0 \underline{\underline{\sigma}} t})_{01} \quad .$$

One gets

$$\frac{\Delta n}{n} \sim \int_0^\infty c_{01}(t)\, dt = \int_0^\infty (e^{-n\langle v\rangle_0 \underline{\underline{\sigma}} t})_{01}\, dt = (\underline{\underline{\sigma}}^{-1})_{01} \quad .$$

From the form of the matrix $\underline{\underline{\sigma}}$ and observing that $\frac{\sigma_{01}\,\sigma_{10}}{\sigma_{00}\,\sigma_{11}} \ll 1$ one gets

$$(\underline{\underline{\sigma}}^{-1})_{01} = \frac{\sigma_{01}}{\sigma_{00}} \, \frac{1}{\tilde{\sigma}} \; .$$

Hence one obtains:

$$\frac{\Delta n}{n} \sim \frac{\sigma_{\eta T}}{\sigma_\eta\, \tilde{\sigma}} \quad .$$

Similarly starting from $\eta = nkT \int_0^\infty \langle \sqrt{2}\, \overline{\underline{W}\,\underline{W}}(0) \sqrt{2}\, \overline{\underline{W}\,\underline{W}}(t)\rangle_0\, dt$ and solving $c_{00} = \langle \sqrt{2}\, \overline{\underline{W}\,\underline{W}}(0) \sqrt{2}\, \overline{\underline{W}\,\underline{W}}(t)$ from the moment equations one gets:

$$\left(\frac{\Delta\eta_{1}^{+}}{\eta}\right)_{sat} = \left(\frac{\Delta\eta_{2}^{+}}{\eta}\right)_{sat} = \frac{(\sigma_{\eta T})^2}{\sigma_\eta\, \tilde{\sigma}} \quad .$$

It is now possible to test this model by comparing $\tilde{\sigma}$ obtained from a combination of $\left(\frac{\Delta\eta}{\eta}\right)_{sat}$ and FBR. Preliminary results of such a comparison is shown in the table:

	$\tilde{\sigma}$ from DPR	$\tilde{\sigma}$ from FE + FBR
N_2	31.6 ± 0.6	31 ± 1
CO	40 ± 1	38 ± 1
CO_2	83 ± 2	140 ± 3

At the time of writing these lectures, investigations along these lines have only just started. So it is still too early to draw more definite conclusions, but the situation looks promising.

IV. THE EFFECTIVE CROSS SECTIONS AND THEIR BEHAVIOUR

We have seen in the foregoing section, in how far the behaviour of the nonequilibrium polarization can be described by effective cross sections $\mathfrak{S}$ that are matrix elements of the Waldmann-Snider collision operator. It is often convenient to use a systematic notation for these quantities. To arrive at such a notation one starts from the form of the expansion of the deviation from equilibrium in terms of irreducible tensors $[\underline{W}]^{(p)}$ and $[\underline{J}]^{(q)}$, while the scalar dependence is expressed using a series expansion in polynomials $S_r^p\ (W^2)$ and $R_s^q\ (J^2)$. $S_0\ (W^2) = 1$, $S_1^1\ (W^2) = -(W^2 - \frac{5}{2})$. Furthermore we take here $R_0^2\ (J^2) = [J^2(J^2 - \frac{3}{4})]^{-\frac{1}{2}}$ to make $R_0^2\ [\underline{J}]^{(2)} = Y^{(2)}$, and $R_0^2 \mathbf{\underline{W}}(\underline{J}^2) = \underline{W}\ Y^{(2)}$.

Finally $R_1^0\ (J^2) = (J^2 - 1)$. In the framework of such a description the matrix elements of the collision operator and hence $\mathfrak{S}$ will be characterised by two sets of 4 indices: pqrs and p'q'r's'. The diagonal cross sections for which p = p' etc., describe the direct decay of the deviation of f considered (e.g. $\mathfrak{S}_T = \mathfrak{S}\binom{0200}{0200}$ describes the direct decay of $\mathbf{Y}^{(2)}$ polarisation by the collisions in the gas). Those quantities are positive. The off-diagonal ones, where at least one pair of indices is unequal, give the strength of the coupling between the different polarizations (e.g. $\mathfrak{S}_{\eta T} = \mathfrak{S}\binom{0200}{2000}$) gives the effective cross section for production of $\mathbf{Y}^{(2)}$ polarisation by the presence of a polarisation in velocity $\overline{\underline{W}\,\underline{W}}$. Such quantities can be both positive and negative. In the example of flow-birefringence, for instance, the sign depends on the direction of the polarisation that is produced. For a general definition of the effective cross sections see [1] . The definition is relatively simple for diagonal and many of the more common off-diagonal elements. The notation discussed above can be simplified when there is no risk of confusion: a) for the diagonal elements one may write only one row of indices, e.g. $\mathfrak{S}(0200)$, b) if no scalar dependence arises, one can omit the last two zeros in the rows, e.g. $\mathfrak{S}\binom{02}{20}$. In the following table we present the most important effective cross sections with the physical processes to which they are related.

Effective cross section	Relaxation (decay) process	Source
$\sigma(0001)$	Rotational energy σ_{rot}	bulk visc η_v
$\sigma(1010)$	$\underline{W}$ $(5/2 - W^2)$, transational heat flow	η, η_v combined
$\sigma(1001)$	$\underline{W}$ $(J^2 - 1)$, rotational heat flow	λ, η, η_v combined
$\sigma(12)$	$\underline{W}$ $Y^{(2)}$, Kagan polarization	$(\frac{H}{p})_{1/2}$ FE λ
$\sigma(20)$	$\overline{\underline{W}\,\underline{W}}$, momentum transport, σ_η	η
$\sigma(02)$	$Y^{(2)}$, tensor polarisation in $\underline{J}$-space, σ_T	$(\frac{H}{p})_{1/2}$ FE η
	Polarizations coupled	
$\sigma\binom{1010}{1200}$	Transational heat flow Kagan polarization	Sat FE η
$\sigma\binom{1001}{1200}$	Rotational heat flow Kagan polarization	Sat FE λ, η combined
$\sigma\binom{02}{20}$	$\overline{W\,W}$ polarization $Y(2)$ polarization, $\sigma_{\eta T}$	Sat FE η

Not all these σ are independent.

Just as for spherical molecules where the effective cross sections for the viscosity and thermal conductivity are related (Eucken factor), so there are relations among these more general cross sections. In fact one has the following relations [2] :

$$\sigma(1010) = \frac{2}{3}\,\sigma(2000) + \frac{5}{6}\,\sigma(0010) \qquad (1)$$

$$\sigma\binom{1010}{1001} = \frac{1}{2}\sqrt{\frac{5k}{2C_{rot}}}\;\sigma(0010) \qquad (2)$$

$$\sigma\binom{1010}{1200} = -\frac{1}{5}\sqrt{5}\;\sigma\binom{0200}{2000} \qquad (3)$$

$$\sigma(0010) = \sqrt{\frac{2C_{rot}}{3k}}\;\sigma\binom{0001}{0010} = \frac{2}{3}\;\frac{C_{rot}}{k}\;\sigma(0001) \qquad (4)$$

$$\sigma\binom{1010}{1100} = 0\ . \qquad (5)$$

From the study of the FE on the viscosity one obtains directly both $\sigma(02)$ and $\sigma\binom{02}{20}$. A considerably more complicated situation arises in the case of the FE on the thermal conductivity. The field free heat conductivity contains already three cross sections: $\sigma(1010)$, $\sigma(1001)$ and $\sigma\binom{1010}{1001}$, describing respectively the behaviour of the translational and the rotational heat flow and their coupling. These quantities can be obtained separately by combining data on λ with those on η and η_v, using also the relation (1). It is then possible to solve the situation for the FE on λ where three additional cross sections occur: $\sigma(1200)$, $\sigma\binom{1200}{1010}$ and $\sigma\binom{1200}{1001}$ describing respectively the decay of the Kagan polarization and its coupling to the translational and rotational heat flow. To do this one combines data on the field effect for λ with similar information on η, making use of relation (3). This situation is summarised in column 3 of table .
Havind reduced the experimental data to effective collision cross sections we can consider in some detail the information that can be obtained in this way. First of all there is the fact that the polarization odd in $\underline{J}$ are absent or very small, this notwithstanding the fact that they occur in a lower order in the $[\underline{W}]^{(p)}[\underline{J}]^{(q)}$ expansion. The fact that $\langle \underline{\underline{A}}_0 R \underline{\underline{B}}_0 \rangle_0 = 0$ can be explained by assuming that the contributions of the loss and gain terms just cancel. This is the case if R_0 is time reversal invariant. The combined behaviour of $\langle \underline{\underline{A}}_0 R \underline{\underline{B}}_0 \rangle_0$ under parity and time reversal will then result in the polarizations odd in $\underline{J}$ being absent.

This gives an answer to the historical problem of inverse collisions in molecules with internal degrees of freedom. In a classical description one has, for spherical molecules, that for every collision $(c_1\ c_2 \rightarrow c_1'\ c_2')$ one can find geometrical conditions such that the inverse process $(c_1'\ c_2' \rightarrow c_1\ c_2)$ takes place. This is in general no longer true in the presence of internal degrees of freedom. This is shown in fig. 18 where a $\rightarrow$ a' while a' will never result in a.

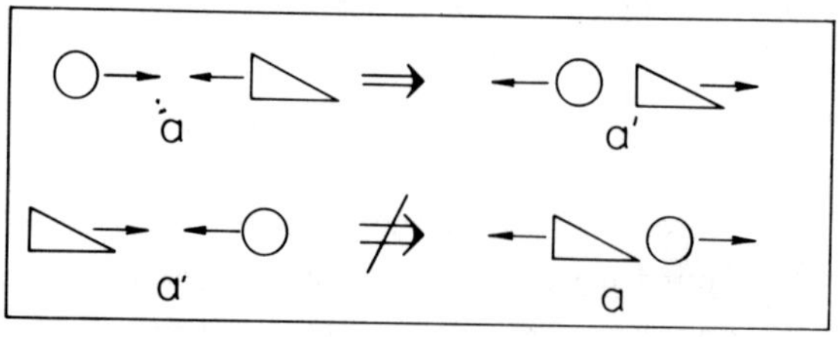

Fig. 18. The problem of inverse collisions.

From the absence of polarizations odd in $\underline{J}$ one can conclude that for diatomic molecules the assumption of the existence of inverse collisions is a good approximation. In a more general forumulation: the collision operator is nearly self adjoint: $\langle \underline{A}\,\underline{\underline{R}}_0\,\underline{B} \rangle_0 = \langle \underline{B}\,\underline{\underline{R}}_0\,\underline{A} \rangle_0$.

One can further pose the question of in how far a perturbation treatment, in which one writes $R_o = R_{spherical} + \varepsilon R_{nonspherical}$, is a promising approach. To investigate this we compare $\sigma(20)$ with $\sigma(02)$. In such a perturbation approach $\sigma(02)$ will be of the order ε and one expects $\sigma(02) \ll \sigma(20)$. This is in general not the case as is shown in the table:

gas \ $\AA^2$	$\sigma(20)$	$\sigma(02)$
N_2	35.0	24
CO	35.5	33
CO_2	52.7	69

An exception is formed by the hydrogen isotopes. Indeed here the combination of small nonsphericity and large rotational level splitting makes $\sigma(02) \ll \sigma(20)$. This is shown in the next table which gives the situation at room temperature.

gas \ $\AA^2$	$\sigma(20)$	$\sigma(02)$
H_2	18.7	0.49
HD	18.7	2.3
D_2	18.7	0.91

One can safely conclude that a small nonsphericity approximation is in general poor. In contrast to this the off diagonal elements of R_0 are always small, see table :

gas \ $\AA^2$	$\sigma(02)$	$\sigma\binom{20}{02}$	$\sigma\binom{1200}{1001}$
N_2	24	1.5	4.6
CO	33	2.0	4.3
HD	2.3	0.29	0.67

This last conclusion remains true even for the case of strongly polar molecules where $\sigma(02) > \sigma(20)$. The exceptional situation for the hydrogen isotopes has as a further consequence that $\sigma(12) \simeq \sigma_D$. In general $\sigma(12)$ will be of the order of $\sigma(02) + \sigma_D$. This situation is illustrated in fig. 19 for CO (note that $\sigma_D \simeq 0.8(\sigma_\eta)$.

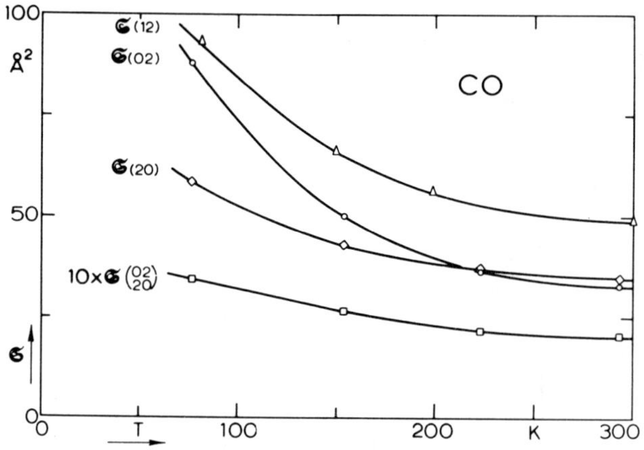

Fig. 19. $\sigma(12)$, $\sigma(02)$, $\sigma(20)$ and $\sigma\binom{02}{20}$ for CO as a function of temperature.

Further conclusions on the properties of the effective cross sections are related to the window that each of the cross sections has on the collision processes. It is immediately clear that diagonal cross sections involving only $\underline{J}$ i.e. $\sigma(02)$ are only sensitive to collisions in which $\underline{J}$ changes in direction ($\Delta m_j \neq 0$). One can further show that in the lowest order DWB approximation the coupling cross section $\sigma\binom{20}{02}$ is caused by energetically inelastic processes ($\Delta j \neq 0$). All the diagonal cross sections involving W are of course also sensitive to changes in $\underline{W}$, while also $\sigma\binom{1200}{1001}$ contains terms of first order in the non-sphericity arising from such collisions. This has an important consequence in testing molecular model calculations. It is far more difficult to fit both magnitude and position of the field effect on η than it is to fit the same quantities for λ [3] . This is illustrated in fig. 20.

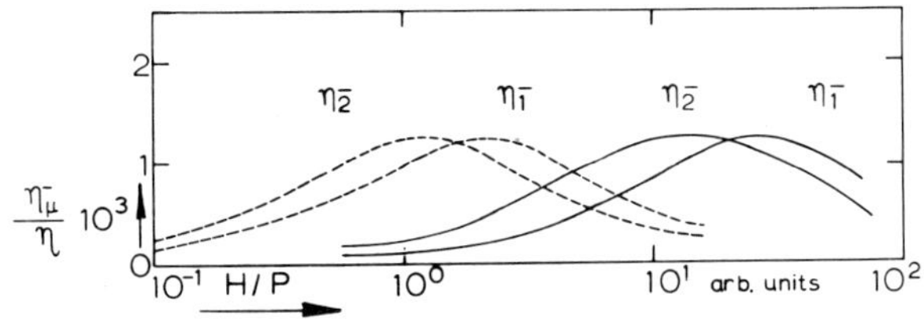

Fig. 20. Comparison between theory (——) and experiment (---) for η_1^- and η_2^- for the case of hard ellipsoids.

In a review article a few years ago we wrote: "The study of the field effects can introduce new perspective (in the study of nonspherical molecules) since it not only allows more collision integrals to be determined, thus increasing largly our source of data, but it also permits the relevant collision integrals to be obtained with good absolute accuracy. In addition, the results can easily be checked for internal consistency and the measurements can be performed over a large range of temperatures without too much difficulty. For these reasons, it is to be expected that a wealth of new information on inelastic collision integrals will become available in the near future. Hopefully, this will contribute greatly to a satisfactory description of the interaction between polyatomic molecules and the way this is reflected in the nonequilibrium behaviour of polyatomic gases". These hopes are by now fulfilled in so far as the collision integrals are concerned. Snider will discuss in his lectures the state of our knowledge of the connection between the collision integrals and the interaction between nonspherical molecules.

V. FIELD EFFECTS IN THE RAREFIED GAS REGIME

Introduction

While in most of the work discussed so far the experiments were performed with a well defined plan in mind, the thermomagnetic torque was discovered by accident. It was present as a large spurious effect in measurements of the Einstein-De Haas effect, where one uses the conservation of angular momentum to determine the angular momentum associated with a magnetic moment in the following way. (see fig. 21)

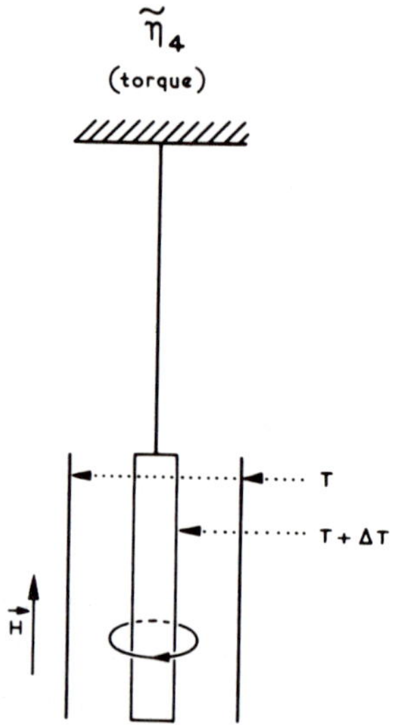

Fig.21. Schematic diagram of Scotts apparatus.

A sample is suspended from a torsion wire inside a solenoid. Upon reversal of the sign of the magnetic field the change in magnetisation of the sample is accompanied by a change of the direction of the angular momentum of the spins. Because of the conservation law this gives rise to an equal and opposite change in the angular momentum of the sample as a whole. So the sample starts to turn till the torsion of the suspension wire compensates for the acquired torque. During such measurements, Scott at General Motors Research Laboratories was troubled by a spurious effect that caused a shift in the zero point of his torsion pendulum by an amount that corresponded to a time-independent torque many orders of magnitude larger than the value caused by the Einstein-De Haas effect. The origin of this torque was not understood but as it was constant, one could easily correct for it. In building a new set up in 1963, Scott and Sturner discovered that the torque was related to the fact that the sample was always slightly heated by the magnetizing coil: it was absent when the sample was at the temperature of the surrounding vessel. They found furthermore that it was also related to the gas residue in the evacuated surrounding vessel. Systematic investigations by Scott, Sturner and Williamson [1] resulted in the following picture

(i) The τ (H) curves behave at constant p approximately like $\frac{H}{1 + aH^2}$.

(ii) The torque is odd in the direction of the temperature difference, and of the magnetic field. It depends further on the sign of the rotational g factor of the gas.

(iii) The torque disappears with increasing pressure, but the behaviour as a function of p is rather complicated.

(iv) The magnitude of the torques is many orders larger than the angular momentum content of the molecules hitting the surface.

The overall behaviour of the Scott effect suggests that it is related to the phenomena discussed earlier in dilute gases. It can however not simply be a transport of internal angular momentum in a temperature gradient and magnetic field as this would be far too small. Its disappearance at higher pressures suggests that it is a property of the rarefied gas regime. In this pressure region the mean free path has become so large that in transport phenomena macroscopic quantities start to vary rapidly over a mean free path. Under such conditions contributions to the transport coefficients occur, that are characterized by an $^1/p$ behaviour. The situation is, however, complicated by the fact that superimposed on these effects there will be the results of the mean free path becoming of the order of the dimensions of the apparatus. This will give rise to so called Knudsen corrections that decrease the measured torque. Only after very carefully correcting for this effect, is the true behaviour of τ (p, H) obtained (see Burgmans and Adair [2]). Such a corrected result is shown in fig. 22.

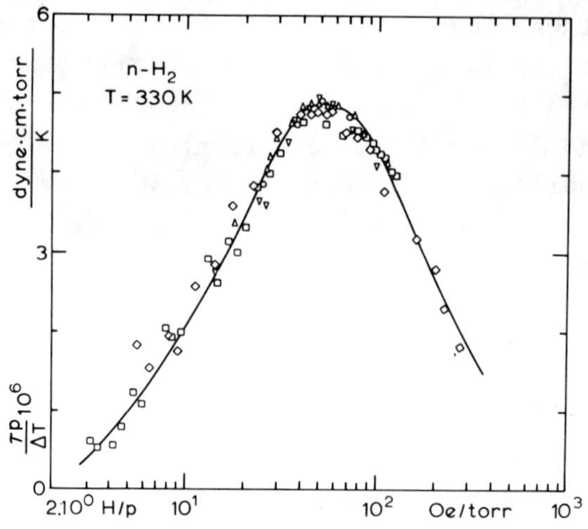

Fig 22. $\frac{\tau p}{\Delta T}$ versus $\frac{H}{p}$ for H_2 at room temperature.

Levi et al. [3] were the first to show how in the presence of a magnetic field the Maxwell stress caused by a $\nabla\nabla T$ gives rise to torque.

To get some insight into the behaviour of transport properties in the rarefied gas regime, we will briefly outline the situation with respect to the Maxwell stress for a monatomic gas. We start with the Boltzmann equation in the steady state:

$$\underline{c} \cdot \underline{\nabla} f = c\ (f\ f_1) \quad .$$

We write for f:

$$f = f^0\ (1 + \epsilon \Phi^{(1)} + \epsilon^2 \Phi^{(2)} + \ldots) \quad .$$

In zeroth order we get

$$0 = c\ (f^0\ f_1^0) \quad , \ f^0\text{: Maxwell distribution} \ .$$

In first order the equation becomes

$$\underline{c} \cdot \underline{\nabla} f^0 = c\ \{f^0\ f_1^0\ (\varphi^{(1)} + \varphi_1^{(1)})\} \equiv -n\ R_0\ \varphi^{(1)}$$

$$\underline{c} \cdot \underline{\nabla} f^0 = -n\ R_0\ \varphi^{(1)}$$

$$\varphi^{(1)} = -\frac{1}{T}\,\underline{A}\cdot\underline{\nabla}T \qquad\qquad \underline{W} = \left(\frac{m}{2kT}\right)^{\frac{1}{2}}\underline{c}$$

$$\left(\frac{2kT}{m}\right)^{\frac{1}{2}}\left(W^2-\frac{5}{2}\right)\underline{W} = n\,R_o\,\underline{A}\ .$$

The second order equation becomes

$$(\underline{c}\cdot\underline{\nabla}f^0)\,\varphi^{(1)} + f^0\,\underline{c}\cdot\nabla\varphi^{(1)} = -n\,R_o\,\varphi^{(2)} + c\,(f^0\,f_1^0\,\varphi^{(1)}\varphi_1^{(1)})\ .$$

Now we will limit our consideration to the case $\frac{q}{q_{random}} \ll 1$, so we neglect $(\varphi^{(1)})^2 :: (\underline{\nabla}T)^2$. Under such conditions it is still possible that $\underline{\nabla}T$ varies rapidly over a mean free path, i.e., it will no longer be possible to neglect $\underline{\nabla}\underline{\nabla}\,T$ and consequently terms containing $\underline{\nabla}\varphi^{(1)}$ should be taken into account. Under these conditions the second order equation becomes

$$\underline{c}\cdot\underline{\nabla}\varphi^{(1)} = -n\,R_o\,\varphi^{(2)}\ .$$

(Terms with $\underline{\nabla}f^0\ \varphi^{(1)} :: (\underline{\nabla}T)^2$ are again neglected). This is known as the linearized Burnett equation. Note that as $\varphi^{(1)} :: \frac{1}{n}$ and $R_o \sim n^0$, one has $\varphi^{(2)} :: \frac{1}{n^2}$ corresponding to a "$\frac{1}{n}$ contribution" to the transport coefficients.

We see further that we get $\varphi^{(2)}$ in terms of $\varphi^{(1)}$, i.e., in terms of the solution of the first order equation. For the second order transport coefficients this statement implies that they can be expressed in terms of the first order transport coefficients. The Maxwell stress corresponds to the $\varphi^{(2)}$ contribution to the pressure tensor. In this way one obtains

$$\underline{\underline{\Pi}}^{(2)} = \int f^0\ \varphi^{(2)}\ \overline{\underline{c}\,\underline{c}}\ d\underline{c}$$

$$2\,\overline{\underline{W}\,\underline{W}} = n\,R_o\,\underline{\underline{B}}$$

$$= 2kT\int f^0\,\varphi^{(2)}\ \overline{\underline{W}\,\underline{W}}\ d\underline{c} = nkT\int f^0\ \varphi^{(2)}\ R_o\,\underline{\underline{B}}\ d\underline{c}$$

$$= nkT\ \ n\left[\varphi^{(2)}\ ,\ \underline{\underline{B}}\right] = nkT\ \ n\left[\underline{\underline{B}},\varphi^{(2)}\right]$$

$$= nkT\int f^0\ \underline{\underline{B}}\ \ R_o\,\varphi^{(2)}\ d\underline{c}$$

$$\underline{c}\cdot\underline{\nabla}\varphi^{(1)} = -n\,R_o\varphi^{(2)}$$

$$= -kT\int f^0\ \ \underline{\underline{B}}\ \underline{c}\cdot\underline{\nabla}\varphi^{(1)}\ d\underline{c}\ .$$

One has $\varphi^{(1)} = -\frac{1}{T}\underline{A}\cdot\underline{\nabla}T$. Neglecting terms like $(\underline{\nabla}T)^2$ and keeping only $\underline{\nabla\nabla}T$, one obtains

$$\overset{\frown}{\underline{\underline{\Pi}}}^{(2)} = kT\int f^0\,\underline{\underline{B}}\,\frac{1}{T}\underline{A}\cdot\underline{\nabla}\,\underline{\nabla}\,T\cdot\underline{c}\,d\,\underline{c}\ .$$

The macroscopic equation reads

$$\overset{\frown}{\underline{\underline{\Pi}}} = 2\,\tilde{\eta}\,\frac{1}{T}\overbrace{\underline{\nabla}\,\underline{\nabla}}T\ .$$

Combination gives the microscopic expression for $\tilde{\eta}$. We have further, in first Sonine approximation:

$$\underline{A} = a_1\,(W^2 - \tfrac{5}{2})\,\underline{W} \qquad\qquad B = b_1\,\overbrace{\underline{W}\,\underline{W}}$$

$$\lambda = n\,\tfrac{5}{2}\sqrt{\frac{2kT}{m}}\;k\,a_1 \qquad\qquad \eta = \tfrac{1}{2}\,nkT\,b_1\ .$$

Using these expressions we obtain:

$$\tilde{\eta} = \frac{2}{5}\,\frac{\lambda\eta}{n\,k}$$

This is known as the Maxwell stress coefficient. Note that $\tilde{\eta}\sim\frac{1}{n}$ as was to be expected. Theory shows further than $\underline{\nabla}\,\underline{\nabla}\,T$ gives no contribution to $\underline{q}$ if no gas flow is present. It was shown by Levi et al. [4,5] that for a gas of rotating molecules the expression becomes

$$\tilde{\eta} = \frac{2}{5}\,\frac{\lambda_{transl}\,\eta}{n\,k}\ ,$$

where λ_{transl} is the coefficient corresponding to the contribution of the translational degrees of freedom to the energy transport. Note that in the lowest order Sonine approximation only the translational heat flux is coupled to the pressure tensor. Retaining only the simplest angular momentum polarization: $(\underline{W}.\underline{J})\,\underline{J}$ and $\overbrace{\underline{J}\,\underline{J}}$ Levi et al. solved the Burnett equation in the presence of a field. $\tilde{\eta}$ will now become a fourth rank tensor $\tilde{\underline{\eta}}$ containing both even and odd in $\underline{H}$. We are here not interested in the even in $\underline{H}$ components as they will be very difficult to measure. The situation for the odd components is more favourable. The expressions for these coefficients in terms of the first order ones become rather complicated mainly because of the complex structure of the energy transport in a gas with internal degrees of freedom. They become much simpler if we consider a hypothetical case where one may neglect the transport of internal energy. In this case the contributions contain four parts: one proportional to $\eta\lambda_{tr}$ and having also the field dependence: $\left(\frac{\xi_\lambda}{1+\xi_\lambda^2} + 2\frac{2\xi_\lambda}{1+4\xi_\lambda^2}\right)$ a second of the type $\lambda_{transl}\bar{\eta}_2$ and behaving like $\bar{\eta}_2$ i.e. $\sim \frac{2\xi_\eta}{1+4\xi_\eta^2}$. The last two contributions are of a mixed nature, they are proportional to $\sqrt{\lambda_{tr}\bar{\eta}_2}$. These contributions occur in two different $\frac{H}{p}$ regions, one with characteristic parameters ξ_λ and $2\xi_\lambda$ and another with the parameters: ξ_η and $2\xi_\eta$. This is a structure one would intuitively expect from the field free behaviour of $\tilde{\eta} \sim \eta\,\lambda_{transl}$. For the case of a torque on a cylinder with radius a and height one gets:

$$\tau = -2\pi a^2 h\, \Pi_{r\theta}$$

with

$$\Pi_{r\theta} = \frac{2}{T}\tilde{\eta}_{r\theta}\frac{d^2T}{dr^2} .$$

The logarithmic temperature field between coaxial cylinders gives:

$$T(r) = T(a) + \frac{\Delta T}{\ln T}\ \ln {}^R/a .$$

In this way we get for the torque:

$$\tau = \frac{4\pi h}{\ln {}^R/a}\frac{\Delta T}{T}\tilde{\eta}_{r\theta} .$$

$\tilde{\eta}_{r\theta}$ is a combination of two Cartesian tensor components of $\tilde{\underline{\eta}}$ appropriate to the cylindrical symmetry. The full expression for the field dependent torque reads

$$\frac{p\tau}{\Delta T}\ln {}^R/a = \frac{8\pi}{5}\lambda_{transl}\,\eta\Big\{-\tfrac{1}{2}A(\bar{f}(\xi_\lambda) + 2\bar{f}(2\xi_\lambda)) - B\bar{f}(2\xi_\lambda) +$$

$$c'_1 = \frac{\xi_\eta}{\xi_\lambda} B \qquad\qquad + c_1\bar{f}(2\xi_\eta) + c_2\bar{f}(2\xi_\eta) .$$

A and C_2 corresponds respectively to the pure λ_{tr} and $\bar{\eta}_2$ terms and B and C_1 resp. to the cross terms in the simplified situation discussed earlier. They are in general rather complicated expressions in the first order field free and field dependent coefficients. The contribution at ξ_η is the dominant one, so the maximum in torque will occur near to the one in the $\bar{\eta}_2$ curve. A detailed comparison is hindered by the fact that there is also a contribution of a boundary layer effect to the Scott torque. While in the bulk $\underline{\nabla}T$ does not couple to

$\underline{\Pi}$, this is no longer true near a wall as there the gas can no longer be considered as an isotropic system. In monatomic gases one has, for example, the phenomenon of thermal creep: i.e. a gas flow along a boundary in the presence of a temperature gradient along the wall. Consider a gas in which there is a temperature gradient parallel to a wall that is at a constant temperature (see fig.23)

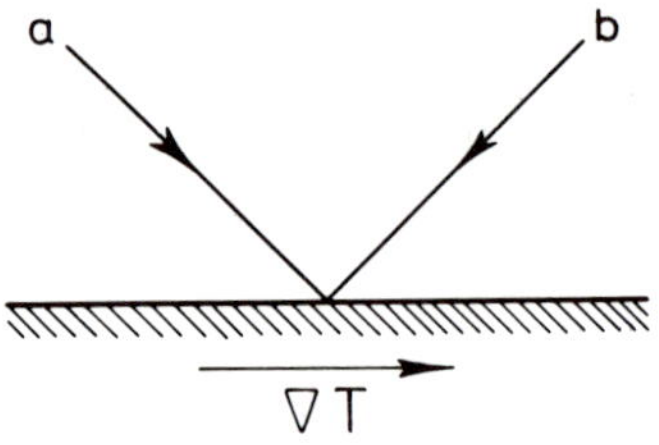

Fig. 23. Schematic diagram illustrating the occurrence of thermal slip.

Let us assume that the accommodation coefficient for translational energy is equal to 1. Thus a molecule hitting the wall leaves the surface with a kinetic energy corresponding to the temperature of the wall and in a random direction. Let us now compare the effect of this accommodation on two molecules coming from resp. a and b. They will both leave the wall with the same average energy, but as $\overline{\frac{1}{2} mv_a^2} < \overline{\frac{1}{2} mv_b^2}$ they gave off a different amount of momentum to the wall in this process. Consequently there is a net loss of momentum by the gas layer near the surface in the $-\underline{\nabla}T$ direction. This will cause the gas to move in the $+\underline{\nabla}T$ direction so that its velocity is such that the ordinary viscous friction will just make up for the momentum lost in the accomodation process. Thus we will have a flow of gas from cold to warm: thermal creep. This flow will under these conditions exert no force on the wall therefore it is also called thermal slip. While monatomic gas slip phenomena have to do with the accommodation of translational energy at the wall, in polyatomic gases also the accommodation of internal energy and the behaviour of the internal angular momentum in its interaction with the wall has to be taken into account. Such boundary layer phenomena can give a contribution to the Scott effect as was shown by Levi and Beenakker [3] and by Waldmann [6]. A discussion of the theory of these effects is, however, outside the scope of these lectures. For our discussion here it is sufficient to realize that it is possible to devise a set-up where ∇T is negligibley small so that only slip effects occur. Such an experiment was suggested by Waldmann and performed at Leiden by Hulsman et al. [7]. From the results of these measurements it is possible to calculate the slip contribution to the Scott torque. It appears to be of opposite sign to the Maxwell stress. Combination of slip and torque measurements allows then a comparison with calculations of the Maxwell stress contributions.

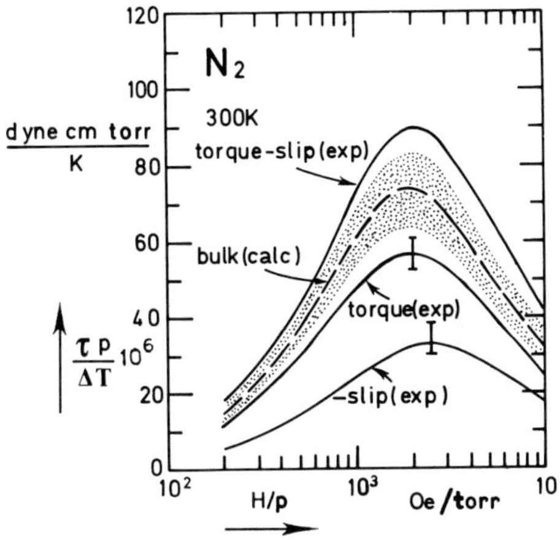

Fig. 24 The boundary-layer (or slip) contribution to the thermomagnetic torque for N_2 as found from the experiment, the total torque as measured by Adair and co-workers and the combination of the two yielding the experimentally determined bulk contribution. The dotted line gives the bulk contribution as calculated from the expression given by Levi et al. The shaded area around the calculated curve reflects the uncertainties in the used cross section values.

Fig. 24 gives the results for N_2. It is, however, still too early to decide whether the discrepancies between theory and experiment are significant.

The transverse viscomagnetic heat flux

When in the Burnett regime a gas flows in such a way that its velocity has non vanishing second spatial derivatives (as, for example, in ordinary Poiseuille flow) a heat flow will be present. In a way similar to the derivation of the expression for the Maxwell stress one gets at uniform temperature

$$\underline{q}^{(2)} = 2\,\tilde{\eta}\,\underline{\nabla}\cdot\overline{\underline{\nabla}\,\underline{v}} \; .$$

This heat flow is, however, difficult to observe in the absence of a field, since the heat flux is parallel to the flow and can hardly be separated from convection and other uninteresting, but much larger phenomena. In a magnetic field, however, the heat flux has a transverse component which can be termed (transverse) visco-magnetic heat flux.

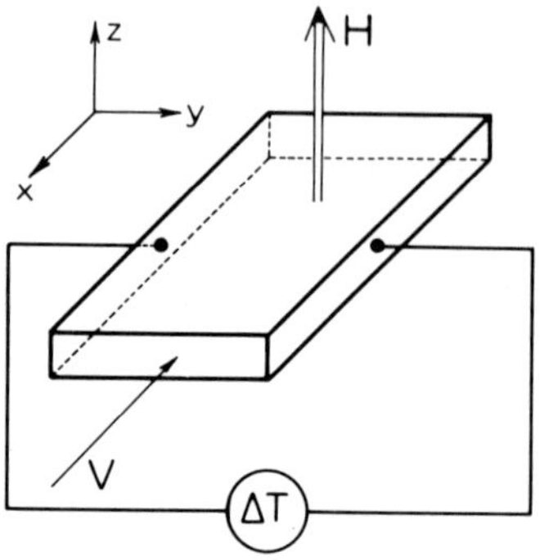

Fig. 25 Schematic diagram for measuring the viscomagnetic heat flux.

In the arrangement shown in the fig. 25 one will have a transverse temperature difference. As shown by Levi et al. [5], its magnitude is directly related to the dilute gas transverse pressure difference by the relation

$$\frac{1}{T}\,\frac{\partial T}{\partial y} = -\,\rho(\xi)\,\frac{1}{p}\,\frac{\partial p}{\partial y}$$

here $\rho(\xi)$ is a slowly varying function of the field with a value around .5. So far no experiments in this direction are reported.

Let us end this discussion by again underlining the importance of the field effect in the rarefied gas regime as being one of the very few methods that give data that are not submerged in larger dilute gas contributions. Through measurements of the Burnett coefficients one has a tool for the direct verification of the kinetic theory in this regime. Other applications are in the study of boundary layer phenomena and in that of Knudsen effects. Recent theoretical work of the Erlangen [8,9] and the Moscow group [10, 11] are in this respect very promising.

Acknowledgements

The collaboration of the members of the Leiden group for molecular physics is gratefully acknowledged.

This work is part of the research program of the "Stichting voor Fundamenteel Onderzoek der Materie (F.O.M.)" and has been made possible by financial support from the "Nederlandse Organisatie voor Zuiver Wetenschappelijk Onderzoek (Z.W.O.)".

REFERENCES

PART I

1) H. SENFTLEBEN, Physik. Z. 31 822. (1930)

2) C.J. GORTER, Naturwizscnschaften 26. 140 (1938)

3) F. ZERNIKE, and C. van LIER, Physica 6 961. (1939)

4) Yu. KAGAN, and L. MAKSIMOV, Sovietу Phys. JETP 14 (1962) 604.

5) F.B. PIDDUCK, Proc. Roy. Soc. A 101 101. (1922)

6) V.D. BORMAN, L.L. GORELIK, B.I. NIKOLAEV and V.V. SINITSYN, JETP Letters 5 (1967) 85.

7) J.J.M. BEENAKKER, G. SCOLES, H.F.P. KNAAP, and R.M. JONKMAN Phys. Lett 2 5. (1962)

8) L.L. GORELIK and V.V. SINITSYN, Soviet Phys. JETP 19 272. (1964)

9) L.J.F. HERMANS, J M KOKS, A f HENGEVELD and H.F.P., Physica 50 410. (1970)

10) H.Van KUIK F.G.HULSMAN K.W.WALSTRA, H.F.P.KNAAP and F.F.M. BEENAKKER, Physica 57. 501. (1972)

11) H. SENFTLEBEN, Ann. Physik 15 273. (1965)

12) G. GALLINARO G. MENEGHETTI, and G. SCOLES, Phys. Lett. 24A 451. (1967)

13) S.R. De GROOT and P. MAZUR, Nonequilibrium Thermodynamics, North Holland Publishing Co., (Amsterdam 1962).

14) J. KORVING, H. HULSMAN. H.F.P. KNAAP, and J.J.M. BEENAKKER, Phys. Lett. 21 5. (1966)

15) L.J.F. HERMANS, P.H. FORTUIN, H.F.P. KNAAP and J.J.M BEENAKKER, Phys. Lett. 25A 81 (1967).

16) L.L. GORELIK, V.G. NIKOLAEVSKII, and V.V. SINITSYN, JETP Letters 4 307 (1966)

17) J.A.R. COOPE, and R.F. SNIDER, J. Chem. Phys 56 2056 (1972)

PART II

1) Yu. KAGAN, and L.A. MAKSIMOV, Soviet Phys. JETP 14 604. (1962)

2) F.R. McCOURT, and SNIDER, R.F., J. Chem. Phys. 47 (1967) 4117.

3) L.J.F. HERMANS, J.M. KOKS, A.F. HENGEVELD, and H.F.P.KNAAP, Physica 50 410. (1970).

4) H. HULSMAN and H.F.P. KNAAP, Physica 50 565 (1970).

5) S. HESS, Phys. Lett. 30A 239. (1969).

6) F. BAAS, Phys. Lett 36A 107. (1971)

7) G. GALLINARO, G. MENEGHETTI, and SCOLES, Phys. Lett. 24A (1967) 451.

8) J. KORVING, Physica 46 619 (1970).

9) J.J. De GROOT L.J.F. HERMANS, C.J.N. van der MEIJDENBERG and J.J.M. BEENAKKER, Phys Lett. 31A 304 (1970)

10) L. WALDMANN, and S. HESS, Z. Naturforsch. 24a 2010 (1969)

PART III

1) S. HESS, Springer Tracts Mod. Phys. 54 136 (1970)

2) R.A.J. KEIJSER, et al. Physica, to be published

3) R. SHAFER, and R.G. GORDON., J. Chem. Phys. 58 5422. (1973)

4) A.G. St. PIERRE, W.E. KÖHLER, S. HESS. Z. Naturforsch. 27a 721. (1972)

PART IV

1) F.R. McCOURT and H. MORAAL, Chem. Phys. Letters 9 39. (1971)

2) H. MORAAL, and R.F. SNIDER, Chem. Phys. Letters 9 401. (1971)

3) W.M. KLEIN, D.K. HOFFMAN, and J.S. DAHLER, J. Chem. Phys. 49 2321. (1968)

PART V

1) G.G. SCOTT, H.W. STURNER and R.M. WILLIAMSON, Phys. Rev. 158 117. (1967).

2) A.L.J. BURGMANS, and T.W. ADAIR, J. Chem. Phys. 59 324 (1973).

3) A.C. LEVI and J.J.M. BEENAKKER, Phys. Lett. 25A 350, (1967)

4) A.C. LEVI, F.R. McCOURT, and J. HAJDU, Physica 42 347. (1969).

5) A.C. LEVI, F.R. McCOURT, and J.J.M. BEENAKKER, Physica 42 363. (1969).

6) L. WALDMANN, Z. Naturforsch 22a 1678 (1967).

7) H. HULSMAN, F.G. van KUIK, H.F.P. KNAAP, and J.J.M. BEENAKKER, Physica 57 522. (1972)

8) H. VESTNER, M. KUBEL, and L. WALDMANN, Proc. Intern Sump. Rarefied Gas Dyn. 7th, Pisa (1970)

9) H. VESTNER, Z. Naturforsch 28a 869. (1973).

10) V.D. BORMAN, V.S. LAZKO and B.I. NIKOLAEV, Soviet Phys. JETP 36 466. (1973).

11) V.D. BORMAN, L.A. MAKSIMOV, B.I. NIKOLAEV, and V.I. TROYAN, Zh. Theor. Fiz. 64. 526 (1973).

SURVEY ARTICLES ON FIELD EFFECTS

I. The influence of electric and magnetic fields on the transport properties of Polyatonic dilute gases. J.J.M. BEENAKKER, 1960 FestKorper problems p: 275-311, Vieweg.

II. Magnetic and electric effects on transport properties J.J.M. BEENAKKER and F/R/ McCOURT. Ann. Rev. Phys. Chem 21 47-72 (1970)

III. Non-equilibrium angular momentum polarisations in rotating molecules. J.J.M. BEENAKKER Heler Physica Austriaca. Suppl.I. (1973) 267-300, Springer.

IV. Experimental information on the ample dependent interaction between polyatomic molecules J.J.M. BEENAKKER, H.F.P. KNAPP and B.C. SANCTUARY AIP Conference Proceedings II (1973) 21-50.

TRANSPORT PROPERTIES OF DILUTE GASES WITH INTERNAL STRUCTURE

R.F. Snider

University of British Columbia
Vancouver, Canada

I. INTRODUCTION

II. THE ROLE OF FREE MOLECULAR MOTION

2.1. Qualitative picture

2.2. Generalization and mathematical formulation

2.3. Implications

III. FIELD DEPENDENCE OF THE VISCOSITY

3.1. Formal expression for the viscosity coefficient

3.2. The simple case of N_2 in a magnetic field

3.3. The treatment of oxygen

3.3.1. Low fields

3.3.2. High fields

IV. ON THE BOLTZMANN EQUATION FOR MOLECULES WITH INTERNAL STRUCTURE

4.1. "Derivation" of the generalized Boltzmann Equation

4.2. Transition operator representation

4.3. Dependence on free motion frequencies

V. COLLISION INTEGRALS OF THE LINEARIZED W-S EQUATION

REFERENCES

I. INTRODUCTION

Professor Beenakker has already given in his lectures [1] a general introduction to the field dependence of gas transport coefficients. In those lectures he has given some of the history of these effects and of their interpretation, as well as the present status of what experiment tells us about what the molecules are doing in these processes. I will briefly sketch some of the theoretical developments as I see them.

Modern theoretical calculations of these effects began in 1961 with the work of Kagan and Maksimov [2] , using a classical Boltzmann equation and a model collision operator. During the next 8 - 9 years, the model dependence was eliminated and a general formulation in terms of quantum mechanics was developed [3]. The central theme that dominated this development was an emphasis on statistical mechanics and a minimal role of individual molecular behaviour. By statistical mechanical emphasis, I mean the ingrained attitude of statistical mechanics: if you don't know what to do, take an average. The only other concept that was retained, was a proper treatment of group theory. This meant that the dependence of the density operator on the velocity and angular momentum directions was carefully considered, but everything else (in particular velocity and angular momentum magnitudes) was averaged. From the directional dependence of the experimental observations, it was deduced that the density operator (or distribution function) of the gaseous system has only a limited dependence on the directions of the internal molecular angular momentum. As Professor Beenakker has discussed in his second lecture, only certain "polarizations" are important. Moreover, the theoretical calculations of this period all involved the ratio of Larmor to collision frequencies, so that the experiments should scale to be functions of $^{H}/p$, and this was usually observed.

Yet some things did not fit, O_2 being the principle counter-example. Improved experimental accuracy showed that the $^{H}/p$ scaling just did not work! This is easily traced to the fact that there is an electron spin angular momentum as well as rotational angular momentum, and the strength of the coupling between the two must be taken into account. Essentially, this can be interpreted as saying that there was too much averaging. The treatment [4] of O_2 in 1970 began a trend to re-emphasize the role of individual molecular behaviour.

An emphasis on molecular motion necessarily complicates the treatment (since more quantities must be taken into account), while averaging simplifies the amount of calculation (fewer quantities). It is also possible to lose the description of experimental detail with too much averaging, while too detailed a description can easily make the calculation intractable. Clearly some compromise is required.

The "Qualitative Picture" is [5] an outgrowth of the detailed treatment of O_2. Besides giving a simple picture by which the field dependence of gas transport properties may be understood, it has the advantage that it exactly parallels the formal solution of the Boltzmann equation and can also be immediately put into correspondence with the formal development that results from any treatment based on time correlation functions. Because of these associations, it serves as an excellent guide both to what is happening in the gas and as to what must be taken into account in any calculation. The first three lectures will describe this "Qualitative Picture" and apply it to a discussion of the field dependence of the viscosity coefficient.

The qualitative picture over-emphasizes the role of the internal state energy level structure. It is also necessary to know something about the collision processes. I will discuss what I consider is the general Boltzmann equation [6] (having only binary collisions) and some of its properties in the fourth lecture, while in the final lecture I will discuss some of the results [7] that can be obtained from an organization and approximate evaluation of the collision rates. The role of collisions can be divided into two classes, the relaxation rates and the production processes. The latter are defined as the collision processes coupling the velocity and angular momentum directions. While in the decade before the "Qualitative Picture", a small number of relaxation times and production rates were used in explaining the experiments, the detailed dependence of the collision rates on, in particular, the angular momentum magnitude j, has recently led to a much better agreement between data obtained from different experiments. Professor Beenakker has discussed [1] some of this work. I will discuss the j dependence of the collision processes from the viewpoint of the Distorted Wave Born Approximation, and this does seem to be in agreement with at least some of the experimental data.

II. THE ROLE OF FREE MOLECULAR MOTION

2.1. Qualitative Picture

Elementary notions of gas phase transport are based on the mean free path picture. In this, transport occurs by free motion between collisions and the collisions act as disruptive processes for the transport. It is also part of the usual treatment, that collisions occur sufficiently often and that the mean free path is short enough, that the gas may be considered to be in local thermodynamic equilibrium. Knudsen effects occur if this latter condition is not fulfilled and I will not discuss such effects.

To calculate the transport due to a mean free flight, it is customary to visualize two planes lying a mean free path "ℓ" apart, with their common normal direction $\hat{u}$ (a unit vector) along the direction of transport. The planes have averages, $\langle B \rangle$ and $\langle B \rangle + \Delta\langle B \rangle$, of the molecular property (quantum mechanical observable) B being transported. Since a third of the molecules can be thought to be moving normal to the planes with an average speed $\bar{c}$, and since there are n molecules per unit volume, the amount of B transported from one plane to the other is

$$\text{B transported} = \frac{1}{3}\, n\bar{c}\, \Delta\langle B \rangle = \frac{1}{3}\, n\bar{c}\, \ell\, (d\langle B \rangle/du) \quad . \tag{1}$$

Expressed in terms of the gradient $d\langle B \rangle/du$ of B, this gives a transport coefficient

$$\eta = \frac{1}{3}\, n\bar{c}\,\ell = \frac{1}{3}\, n\bar{c}^2 \tau \, . \tag{2}$$

which is more conveniently expressed in terms of the mean free time $\tau \equiv \ell/\bar{c}$ rather than in terms of ℓ . Most transport coefficients (the Diffusion coefficient is the major exception) are density independent because of two compensating effects. With increased density, more molecules are doing the transporting, while each molecule travels a smaller distance because of the increased frequency of collisions (more disruptions); mathematically this is expressed by relating τ^{-1} to an effective collision cross-section σ according to

$$\tau^{-1} = n\,\langle v \rangle_{rel}\, \sigma \tag{3}$$

where $\langle v \rangle_{rel}$ is the average relative speed of two molecules. For the viscosity, $\bar{c}^2$ should be replaced by the root mean square speed, compare Eq. (48).

It is a tacit assumption in the derivation of Eq. (2) that a molecule, on moving from one plane to the other, decreases the amount of B in one plane by $\Delta\langle B\rangle$ and increases the other by $\Delta\langle B\rangle$. That is, B is a constant of the free motion and there is thus a transport of $\Delta\langle B\rangle$ between the planes. This is true for mass, momentum and energy but not true for most properties - in particular for an angular momentum (vector) $\underset{\sim}{J}$ in a magnetic field $\underset{\sim}{H} = H\hat{z}$ directed, for convenience of presentation, along the $\hat{z}$ axis. If, in particular, the average angular momentum $\langle \underset{\sim}{J} \rangle$ is directed along the $\hat{x}$ axis in one plane, a molecule having this average angular momentum will have, at time t later during its free flight, the precessed angular momentum

$$\langle \underset{\sim}{J} \rangle (t) = \langle J_x \rangle (0) \left[\hat{x} \cos \omega t + \hat{y} \sin \omega t \right] \tag{4}$$

where ω is the Larmor precession frequency $-\gamma H$, γ being the gyromagnetic ratio. For planes a distance "ℓ" apart, the angular momentum reaching the second plane has the same initial magnitude but has been rotated by an angle $\omega\tau$ around the field direction. Thus the transport of angular momentum should be modified by the directional factor in Eq. (4), in order to take into account the precessional motion. Note firstly that these effects depend on direction, for example, the $\hat{z}$ component of the angular momentum does not precess, but secondly, that if $\omega\tau >> 1$, then the exact value of τ is very important to determine exactly what direction the angular momentum has when it arrives at the second plane. Such a sensitive behaviour is contrary to the statistical nature of the treatment. There is an easy remedy for this latter defect once it is realized that τ is a mean free time and after all, different molecules have different free times between collisions, and hence different amounts of precession. One should take the average effect of the precession obtained, by averaging over the distribution of free times. Elementary arguments lead to a Poisson distribution [8]

$$P(t) = \tau^{-1} \exp(-t/\tau) \tag{5}$$

of free times which implies that an $\hat{x}$ directed angular momentum in one plane give rise to a resultant arriving angular momentum of

$$\langle \underset{\sim}{J} \rangle_{\text{arrival}} = \langle J_x \rangle_{\text{leaving}} \int_0^{\tau} P(t) \left[\hat{x} \cos \omega t + \hat{y} \sin \omega t\right] dt$$

$$= \langle J_x \rangle_{\text{leaving}} \left[\frac{\hat{x}}{1+\omega^2\tau^2} + \frac{\hat{y}\,\omega\tau}{1+\omega^2\tau^2} \right] \qquad (6)$$

The averaging of the precessional motion does two things. First of all, the total magnitude of the arriving angular momentum is less than that leaving the first plane according to

$$|\langle \underset{\sim}{J} \rangle_{\text{arrival}}| = \left[\langle J_x \rangle_{\text{arr.}}^2 + \langle J_y \rangle_{\text{arr.}}^2 \right]^{1/2}$$

$$= |\langle J_x \rangle_{\text{leaving}}| \; (1+\omega^2\tau^2)^{-1/2}, \qquad (7)$$

while secondly, the effective precession angle α_{prec} is

$$\alpha_{\text{prec.}} = \arctan \omega\tau \qquad (8)$$

and has a maximum of 90° in magnitude. These effects are demonstrated in Table I.

TABLE I

EFFECT OF AVERAGING THE PRECESSION

$\langle \underset{\sim}{J} \rangle_{\text{leaving}} = \hat{x}$ (unit angular momentum in the $\hat{x}$ direction)

$\omega\tau$	$\langle J_x \rangle$ arr.	$\langle J_y \rangle$ arr.	$\lvert\langle \underset{\sim}{J} \rangle$ arr.$\rvert$	$\alpha_{\text{prec.}}$
0	1	0	1	0
1	½	½	$(2)^{-1/2}$	45°
10	$^1/101 \sim 10^{-2}$	$^{10}/101 \sim 10^{-1}$	$(101)^{-1/2} \sim 10^{-1}$	84°
∞	0	0	0	90°

I could spend a great deal of time discussing the implications of this simple result but let me comment at this stage on only four things:

(i) The precessional effects are governed by the quantity $\omega\tau$ which is the ratio of Larmor to collision frequencies and thus experimentally dependent on the ratio $^{H}/p$, first observed by Senftleben in 1930.

(ii) The direction of precession depends on the sign of ω , hence on the sign of the gyromagnetic ratio. This has been used to experimentally [9] verify the theoretically predicted signs of γ.

(iii) The decrease in angular momentum magnitude is not a relaxation effect but rather due to a randomization or dephasing of the angular momentum directions. We prefer to call it "phase randomization". This association arises from the mathematical treatment which I will discuss presently.

(iv) In this picture, the collisions are field independent and merely act as disruptions of the transport processes. This is in contrast to the earlier literature (in particular Gorter's picture [10]), in which stress is given to the orientation of the colliding molecules. The present picture is in agreement with the formal methods of treating these phenomena and the rationalization of field independence of collision processes is usually based on the fact that the Zeeman energies associated with the Larmor precession are negligible in comparison with the thermal energies that are available during collisions.

2.2 Generalization and Mathematical Formulation

For dilute gases, all observables of interest are one-molecule observables B and their expectation values are obtained by tracing with the (time dependent - that is - Schrödinger picture) singlet density operator $\rho(t)$, namely

$$\langle B \rangle (t) = \mathrm{Tr}\, \rho(t)\, B \quad . \tag{9}$$

Neglecting collisions for the present discussion, $\rho(t)$ evolves in time according to the von Neumann (or quantum Liouville) equation

$$i \frac{\partial \rho}{\partial t} = \mathcal{L}\rho \equiv \frac{1}{\hbar}(\mathcal{H}\rho - \rho\mathcal{H}) \tag{10}$$

Here $\mathcal{H}$ is the hamiltonian for one molecule and $\mathcal{L}$ is the superoperator [11] (that is, operator acting on operators) defined as the "commutator of $\mathcal{H}/\hbar$ with", namely

$$\mathcal{L}A \equiv (\mathcal{H}A - A\mathcal{H})/\hbar \quad . \tag{11}$$

The insertion of $\hbar^{-1}$ at this point is convenient in that $\mathcal{L}$ now has the units of inverse time.

If one now visualizes one plane of our gas as having expectation value $\langle B \rangle$, then a molecule leaving this plane with this value of B will arrive at a plane a mean free path away with $\langle B \rangle$ modified by free motion according to

$$\int_0^\infty P(t)\langle B \rangle(t)\, dt = \int_0^\infty P(t) \text{ Tr } \rho(t) \text{ B } dt$$

$$= \int_0^\infty P(t) \text{ Tr } \left[\exp(-i\mathcal{L}t)\rho\right] \text{ B } dt$$

$$= \int_0^\infty P(t) \text{ Tr } \rho \exp(i\mathcal{L}t) \text{ B } dt$$

$$= \langle (1 - i\mathcal{L}\tau)^{-1} B \rangle . \tag{12}$$

I will thus interpret the factor $(1 - i\mathcal{L}\tau)^{-1}$, which arises in calculating transport properties by formally solving the Boltzmann equation, as arising from free motion between collisions. Here, I first want to discuss the properties of this "phase randomization factor".

It is convenient to consider operators which are eigenoperators of $\mathcal{L}$. If $|m\rangle$, $|n\rangle$ are eigenvectors of $\mathcal{H}$, with energies $\mathcal{E}_m$ and $\mathcal{E}_n$ respectively, then $\Theta_{mn} = |m\rangle\langle n|$ is an eigenoperator of $\mathcal{L}$ with eigenvalue $\omega_{mn} = \hbar^{-1}(\mathcal{E}_m - \mathcal{E}_n)$ according to

$$\mathcal{L}\Theta_{mn} = \hbar^{-1}\left[\mathcal{H}|m\rangle\langle n| - |m\rangle\langle n|\mathcal{H}\right] = \omega_{mn}\,\Theta_{mn} \quad . \tag{13}$$

Any observable can be expanded in terms of this set of operators Θ_{mn}. The expectation value of Θ_{mn} is the matrix element

$$\langle \Theta_{mn} \rangle = \text{Tr } \rho\Theta_{mn} = \langle n|\rho|m\rangle = \rho_{nm} \tag{14}$$

of the density operator, so that free motion modifies the matrix elements of ρ according to

$$\rho_{nm} = \langle \Theta_{mn} \rangle \rightarrow \langle (1 - i\mathcal{L}\tau)^{-1}\Theta_{mn} \rangle = (i - i\omega_{mn}\tau)^{-1}\rho_{nm} . \tag{15}$$

Note firstly the order of indices, ω_{mn} versus ρ_{nm} and secondly that if m = n, then $\omega_{nn} = 0$ and thus that the diagonal elements of ρ are unaffected by the free motion. If the energy level spacing is large enough so that $\omega_{mn}\tau \gg 1$ for all $n \neq m$, then

$$\left|(1 - i\omega_{mn}\tau)^{-1} \rho_{nm}\right| = \left[1 + (\omega_{mn}\tau)^2\right]^{-1/2} |\rho_{nm}| \underset{n \neq m}{\longrightarrow} 0 \qquad (16)$$

and the density operator becomes diagonal. This is the usual statement of phase randomization. This is especially emphasized if the "initial" ρ is a pure state with different energy components, $\rho_a = |a\rangle\langle a|$ where

$$|a\rangle = \sum_m |m\rangle\, c_m \quad . \qquad (17)$$

Now free motion modifies ρ_a according to

$$\rho_a \longrightarrow \sum_{mn} |m\rangle\, c_m\, (1 - i\omega_{nm}\tau)^{-1}\, c_n^* \langle n|$$

$$\underset{\tau\to\infty}{\longrightarrow} \sum_m |m\rangle\, |c_m|^2 \langle m| \qquad (18)$$

with the extreme case given when $\tau \to \infty$ and all states are non-degenerate. This is a diagonal density operator and is necessarily mixed rather than a pure state. The off-diagonal elements of ρ_a, in the m representation, involve phase relations between the c_m's (another way is to say that these are coherences). The free motion randomizes these and leaves only the incoherent part of the density operator, namely the diagonal part. The intermediate stage of phase randomization, when $\omega_{mn}\tau$ is neither 0 nor very large, will be referred to as partial phase randomization.

Now the adjoint $\mathcal{O}^{+}_{mn} = \mathcal{O}_{nm}$ of $\mathcal{O}_{mn}$ has the negative frequency $\omega_{nm} = -\omega_{mn}$ and the commutator

$$\left[\mathcal{O}_{mn}, \mathcal{O}^{+}_{mn}\right]_- = \mathcal{O}_{mm} - \mathcal{O}_{nn} \qquad (19)$$

is a constant of the (free) motion. The time evolution of $\mathcal{O}_{mn}$ involves an interconversion of the hermitian and antihermitian parts of $\mathcal{O}_{mn}$, or of their expectation values

$$P_{mn} \equiv \tfrac{1}{2}(\rho_{mn} + \rho_{nm}) \quad ; \quad I_{mn} = -\tfrac{1}{2} i\, (\rho_{mn} - \rho_{nm}) \quad . \qquad (20)$$

The average effect of free motion is to lead to the relation

$$\begin{pmatrix} P_{mn} \\ \\ I_{mn} \end{pmatrix}_{\text{arrival}} = \begin{pmatrix} f^{+}(\omega\tau) & f^{-}(\omega\tau) \\ \\ -f^{-}(\omega\tau) & f^{+}(\omega\tau) \end{pmatrix} \begin{pmatrix} P_{mn} \\ \\ I_{mn} \end{pmatrix}_{\text{leaving}} \tag{21}$$

in terms of the Lorentz-Debye absorption and dispersion line shapes

$$f^{+}(\omega\tau) = \left[1 + (\omega\tau)^2\right]^{-1} \quad \text{and} \quad f^{-}(\omega\tau) = \omega\tau\left[1 + (\omega\tau)^2\right]^{-1} . \tag{22}$$

This is the generalization of the angular momentum, Larmor precession frequency case that was considered at the beginning of the discussion of the qualitative picture, see Eq. (6). Since J_+ and J_- are eigen-operators of the Zeeman Liouville superoperator, J_x is the hermitian and J_y the antihermitian part of J_+. It is coincidental to the phase randomization that J_x and J_y have directional properties, but of course of fundamental significance experimentally. What must be stressed is that the effects of free particle motion are very far reaching and are not in general connected to any directional properties of the gas.

To continue the discussion of angular momentum, consider a quantum mechanical pure state having the angular momentum pointing along the $\hat{u}$ direction given by the spherical coordinate angles θ and ϕ. It is easily shown that the c_m of Eq. (17) are

$$c_m = \left[(2J)!/(J+m)!(J-m)!\right]^{1/2} (\sin\tfrac{1}{2}\theta)^{J-m}(\cos\tfrac{1}{2}\theta)^{J+m} \exp(-im\phi) \tag{23}$$

in the equation

$$|J\,\theta\,\phi\rangle = \sum_{m=-J}^{J} |J\,m\rangle\, c_m(\theta,\phi) . \tag{24}$$

Complete phase randomization leads to

$$\rho_{\text{phase random.}} = \sum_{m=-J}^{J} |Jm\rangle\, |c_m|^2 \langle Jm| \tag{25}$$

with angular momentum expectation value

$$\begin{aligned} \langle \underset{\sim}{J} \rangle &= \mathrm{Tr}\, \underset{\sim}{J}\, \rho_{\text{phase random.}} \\ &= \hat{z} \sum_{m=-J}^{J} m\, |c_m|^2 \\ &= J \cos\ \hat{z} . \end{aligned} \tag{26}$$

That is, only the component of the angular momentum parallel to the field axis has remained after complete phase randomization; this is the only part that has zero frequency.

2.3. Implications

Two important consequences are obtained from this discussion. First, free motion and partial phase randomization can give rise to external field and pressure dependent phenomena, as is exemplified by the Senftleben effects. This is discussed in my subsequent lecture. Secondly, if the separation of two energy states is large compared to $\hbar/\tau$, then (off-diagonal) matrix elements of ρ between these two states will be relatively unimportant. This has very important consequences in any theoretical calculation, for it serves as a guide to what terms one can ignore and so increase the efficiency of one's calculation.

III. FIELD DEPENDENCE OF THE VISCOSITY

The discussion here is limited to situations in which the Navier Stokes equations and the Chapman Enskog method are applicable. This means that the gas is, first of all, dilute enough that only binary collisions are important, and secondly, dense enough so that there is a sufficient collision rate to maintain local equilibrium. Lack of the first would mean that there is little if any free motion and we could not use the Boltzmann equation as the starting point, while lack of the second means that a separation into kinetic and hydrodynamic effects is not possible and consequently that the properties of the whole gas must be solved all together.

Between these two extremes there is a region in which we can treat the gas as linearly perturbed from local equilibrium and, because of this linearity, the results can be compared with those of linear response theory.

3.1. Formal Expression for the Viscosity Coefficient

Local equilibrium means that in each macroscopically small part of the gas, the properties of the gas are describable in terms of a local temperature T ($\underset{\sim}{r}$, t), number density n ($\underset{\sim}{r}$, t) and stream velocity $\underset{\sim}{v}_0$ ($\underset{\sim}{r}$, t) at position $\underset{\sim}{r}$ and time t. Correspondingly, there is a local equilibrium (Maxwell-Boltzmann) density operator for the system

$$f^{(0)} = \frac{n}{(2\pi\ mkT)^{3/2}\ Q} \exp\left(-W^2 - \mathcal{H}_{int}/kT\right) \quad . \tag{27}$$

Here k is Boltzmann's constant, Q is the internal state partition function $Q = \mathrm{Tr}\exp(-\mathcal{H}_{int}/kT)$ defined in terms of the internal state hamiltonian $\mathcal{H}_{int}$ while $\underset{\sim}{W}$ is the reduced peculiar velocity of a molecule

$$\underset{\sim}{W} = (m/2kT)^{1/2}\ (\underset{\sim}{v} - \underset{\sim}{v}_0) \equiv (m/2kT)^{1/2}\ \underset{\sim}{V} \quad . \tag{28}$$

Actually, as given, $f^{(0)}$ is a density operator in internal states but a Wigner [12] distribution function in the translational degrees of freedom, see also part IV of this set of lectures. One has the immediate relations ($\underset{\sim}{p} = m\underset{\sim}{v}$)

$$n(\underset{\sim}{r}, t) = \mathrm{Tr} \int f^{(0)}\ d\underset{\sim}{p} \tag{29a}$$

$$\underset{\sim}{v}_0\ (\underset{\sim}{r}, t) = n^{-1}\ \mathrm{Tr} \int \underset{\sim}{v}\ f^{(0)}\ d\underset{\sim}{p} \tag{29b}$$

$$\frac{3}{2}\ n\ (\underset{\sim}{r}, t)\ kT\ (\underset{\sim}{r}, t) = \mathrm{Tr} \int \tfrac{1}{2} mV^2\ f^{(0)}\ d\underset{\sim}{p} \tag{29c}$$

which also serves to define the normalization, the trace being over the internal states. It is also required that the true Wigner distribution function-density operator f satisfy these three equations, thus that n, $\underset{\sim}{v}_0$ and T <u>are</u> the local density, stream velocity and temperature.

Now f evolves in time due to the free motion of the molecules between collisions, and also due to binary collisions. This can be formulated as

$$\frac{\partial f}{\partial t} + \underset{\sim}{v} \cdot \frac{\partial f}{\partial \underset{\sim}{r}} + i\mathcal{L}f = \frac{\partial f}{\partial t}_{\mathrm{coll.}} \quad . \tag{30}$$

The form of the collision term is left unspecified at this stage but will be discussed in part IV. Otherwise, there are two types of free particle motion, one is the streaming due to the molecular velocities while the other is due to the internal state changes. Here $\mathcal{L}$ is the Liouville operator, contrast Eq. (11), associated with the internal state hamiltonian

$$\mathcal{L}A = (\mathcal{H}_{int} A - A\mathcal{H}_{int})\hbar^{-1} , \qquad (31)$$

while the translational hamiltonian has led to the streaming term. Since particle density (consider only a one component gas), momentum and energy are collisionally conserved quantities, their expectation values evolve according to the equations of:

continuity

$$\partial n/\partial t + \nabla \cdot (n \underset{\sim}{v}_0) = 0 \quad ; \qquad (32)$$

momentum balance

$$\partial(nm \underset{\sim}{v}_0)/\partial t + \nabla \cdot (nm \underset{\sim}{v}_0 \underset{\sim}{v}_0 + \underset{\approx}{P}) = 0 \qquad (33)$$

where $\underset{\approx}{P}$ is the conductive momentum flux or pressure tensor

$$\underset{\approx}{P} = \mathrm{Tr} \int m \underset{\sim}{V} \underset{\sim}{V} f \, d\underset{\sim}{p}$$

$$= 2kT \,\mathrm{Tr} \int \underset{\sim}{W} \underset{\sim}{W} f \, d\underset{\sim}{p} \quad ; \qquad (34)$$

and an energy balance equation which I will not display.

It follows from the form of the streaming term in the Boltzmann equation, that if f is position dependent (that is, the gas is inhomogeneous), then f cannot be of Maxwell-Boltzmann form. Hence inhomogeneities attempt to make f to be of non Maxwell-Boltzmann form, while collisions try to bring everything back to local equilibrium. A steady state is reached with f deviating fractionally from $f^{(0)}$ according to

$$f = f^{(0)} (1 + \phi) \qquad (35)$$

and ϕ is linear in the macroscopic gradients, namely ∇n, $\nabla \underset{\sim}{v}_0$ and ∇T. In this approximation, the pressure tensor is given by

$$\underset{\approx}{P} = nkT \underset{\approx}{U} + \underset{\approx}{\Pi} \qquad (36)$$

where

$$\underset{\approx}{\Pi} = 2kT \,\mathrm{Tr} \int \underset{\sim}{W} \underset{\sim}{W} f^{(0)} \phi \, d\underset{\sim}{p} \qquad (37)$$

is the viscous pressure tensor, also linear in the gradients, $\underset{\approx}{U}$ is the unit second rank tensor, $U_{xx} = U_{yy} = U_{zz} = 1$, all the other six components being zero and nkT is the local equilibrium pressure.

A consistent expansion of the Boltzmann equation to terms linear in the spatial gradients leads to the equation

$$X \equiv 2[\underset{\sim}{W}]^{(2)} : [-\nabla \underset{\sim}{v}_0]^2 - \left[\left(\frac{2}{3} - \frac{k}{C_v}\right) \left(W^2 - \frac{3}{2}\right) - \frac{\mathcal{H}_{int} - \langle \mathcal{H}_{int} \rangle}{C_v T} \right] \nabla \cdot \underset{\sim}{v}_0$$

$$+ \underset{\sim}{V} \left[W^2 - \frac{5}{2} + \frac{\mathcal{H}_{int} - \langle \mathcal{H}_{int} \rangle}{kT} \right] \cdot (-\nabla \ln T) = (\mathcal{R} + i\mathcal{L})\, \phi \, . \qquad (38)$$

Here $\mathcal{R}$ is a linearized collision superoperator whose detailed form is discussed in part V while C_v is the molecular heat capacity per unit volume. It should be noted that since both ϕ and the streaming term $\underset{\sim}{v} \cdot \nabla f$ are linear in the gradients, there is no term in Eq. (38) involving $\underset{\sim}{v} \cdot \nabla \phi$. In contrast, $\mathcal{L}$ is a free motion term that does not involve gradients so $\mathcal{L}\phi$ is linear in the spatial gradients and does appear in this equation. The symmetric traceless part $[\ldots]^{(p)}$ of a pth rank tensor has been used previously by Professor Beenakker in his lectures. This division of the dyad $\underset{\sim}{W}\,\underset{\sim}{W}$ into traceless

$$[\underset{\sim}{W}]^{(2)} = \underset{\sim}{W}\,\underset{\sim}{W} - \frac{1}{3} W^2 \underset{\approx}{U} \qquad (39)$$

and trace $\frac{1}{3} W^2 \underset{\approx}{U}$ parts corresponds physically to the distinction between shear and bulk viscosity, while mathematically it is associated with the reduction under the 3-dimensional rotation group SO(3) to irreducible representations[13] of dimensions 5 and 1 respectively - note that of the 9 components in a second rank symmetric traceless tensor only 5 are independent.

The formal solution of Eq. (38) is

$$\phi = (\mathcal{R} + i\mathcal{L})^{-1} X \qquad (40)$$

provided that the inverse exists. In general it does not, since mass, momentum and energy are null eigenoperators of both $\mathcal{R}$ and $\mathcal{L}$. However, within the requirement that ϕ does not contribute to n, $\underset{\sim}{v}_0$ and T [Eqs. (29) are satisfied by both $f^{(0)}$ and f] the inverse $(\mathcal{R} + i\mathcal{L})^{-1}$ exists in the subspace orthogonal to these collisionally conserved quantities (usually referred to as the summational invariants). On dividing up the viscous pressure tensor, $\underset{\approx}{\Pi}$ of Eq. (37), into traceless and trace parts, the former is given by

$$\pi^{(2)} = 2kT\ \mathrm{Tr} \int [\underset{\sim}{W}]^{(2)}\ f^{(0)} \phi\ d\underset{\sim}{p}$$

$$= 2nkT \langle\langle [\underset{\sim}{W}]^{(2)} | \phi \rangle\rangle$$

$$= 2\eta : [-\nabla \underset{\sim}{v}_0]^{(2)} + \mathcal{L}^{(20)}_{\pi\pi} (-\nabla \cdot \underset{\sim}{v}_0)$$

$$\mathcal{L}^{(2)}_{\pi q} \cdot (-\nabla \ln T) \tag{41}$$

in terms of a fourth rank shear viscosity coefficient

$$\underset{\sim}{\eta} = nkT \langle\langle \sqrt{2}[W]^{(2)} | (\mathcal{R} + i\mathcal{L})^{-1} | \sqrt{2}[W]^{(2)} \rangle\rangle \tag{42}$$

and second and third rank tensors [14] $\mathcal{L}^{(20)}_{\pi\pi}$, $\mathcal{L}^{(2)}_{\pi q}$ for the cross effects between bulk and shear viscosities and between heat conductivity and shear viscosity. These latter vanish in the absence of a field since the gas is then isotropic. Parity forbids a magnetic field to couple heat conductivity and viscosity but this coupling is possible in an electric field. Only the shear viscosity will be discussed further so only its detailed formula has been given.

It is convenient at this stage to introduce the inner product of two operators

$$\langle\langle A | B \rangle\rangle = \mathrm{Tr} \int A^{\dagger t}\ n^{-1}\ f^{(0)}\ B\ d\underset{\sim}{p} \quad . \tag{43}$$

This requires the operator adjoint † and tensor transpose ^{t} of the quantity A so that this definition is general enough to treat tensor valued quantities A, B, which are simultaneously operators in internal state space and functions of velocity. However the form of inner product given assumes that A and/or B commutes with $f^{(0)}$ (as internal state operators) which can be shown to be sufficient for the present purpose but can easily be (and must be) generalized if this condition is not satisfied. The double ket $|\rangle\rangle$ notation was, as far as I know, first introduced by Baranger [15] in pressure broadening studies. The quantity $\sqrt{2}\,[\underset{\sim}{W}]^{(2)}$ is normalized in the sense that

$$\langle\langle \sqrt{2}\,[\underset{\sim}{W}]^{(2)} | \sqrt{2}\,[\underset{\sim}{W}]^{(2)} \rangle\rangle = E^{(2)} \tag{44}$$

where $E^{(2)}$ is the fourth rank tensor which acts as the identity (projection operator) for symmetric traceless second rank tensors, that is, under double-dot contraction (always of adjacent indices), any symmetric traceless second rank tensor $T^{(2)}$ satisfies

$$E^{(2)} : T^{(2)} = T^{(2)} \tag{45}$$

which requires in particular that $E^{(2)}$ is idempotent

$$E^{(2)} : E^{(2)} = E^{(2)} \quad . \tag{46}$$

This is sufficient information for its explicit calculation which implies that it can be written

$$(E^{(2)})_{ijk\ell} = \frac{1}{2}(\delta_{ik}\,\delta_{j\ell} + \delta_{i\ell}\,\delta_{jk}) - \frac{1}{3}\delta_{ij}\,\delta_{k\ell} \; . \tag{47}$$

If $\mathcal{L} = 0$, this would be expected for a gas of monatomic molecules, and $\mathcal{R}^{-1}$ is written as a collision time τ , then the viscosity tensor is

$$\begin{aligned} \underset{\sim}{\eta} &= nkT \left\langle\!\left\langle \sqrt{2}\,[\underset{\sim}{W}]^{(2)} \middle| \tau \middle| \sqrt{2}\,[\underset{\sim}{W}]^{(2)} \right\rangle\!\right\rangle \\ &= nkT\,\tau\,E^{(2)} = \bar{\eta}\,E^{(2)} \end{aligned} \tag{48}$$

and the shear pressure tensor is

$$\Pi^{(2)} = 2\,\bar{\eta}[-\nabla \underset{\sim}{v}_0]^{(2)} \tag{49}$$

in terms of one scalar shear viscosity coefficient $\bar{\eta} = nkT\tau$. An estimate of τ , as calculated from intermolecular forces and binary collisions, is that $^1/\tau$ is a matrix element of $\mathcal{R}$, namely

$$^1/\tau \;\; E^{(2)} = \left\langle\!\left\langle \sqrt{2}\,[\underset{\sim}{W}]^{(2)} \middle| \mathcal{R} \middle| \sqrt{2}\,[\underset{\sim}{W}]^{(2)} \right\rangle\!\right\rangle \tag{50}$$

or on contracting the tensorial indices

$$^1/\tau = \frac{1}{5} \left\langle\!\left\langle \sqrt{2}\,[\underset{\sim}{W}]^{(2)} \middle| : \mathcal{R} \middle| \sqrt{2}\,[\underset{\sim}{W}]^{(2)} \right\rangle\!\right\rangle , \tag{51}$$

which indicates a double-dot contraction between the two velocity tensors after calculating the matrix element of $\mathcal{R}$. The approximation made in going from Eq. (48) to Eq. (50) is to demand the equality

$$\begin{aligned} &\frac{1}{5} \left\langle\!\left\langle \sqrt{2}\,[\underset{\sim}{W}]^{(2)} \middle| : \mathcal{R}^{-1} \middle| \sqrt{2}\,[\underset{\sim}{W}]^{(2)} \right\rangle\!\right\rangle = \\ &\frac{1}{5} \left\langle\!\left\langle \sqrt{2}\,[\underset{\sim}{W}]^{(2)} \middle| : \mathcal{R} \middle| \sqrt{2}\,[\underset{\sim}{W}]^{(2)} \right\rangle\!\right\rangle^{-1} \end{aligned} \tag{52}$$

which is an interchange of the order of taking the inverse and matrix element of $\mathcal{R}$. That this is an excellent approximation ($\simeq 5\%$ error) when calculating the viscosity of monatomic gases, is well known [16].

This is equivalent to a one-moment approximation to the perturbation function ϕ, see also the discussion by Professor Beenakker.

Before ending this discussion of the formal expression for the viscosity, I will make some connection with time correlation function theory. Within a dilute gas framework and a one-molecule picture, an observable evolves in time according to

$$B(t) = \exp\left[-(\mathcal{R} - i\mathcal{L})\, t\right] B(0) \qquad (53)$$

which accounts for its free particle evolution (given by $\mathcal{L}$), and due to collisions. Eq. (42) for the viscosity can be recognized as the time integral

$$\begin{aligned} \underset{\sim}{\eta} &= nkT \left\langle\!\left\langle \sqrt{2}[\underset{\sim}{W}]^{(2)} \,\middle|\, (\mathcal{R} + i\mathcal{L})^{-1} \,\middle|\, \sqrt{2}\,[\underset{\sim}{W}]^{(2)} \right\rangle\!\right\rangle \\ &= nkT \left\langle\!\left\langle (\mathcal{R} - i\mathcal{L})^{-1} \sqrt{2}\,[\underset{\sim}{W}]^{(2)} \,\middle|\, \sqrt{2}[\underset{\sim}{W}]^{(2)} \right\rangle\!\right\rangle \\ &= nkT \int_0^{\infty} \left\langle \sqrt{2}\,[\underset{\sim}{W}]^{(2)}\,(t)\; \sqrt{2}\,[\underset{\sim}{W}]^{(2)}\,(0) \right\rangle dt \end{aligned} \qquad (54)$$

of nkT times the autocorrelation function of $\sqrt{2}[\underset{\sim}{W}]^{(2)}$. The same result can be obtained if one starts from an N-particle picture and makes a binary collision expansion of the resulting resolvents. Note also the interplay of Heisenberg and Schrödinger pictures with operator adjoints to get the phase factors correct. $(\mathcal{R} + i\mathcal{L})^{-1}$ appears in Eq. (42), but $(1 - i\mathcal{L}\tau)^{-1}$ appears in Eq. (12). $\mathcal{L}$ is automatically a self-adjoint superoperator in the inner product defined by Eq. (43), while to obtain Eq. (54), it has tacitly been assumed that $\mathcal{R}$ is self-adjoint. This is not strictly valid, as will be mentioned later.

3.2. The Simple Case of N_2 in a Magnetic Field

At room temperature N_2 can be considered to be in its ground electronic and vibrational state. On the basis that the nuclear spins are only very weakly coupled to the rotational states, these will be ignored. The remaining internal state Hamiltonian is for rotational and Zeeman effects, and is written as

$$\mathcal{H}_{int} = Bj\,(j + 1) - \gamma \underset{\sim}{H} \cdot \underset{\sim}{J}\hbar \;, \qquad (55)$$

where the possible consequences of centrifugal distortion, etc. are here ignored. As a matter of fact they would have only secondary effects. Since $B\langle j\rangle\hbar^{-1} \simeq 3 \times 10^{14}$ sec^{-1} for N_2 while the mean free time is roughly 6×10^{-13} sec. at atmospheric pressure, it is seen that any frequency associated with different j levels will be completely phase randomized at all pressures less than atmospheric. This means that ϕ may be accurately approximated as being diagonal in j. Now $\mathcal{L}$ consists of two terms, one due to $Bj(j+1)$ and the other associated with the Zeeman hamiltonian. If ϕ is diagonal in j, then the first part of $\mathcal{L}$ does not affect ϕ, see Eq. (38), so that in the expression, Eq. (42), for the viscosity, only the Zeeman part, $\mathcal{L}_Z$ of $\mathcal{L}$ needs to be retained, but there is then the implicit requirement that $(\mathcal{R} + i\mathcal{L}_Z)^{-1}\ 2\,[\underset{\sim}{W}]^{(2)} : [-\nabla \underset{\sim}{v}_0]^{(2)} = \phi$ be diagonal in J magnitude. The important fact is that there is no B dependence in the resolvent $(\mathcal{R} + i\mathcal{L}_Z)^{-1}$. If $\mathcal{R}$ is now approximated as an average collision rate $^1/\tau$ times the identity, as in Eq. (50), then the viscosity tensor is given by

$$\underset{\sim}{\eta} = nkT\tau \left\langle\!\left\langle \sqrt{2}[\underset{\sim}{W}]^{(2)} \right| (1 + i\mathcal{L}_Z\tau)^{-1} \left| \sqrt{2}\,[\underset{\sim}{W}]^{(2)} \right\rangle\!\right\rangle \tag{56}$$

which shows a phase randomization factor. Moreover, since $\mathcal{L}_Z$ is proportional to H and τ to $1/p$, it is seen that the viscosity is dependent on pressure (or density) only in the combination of H/p, namely $\underset{\sim}{\eta}$ (H/p). This should be contrasted to the possible dependence on H/p and B/p if the $Bj(j+1)$ part of $\mathcal{L}$ were present. The dependence on B/p has been eliminated on the basis of phase randomization! It is noticed that it is the large energy level spacings that are eliminated from the resolvent. In contrast, the Boltzmann factors in $f^{(0)}$, Eq. (27), are insensitive to the small Zeeman energies and so the Zeeman term will subsequently be eliminated from $f^{(0)}$. This has the important mathematical advantage that now $f^{(0)}$ is field independent and hence isotropic, consequently the inner product, Eq. (43) is also isotropic and this makes (3-dimensional rotation) group theoretical arguments very simple.

Although the deductions that I have stressed are correct, the simple procedure adopted in getting these from Eq. (56) just doesn't work! In Eq. (56) τ is a constant and $\mathcal{L}_Z[\underset{\sim}{W}]^{(2)} = 0$, so the phase randomization factor vanishes and there is no field dependence of the viscosity! The replacement of $\mathcal{R}$ by $1/\tau$ is too naive! It is important that the velocity and internal state spaces be coupled and this can only occur by means of collisions. A perturbational expansion in this coupling appears to be consistent with the small (0.5 - 1%) field dependence of the transport coefficients and more detailed deductions, arising from a perturbational treatment, also appear to be borne out. Thus $\mathcal{R}$ is separated into a part $\mathcal{R}_d$ diagonal in <u>either</u> velocity or internal state space, and a non-diagonal part $\mathcal{R}_{nd}$ that causes the coupling. An exact prescription about what separation is physically significant, has not, and most likely can never, be precisely made. The division that seems to work is based on the <u>directional</u> independence versus dependence, of the internal state angular momenta. There is obviously a great need for a better understanding of collision processes in order to improve our present understnading of this problem.

First order terms in $\mathcal{R}_{nd}$ do not contribute to the viscosity since one must go from velocity space to velocity space. Hence to second order in $\mathcal{R}_{nd}$, the viscosity tensor is

$$\underset{\sim}{\eta} = nkT \langle\langle \sqrt{2}[\underset{\sim}{W}]^{(2)} | \mathcal{R}_d^{-1} + \mathcal{R}_d^{-1}\mathcal{R}_{nd}(\mathcal{R}_d + i\mathcal{L})^{-1}\mathcal{R}_{nd}\mathcal{R}_d^{-1} | \sqrt{2}[\underset{\sim}{W}]^{(2)} \rangle\rangle$$

$$= \bar{\eta} E^{(2)} + \Delta\eta^{int} \qquad (57)$$

where $\bar{\eta} = nkT\,\tau$ is the viscosity due to pure velocity polarizations following Eq. (48), while the contribution from internal state polarizations is

$$\Delta\underset{\sim}{\eta}^{int} = nkT \langle\langle \sqrt{2}[\underset{\sim}{W}]^{(2)} | \mathcal{R}_d^{-1}\mathcal{R}_{nd}(\mathcal{R}_d + i\mathcal{L})^{-1}\mathcal{R}_{nd}\mathcal{R}_d^{-1} | \sqrt{2}\, W^{(2)} \rangle\rangle$$

$$= nkT\,\varphi^2 \langle\langle B_T^{int} | (\mathcal{R}_d + i\mathcal{L})^{-1} | B^{int} \rangle\rangle \; . \qquad (58)$$

There are several technicalities that have been introduced here. First, when acting in velocity space, $\mathcal{L}$ vanishes so that only $\mathcal{R}_d^{-1}$ appears when acting on $\sqrt{2}[\underset{\sim}{W}]^{(2)}$. Second, it is $\mathcal{R}_d^{-1}$ rather than $\mathcal{R}^{-1}$ that is approximated by τ when acting on $\sqrt{2}[\underset{\sim}{W}]^{(2)}$. Third, the internal state polarization is defined by

$$\varphi B^{int} = \mathcal{R}_{nd}\mathcal{R}_d^{-1}\sqrt{2}[\underset{\sim}{W}]^{(2)} \qquad (59)$$

with φ to be a positive normalization constant, so that

$$\langle\langle B^{int} | B^{int} \rangle\rangle = E^{(2)} \quad . \qquad (60)$$

φ thus plays the role of a coupling constant between internal state and velocity polarizations. Since the field dependence is about 1%, φ will have approximately the value 1/10. Lastly, it is a property of the collision <u>superoperator</u> $\mathcal{R}$ that its superoperator adjoint $\mathcal{R}^{\ddagger}$

is the same as the time reversed collision superoperator $\mathcal{R}_T$. Hence it follows that $\Delta\eta^{int}$ involves a matrix element between B^{int} and its time resersed polarization B_T^{int}. The consequence of this is mentioned later.

It was stated earlier that phase randomization requires that only that part of (and of B^{int}) that is diagonal in J magnitude, is of appreciable size. It is convenient to take this explicitly into account in any assumed form for B^{int}. Consistent with this, let me first consider the consequences of assuming

$$B^{int} \simeq [\underset{\sim}{J}]^{(2)} \left\langle \frac{2}{3} \underset{\sim}{J}^2 (\underset{\sim}{J}^2 - \tfrac{3}{4}) \right\rangle^{-1/2} \tag{61}$$

as has standardly been done in most of the work up to now. At the same time, $\mathcal{L}$ can be replaced by $\mathcal{L}_Z$ in Eq. (58). If $\mathcal{R}_d$, acting on B^{int}, is approximated by a collision rate $1/\tau_{int}$ (τ_{int} is not the same as τ), then one has the contribution

$$\Delta\underset{\sim}{\eta}^{int} = nkT\, \tau_{int} \varphi^2 \langle\langle B_T^{int} |\, (1 + i\mathcal{L}_Z \tau_{int})^{-1} | B^{int} \rangle\rangle \tag{62}$$

to the viscosity. The field strength H and the pressure p enter only in the combination H/p. In the absence of a field, the total viscosity tensor becomes scalar ($\underset{\sim}{\eta} = \eta E^{(2)}$) with scalar viscosity coefficient

$$\eta = \bar{\eta} + nkT\, \tau_{int} \varphi^2 = nkT\, (\tau + \tau_{int} \varphi^2) \quad . \tag{63}$$

The field dependence of the viscosity is governed by the properties of the Zeeman superoperator $\mathcal{L}_Z$. Let us now look at its structure. $\mathcal{L}_Z$ is effectively the commutator $\mathcal{J}_z$ of J_z, namely

$$\mathcal{L}_Z A = -\gamma H [J_z , A]_- = \omega [J_z, A]_- \equiv \omega \mathcal{J}_z A \tag{64}$$

and it is convenient to recognize that the superoperator $\mathcal{J}_z$ is the rotation generator, about the $\hat{z}$-axis (field direction), for angular momentum operators. In a magnetic field, the symmetry group of the system is C_∞ (actually $C_{\infty h}$) and the viscosity should be classified under this group rather than under the full rotation group SO(3). The 5-dimensional irreducible representation of SO(3) to which $[\underset{\sim}{W}]^{(2)}$, $[\underset{\sim}{J}]^{(2)}$ and the shear viscosity belong are completely reduced to five 1-dimensional irreducible representations of C_∞. This is of course exactly equivalent to the five spherical harmonics $Y^{(2)\mu}$, but in Cartesian form [13]. The effect of $\mathcal{J}_z$ on an irreducible component $[\underset{\sim}{J}]^{(2)\mu}$ of $[\underset{\sim}{J}]^{(2)}$ is

$$\mathcal{J}_z \, [\underset{\sim}{J}]^{(2)\mu} = \mu [\underset{\sim}{J}]^{(2)\mu} \tag{65}$$

while there exists a basis $\underset{\sim}{e}^{(2)\mu}$ of second rank symmetric traceless tensors satisfying

$$[\underset{\sim}{J}]^{(2)\mu} = [\underset{\sim}{J}]^{(2)} : \underset{\sim}{e}^{(2)\mu} . \tag{66}$$

Explicitly these are

$$\underset{\sim}{e}^{(2)\pm 2} = -\frac{1}{2}(\hat{x} \pm i\hat{y})\,(\hat{x} \pm i\hat{y}) , \tag{67a}$$

$$\underset{\sim}{e}^{(2)\pm 1} = \pm\frac{1}{2}\left[(\hat{x} \pm i\hat{y})\,\hat{z} + \hat{z}\,(\hat{x} \pm i\hat{y})\right], \tag{67b}$$

and

$$\underset{\sim}{e}^{(2)0} = -\left(\frac{3}{2}\right)^{1/2} \left(\hat{z}\,\hat{z} - \frac{1}{3}\underset{\approx}{U}\right) , \tag{67c}$$

which are in natural phase according to Condon and Shortley [17] and obey the relations

$$(\underset{\sim}{e}^{(2)\mu})^{*} = (-1)^{\mu} \, \underset{\sim}{e}^{(2)-\mu} , \tag{68a}$$

$$\underset{\sim}{e}^{(2)\mu} : \underset{\sim}{e}^{(2)\nu} = (-1)^{\mu} \delta_{\nu,-\mu} , \tag{68b}$$

and

$$\sum_{\mu} \underset{\sim}{e}^{(2)\mu} (-1)^{\mu} \, \underset{\sim}{e}^{(2)-\mu} = E^{(2)} . \tag{68c}$$

If B^{int} is now expanded in the $e^{(2)\mu}$ basis, then the internal state viscosity contribution of Eq. (62) can be written as

$$\Delta \underset{\sim}{\eta}^{int} = \sum_{\mu} e^{(2)-\mu} \Delta \eta_{\mu}^{int} (-1)^{\mu} \, e^{(2)\mu} \tag{69}$$

with

$$\Delta \eta_{\mu}^{int} = nkT \, \tau_{int} \phi^2 \, (1 - i\mu\omega\tau_{int})^{-1} . \tag{70}$$

The phase randomization factor and H/p dependence is clear from this equation while the directional dependence of the viscosity can be sorted out in conjunction with Eqs. (67). The total viscosity tensor $\underset{\sim}{\eta}$ can be written in the same way as Eq. (69) and it is now customary to separate the η_{μ} into real η_{μ}^{+} and imaginary η_{μ}^{-} parts. This has already been discussed by Professor Beenakker. Note that for zero field, $\Delta \eta_\mu^{int}$ is an additive (real) contribution to $\bar{\eta}$ but that phase randomization decreases the internal state contribution. Since experimentally it is the field dependence that is measured, η of Eq. (63) is the reference viscosity at zero field and deviations from that,

$$\eta_\mu^{field} \equiv \eta_\mu - \eta = nkT\, \tau_{int} \phi^2\, i\mu\omega\, \tau_{int} (1 - i\mu\omega\, \tau_{int})^{-1} \,, \tag{71}$$

are, for the real parts, negative. This is the decrease in viscosity with increasing field that is usually observed. Of course the $\mu = 0$ component of $\underset{\sim}{\eta}$ does not phase randomize.

It was tacitly assumed that B^{int}, Eq. (61), is velocity independent, so that tensorially, B^{int} must depend on the directions of $\underset{\sim}{J}$ as indicated. The deductions based on the tensorial behaviour of $\underset{\sim}{\eta}$ are then valid, for instance, that η_0 does not phase randomize. A possible choice of velocity dependence of B^{int} is of the form $[\underset{\sim}{W}]^{(2)} \underset{\sim}{J}$. Such a B^{int} is odd to time reversal so that such a contribution would increase the real parts of η_μ^{field}. So far, the experimental evidence indicates that B^{int} is velocity independent for N_2.

Finally, even if B^{int} is velocity independent and diagonal in j magnitude, there is no requirement that all j states contribute proportionately to $[\underset{\sim}{J}]^{(2)}$ as given in Eq. (61). In fact, recent evidence is to the contrary. The generalization to allow an arbitrary dependence on j is to set

$$\mathcal{R}_{nd} \mathcal{R}_d^{-1} \sqrt{2}[\underset{\sim}{W}]^{(2)} \simeq \sum_j \phi_j \mathcal{Y}^{(2)}(\underset{\sim}{J})\, P_j (p_j)^{-1/2} \equiv \sum_j \phi_j A_{020j} \,. \tag{72}$$

Here P_j is a projection operator onto j magnitude quantum states, $\mathcal{Y}^{(2)}(\underset{\sim}{J})$ is explicitly

$$\mathcal{Y}^{(2)}(\underset{\sim}{J}) = \left[15/2\, \underset{\sim}{J}^2 (\underset{\sim}{J}^2 - 3/4)\right]^{-1/2} [\underset{\sim}{J}]^{(2)} \,, \tag{73}$$

so that $A_{020j} \equiv \mathcal{Y}^{(2)}(\underset{\sim}{J})\, P_j\, (p_j)^{-1/2}$ is normalized,

$$\langle\langle \mathcal{Y}^{(2)}(\underset{\sim}{J})\, P_j (p_j)^{-1/2} | \mathcal{Y}^{(2)}(\underset{\sim}{J})\, P_{j'} (p_{j'})^{-1/2} \rangle\rangle = \delta_{jj'}\, E^{(2)} \tag{74}$$

with Boltzmann factor

$$p_j = (2j + 1)\, Q^{-1} \exp\left[-Bj\,(j + 1)/kT\right] \,. \tag{75}$$

The general notation A_{020j} will be used again in Part V. Lastly, ϕ_j is a coupling strength. This allows for the possibility that τ_{int}

is j dependent, giving a τ_j and

$$\Delta\eta_\mu^{int} = \sum_j nkT\ \tau_j\ \phi_j^2\ (1 - i\mu\omega\,\tau_j)^{-1} \quad . \tag{76}$$

This is called an uncoupled model [18]. It seems that this is still not general enough and that it is necessary to allow collisional coupling between the different j-shells. That is, one needs to allow for a relaxation matrix

$$\langle\langle A_{020j} | \mathcal{R}_d | A_{020j'} \rangle\rangle \equiv n \langle v \rangle_{rel} \mathfrak{S}(0\hat{2})_{jj'}\ E^{(2)} \ . \tag{77}$$

In particular the zero field internal state contribution to the viscosity, which is identical to the saturation of the field dependence of the viscosity is

$$\Delta\eta_\mu^{int}\Big|_{H=0} = kT \langle v \rangle_{rel.}^{-1} \sum_{jj'} \phi_j \left[\mathfrak{S}(0\hat{2})^{-1}\right]_{jj'} \phi_{j'} \quad . \tag{78}$$

The distorted wave Born approximation (Part V) indicates that for large j, $\phi_j = (p_j)^{1/2}\phi$, ϕ independent of j, so that the same weighting

$$\mathfrak{S}(0\hat{2})^{-1} = \sum_{jj'} (p_j)^{1/2} \left[\mathfrak{S}(0\hat{2})^{-1}\right]_{jj'} (p_{j'})^{1/2} \tag{79}$$

of the matrix inverse of $\mathfrak{S}(0\hat{2})$ appears in depolarized Rayleigh scattering as in the viscosity. Professor Beenakker has already indicated some of the initial successes in following this comparison of experimental data.

3.3. The Treatment of Oxygen

Oxygen is a paramagnetic molecule and was the gas used in the first experiments on the field dependence of gas transport properties. Nevertheless, it does fit into as simple a scheme as does N_2, as was, in particular, discovered by Kikoin et al. [19] , and more accurately by Hulsman et al. [20] . This is easily traced to the relatively complex internal state energy level structure, being finally theoretically understood [4] in 1971, and in so doing, led to the qualitative picture.

O_2 has an electronic spin S of 1 which is, because of the enormous difference in gyromagnetic ratios, much more affected by a magnetic field than is the rotational angular momentum, here denoted by $\underset{\sim}{N}$. The resultant $\underset{\sim}{J} = \underset{\sim}{N} + \underset{\sim}{S}$ is obviously denoted by $\underset{\sim}{J}$. The internal state hamiltonian of oxygen is given by

$$\mathcal{H}_{int} = B\,\underset{\sim}{N}^2 + c_{sr}\,\underset{\sim}{S}\cdot\underset{\sim}{N} + c_d'\,[\underset{\sim}{S}]^{(2)} : [\hat{u}]^{(2)} - \gamma_S\,\underset{\sim}{H}\cdot\underset{\sim}{S} - \gamma_N\,\underset{\sim}{H}\cdot\underset{\sim}{N} \tag{80}$$

with spin rotation and dipolar coupling constants c_{sr} and c_d'. $\hat{u}$ is the unit vector parallel to the internuclear axis while γ_S and γ_N are the electronic and nuclear gyromagnetic ratios. Only the rotational energy level splittings associated with $B\underset{\sim}{N}^2$ can become comparable with kT, so in $f^{(0)}$ only this term needs to be retained, just as in N_2. Moreover, the rotational energy level splitting due to $B\underset{\sim}{N}^2$ is much larger than τ^{-1} so phase randomization implies that ϕ is diagonal in the magnitude of N, we say that N is a "good quantum number". This also implies that we need only consider that part of $\mathcal{H}_{int}$ which is diagonal in N. This only affects $[\hat{u}]^{(2)}$ which must now be proportional to $[\underset{\sim}{N}]^{(2)}$, in fact

$$[\hat{u}]^{(2)}\Big|_{\text{diagonal in N}} = -\,[\underset{\sim}{N}]^{(2)}\left[2\,(\underset{\sim}{N}^2 - \tfrac{3}{4})\right]^{-1} \tag{81}$$

and $\mathcal{H}_{int}$ can effectively be replaced by

$$\mathcal{H}_{int} = B\,\underset{\sim}{N}^2 + c_{sr}\,\underset{\sim}{S}\cdot\underset{\sim}{N} + c_d\,(\underset{\sim}{N}^2)\,[\underset{\sim}{S}]^{(2)} : [\underset{\sim}{N}]^{(2)} - \gamma_S\,\underset{\sim}{H}\cdot\underset{\sim}{S} - \gamma_N\,\underset{\sim}{H}\cdot\underset{\sim}{N} \tag{82}$$

with

$$c_d\,(\underset{\sim}{N}^2) = -\,c_d'\left[2\,(\underset{\sim}{N}^2 - \tfrac{3}{4})\right]^{-1} . \tag{83}$$

The treatment so far requires that $B\,\underset{\sim}{N}^2 \gg c_d'\,[S]^{(2)} : [\hat{u}]^{(2)}$ which is the Hund's case (b) coupling scheme that is valid for the thermally most populated rotational levels $N \sim 10$, of O_2 at room temperature. Now the spin rotation and dipolar couplings both work to couple $\underset{\sim}{N}$ and $\underset{\sim}{S}$ to a resultant $\underset{\sim}{J}$, since, e.g.

$$\underset{\sim}{S}\cdot\underset{\sim}{N} = \tfrac{1}{2}\left[J\,(J+1) - S\,(S+1) - N(N+1)\right] , \tag{84}$$

while the Zeeman terms separately work to quantize S_z and N_z. It depends on which effect is stronger as to which quantization scheme dominates, and of course there are intermediate cases. I will briefly discuss both extremes.

3.3.1. Low Fields

Since the field dependence of the transport coefficients is to be observed, we are interested predominantly in the situation where $\gamma H \tau$ is approximately 1 and thus all energy level splittings that are larger than γH will be phase randomized. For low fields, this implies that J magnitude is a good quantum number and consequently $\mathcal{H}_{int}$ should also be diagonal in J. That is, the Zeeman terms should be truncated so they both depend only on $\underset{\sim}{H} \cdot \underset{\sim}{J}$ and this gives the effective hamiltonian

$$\tilde{\mathcal{H}} = -\gamma_{NJ}\, \underset{\sim}{H} \cdot \underset{\sim}{J} \tag{85}$$

with an N and J dependent effective gyromagnetic ratio

$$\begin{aligned}\gamma_{NJ} &= (\gamma_S \underset{\sim}{S} + \gamma_N \underset{\sim}{N}) \cdot \underset{\sim}{J} / \underset{\sim}{J}^2 \\ &= \tfrac{1}{2}\left[\gamma_S (\underset{\sim}{J}^2 + \underset{\sim}{S}^2 - \underset{\sim}{N}^2) + \gamma_N (\underset{\sim}{J}^2 + \underset{\sim}{N}^2 - \underset{\sim}{S}^2)\right] / \underset{\sim}{J}^2 \,.\end{aligned} \tag{86}$$

For O_2, the magnitude of S is 1, and $J - N \equiv \sigma$ can take the values -1, 0 and 1; moreover $\gamma_S \sim 2000\, \gamma_N$ so the γ_N term can be ignored. The result is that for $\sigma = \pm 1$, the effective gyromagnetic ratio is $\simeq \pm \gamma_s / N$ (N = 10 is the thermal average) while the $\sigma = 0$ states have a γ_{NN} of $\gamma_S / N(N+1)$, roughly a factor of 10 smaller.

An uncoupled calculation with each multiplet NJ giving its own contribution is

$$\Delta \eta_\mu^{int} = nkT \sum_{NJ} \tau_{NJ}\, \phi_{NJ}^2\, (1 + i\mu \gamma_{NJ} H \tau_{NJ})^{-1} \,. \tag{87}$$

Because of the disparity of magnitudes of γ_{NJ}, there are two ratios of H/p at which phase randomization effects set in. The first for $\sigma = \pm 1$ and the second occurring at higher H/p values (roughly a factor of 10) for $\sigma = 0$. An interesting effect now occurs. For the $\sigma = 0$ contributions, the higher field and lower gyromagnetic ratio make it very easy for a breakdown of the low field coupling to J to be observed. This manifests itself first as a non-linear dependence of the precession frequency on field and shows up experimentally as an extra pressure dependence of the viscosity. Fig. 1 shows the experimental and theoretical η_1^- viscosity. At low pressures one has only the double peaked curve appropriate to the low field case, but at higher pressures (and a constant H/p ratio which means higher fields), it is found that many of the energy level spacings become smaller and actually cross. Hence there is a cancellation of contributions to η_1^- and in fact this quantity is negative for certain values of H and p.

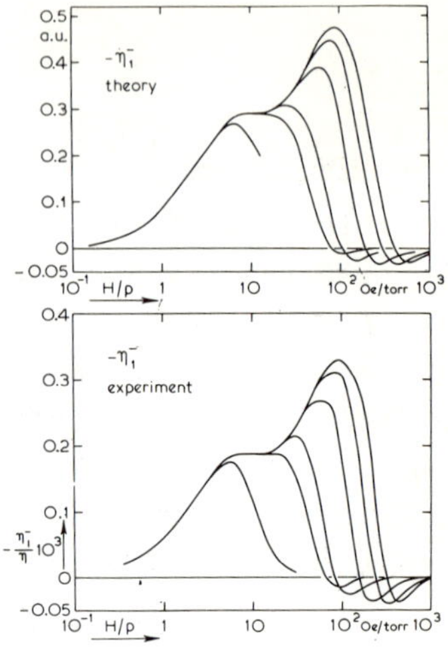

The theoretical calculation[4] was based on the polarizations

$$\mathcal{R}_{nd}\,\mathcal{R}_d^{-1}\,\sqrt{2}\,[\underset{\sim}{W}]^{(2)} \simeq \sum_{NJ} \varphi_{NJ}\,\mathcal{Y}^{(2)}(\underset{\sim}{J})\,P_{NJ} \tag{88}$$

with P_{NJ} being the projection operator onto an NJ multiplet. The uncoupled model formula is then

$$\Delta \underset{\sim}{\eta}^{int} = nkT \sum_{NJ} \varphi_{NJ}^{\;2}\,\tau_{NJ}$$

$$\langle\langle \mathcal{Y}^{(2)}(\underset{\sim}{J})\,|\,(1 + i\mathcal{L}^{(1)}\,\tau_{NJ})^{-1}\,P_{NJ}\,|\mathcal{Y}^{(2)}(\underset{\sim}{J})\rangle\rangle \tag{89}$$

in which $\mathcal{L}^{(1)}$ includes both linear and quadratic Zeeman terms for the energy level structure. Assuming that $\varphi_{NJ} = \varphi$ is independent of N and J and $\tau_{NJ} = J\,(J+1)\,\tau_0$, implies that there are only two adjustable parameters. These are an overall magnitude factor φ and <u>one</u> reference collision time τ_0. The latter was picked so that the peak of the $\sigma = 0$ contribution in the limit of zero pressure fitted experiment. The success of the fit is that <u>all</u> curves depend on the <u>same</u> two parameters and the relative <u>shapes</u> of the curves are dependent on the assumed N and J dependence of φ_{NJ} and τ_{NJ}. These assumed dependences are in fair agreement with collision theory results based on the distorted wave Born approximation.

3.3.2. High Fields

At high fields, the Zeeman energies $-\gamma_S \underset{\sim}{H} \cdot \underset{\sim}{S}$ will dominate over the coupling of $\underset{\sim}{S}$ and $\underset{\sim}{N}$ to give $\underset{\sim}{J}$. At the same time, the rotational Zeeman term $-\gamma_N \underset{\sim}{H} \cdot \underset{\sim}{N}$ is comparatively small because of the difference in gyromagnetic ratios. I will discuss only the limiting case to indicate the type of field dependence that one expects.

Now phase randomization of the electronic spin (Zeeman) frequencies implies that S_z is a good quantum number. That is ϕ and B^{int} are to be diagonal in S_z. Truncating the hamiltonian in the same way gives

$$\mathcal{H}_{int} = B \underset{\sim}{N}^2 - \gamma_S H m_S + (c_{sr} m_S - \gamma_N H) N_z$$

$$+ \frac{3}{2}\left[m_S{}^2 - \frac{1}{2} S (S+1)\right] c_d (\underset{\sim}{N}^2) (N_z{}^2 - \frac{1}{3} \underset{\sim}{N}^2) \quad . \tag{90}$$

This is obtained by replacing $\underset{\sim}{S}$ by $m_S \hat{z}$ everywhere that $\underset{\sim}{S}$ occurs. If $\gamma_N H$ is ignored as being small compared to $c_{sr} m_S$, then it is noted that the rotational states are unaffected by the magnitude of the field. In this case, the spin rotation term $c_{sr} m_S N_z$ implies that the rotational angular momentum $\underset{\sim}{N}$ will precess about the spin which is quantized parallel to the field. The dipolar term gives a much more complicated motion which is absent if the electronic spin is ½ rather than 1. The appropriate polarizations are given by

$$\mathcal{R}_{nd} \mathcal{R}_d^{-1} \sqrt{2}\, [\underset{\sim}{W}]^{(2)} \simeq \sum_{N,m_S} \varphi_N \mathcal{Y}^{(2)} (\underset{\sim}{N})\, P_{N,m_S} \tag{91}$$

with projection P_{N,m_S} onto N, m_S states. φ_N is independent of m_S on the basis that collisions affect the molecular orientation rather than electronic spin. For spin ½ molecules, an uncoupled model calculation gives

$$\Delta\eta_\mu^{int} = nkT \sum_N \varphi_N^2 \tau_N\, p_N\, 2 \left[1 + \frac{1}{4} (\mu\, c_{sr} \tau_N)^2\right]^{-1} \tag{92}$$

on the basis that τ_N is again independent of m_S. p_N is given by Eq. (75) and distorted wave Born approximations can be made for φ_N and τ_N. It is to be noted that the $m_S = \pm \frac{1}{2}$ contributions have been formally added to give a result that is real. Thus there are no transverse viscosity effects. This is because there is just as much precession in one direction as the other. Moreover, the "field dependence" is independent of the magnitude of the field, but of course the direction of the field determines the axis of precession. The effects are then characterized as dependent on the ratio c_{sr}/p which appears as a pressure dependence since c_{sr} is a constant. Professor Coope [21] used these formulae to explain data which was gathered [22] in 1938 on NO_2 but had remained unexplained until his work in 1971.

IV. ON THE BOLTZMANN EQUATION FOR MOLECULES WITH INTERNAL STRUCTURE

This year the Boltzmann equation is 102 years old [23]. I think it is safe to say that no other equation and its consequences was greeted with such criticism and only the concept of entropy has been (and still is) the center of more controversy.

As I see it, there are three, what we would now call statistical assumptions at the root of Boltzmann's original equation. These are:

i) Only binary collisions are important;

ii) Before a collision the molecules are statistically independent, and

iii) A collision occurs at one point in space.

There has been much work during the last 28 years [24] in generalizing the first, which has necessitated generalizing the second and also the third. This I will not do since for a dilute gas, i) should be a good approximation. In fact, I consider the first two to constitute the "philosophy of the Boltzmann equation" while the third is unnecessary and for certain considerations even in the dilute gas case, incorrect. To give Boltzmann his due respect, iii) is a property of the special collision processes that he was considering.

Except for special molecular models [25], the first Boltzmann equation valid for molecules with internal structure was introduced by Wang Chang and Uhlenbeck [26]. Essentially, their only change from the monatomic case is to introduce a set of distribution functions $f_i(\underset{\sim}{r}, \underset{\sim}{p}, t)$, one for each internal state, and a cross-section matrix which describes the transitions from one internal state to another. This is obviously not general enough to account for the phenomena that I have been discussing in my previous lectures, since there the density matrix must be non-diagonal. A Boltzmann equation of sufficient generality to allow for degenerate internal states was first derived by Waldmann [27] and independently, by myself [28]. Since the emphasis at that time was to obtain a collision operator in as close agreement as possible with the classic form, the result was specialized too much. Rather, I now prefer to leave the form of the equation as general as possible [29] and to see how it reduces in special cases. I will "derive" the general form and discuss some of its properties.

4.1. "Derivation" of the Generalized Boltzmann Equation

An N-molecule isolated system obeys the von Neumann equation

$$i\,\partial\rho^{(N)}/\partial t = \mathcal{H}^{(N)}\,\rho^{(N)} - \rho^{(N)}\mathcal{H}^{(N)} \equiv \mathcal{L}^{(N)}\,\rho^{(N)}\,. \tag{93}$$

The N-molecule density operator has reduced density operators

$$\rho^{(n)}_{1\ldots n} = \left[(N-n)!\right]^{-1}\operatorname*{Tr}_{n+1\ldots N}\rho^{(N)} \tag{94}$$

with "convenient" normalizations (n = 1, ..., N)

$$\mathrm{Tr}\,\rho^{(n)} = N!/(N-n)!\quad . \tag{95}$$

By successively tracing Eq. (93) over more and more molecules, the quantum version of the BBGKY hierarchy is obtained. The first two of these are

$$i\,\partial\rho_1/\partial t = \mathcal{L}_1^{(1)}\,\rho_1 + \mathrm{Tr}_2\,\mathcal{V}_{12}\,\rho^{(2)}_{12} \tag{96}$$

and

$$i\,\partial\rho^{(2)}_{12}/\partial t = \mathcal{L}^{(2)}_{12}\,\rho^{(2)}_{12} + \mathrm{Tr}_3(\mathcal{V}_{13}+\mathcal{V}_{23})\,\rho^{(3)}_{123}\,. \tag{97}$$

Here $\rho^{(1)}_1$ has been abbreviated as ρ_1, $\mathcal{L}^{(n)}_{1\ldots n}$ is the Liouville operator for molecules 1...n, $\mathcal{V}_{ij}$ is the commutator superoperator

$$\mathcal{V}_{ij}\,A \equiv V_{ij}\,A - A\,V_{ij} \tag{98}$$

for the intermolecular potential V_{ij}, and it has been assumed that

$\mathcal{H}^{(N)}$ consists of only one-molecule $\mathcal{H}_i^{(1)}$ and pair V_{ij} interactions,

$$\mathcal{H}^{(N)} = \sum_i \mathcal{H}_i^{(1)} + \sum_{i<j} V_{ij} \quad . \tag{99}$$

As is well known, the BBGKY hierarchy is not a closed set of equations and some closure condition must be applied. For dilute gases whose molecules have potentials of short range, this closure can be based on binary collisions and statistical independence <u>before</u> collision, namely the two assumptions that I have said constitute the philosophy of the Boltzmann equation.

The trace term in Eq. (96) vanishes if V_{12} is zero and so requires the two molecules to be colliding. It is <u>only</u> in this term that $\rho^{(2)}$ is needed. Hence for molecules 1 and 2 <u>in the process of a collision</u>, the binary collision approximation means that the $\mathcal{V}_{13} + \mathcal{V}_{23}$ term in the second BBGKY equation can be ignored, that is, <u>during a collision</u>

$$i \, \partial \rho_{12}^{(2)} / \partial t = \mathcal{L}_{12}^{(2)} \rho_{12}^{(2)} \quad . \tag{100}$$

This has the formal solution

$$\rho_{12}^{(2)} (t) = \exp \left[-i \mathcal{L}_{12}^{(2)} (t - t_0) \right] \rho_{12}^{(2)} (t_0) \quad . \tag{101}$$

If now, t_0 is a time <u>before</u> the collisions occurred, then the two molecules are statistically independent, namely

$$\rho_{12}^{(2)} (t_0) = \rho_1 (t_0) \rho_2 (t_0) \quad . \tag{102}$$

The time t_0 must be understood as some time <u>between</u> collisions. Molecules 1 and 2 have come from collisions with other unspecified molecules and move in free motion towards each other till they begin to collide. t_0 is any time during that free motion and it is a necessary requirement for such arguments to be correct, that the free flight time be large compared to the time of duration of a collision. This <u>is</u> true if the density is low and the intermolecular potential of short range.

Free molecular motion for the pair is governed by the Liouville superoperator

$$\mathcal{K} \equiv \mathcal{L}_1^{(1)} + \mathcal{L}_2^{(1)} \tag{103}$$

and one can write

$$\rho_1(t_0)\, \rho_2(t_0) = \exp\left[-i\mathcal{K}(t_0 - t)\right] \rho_1(t)\, \rho_2(t) \tag{104}$$

which connects $\rho_1\ \rho_2$ evaluated at a time t_0 before the collision to the product density operator $\rho_1 \rho_2$ that these molecules would have had, if no collision had occurred. The combination of Eqs. (101), (102) and (104) give

$$\rho_{12}^{(2)}(t) = \exp\left[-i\mathcal{L}_{12}^{(2)}(t-t_0)\right] \exp\left[-i\mathcal{K}(t_0-t)\right] \rho_1(t)\, \rho_2(t)$$

$$\xrightarrow[\lim t_0 \to -\infty]{} \underset{\sim}{\Omega}\, \rho_1(t)\, \rho_2(t)\ . \tag{105}$$

This is a relation between the density operator $\rho_{12}^{(2)}$ of a pair of colliding molecules and the product density operator $\rho_1 \rho_2$ that they would have had if there was no collision. Another way of viewing the two exponentials is that, starting from $\rho_1(t)\, \rho_2(t)$, one goes <u>back</u> on free particle trajectories to a time t_0 <u>before</u> the collision, and then forward in time (to t) along a collision trajectory to get to $\rho^{(2)}$. In this way, free particle motion is changed into collisional motion, a result which will be made precise in the following. Since free and collisional trajectories are identical at t_0, the precise value of t_0 is unimportant and so can be taken to $-\infty$ to define the Møller superoperator $\underset{\sim}{\Omega}$. Actually, real intermolecular potentials are infinite in range so that this limit is required in order to have a mathematical meaning - following Jauch, Misra and Gibson [30], the limit should be taken in the trace norm - but in practise $t_0 \to -\infty$ is to be on a microscopic time scale related to the duration of a collision while t_0 is still small on a macroscopic time scale related to the free flight time. Again, one needs low densities and short ranged potentials.

The generalized Boltzmann equation [29] is obtained by inserting Eq. (105) into the first BBGKY equation, namely

$$i\, \partial\rho_1 / \partial t = \mathcal{L}_1^{(1)} \rho_1 + \mathrm{Tr}_2\, \mathcal{J}_{12}\, \rho_1 \rho_2 \tag{106}$$

where the transition superoperator

$$\mathcal{J}_{12} \equiv \mathcal{V}_{12}\, \underset{\sim}{\Omega} \tag{107}$$

contains all the effects of the collisions. In my view, this equation is the most general mathematical formulation of the first two statistical assumptions i) and ii) and moreover, has used <u>only</u> these two assumptions. There are various alternate forms for this equation and many special cases. I will discuss some important ones.

4.2. Transition Operator Representation

The scattering of a (Dirac) state vector $|\rangle$ is governed by the Møller wave operator ($K \equiv \mathcal{H}_1^{(1)} + \mathcal{H}_2^{(1)}$)

$$\Omega \equiv \underset{t_0 \to -\infty}{\text{st-lim}} \exp\,(i\mathcal{H}^{(2)}\, t_0)\, \exp\,(-i\, K\, t_0) \quad , \tag{108}$$

defined as a strong operator limit of the product of the colliding and free motion evolution operators. This satisfies the intertwining relation

$$\mathcal{H}^{(2)}\Omega = \Omega K \tag{109}$$

which means that it is capable of making eigenfunctions of $\mathcal{H}^{(2)}$ out of eigenfunctions of K, having the same energy. Thus, if $|E\rangle$ is an eigenfunction of K (these do not exist in the Hilbert space of state vectors but, following Dirac, I will use them for ease of presentation and thus avoid problems of spectral measures, etc.), then

$$|E\rangle^{+} \equiv \Omega|E\rangle = \Omega(E)\,|E\rangle \tag{110}$$

can be given in terms of the energy dependent Møller wave operator defined by the Lippmann-Schwinger equation [31]

$$\Omega(E) = 1 + G\,(E)\; V\Omega(E) \quad , \tag{111}$$

and the Green's"function"

$$G\,(E) = \underset{\epsilon \to 0^+}{\text{st-lim}} \;(\,E - K + i\,\epsilon\,)^{-1} \quad . \tag{112}$$

If Ω exists for the given intermolecular potential, and the transition operator is defined by

$$t \equiv V\Omega \quad , \tag{113}$$

then [30] (operator adjoint)

$$\underset{\sim}{\Omega} A = \Omega A \Omega^{\dagger} \tag{114}$$

and

$$\mathcal{T} A = \mathcal{V}\underset{\sim}{\Omega} A = t\, A\Omega^{\dagger} - \Omega A\, t^{\dagger} \quad . \tag{115}$$

From the Lippmann-Schwinger equation, it is seen that $\mathcal{T}$ involves t both linearly and quadratically. Rather than exploring this form in more detail at this stage, it is convenient to first make a Weyl correspondence [32] transformation of the Boltzmann equation so that the position and momentum dependence of the collision operator is more explicit.

The Wigner distribution function [12] is defined by means of the unitary transformation on the translational part of the density operator according to

$$f\,(\underset{\sim}{r},\ \underset{\sim}{p},\ t) \equiv h^{-3}\int \exp\,(i\underset{\sim}{q}\cdot\underset{\sim}{r}/\hbar)\langle\,\underset{\sim}{p}+\tfrac{1}{2}\underset{\sim}{q}|\rho|\underset{\sim}{p}-\tfrac{1}{2}\underset{\sim}{q}\,\rangle\,d\underset{\sim}{q} \tag{116}$$

with the density operator here expressed in momentum representation of the translational states while both f and $\langle p|\rho|p'\rangle$ are operators in internal states. The free motion as given by $\mathcal{L}_1^{(1)}$ gives rise both to a drift term from the translational states and an internal state Liouville operator which will be simply denoted by $\mathcal{L}$, with no subscript. This corresponds to the notation in Eq. (31). Since the collision term involves three different positions, there are a multitude of ways of writing down the result. A convenient method was introduced by Baerwinkel and Grossmann [33] and consists essentially of expressing all momenta in terms of sums and differences. Written out, this is [34]

$$\frac{\partial f}{\partial t}+\frac{\underset{\sim}{p}}{m}\cdot\frac{\partial f}{\partial \underset{\sim}{r}}+i\mathcal{L}f = J\,(\underset{\sim}{r},\ \underset{\sim}{p},\ t) \tag{117}$$

where

$$J(\underset{\sim}{r},\underset{\sim}{p}) \equiv \frac{-(2)^6\, i}{\hbar}\left(\frac{2}{h}\right)^3 \mathrm{Tr}_2\int \exp\left[\frac{-2i}{\hbar}\,(\underset{\sim}{x}\cdot\underset{\sim}{\kappa}\ -\ \underset{\sim}{q}\cdot\underset{\sim}{y})\right]\mathcal{J}(\underset{\sim}{\beta}\underset{\sim}{q}\ \underset{\sim}{k}\underset{\sim}{\kappa})$$

$$f_1\ (\underset{\sim}{r}+\tfrac{1}{2}[\underset{\sim}{y}-\underset{\sim}{x}],\ \underset{\sim}{p}+\underset{\sim}{\beta}-\underset{\sim}{k})\ f_2\ (\underset{\sim}{r}+\tfrac{1}{2}[\underset{\sim}{y}+\underset{\sim}{x}],\ \underset{\sim}{p}+\underset{\sim}{\beta}+\underset{\sim}{k})\ d\underset{\sim}{k}\ d\underset{\sim}{\kappa}\ d\underset{\sim}{\beta}\ d\underset{\sim}{q}\ \ d\underset{\sim}{y}\ d\underset{\sim}{x} \tag{118}$$

and $\mathcal{J}\ (\underset{\sim}{\beta}\underset{\sim}{q}\ \underset{\sim}{k}\underset{\sim}{\kappa})$ is a superoperator acting on internal state operators according to

$$\begin{aligned}\mathcal{J}(\underset{\sim}{\beta}\underset{\sim}{q}\ \underset{\sim}{k}\underset{\sim}{\kappa})\ A \equiv &\langle\underset{\sim}{\beta}+\underset{\sim}{q}|\ t\ |\underset{\sim}{k}+\underset{\sim}{\kappa}\rangle\ \ A\,\langle \underset{\sim}{k}-\underset{\sim}{\kappa}|\Omega^{\dagger}|\underset{\sim}{\beta}-\underset{\sim}{q}\,\rangle \\ &-\langle\underset{\sim}{\beta}+\underset{\sim}{q}|\Omega|\ \underset{\sim}{k}+\underset{\sim}{\kappa}\rangle\ \ A\,\langle\underset{\sim}{k}-\underset{\sim}{\kappa}|\ t^{\dagger}|\underset{\sim}{\beta}-\underset{\sim}{q}\,\rangle\end{aligned} \tag{119}$$

This can be identified as a momentum representation of $\mathcal{J}$. One immediately sees that this collision operator is non-local, the colliding molecules are <u>not</u> at the same position nor at the position of the f appearing in the drift term.

Since $\underset{\sim}{x}$ and $\underset{\sim}{y}$ arise from the non-locality of the collisions, one expects that their appearance in f_1 and f_2 can be ignored if f is sufficiently homogeneous. This gives rise to δ-functions in $\underset{\sim}{q}$

and $\underset{\sim}{K}$ so that

$$J(\underset{\sim}{r}, \underset{\sim}{p}) \rightarrow J_0(\underset{\sim}{r}, \underset{\sim}{p}) \equiv -64\pi^3 i \hbar^2 \, \mathrm{Tr}_2 \int \mathcal{T}(\underset{\sim}{\beta} o \underset{\sim}{k} o) \, f_1(\underset{\sim}{r}, \underset{\sim}{p} + \underset{\sim}{\beta} - \underset{\sim}{k})$$

$$f_2(\underset{\sim}{r}, \underset{\sim}{p} + \underset{\sim}{\beta} + \underset{\sim}{k}) \, d\underset{\sim}{k} \, d\underset{\sim}{\beta} \qquad (120)$$

is a local collision operator. If, moreover, f is sufficiently phase randomized that one can treat f as being diagonal in (internal state) energy in the collision operator, then this collision term is the one first derived by Waldmann [27] and myself [28] , but of course in a different notation. To specialize further, if one assumes that f is diagonal in internal states,

$$f(\underset{\sim}{r}, \underset{\sim}{p}, t) = \sum_a |a\rangle f_a(\underset{\sim}{r}, \underset{\sim}{p}, t) \langle a| \quad , \qquad (121)$$

and uses the relation

$$(x - i\epsilon)^{-1} - (x + i\epsilon)^{-1} \underset{\epsilon \to 0^+}{\longrightarrow} 2\pi i \; \delta(x) \quad , \qquad (122)$$

the cross-section

$$\sigma_{cd \to ab} \equiv (2\pi)^4 \hbar^2 \mu_{ab} \mu_{cd} (p_{ab}/p_{cd}) \, |\langle ab \, \underset{\sim}{p}_{ab} | t | cd \, \underset{\sim}{p}_{cd} \rangle|^2 \; , \qquad (123)$$

and the optical theorem

$$\langle ab \, \underset{\sim}{p}_{ab} | t^{\dagger} - t | ab \, \underset{\sim}{p}_{ab} \rangle = 2\pi i \sum_{cd} \int d\underset{\sim}{p}_{cd}$$

$$\delta(E_{ab} + \tfrac{1}{2} p_{ab}^2/\mu_{ab} - E_{cd} - \tfrac{1}{2} p_{cd}^2/\mu_{cd}) |\langle ab \, \underset{\sim}{p}_{ab} | t | cd \, \underset{\sim}{p}_{cd} \rangle|^2 \; , \qquad (124)$$

then the Boltzmann equation becomes

$$\frac{\partial f_a}{\partial t} + \frac{\underset{\sim}{p}}{m} \cdot \nabla f_a = \sum_{bcd} \iiint \Big[f_c(\underset{\sim}{r}, \underset{\sim}{p}', t) \, f_d(\underset{\sim}{r}, \underset{\sim}{p}_2', t) \qquad (125)$$

$$- f_a(\underset{\sim}{r}, \underset{\sim}{p}, t) \, f_b(\underset{\sim}{r}, \underset{\sim}{p}_2, t) \Big] \frac{p_{cd} \, \sigma_{cd \to ab}}{p_{ab} \, \mu_{ab} \, \mu_{cd}} \, \delta(\text{energy})$$

$$\delta(\underset{\sim}{p}' + \underset{\sim}{p}_2' - \underset{\sim}{p} - \underset{\sim}{p}_2) \, d\underset{\sim}{p}' \, d\underset{\sim}{p}_2 \, d\underset{\sim}{p}_2'$$

where δ(energy) is the same as in Eq. (124) and p_{ab} is the relative momentum associated with $\underset{\sim}{p}$ and $\underset{\sim}{p}_2$. This is exactly the Wang Chang Uhlenbeck equation [26] . If there is no position dependence, then this is of Master equation form with uncorrelated pair probabilities while if there is only one internal state, one has the usual monatomic gas Boltzmann equation [23] but with quantum mechanical cross section.

Now angular momentum is a conserved quantity. If the internal states lose angular momentum during a collision, then the translational states must gain the same amount of angular momentum, about the center of mass. If the collision is local, all molecules are at the same position and no torque can be exerted to create translational angular momentum. On the other hand, localizing the collision operator does not affect the internal state operators, so a collision can change the internal state angular momentum. In this way, the localized collision operator violates angular momentum conservation. It is interesting to note [34] that if f is not diagonal in energy, then the localized collision operator also violates energy conservation. This can be remedied [34] by adding a potential energy density calculated according to

$$nm\, E_V(\underset{\sim}{r}, t) \cong \mathrm{Tr}_{1,2}\, \delta(\underset{\sim}{r} - \underset{\sim}{r}_1)\, V \Omega \rho_1 \rho_2 \Omega^{\dagger}, \tag{126}$$

to the usual kinetic (translational and internal) energy density. Essentially a non-local collision allows the molecules to separate and acquire potential energy. By retaining part of the non-locality of the collision operator, a scheme has been given [34] which is consistent with all conservation laws but I do not have time to discuss that here.

4.3. Dependence on Free Motion Frequencies

Let $|\omega\rangle\rangle$ be an eigenoperator of the free (pair) particle Liouville superoperator $\mathcal{K}$,

$$\mathcal{K}|\omega\rangle\rangle = \omega\,|\omega\rangle\rangle \tag{127}$$

with frequency ω. Acting on such an operator, the transition superoperator $\mathcal{T}$ is equivalent to [29]

$$\mathcal{T}(\omega) \equiv \mathcal{V} + \mathcal{V}(\omega - \mathcal{L}^{(2)} + i\epsilon)^{-1}\mathcal{V}\ , \tag{128}$$

compare the Lippmann-Schwinger Eq. (111). This quantity is identical to the m (ω) operator defined by Fano [35] in his theory of pressure broadening.

A matrix element of $\mathcal{T}$ is then

$$\langle\langle \omega' | \mathcal{T} | \omega \rangle\rangle = \langle\langle \omega' | \mathcal{T}(\omega) | \omega \rangle\rangle\ . \tag{129}$$

It is ω, not ω', that appears in $\mathcal{J}(\omega)$. A diagonal matrix element is complex and is written

$$\langle\langle \omega | \mathcal{J} | \omega \rangle\rangle = \langle\langle \omega | \mathcal{J}(\omega) | \omega \rangle\rangle = \delta - i/\tau \quad . \tag{130}$$

If the collision partner is ignored and one considers only the component ρ_ω of the density operator for the molecule of interest that has a free motion frequency ω $(\mathcal{L}\rho_\omega = \omega\rho_\omega)$, then this part of the Boltzmann equation becomes

$$i\, \partial\rho_\omega/\partial\tau = \mathcal{L}\rho_\omega + \mathcal{J}(\omega)\rho_\omega \cong (\omega + \delta - i/\tau)\rho_\omega \tag{131}$$

This demonstrates that δ is a frequency shift and τ a relaxation time for ρ_ω. It is easily shown that $1/\tau \geq 0$. In this way, $\mathcal{J}$ describes collisional frequency shifts and relaxation rates, Eq. (130), as well as frequency couplings, Eq. (129). It must be noted that the frequencies discussed here refer to the eigenvalues of $\mathcal{L}$ and have nothing to do with the true time evolution of ρ, in particular, ρ_ω is not the time Fourier transform of $\rho(t)$.

It has been argued - in the first lecture of this series - that long free flight times lead to phase randomization and diagonalization(in energy) of the density operator. Complete phase randomization thus means that only zero frequency terms remain. At an intermediate stage of phase randomization one can expect that all the important energy off-diagonalities have frequencies small compared to $kT/\hbar$. Since thermal energies are averaged over, I expect that at this stage, the frequency dependence of $\mathcal{J}(\omega)$ is negligible and one can approximate the collision operator with $\mathcal{J}(0)$. This means that not only is ρ "almost" diagonal in internal energy but also that the translational energy off-diagonality is small, which is reflected in the slow position dependence of $f(\underset{\sim}{r}, \underset{\sim}{p}, t)$. This is again the realm of the Waldmann-Snider collision operator. These arguments have never been fully investigated and much work needs to be done on understanding all the properties of $\mathcal{J}$.

The concept of entropy increase, the H-theorem, is central to the history of the Boltzmann equation. Here this is generally not provable. For a general singlet density operator, many frequency components are present so that any expression for the entropy production involves matrix elements of $\mathcal{J}$ reflecting the collisional coupling of different free motion frequencies. The signs of these terms are unknown, nor have the bounds on the off-diagonal frequency components of $\mathcal{J}$ been investigated. Since the H-theorem requires the negative definiteness of the anti-hermitian part of $\mathcal{J}$, it follows that the entropy need not monotonically increase. The rationale [29], is that as time progresses, phase randomization destroys the higher frequency components of ρ until ρ is sufficiently diagonal in energy, so that only $\mathcal{J}(0)$ is important and entropy increases. For short times, all kinds of oscillatory motion are possible, which reflects the complicated interplay of free molecule (reversible) motion and decay. Detailed calculations to verify these ideas have so far not been carried out.

V. COLLISION INTEGRALS OF THE LINEARIZED W-S EQUATION

In the regime of validity of the W-S equation, the Wigner distribution function-density operator $f(\underset{\sim}{r}, \underset{\sim}{p}, t)$, Eq. (116), is sufficiently diagonal in energy that as far as the collision processes are concerned, f is diagonal in internal energy, and moreover f commutes with the effective Boltzmann factor exp $(-\mathcal{H}_{int}/kT)$. For example, in studying the field dependence of the transport coefficients of diamagnetic diatomic molecules, the Zeeman terms are dropped from $\mathcal{H}_{int}$ in the Boltzmann factors and from the evolution operators $\mathcal{L}^{(2)}$ and $\mathcal{K}$. Thus m states are degenerate as far as the collisions and the inner product is concerned.

If the gas is close to (local) equilibrium, then one can write

$$f = f^{(0)} (1 + \phi) \quad , \tag{132}$$

compare Eq. (35), and the W-S collision operator becomes, Eq. (119),

$$J_0 (\underset{\sim}{r}, \underset{\sim}{p}) \simeq - 64\pi^3 i\hbar^2 \, Tr_2 \int \mathcal{J}(\underset{\sim}{\beta} \, o \, \underset{\sim}{k} \, o)$$

$$f_1^{(0)}(\underset{\sim}{p} + \underset{\sim}{\beta} - \underset{\sim}{k}) \, f_2^{(0)}(\underset{\sim}{p} + \underset{\sim}{\beta} + \underset{\sim}{k}) \left[\phi_1(\underset{\sim}{p} + \underset{\sim}{\beta} - \underset{\sim}{k}) + \phi_2 (\underset{\sim}{p} + \underset{\sim}{\beta} + \underset{\sim}{k})\right] d\underset{\sim}{k} \, d\underset{\sim}{\beta}$$

$$\equiv - f^{(0)} (\underset{\sim}{r}, \underset{\sim}{p}) \mathcal{R} \phi \, . \tag{133}$$

The linearized W-S collision operator is usually written in terms of the transition operator t as

$$(\mathcal{R}\phi)_1 = -(2\pi)^4 \hbar^2 \, \mathrm{Tr}_2 \int f_2^{(0)}$$

$$\left\{ \int t^{g}_{g'} \, (\phi_1' + \phi_2') \, t^{g\dagger}_{g'} \, \delta(E) \, d(\mu \underset{\sim}{g}') \right.$$

$$\left. + (2\pi i)^{-1} \left[t^{g}_{g} \, (\phi_1 + \phi_2) - (\phi_1 + \phi_2) \, t^{g\dagger}_{g} \right] \right\} d\underset{\sim}{p}_2 \qquad (134)$$

Here $\underset{\sim}{g}$ and $\underset{\sim}{g}'$ are the relative velocities after and before collision and $t^{g}_{g'}$ is an operator in internal states, being a partial matrix element of t defined as

$$\langle a | t^{g}_{g'} | b \rangle \equiv \langle a, \mu \underset{\sim}{g} | t | b, \mu \underset{\sim}{g}' \rangle \qquad (135)$$

for internal states a and b (for the pair of molecules) and momentum kets normalized according to $\langle \mu g | \mu g' \rangle = \delta(\mu \underset{\sim}{g} - \mu \underset{\sim}{g}')$. ϕ_1' is short for $\phi_1(\underset{\sim}{W}_1')$, the linear perturbation function for molecule 1 evaluated at reduced velocity, Eq. (28), $\underset{\sim}{W}_1'$ appropriate to molecule 1 before collision. It is understood that the velocities are related by

$$\underset{\sim}{W}_1 = (2)^{-1/2} (\underset{\sim}{\mathcal{G}} - \underset{\sim}{\Upsilon}) \quad , \quad \underset{\sim}{W}_2 = (2)^{-1/2} (\underset{\sim}{\mathcal{G}} + \underset{\sim}{\Upsilon})$$

$$\underset{\sim}{W}_1' = (2)^{-1/2} (\underset{\sim}{\mathcal{G}} - \underset{\sim}{\Upsilon}') \quad , \quad \underset{\sim}{W}_2' = (2)^{-1/2} (\underset{\sim}{\mathcal{G}} + \underset{\sim}{\Upsilon}') \qquad (136)$$

where $\underset{\sim}{\mathcal{G}}$ is the reduced center-of-mass velocity and $\underset{\sim}{\Upsilon} = (m/4kT)^{1/2} \underset{\sim}{g}$ is the reduced relative velocity. Lastly the magnitude of $\underset{\sim}{\Upsilon}'$ is restricted by energy conservation as denoted by $\delta(E)$ which is shorthand for

$$\delta(E) \rightarrow \delta(E_a + \tfrac{1}{2}\mu g^2 - E_b - \tfrac{1}{2}\mu g'^2) \quad ,$$

if the pair molecular internal states a and b appear as in Eq. (135). The dagger, $\dagger$, here acts only on internal states.

This equation is of much more complex form than the Wang Chang Uhlenbeck equation, Eq. (125), since both ϕ and $\mathcal{R}\phi$ are generally non-diagonal in internal state labels. It is seen that this is not written in terms of cross sections and in fact t appears linearly as well as quadratically. If ϕ and $\mathcal{R}\phi$ are diagonal then the optical theorem, Eq. (124), reduces Eq. (134) to the linearized version of the WCU, Eq. (125)

Matrix elements of $\mathcal{R}$ are conveniently expressed as kinetic theory cross sections

$$\sigma(A|B) = \langle\langle A|\mathcal{R}|B\rangle\rangle \; n^{-1} \; (\pi\mu/8kT)^{1/2} \; , \tag{137}$$

it being understood that A and B are normalized according to the inner product of Eq. (43). It is necessary to pick a basis in which to expand ϕ and for which all the collision integrals can be expressed. Now ϕ is a function of velocity and an operator in internal states, moreover $\mathcal{R}$ is rotationally invariant. These considerations suggest that a convenient basis is one in which velocity polarizations and internal state polarizations are separately irreducible representations of the 3-dimensional rotation group. This also is motivated by the experimental fact that velocity angular momentum directional coupling appears to be weak. It is thus convenient [7] to use the normalized velocity polarizations [7] $\mathcal{L}^{ps}\,(\underset{\sim}{W})$,

$$\langle\langle \mathcal{L}^{ps}\,(\underset{\sim}{W})|\mathcal{L}^{p's'}\,(\underset{\sim}{W})\rangle\rangle = \delta_{pp'}\,\delta_{ss'}\,E^{(p)} \; , \tag{138}$$

which are proportional to a product of irreducible tensors $[\underset{\sim}{W}]^{(p)}$ and associated Laguerre polynomials $L_s^{p+\frac{1}{2}}(W^2)$ together with orthonormal internal state polarizations $\mathcal{Y}^{(q)}\,(\underset{\sim}{J})\;P^j\,(\bar{v}v)(p_{jv})^{-1/2}$, where

$$P^j\,(\bar{v}v) \equiv \sum_m |jm\bar{v}\rangle\langle jmv| \tag{139}$$

and $\mathcal{Y}^{(q)}\,(\underset{\sim}{J})$ is proportional to $[\underset{\sim}{J}]^{(q)}$, all being normalized such that

$$\langle\langle \mathcal{Y}^{(q)}\,(\underset{\sim}{J})\;P^j\,(\bar{v}v)\;(p_{jv})^{-1/2}|\mathcal{Y}^{(q')}(\underset{\sim}{J})\;P^{j'}\,(\bar{v}'v')\;(p_{j'v'})^{-1/2}\rangle\rangle$$

$$= \delta_{qq'}\,\delta_{jj'}\,\delta_{\bar{v}\bar{v}'}\,\delta_{vv'}\,E^{(q)} \; . \tag{140}$$

The $\bar{v}$, v indices label degenerate internal states other than the angular momentum while the m dependence of a polarization has been recoupled to a spherical harmonic $\mathcal{Y}^{(q)\nu}(\underset{\sim}{J})$ dependence on angular momentum according to

$$\mathcal{Y}^{(q)\nu}(\underset{\sim}{J})\;P^j\,(\bar{v}v) = (i)^q\left[(2j+1)\;(2q+1)\right]^{1/2}\sum_{mm'}(-1)^{j-m}\begin{pmatrix} j & q & j \\ -m & \nu & m' \end{pmatrix}|jm\bar{v}\rangle\langle jm'v| \tag{141}$$

(...) being a 3-j symbol [36]. The product basis

$$A_{pq\;sj\;\bar{v}v} \equiv \mathcal{L}^{ps}\,(\underset{\sim}{W})\;\mathcal{Y}^{(q)}\,(\underset{\sim}{J})\;P^j\,(\bar{v}v)\;(p_{jv})^{-1/2} \tag{142}$$

is thus tensor valued in both velocity and internal states and reduces for diamagnetic diatomic molecules (e.g. N_2) to $A_{pq\;sj}$, there being no v states.

A kinetic cross section, Eq. (137), arising from the use of such polarizations is a tensor that is symmetric traceless in four sets of indices, labelled pqp' and q'. Since $\mathcal{R}$ is rotationally invariant, such a quantity must be expressible in terms of one or more scalars. The choice of coupling scheme for expressing the results is motivated by the smallness of the directional coupling of velocity and angular momentum. In terms of 3-j symbols, this is expressed as

$$\mathfrak{S}(A_{pq\,sj\,\bar{v}v}{}^{\mu\nu}|A_{p'q's'j'\bar{v}'v'}{}^{\mu'\nu'}) = \sum_{k\kappa} (-1)^{k+p+q+\mu+\nu'}\Omega(kq'q)^{1/2}$$

$$\Omega(kpp')^{1/2} \begin{pmatrix} k & q & q' \\ \kappa & -\nu & \nu' \end{pmatrix} \begin{pmatrix} k & p & p' \\ -\kappa & -\mu & \mu' \end{pmatrix} \mathfrak{S}(pqsj\bar{v}v|p'q's'j'\bar{v}'v')_k \tag{143}$$

with scalar cross sections $(\ldots|\ldots)_k$ and normalization constant [13] $(\mathcal{L} = \ell_1 + \ell_2 + \ell_3)$

$$\Omega(\ell_1\ell_2\ell_3) = \frac{(\mathcal{L}+1)!\,(\mathcal{L}-2\ell_1)!\,(\mathcal{L}-2\ell_2)!\,(\mathcal{L}-2\ell_3)!}{2\,(2\ell_1)!\,(2\ell_2)!\,(2\ell_3)!}\left[3-(-1)^{\mathcal{L}}\right]. \tag{144}$$

It is seen that k is a measure of the directional coupling between velocities and angular momenta. For $k = 0$, this coupling vanishes and I will refer to such cross sections as relaxation terms while $k \neq 0$ terms necessarily arise in the field dependence of transport coefficients, compare Eq. (57), to produce polarizations that are phase randomized by the field. Hence these will be called production cross sections.

The polarizations $A_{pqsj\bar{v}v}$ are very useful for uncoupled model calculations and for cataloguing the results of collision calculations. On the other hand, the laws of conservation of mass, momentum and energy involve combinations of these quantities. The flux terms that naturally arise in hydrodynamics are also combinations of the A's. The six quantities of hydrodynamic interest are:

$$\text{mass} \qquad B_{0000} \equiv 1 = \sum_{jv} (p_{jv})^{1/2}\, A_{000jvv} \tag{145}$$

$$\text{momentum} \quad (2mkT)^{1/2}\, B_{1000} \equiv m\underset{\sim}{v} = (2mkT)^{1/2} \sum_{jv} (p_{jv})^{1/2}\, A_{100jvv} \tag{146}$$

reduced non-equilibrium energy

$$(C_{int}/k)^{1/2} B_{0001} - (3/2)^{1/2} B_{0010} \equiv \mathcal{H}_{int} + W^2 - \langle \mathcal{H}_{int} \rangle - 3/2$$

$$= \sum_{jv} (p_{jv})^{1/2} \left[(\epsilon_{jv} - \langle \mathcal{H}_{int} \rangle) A_{000jvv} - (3/2)^{1/2} A_{001jvv} \right] \qquad (147)$$

reduced shear momentum flux

$$B_{2000} \equiv (2)^{1/2} [\underset{\sim}{W}]^{(2)} = \sum_{jv} (p_{jv})^{1/2} A_{200jvv} \qquad (148)$$

translational heat conductivity

$$B_{1010} \equiv -(4/5)^{1/2} (W^2 - 5/2) \underset{\sim}{W} = \sum_{jv} (p_{jv})^{1/2} A_{101jvv} \qquad (149)$$

and internal heat conductivity

$$B_{1001} \equiv (2)^{1/2} \underset{\sim}{W} (\mathcal{H}_{int} - \langle \mathcal{H}_{int} \rangle) (kT^2 C_{int})^{-1/2} \qquad (150)$$

$$= \sum_{jv} (p_{jv})^{1/2} A_{100jvv} (\epsilon_{jv} - \langle \mathcal{H}_{int} \rangle) (kT^2 C_{int})^{-1/2} .$$

Here the B_{pqst} notation is one that was used in the earlier literature.[3] . Cross sections in mixed representation will be denoted

$$\mathfrak{S}(pqsj\bar{v}v | |p'q's't')_k \equiv \mathfrak{S}(A_{pqsj\bar{v}v} | B_{p'q's't'})_k \qquad (151)$$

with the same tensorial reduction as given above in Eq. (143).

Collisions are unaffected by the center of mass motion. Hence it is possible to explicitly carry out the integral over $\underset{\sim}{\mathcal{G}}$ leaving integrals over the relative momenta of the pair. Two types of relative coordinate integrals arise, $\mathfrak{S}'$ in which both internal state polarizations in $\mathfrak{S}$ are carried by the same molecule, and $\mathfrak{S}''$ for different molecules. The result of the $\underset{\sim}{\mathcal{G}}$ integration is expressed as

$$\mathfrak{S}(pq\ sj\ \bar{v}v | p'q's'j'\bar{v}'v')_k$$

$$= (2k+1) \Omega(kpp')^{-1/2} \sum_{\ell n \ell' n'} I^{(k)}{}_{\ell n \ell' n';\, psp's'} \Omega(k \ell \ell')^{1/2}$$

$$\Big\{ \sum_{j_2 v_2 j_2' v_2'} (p_{j_2' v_2'} p_{j_2 v_2})^{1/2} \mathfrak{S}'(\ell n j \bar{v} v j_2 v_2 q | \ell' n' j' \bar{v}' v' j_2' v_2' q')_k \qquad (152)$$

$$+ (-1)^{\ell'} \sum_{j_1' v_1' j_2 v_2} (p_{j_1' v_1'} p_{j_2 v_2})^{1/2} \mathfrak{S}''(\ell n j \bar{v} v j_2 v_2 q | \ell' n' j_1' v_1' j' \bar{v}' v' q')_k \Big\}$$

and involves a complicated expansion coefficient $I^{(k)}_{\ell n \ell' n' ;psp's'}$. This involves [37] a 9-j symbol and will not be reproduced here. An examination of these results shows that there are various exact relations between cross sections, exemplified by

$$\mathfrak{S}(\,120\;\; j\bar{v}v\,\|\,1010)_2 = -\,(5)^{-1/2}\,\mathfrak{S}(020\;\; j\bar{v}v\,\|\,2000)_2 \qquad (153)$$

and others have been mentioned by Professor Beenakker [1]. A study of $I^{(k)}_{\cdots}$ limits what polarizations can be produced from the hydrodynamic fluxes and these are summarized in Table II.

Further progress can be made within a Distorted Wave Born Approximation (DWBA). The intermolecular potential is considered as a sum of spherical V_0 and non-spherical V_1 parts with the collisional effects of the first dominating the second. If there was only the spherical potential, only relaxation (k = 0) cross sections occur and these are explicitly given by

$$\mathfrak{S}^{(0)}\,(pqsj\bar{v}v\,|\,p'q's'j'\bar{v}'v')_k = \delta_{ko}\;\delta_{q'q}\;\delta_{p'p}$$
$$\delta_{\bar{v}v}\;\delta_{\bar{v}'v'}\;\sum_{\ell n n'}(2p+1)^{-\frac{1}{2}}\;(2\ell+1)^{\frac{1}{2}}\;I_{\ell n \ell n';psps'}{}^{(0)}$$
$$\left[\delta_{jj'}\;\delta_{vv'} + \delta_{qo}\;(-1)^{\ell}\;(p_{jv}p_{j'v'})^{\frac{1}{2}}\right]\mathfrak{S}^{(1)}_{v}\;(\ell nooo\,|\,\ell n'ooo\,|\,o)_0\,. \qquad (154)$$

The velocity cross section $\mathfrak{S}^{(1)}_v$ is in turn expressible according to

$$\mathfrak{S}^{(1)}_{v}\;(\ell nooo\,|\,\ell n'ooo\,|\,o)_0 = \tfrac{1}{2}\,(\pi)^{\frac{1}{2}}\sum_{r} B\,(n\ell,\,n'\ell,\,r)\;I^{\ell}_{r}\;(T) \qquad (155)$$

in terms of Brody-Moshinsky coefficients [38] and Talmi integrals [39]

$$I^{\ell}_{r}\;(T) \equiv 4\pi\left[\Gamma(r+\tfrac{3}{2})\right]^{-1}\int_0^{\infty}\int_{-1}^{1}\exp(-\gamma^2)\;\gamma^{2r+3}\left[1-P_{\ell}\,(\cos\chi)\right]\sigma\,(g,\chi)\;d\,\cos\chi\,d\gamma \qquad (156)$$

of the differential cross section $\sigma\,(g,\chi)$ at deflection angle χ and Legendre polynomial $P_\ell\,(\cos\chi)$. An alternative form in terms of Chapman Cowling [25] $\Omega^{(\ell,s)}$ integrals is also given in reference 7. If p and/or s is non-zero, it might be expected that $\mathfrak{S}^{(0)}(\ldots|\ldots)_0$ dominates $\mathfrak{S}(\ldots|\ldots)_0$ but in practise this often appears to be only 50% true, see in particular the $\mathfrak{S}(1200)$ data of N_2 appearing in reference 40.

A perturbation expansion of the transition operator $t = t_0 + t_1 + \ldots$ in powers of V_1 gives for t_1 the expression

TABLE II

The possible Production $\mathfrak{S}$'s-$\mathfrak{S}(pqsj\bar{v}v \| p'os't')_q$

Type	p′q′s′t′	ℓn	ℓ′n′	Conditions on q	Parity of $\pi_{\bar{v}}\pi_v$
Viscosity	2000	ps	20	triangle p, q, 2	$(-1)^p$
Translational	1010	p-1, s	01	p-1	$(-1)^{p+1}$
Heat Cond.		p+1, s-1,	01	p+1	$(-1^{p+1}$
		p-1, s,	20	triangle p, q, 1	$(-1)^{p+1}$
		p+1, s-1,	20	triangle p, q, 1	$(-1^{p+1}$
Internal	1001	p-1, s ,	00	p-1	$(-1)^{p+1}$
Heat Cond.		p+1, s-1,	00	p+2	$(-1)^{p+1}$
		p s	10	triangle p, q, 1	$(-1)^{p+1}$

$$t_1 = \tilde{\Omega}_0 V_1 \Omega_0 \tag{157}$$

where Ω_0 is the Møller wave operator for the spherical potential and $\tilde{\Omega}_0$ is its operator transpose (time reverse of the adjoint - also called [41] the L conjugate). If V_1 is written as

$$V_1 = \sum_{\ell_1 \ell_2 L} \mathcal{J}_1^{(\ell_1)} \mathcal{J}_2^{(\ell_2)} \odot^{\ell_1+\ell_2} V(\ell_2 \ell_1 L) \odot^L \mathcal{Y}^{(L)}(\hat{R})\, b_{\ell_1 \ell_2 L}(R) \tag{158}$$

Where $\underset{\sim}{R} = R\hat{R}$ is the position of molecule 2 relative to molecule 1 and $\mathcal{J}_i^{(\ell)}$ is an ℓth rank tensor operator (assumed hermitian) on molecule i, then the DWBA transition operator is

$$t^{g}_{1g'} \equiv \langle \mu \underset{\sim}{g} | t_1 | \mu \underset{\sim}{g}' \rangle =$$

$$= \sum_{\ell_1 \ell_2 L} \mathcal{J}_1^{(\ell_1)} \mathcal{J}_2^{(\ell_2)} \odot^{\ell_1+\ell_2} V(\ell_2 \ell_1 L) \odot^L A^{(L)}_{\ell_1 \ell_2}(\underset{\sim}{g}, \underset{\sim}{g}') \tag{159}$$

where the translational dependence is contained entirely in the quantity

$$A^{(L)}_{\ell_1 \ell_2}(\underset{\sim}{g}, \underset{\sim}{g}') \equiv \langle \mu \underset{\sim}{g} | \tilde{\Omega}_0 \mathcal{Y}^{(L)}(\hat{R})\, b_{\ell_1 \ell_2 L}(R) \Omega_0 | \mu \underset{\sim}{g}' \rangle \; . \tag{160}$$

The evaluation of $A^{(L)}_{\ell_1 \ell_2}(\underset{\sim}{g}, \underset{\sim}{g}')$ requires a knowledge of the complete scattering wavefunctions for V_0 while the internal state dependence of t_1 is entirely separated from the translational motion. It is now useful to express the $\mathcal{J}^{(\ell)}$ in terms of reduced (group theoretical reduction, see e.g. Edmonds [36]) matrix elements

$$\langle jmv | \mathcal{J}^{(\ell)\mu} | j'm'v' \rangle \equiv (-1)^{j-m} \begin{pmatrix} j & \ell & j' \\ -m & \mu & m' \end{pmatrix} \langle jv || \mathcal{J}^{(\ell)} || j'v' \rangle \; . \tag{161}$$

With these definitions, it is possible to say a little bit about the kind of results that one can obtain while the detailed treatment and a complete definition of the quantities used are given in reference 7.

Relaxation (k = 0) collision integrals are exemplified by the special case $p = s = p' = s' = 0$, these collision integrals (for $q = q' = 2$) being important in viscosity, depolarized Rayleigh, Raman and flow birefringence studies. From Eq. (152) there are contributions from both $\mathfrak{S}'$ and $\mathfrak{S}''$ collision integrals. If many different j levels are occupied, it turns out that the dependence on Boltzmann factors p_j, Eq. (75), implies that the $\mathfrak{S}''$ contributions are small. An explicit formula for the $\mathfrak{S}'$ contribution, within the DWBA, is

$$\mathfrak{S}'(oqoj\bar{v}v|oqoj'\bar{v}'v')_0 =$$

$$= \sum_{\ell_1\ell_2 L j_2 v_2 j_2' v_2'} \left[\frac{p_{j_2v_2}\, p_{j_2'v_2'}}{(2j_2+1)\,(2j_2'+1)}\right]^{\frac{1}{2}} \frac{|\langle j_2v_2\| \mathscr{d}^{(\ell_2)} \|j_2'\, v_2'\rangle|^2}{(2\ell_1+1)(2\ell_2+1)(2L+1)^{\frac{1}{2}}} \left[\frac{1}{2}\sum_{j''v''}\left(\frac{p_{j''v''}}{p_{jv}}\right)^{\frac{1}{2}}\right.$$

$$\delta_{jj'}\; (2j+1)^{-\frac{1}{2}}\; (2j''+1)^{-\frac{1}{2}}\; (-1)^{\ell_1+j+j''}$$

$$\Big(\langle j\bar{v}\| \mathscr{d}^{(\ell_1)} \|j''v''\rangle\langle j''v''\| \mathscr{d}^{(\ell_1)} \|j\bar{v}'\rangle\delta_{v'v}$$

$$+\langle jv'\| \mathscr{d}^{(\ell_1)} \|j''v''\rangle\langle j''v''\|\mathscr{d}^{(\ell_1)}\|jv\rangle\delta_{\bar{v}\bar{v}'}\Big)\mathfrak{S}_p^{(1)}(00L\ell_1\ell_2|00L\ell_1\ell_2|\epsilon_{j''v''}$$

$$+\epsilon_{j_2'v_2'} - \epsilon_{jv} - \epsilon_{j_2v_2})_0$$

$$-(-1)^q\begin{Bmatrix}\ell_1 & j & j'\\ q & j' & j\end{Bmatrix}\langle j\bar{v}\| \mathscr{d}^{(\ell_1)} \| j'\bar{v}'\rangle\langle j'v'\|\mathscr{d}^{(\ell_1)}\| jv\rangle$$

$$\left.\mathfrak{S}_p(1)\;(00L\ell_1\ell_2|00L\ell_1\ell_2|\epsilon_{j'v'} + \epsilon_{j_2'v_2'} - \epsilon_{jv} - \epsilon_{j_2v_2})\;0\right]. \tag{162}$$

This involves an integral over velocities of the square of $A^{(L)}_{\ell_1\ell_2}$, namely

$$\mathfrak{S}_p^{(1)}(00L\ell_1\ell_2|00L\ell_1\ell_2|x)_0 \equiv (2\pi)^4\hbar^2(\pi\mu/8kT)^{\frac{1}{2}}\pi^{-3/2}\iint \exp\left[-\tfrac{1}{2}(\gamma^2+\gamma'^2)\right.$$

$$(2L+1)^{-\frac{1}{2}}\, A^{(L)}_{\ell_1\ell_2}(\underset{\sim}{g},\underset{\sim}{g}')\odot^L A^{(L)}_{\ell_1\ell_2}(\underset{\sim}{g},\underset{\sim}{g}')^*\,\delta(\tfrac{1}{2}\mu g'^2-\tfrac{1}{2}\mu g^2+x)\,d(\mu\underset{\sim}{g}')\,d\gamma. \tag{163}$$

If the energy inelasticity of the collisions is ignored, x = 0, then one has a <u>complete</u> factorization of internal and translational state dependence with the $\mathfrak{S}_p^{(1)}$ integral acting as a scale factor for all these relaxation cross sections. So far, only the j diagonal contributions to the relaxation matrix, Eq. (162), have been explored and these do not agree with the DPR experimental data [42]. The implications of using the full matrix has still to be explored. For diamagnetic diatomics (no v states) the j dependence of the diagonal $\mathfrak{S}$'s is given by

$$\mathfrak{S}'(oqoj|oqoj)_0 = \sum_{\ell_1} \begin{pmatrix} j & \ell_1 & j \\ o & o & o \end{pmatrix}^2 \left[1-(-1)^{\ell_1+q} (2j+1) \begin{Bmatrix} j & j & q \\ j & j & \ell_1 \end{Bmatrix}\right] \times$$

$$\times \text{ function } (\ell_1, T)$$

$$\simeq (2j+1)^{-2} \times \text{ function } (T) \qquad (164)$$

where the last form is valid for large values of j. This is the basis of the dependence that was assumed in fitting the O_2 viscosity data, Eq. (89).

A typical production integral is

$$\mathfrak{S}'(020\ j\bar{v}v\|\ 2000)_2 = \sum_{j_1' v_1' j_2' v_2' j_2 v_2} \left[\frac{p_{j_1' v_1'} \, p_{j_2 v_2} \, p_{j_2' v_2'}}{(2j_1'+1)(2j_2+1)(2j_2'+1)}\right]^{1/2}$$

$$\sum_{\ell_1 \ell_1' \ell_2 LL'} \begin{Bmatrix} 2 & \ell_1 & \ell_1' \\ j_1' & j & j \end{Bmatrix} \begin{Bmatrix} 2 & L & L' \\ \ell_2 & \ell_1' & \ell_1 \end{Bmatrix} \frac{|\langle j_2 v_2 \| \mathcal{d}^{(\ell_2)} \| j_2' v_2' \rangle|^2}{2\ell_2+1}$$

$$(5)^{\frac{1}{2}} (-1)^{\ell_1+\ell_2+L'+1} \langle j\bar{v} \| \mathcal{d}^{(\ell_1)} \| j_1' v_1' \rangle \langle j_1' v_1' \| \mathcal{d}^{(\ell_1)} \| jv \rangle$$

$$\mathfrak{S}_v^{(1)} (00L\ell_1\ell_2|20L'\ell_1'\ell_2|\epsilon_{j_1' v_1'} + \epsilon_{j_2' v_2'} - \epsilon_{jv} - \epsilon_{j_2 v_2})_2 \qquad (165)$$

where $\mathfrak{S}_v^{(1)}$ is a translational collision integral of slightly more complicated structure that is $\mathfrak{S}_p^{(1)}$, Eq. (163). However it can be shown that this $\mathfrak{S}_v^{(1)}$ vanishes if the energy inelasticity is set equal to zero. It is thus reasonable to consider the limit

$$\lim_{x\to 0} (kT/x)\, \mathfrak{S}_v^{(1)} (00L\ell_1\ell_2|\ 20\ L'\ell_1'\ell_2|x)_2 \equiv \lim \tilde{\mathfrak{S}} \qquad (166)$$

which can be shown to be non-zero, at least in one approximation [7,43]. To first order in x, the j dependence of the production integral is then given by [7]

$$\mathfrak{S}(020\ j\bar{v}v\|\ 2000)_2 = \sum_{j'\,v'\,\ell_1\,\ell_1'} \left(\frac{p_{j'v'}}{2j'+1}\right)^{\frac{1}{2}} \begin{Bmatrix} 2 & \ell_1 & \ell_1' \\ j' & j & j \end{Bmatrix} \left(\frac{\epsilon_{jv}-\epsilon_{j'v'}}{kT}\right)$$

$$\langle j\bar{v}\| \mathscr{L}^{(\ell_1)} \| j'v'\rangle \langle j'v'\| \mathscr{L}^{(\ell_1)} \| jv\rangle \times \text{function}\ (\ell_1', \ell_1, T) \quad . \qquad (167)$$

For diamagnetic diatomic molecules in the large j approximation, this is proportional to

$$\mathfrak{S}(020\ j\|\ 2000)_2 \cong (p_j)^{\frac{1}{2}} \times \text{function}\ (T) \qquad (168)$$

which is the j dependence that is now being used in correlating theory and experiment.

If it is further assumed that $A^{(L)}_{\ell_1 \ell_2}(\underset{\sim}{g}, \underset{\sim}{g}')$ is, tensorially, only a function of the momentum transfer

$$\underset{\sim}{K} \equiv \mu(\underset{\sim}{g}' - \underset{\sim}{g}) \quad , \qquad (169)$$

then it can be shown that for $L' = L$,

$$\lim \mathfrak{S} = \frac{2}{5}\left[\frac{L\ (L+1)}{3\ (2L+3)\ (2L-1)}\right]^{\frac{1}{2}} \mathfrak{S}_p^{(1)}\ (00L\ell_1\ell_2|00L\ell_1'\ell_2|0)_0 \qquad (170)$$

which allows one to obtain the approximate relation

$$\mathfrak{S}\begin{pmatrix}0200\\2000\end{pmatrix}_2 \cong \sum_j \frac{(p_j)^{\frac{1}{2}}[j(j+1)\ (2j+3)\ (2j-1)]^{\frac{1}{2}}}{\langle j\ (j+1)\ (2j+3)\ (2j-1)\rangle^{\frac{1}{2}}}\ \mathfrak{S}(020j\,\|\ 2000)_2$$

$$\simeq \frac{kT\ (-1)^{\ell_1+\ell_2+L}}{B\langle j\ (j+1)\ (2j+3)\ (2j-1)\rangle^{\frac{1}{2}}}$$

$$\left[\frac{3\ L(L+1)\ (2L+1)\ (2\ell_1+1)\ (2\ell_1+3)\ (2\ell_1-1)}{5\ (2L+3)\ (2L-1)\ \ell_1\ (\ell_1+1)}\right]^{\frac{1}{2}}$$

$$\begin{Bmatrix} \ell_2 & \ell_1 & L \\ 2 & L & \ell_1 \end{Bmatrix} \mathfrak{S}\begin{pmatrix}0010\\0010\end{pmatrix} \qquad (171)$$

where B is the rotational constant and it is assumed that only one anisotropic potential, specified by $\ell_1\ell_2L$ is present. This approximate relation between <u>an</u> effective viscosity production integral and a rotational energy relaxation rate constant has had some measure of success in comparison with experiment [44].

REFERENCES

1) J.J.M. BEENAKKER, Transport Properties in Gases in the Presence of External Fields, Lecture notes, this volume, p.

2) Yu. KAGAN and L. MAKSIMOV. Zh. Eksp. Teor. Fiz. 41, 842 (1961) (Sov. Phys. JETP 14, 604 (1962)).

3) For references and a discussion of this work, see the review by J.J.M BEENAKKER and F.R. McCOURT, Ann. Rev. Phys. Chem 21 47 (1970).

4) J.J.M. BEENAKKER, J.A.R. COOPE and R.F. SNIDER, Phys. Rev. A4, 788 (1971). This was motivated by the earlier work by J.A.R. Coope, R.F. SNIDER and F.R. McCOURT, J. Chem. Phys. 53. 3358 (1970)but the Phys. Rev. article completely supercedes the earlier work.

5) J.A.R. COOPE and R.F. SNIDER, J. Chem. Phys. 57, 4266 (1972).

6) R.F. SNIDER and B.C. SANCTUARY, J. Chem. Phys. 55, 1555 (1971).

7) F.M. CHEN, H. MORAAL and R.F. SNIDER, J. Chem. Phys. 57, 542 (1972) R.F. SNIDER, Physica (to be published).

8) See e.g. J.H. JEANS, The Dynamical Theory of Gases (Dover, New York, 1954), 4th ed., p. 257.

9) L.J.F. HERMANS , P.H. FORTUIN, H.F.P. KNAAP and J.J.M. BEENAKKER, Phys. Lett. 25a, 81 (1967).

10) C.J. GORTER, NATURWISS. 26, 140 (1938).

11) This convention was introduced by J.A. CRAWFORD, Nuovo Cimento 10, 698 (1958) and propagated into the present field through the work of H. PRIMAS, Mol. Phys. 6, 225 (1963) and Rev. Mod. Phys. 35, 710 (1963).

12) E. WIGNER, Phys. Rev. 40, 479 (1932)

13) Irreducible Cartesian tensors are extensively treated in the three articles: J.A.R. COOPE, R.F. SNIDER and F.R. McCOURT, J. Chem. Phys. 43, 2269 (1965); J.A.R. COOPE and R.F. SNIDER, J. Math. Phys. 11, 1003 (1970); J.A.R. COOPE, ibid 11, 1591 (1970) while many of the properties and the phases used in the present discussion is given in the Appendix of reference 7.

14) This is an adaption of the notation used in R.F. SNIDER and K. S. LEWCHUK, J. Chem. Phys 46, 3163 (1967).

15) M. BARANGER, Phys. Rev. 111, 494 (1958).

16) See e.g. J.O. HIRSCHFELDER, C.F. CURTISS and R.B. BIRD, The Molecular Theory of Gases and Liquids (Wiley, New York, 1954).

17) E.U. CONDON and G.H. SHORTLEY, Theory of Atomic Spectra (Cambridge U.P., London 1935).

18) J.A.R. COOPE and R.F. SNIDER, J. Chem. Phys. 56, 2049, 2056 (1972).

19) I.K. KIKOIN, K.I. BALASHOV, S.D. LASAREV and R.E. NEUSHTADT, Phys. Letters A24, 165 (1967).

20) H.HULSMAN, A.L.J. BURGMANS, E.J. van WAASDIJK and H.F.P. KNAAP, Physica 50, 558 (1970).

21) J.A.R. COOPE, Mol. Phys. 21, 217 (1971)

22) H. TORWEGGE, Ann. Physik 33, 459 (1938).

23) L. BOLTZMANN, Wein. Ber. 66, 275 (1872).

24) I am setting this date according to the work that has had the most influence on the subsequent development, namely that of N. BOGOLIUBOV, J. Phys. (U.S.S.R.) 10, 265 (1946).

25) See e.g. S. CHAPMAN and T.G. COWLING, The Mathematical Theory of Non-Uniform Gases (Cambridge U.P., Cambridge, England, 1970), 3rd ed.

26) C.S. WANG CHANG and G.E. UHLENBECK, Univ. of Michigan Rept. CM-681, 1951. See also C.S. WANG CHANG, G.E. UHLENBECK and J. de BOER, in Studies in Statistical Mechanics, edited by J. de Boer and G.E. UHLENBECK (North-Holland, Amsterdam, 1964) Vol. 2.

27) L. WALDMANN, Z. NATURFORSCH. 12a 660 (1957), See also Handbuch der Physik, edited by S. FLUGGE (Springer, Berlin, 1958) Vol.12.

28) R.F. SNIDER, J. Chem.Phys. 32, 1051 (1960)

29) R.F. SNIDER and B.C. SANCTUARY, J. Chem. Phys. 55, 1555 (1971).

30) J.M. JAUCH, B. MISRA and A.G. GIBSON, Helv. Phys. Acta 41, 513 (1968).

31) B.A. LIPPMANN and J. SCHWINGER, Phys. Rev. 79, 469 (1950).

32) H. WEYL, Group Theory and Quantum Mechanics (Methuen, 1931).

33) K. BAERWINKEL and S. GROSSMANN, Z. Physik 198, 277 (1967).

34) M.W. THOMAS and R.F. SNIDER, J. Stat. Phys. 2, 61 (1970).

35) U. FANO, Phys Rev.131, 259 (1963).

36) A.R. EDMONDS, Angular Momentum in Quantum Mechanics (Princeton U.P., Princeton, 1960).

37) Given as Eq. (B16) in reference 7.

38) T.A. BRODY and M. MOSHINSKY, Tables of Transformation Brackets (Monografias de Instituto de Fisica, Mexico, 1960). See also T.A. BRODY, G. JACOB and M. MOSHINSKY, Nucl. Phys. 17, 16 (1960).

39) I. TALMI, Helv. Phys. Acta 25, 185 (1952).

40) J.P.J. HEEMSKERK, G.F. BULSING and H.F.P. KNAAP, Physica, to be published.

41) B.A. LIPPMANN, Ann. Phys. (N.Y.) 1, 113 (1957). See also R.D. LEVINE, Quantum Mechanics of Molecular Rate Processes (Oxford U.P., London, 1969).

42) R.A.J. KEIJSER, K.D. van den HOUT and H.F.P. KNAAP, Physica, to be published.

43) H. MORAAL, Z. Naturforsch. 28a 824 (1973).

44) G.J. PRANGSMA, A.L.J. BURGMANS, H.F.P. KNAAP and J.J.M. BEENAKKER, Physica 65, 579 (1973).

Lecture Notes in Physics

Bisher erschienen / Already published

Vol. 1: J. C. Erdmann, Wärmeleitung in Kristallen, theoretische Grundlagen und fortgeschrittene experimentelle Methoden. 1969. DM 20,–

Vol. 2: K. Hepp, Théorie de la renormalisation. 1969. DM 18,–

Vol. 3: A. Martin, Scattering Theory: Unitarity, Analyticity and Crossing. 1969. DM 16,–

Vol. 4: G. Ludwig, Deutung des Begriffs physikalische Theorie und axiomatische Grundlegung der Hilbertraumstruktur der Quantenmechanik durch Hauptsätze des Messens. 1970. DM 28,–

Vol. 5: M. Schaaf, The Reduction of the Product of Two Irreducible Unitary Representations of the Proper Orthochronous Quantummechanical Poincaré Group. 1970. DM 16,–

Vol. 6: Group Representations in Mathematics and Physics. Edited by V. Bargmann. 1970. DM 24,–

Vol. 7: R. Balescu, J. L. Lebowitz, I. Prigogine, P. Résibois, Z. W. Salsburg, Lectures in Statistical Physics. 1971. DM 18,–

Vol. 8: Proceedings of the Second International Conference on Numerical Methods in Fluid Dynamics. Edited by M. Holt. 1971. DM 28,–

Vol. 9: D. W. Robinson, The Thermodynamic Pressure in Quantum Statistical Mechanics. 1971. DM 16,–

Vol. 10: J. M. Stewart, Non-Equilibrium Relativistic Kinetic Theory. 1971. DM 16,–

Vol. 11: O. Steinmann, Perturbation Expansions in Axiomatic Field Theory. 1971. DM 16,–

Vol. 12: Statistical Models and Turbulence. Edited by M. Rosenblatt and C. Van Atta. 1972. DM 28,–

Vol. 13: M. Ryan, Hamiltonian Cosmology. 1972. DM 18,–

Vol. 14: Methods of Local and Global Differential Geometry in General Relativity. Edited by D. Farnsworth, J. Fink, J. Porter and A. Thompson. 1972. DM 18,–

Vol. 15: M. Fierz, Vorlesungen zur Entwicklungsgeschichte der Mechanik. 1972. DM 16,–

Vol. 16: H.-O. Georgii, Phasenübergang 1. Art bei Gittergasmodellen. 1972. DM 18,–

Vol. 17: Strong Interaction Physics. Edited by W. Rühl and A. Vancura. 1973. DM 28,–

Vol. 18: Proceedings of the Third International Conference on Numerical Methods in Fluid Mechanics, Vol. I. Edited by H. Cabannes and R. Temam. 1973. DM 18,–

Vol. 19: Proceedings of the Third International Conference on Numerical Methods in Fluid Mechanics, Vol. II. Edited by H. Cabannes and R. Temam. 1973. DM 26,–

Vol. 20: Statistical Mechanics and Mathematical Problems. Edited by A. Lenard. 1973. DM 22,–

Vol. 21: Optimization and Stability Problems in Continuum Mechanics. Edited by P. K. C. Wang. 1973. DM 16,–

Vol. 22: Proceedings of the Europhysics Study Conference on Intermediate Processes in Nuclear Reactions. Edited by N. Cindro, P. Kulišić and Th. Mayer-Kuckuk. 1973. DM 26,–

Vol. 23: Nuclear Structure Physics. Proceedings of the Minerva Symposium on Physics. Edited by U. Smilansky, I. Talmi, and H. A. Weidenmüller. 1973. DM 26,–

Vol. 24: R. F. Snipes, Statistical Mechanical Theory of the Electrolytic Transport of Non-electrolytes. 1973. DM 20,–

Vol. 25: Constructive Quantum Field Theory. The 1973 "Ettore Majorana" International School of Mathematical Physics. Edited by G. Velo and A. Wightman. 1973. DM 26,–

Vol. 26: A. Hubert, Theorie der Domänenwände in geordneten Medien. 1974. DM 28,–

Vol. 27: R. Kh. Zeytounian, Notes sur les Ecoulements Rotationnels de Fluides Parfaits. 1974. DM 28,–

Vol. 28: Lectures in Statistical Physics. Edited by W. C. Schieve and J. S. Turner. 1974. DM 24,–

Vol. 29: Foundations of Quantum Mechanics and Ordered Linear Spaces. Advanced Study Institute Held in Marburg 1973. Edited by A. Hartkämper and H. Neumann. 1974. DM 26,–

Vol. 30: Polarization Nuclear Physics. Proceedings of a Meeting held at Ebermannstadt October 1–5, 1973. Edited by D. Fick. 1974. DM 24,–

Vol. 31: Transport Phenomena. Sitges International School of Statistical Mechanics, June 1974. Edited by G. Kirczenow and J. Marro. 1974. DM 39,–